T0314140

CIVIL AVIONICS SYSTEMS

Aerospace Series List

Civil Avionics Systems, Second Edition	Moir, Seabridge and Jukes	August 2013
Modelling and Managing Airport Performance	Zografos	July 2013
Advanced Aircraft Design: Conceptual Design, Analysis and Optimization of Subsonic Civil Airplanes	Torenbeek	June 2013
Design and Analysis of Composite Structures: With Applications to Aerospace Structures, Second Edition	Kassapoglou	April 2013
Aircraft Systems Integration of Air-Launched Weapons	Rigby	April 2013
Design and Development of Aircraft Systems, Second Edition	Moir and Seabridge	November 2012
Understanding Aerodynamics: Arguing from the Real Physics	McLean	November 2012
Aircraft Design: A Systems Engineering Approach	Sadraey	October 2012
Introduction to UAV Systems, Fourth Edition	Fahlstrom and Gleason	August 2012
Theory of Lift: Introductory Computational Aerodynamics with MATLAB and Octave	McBain	August 2012
Sense and Avoid in UAS: Research and Applications	Angelov	April 2012
Morphing Aerospace Vehicles and Structures	Valasek	April 2012
Gas Turbine Propulsion Systems	MacIsaac and Langton	July 2011
Basic Helicopter Aerodynamics, Third Edition	Seddon and Newman	July 2011
Advanced Control of Aircraft, Spacecraft and Rockets	Tewari	July 2011
Cooperative Path Planning of Unmanned Aerial Vehicles	Tsourdos et al.	November 2010
Principles of Flight for Pilots	Swatton	October 2010
Air Travel and Health: A Systems Perspective	Seabridge et al.	September 2010
Unmanned Aircraft Systems: UAVS Design, Development and Deployment	Austin	April 2010
Introduction to Antenna Placement and Installations	Macnamara	April 2010
Principles of Flight Simulation	Allerton	October 2009
Aircraft Fuel Systems	Langton et al.	May 2009
The Global Airline Industry	Belobaba	April 2009
Computational Modelling and Simulation of Aircraft and the Environment: Volume 1 – Platform Kinematics and Synthetic Environment	Diston	April 2009
Handbook of Space Technology	Ley, Wittmann Hallmann	April 2009
Aircraft Performance Theory and Practice for Pilots	Swatton	August 2008
Aircraft Systems, Third Edition	Moir and Seabridge	March 2008
Introduction to Aircraft Aeroelasticity and Loads	Wright and Cooper	December 2007
Stability and Control of Aircraft Systems	Langton	September 2006
Military Avionics Systems	Moir and Seabridge	February 2006
Design and Development of Aircraft Systems	Moir and Seabridge	June 2004
Aircraft Loading and Structural Layout	Howe	May 2004
Aircraft Display Systems	Jukes	December 2003
Civil Avionics Systems	Moir and Seabridge	December 2002

CIVIL AVIONICS SYSTEMS

Second Edition

Ian Moir
Aerospace Consultant, UK

Allan Seabridge
Aerospace Consultant, UK

Malcolm Jukes
Aerospace Consultant, UK

This edition was published in 2013
© 2013 John Wiley & Sons, Ltd

First Edition published in 2003
© 2003 John Wiley & Sons, Ltd

Registered office
John Wiley & Sons Ltd, The Atrium, Southern Gate, Chichester, West Sussex, PO19 8SQ, United Kingdom

For detailsof our global editorial offices, for customer services and for information about how to apply for permission to reuse the copyright material in this book please see our website at www.wiley.com.

Library of Congress Cataloging-in-Publication Data

Moir, I. (Ian)
  Civil avionic systems / Ian Moir, Allan Seabridge, Malcolm Jukes. – 2nd edition.
    1 online resource.
  Some parts of ECIP data have title: Civil avionics systems
  Includes bibliographical references and index.
  Description based on print version record and CIP data provided by publisher; resource not viewed.
    ISBN 978-1-118-53672-8 (ePub) – ISBN 978-1-118-53673-5 (Adobe PDF) – ISBN 978-1-118-53674-2 (MobiPocket) – ISBN 978-1-118-34180-3 (cloth)   1. Avionics.   I. Seabridge, A. G. (Allan G.)   II. Jukes, Malcolm.   III. Title.   IV. Title: Civil avionics systems.
  TL695
  629.135–dc23

                                                                                    2013023778

A catalogue record for this book is available from the British Library

ISBN: 978-1-118-34180-3

Typeset in 10/12pt Times by Aptara Inc., New Delhi, India

1   2013

This book is dedicated to Sheena, Sue and Marianne who once again allowed us to indulge our passion for aircraft engineering.

We also wish to acknowledge the passing of a friend, colleague, fellow author, and Series Editor: a major contributor to the Aerospace Series. A vital member of the global aerospace engineering community who passed away on 22 November 2012.

An aerospace systems engineer *'par excellence'*

Roy Langton, 1939 to 2012

Contents

About the Authors

Ian Moir, after 20 years in the Royal Air Force as an engineering officer, went on to Smiths Industries in the UK where he was involved in a number of advanced projects. Since retiring from Smiths (now GE aviation), he is now in demand as a highly respected consultant. Ian has a broad and detailed experience working in aircraft avionics systems in both military and civil aircraft. From the RAF Tornado and Army Apache helicopter to the Boeing 777 electrical load management system (ELMS), Ian's work has kept him at the forefront of new system developments and integrated systems in the areas of more-electric technology and system implementations. With over 50 years of experience, Ian has a special interest in fostering training and education and further professional development in aerospace engineering.

Allan Seabridge was until 2006 the Chief Flight Systems Engineer at BAE Systems at Warton in Lancashire in the UK. In over 45 years in the aerospace industry, his work has included the opportunity to work on a wide range of BAE Systems projects including Canberra, Jaguar, Tornado, EAP, Typhoon, Nimrod, and an opportunity for act as reviewer for Hawk, Typhoon and Joint Strike Fighter, as well being involved in project management, research and development, and business development. In addition, Allan has been involved in the development of a range of flight and avionics systems on a wide range of fast jets, training aircraft, and ground and maritime surveillance projects. From experience in BAE Systems with Systems Engineering education, he is keen to encourage a further understanding of integrated engineering systems. An interest in engineering education continues since retirement with the design and delivery of systems and engineering courses at a number of UK universities at undergraduate and postgraduate level. Allan has been involved at Cranfield University for many years and has recently started a three-year period as External Examiner for the MSc course in Aerospace Vehicle Design.

Malcolm Jukes has over 35 years of experience in the aerospace industry, mostly working for Smiths Aerospace at Cheltenham, UK. Among his many responsibilities as Chief Engineer for Defence Systems Cheltenham, Malcolm managed the design and experimental flight trials of the first UK electronic flight instrument system (EFIS) and the development and application of head-up displays, multifunction head-down displays, and mission computing on the F/A-18, AV8B, Eurofighter Typhoon, Hawk and EH101 aircraft. In this role, and subsequently as Technology Director, he was responsible for product technical strategy and the acquisition of new technology for Smiths UK aerospace products in the areas of displays and controls, electrical power management systems, fuel gauging and management systems, and health and usage monitoring systems. One of his most significant activities was the application of AMLCD technology to civil and military aerospace applications. Malcolm was also a member of the UK Industrial Avionics Working Group (IAWG), and is now an aerospace consultant and university lecturer operating in the areas of displays, display systems, and mission computing.

Between them the authors have been actively involved in undergraduate, postgraduate and supervisory duties in aerospace at the Universities of Bristol, Bath, City, Cranfield, Lancaster, Loughborough, Imperial, Manchester, and the University of the West of England. The authors are course leaders for the postgraduate Avionics Systems and Aircraft Systems modules for the Continuous Professional Development in Aerospace (CPDA) course delivered by a consortium of the Universities of Bristol, Bath and the West of England to UK aerospace companies including BAE Systems, Airbus UK and Augusta Westland.

Series Preface

The field of aerospace is wide ranging and covers a variety of products, disciplines and domains, not merely in engineering but in many related supporting activities. These combine to enable the aerospace industry to produce exciting and technologically challenging products. A wealth of knowledge is retained by practitioners and professionals in the aerospace fields that is of benefit to other practitioners in the industry, and to those entering the industry from University.

The Aerospace Series aims to be a practical and topical series of books aimed at engineering professionals, operators, users and allied professions such as commercial and legal executives in the aerospace industry. The range of topics is intended to be wide ranging, covering design and development, manufacture, operation and support of aircraft as well as topics such as infrastructure operations, and developments in research and technology. The intention is to provide a source of relevant information that will be of interest and benefit to all those people working in aerospace.

Avionic systems are an essential and key component of modern aircraft that control all vital functions, including navigation, traffic collision avoidance, flight control, data display and communications. It would not be possible to fly today's advanced aircraft designs without such sophisticated systems.

This 2nd edition of *Civil Avionics Systems* provides many additions to the original edition, taking into account many of the innovations that have appeared over the past decade in this rapidly advancing field. The book follows the same successful format of the first edition, and is recommended for those wishing to obtain either a top-level overview of avionic systems or a more in-depth description of the wide range of systems used in today's aircraft.

Peter Belobaba, Jonathan Cooper and Allan Seabridge

Preface to Second Edition

It has been over ten years since the first edition of Civil Avionics Systems was published. The book has been in print since that time and it is used as a course text book for a number of university undergraduate and postgraduate courses. It continues to be popular with students and practitioners, if the sales are anything to go by, and the authors continue to use it as the basis of lectures whilst continuously updating and improving the content.

However, much has happened in the world of commercial aviation and in the technological world of avionics since the first publication, prompting a serious update to the book. Despite worldwide economic recession, people still feel a need to fly for business and leisure purposes. Airlines have introduced new and larger aircraft and also introduced more classes to improve on the basic economy class, with more people choosing premium economy and even business class for their holiday flights. This has seen the introduction of the world's largest airliner, the Airbus A380, and an airliner seriously tackling some of the environmental issues in the form of the Boeing B787.

In the field of avionics there have been many advances in the application of commercial data bus networks and modular avionic systems to reduce the risk of obsolescence. Global navigation systems including interoperability of European, US, Russian and Chinese systems and associated standards will seek to improve the ability of aircraft to navigate throughout the world, maybe leading to more 'relaxed' rules on navigation and landing approaches. The crew have been served well with ergonomically improved flight decks providing improved situational awareness through larger, clearer, head-down displays and the addition of head-up displays, with enhanced flight vision and synthetic vision systems.

Propulsion systems have improved in the provision of thrust, reduced noise, improved availability and economic operation. Modern airliners are beginning to move towards more-electric operation.

All these topics and more are covered in this new edition, at considerable effort to keep the book to a reasonable number of pages.

Preface to First Edition

This book on '*Civil Avionic Systems*' is a companion to our book on '*Aircraft Systems*'. Together the books describe the complete set of systems that form an essential part of the modern military and commercial aircraft. There is much read across – many basic aircraft systems such as fuel, air, flight control and hydraulics are common to both types, and modern military aircraft are incorporating commercially available avionic systems such as liquid crystal cockpit displays and flight management systems.

Avionics is an acronym which broadly applies to AVIation (and space) electrONICS. Civil avionic systems are a key component of the modern airliner and business jet. They provide the essential aspects of navigation, human machine interface and external communications for operation in the busy commercial airways. The civil avionic industry, like the commercial aircraft industry it serves, is driven by regulatory, business, commercial and technology pressures and it is a dynamic environment in which risk must be carefully managed and balanced against performance improvement. The result of many years of improvement by systems engineers is better performance, improved safety and improved passenger facilities.

'Civil Avionic Systems' provides an explanation of avionic systems used in modern aircraft, together with an understanding of the technology and the design process involved. The explanation is aimed at workers in the aerospace environment – researchers, engineers, designers, maintainers and operators. It is, however, aimed at a wider audience than the engineering population, it will be of interest to people working in marketing, procurement, manufacturing, commercial, financial and legal departments. Furthermore it is intended to complement undergraduate and post graduate courses in aerospace systems to provide a path to an exciting career in aerospace engineering. Throughout the book 'industry standard' units have been used, there is therefore a mix of metric and Imperial units which reflects normal parlance in the industry

The book is intended to operate at a number of levels:

- Providing a top level overview of avionic systems with some historical background.
- Providing a more in-depth description of individual systems and integrated systems for practitioners.
- Providing references and suggestions for further reading for those who wish to develop their knowledge further.

We have tried to deal with a complex subject in a straightforward descriptive manner. We have included aspects of technology and development to put the systems into a rapidly changing context. To fully understand the individual systems and integrated architectures of systems to meet specific customer requirements is a long and complicated business. We hope that this book makes a contribution to that understanding.

Ian Moir and Allan Seabridge 2002

Acknowledgements

Many people have helped us with this book, albeit unknowingly in a lot of cases. Some of the material has come from our lecturing to classes of short-course delegates and continuing professional development students. The resulting questions and discussions inevitably help to develop and improve the material. Thanks are due to all those people who patiently listened to us and stayed awake.

Colleagues in industry have also helped us in the preparation. Mike Hirst critiqued a number of chapters, and Brian Rawnsley of GE Aviation reviewed and advised upon the latest regulatory issues. Our Airbus UK course mentors Barry Camwell, Martin Rowlands and Martin Lee provided invaluable advice and really gave a stimulus to generating a lot of new material. We have also been helped by Leon Skorczewski and Dave Holding who have joined in the avionics courses by providing material and lectures.

BAE Systems, Cranfield University and the University of the West of England have invited us to lecture on their continuing professional development courses, which opens the door to discussions with many mature students. We wish to thank the organisers of the courses and also the students.

We have been guided throughout the preparation of the manuscript by Anne Hunt, Tom Carter and Eric Willner at John Wiley's at Chichester, and also to Samantha Jones, Shikha Jain from Aptara Delhi and Wahidah Abdul Wahid from Wiley Singapore for the proof-reading, copy-editing and publishing stages of production. Their guidance and patience is, as always, gratefully received.

Ian Moir, Allan Seabridge and Malcolm Jukes
January 2013

List of Abbreviations

3-D	three-dimensional
4-D	four-dimensional
ABS	automatic braking system
AC	alternating current
AC	Advisory Circular
ACARS	ARINC Communications and Reporting System
ACE	actuator control electronics
ACK	receiver acknowledge
ACFD	Advanced Civil Flight Deck
ACP	audio control panel
ADC	air data computer
ADC	analogue to digital conversion/converter
ADD	airstream direction detector
ADF	automatic direction finding
ADI	attitude director indicator
ADI	aircraft direction indicator
ADIRS	Air Data & Inertial Reference System
ADIRU	Air Data and Inertial Reference Unit (B777)
ADM	air data module
ADP	air-driven pump
ADS-A	automatic dependent surveillance – address
ADS-B	automatic dependent surveillance – broadcast
AEW	airborne early warning
AEW&C	Airborne Early Warning and Control
AFDC	autopilot flight director computer
AFDS	autopilot flight director system
AFDX	Aviation Full Duplex
AH	artificial horizon
AHRS	attitude and heading reference system
AIM	Apple–IBM–Motorola alliance
AIMS	Aircraft Information Management System (B777)
Al	aluminium
ALARP	As Low as Reasonably Practical
ALT	barometric altitude

ALU	arithmetic logic unit
AM	amplitude modulation
AMCC	Applied Micro Circuits Corporation
AMLCD	active matrix liquid crystal display
ANO	Air Navigation Order
ANP	actual navigation performance
AoA	angle of attack
AOC	airline operation communication
AOR-E	Azores Oceanic Region – East
AOR-W	Azores Oceanic Region – West
APEX	Application Executive
API	Application Programming Interface
APU	auxiliary power unit
AR	Authorisation Required
ARINC	Air Radio Inc.
ARM	Advanced RISC machine
ASCB	Avionics Standard Communications Bus (Honeywell)
ASCII	American Standard Code for Information Interchange
ASI	airspeed indicator
ASIC	application-specific integrated circuit
ASPCU	air supply and pressure control unit
ASTOR	Airborne Stand-off Radar
ATA	Air Transport Association
ATC	air traffic control
ATI	air transport indicator
A to D	analogue to digital
ATM	air traffic management
ATN	aeronautical telecommunications network
ATR	Air Transport Radio
ATS	air traffic services
ATSU	Air Traffic Service Unit – Airbus unit to support FANS
AWACS	Airborne Warning and Control System
AWG	American Wire Gauge
B	Blue Channel (hydraulics) Airbus
BAG	bandwidth allocation gap
BAT	battery
BC	bus controller
BCD	binary coded decimal
BGA	ball grid array
BGAN	Broadcast Global Area Network
BIT	built-in-test
BLC	battery line contactors
BPCU	bus power control unit
BPCU	brake power control unit

bps	bits per second
BRNAV	basic area navigation
BSCU	brake system control unit
BTB	bus tie breaker
BTC	bus tie contactor
BTMU	brake temperature monitoring unit
C	Centre
C	Centre Channel (hydraulic) Airbus
C	C Band (3.90 to 6.20 GHz)
C1	Centre 1 (Boeing 777)
C2	Centre 2 (Boeing 777)
CA	Course/Acquisition – GPS Operational Mode
CAA	Civil Airworthiness Authority
CANbus	a widely used industrial data bus developed by Bosch
CAS	calibrated air speed
CAST	Certification Authorities Software Team
Cat I	Automatic Approach Category I
Cat II	Automatic Approach Category II
Cat III	Automatic Approach Category III
Cat I	Category I Autoland
Cat II	Category II Autoland
Cat IIIA	Category IIIA Autoland
Cat IIIB	Category IIIB Autoland
CCA	common cause analysis
CCR	common computing resource
CCS	communications control system
CD	collision detection
Cd/m^2	candela per square metre
CDU	control and display unit
CDR	critical design review
CF	constant frequency
CF	course to a fix
CFIT	controlled flight into terrain
CFR	Code of Federal Regulations
CLB	configurable logic block
CMA	common mode analysis
CMCS	Central Maintenance Computing System (Boeing)
C-MOS	complementary metal-oxide semiconductor
CMS	Central Maintenance System (Airbus)
CNS	Communications, Navigation, Surveillance
CO_2	carbon dioxide
C of G	centre of gravity
COM	command
COMMS	communications mode

COMPASS	Chinese equivalent of GPS (Bei Dou)
COTS	commercial off-the-shelf systems
CPIOM	central processor input/output module
CPU	central processing unit
CRI	configuration reference item
CRC	cyclic redundancy check
CRDC	common remote data concentrator (A350)
CRT	cathode ray tube
CS	certification specification
CSD	constant speed drive
CSDB	Commercial Standard Data Bus
CSMA	carrier sense multiple access
CSMA/CD	carrier sense multiple access/collision detection
CTC	cabin temperature controller
Cu	copper
CVR	cockpit voice recorder
CVS	combined vision system
CW	continuous wave
CW/FM	continuous wave/frequency modulated
DA	decision altitude
DAC	digital to analogue conversion/converter
DAL	design assurance level
dB	decibel
DC	direct current
DCDU	Data-Link Control & Display Unit (Airbus)
DC TIE CONT	DC tie contactor
Def Stan	Defence Standard
DF	direct to a fix
DFDAU	digital flight data acquisition unit
DFDR	digital flight data recorder
DFDRS	digital flight data recording system
DG	directional gyro
DGPS	Differential GPS
DH	decision height
DIP	dual in-line package
DLP	digital light projector
DMD	digital micro-mirror
DME	distance measuring equipment
DoD	Department of Defense (US)
D-RAM	dynamic random access memory
DTED	Digital Terrain Elevation Data
DTI	Department of Trade and Industry
D to A	digital to analogue
DTSA	dynamic time-slot allocation

DU	display unit
DVOR	Doppler VOR
E	east
EADI	electronic ADI
EAS	equivalent airspeed
EASA	European Aviation Safety Authority
EBHA	electrical backup hydraulic actuator
EC	European Community
ECAM	Electronic Centralised Aircraft Monitor (Airbus)
ECB	external power contactor
ECC	error correcting code
ECCM	electronic counter-counter measures
ECM	electronic counter measures
ECS	environmental control system
EDP	engine-driven pump
EE	electrical equipment
EEPROM	electrically erasable programmable read only memory
EFB	electronic flight bag
EFIS	electronic flight instrument system
EFVS	enhanced flight vision system
EGPWS	enhanced ground proximity warning system
EHA	electro-hydrostatic actuator
EHF	extremely high frequency
EHSI	electronic HSI
EICAS	Engine Indicating & Crew Alerting System (Boeing)
EIS	electronic instrument system
ELAC	elevator/aileron computer (A320)
ELCU	electrical load control unit
ELINT	electronic intelligence
ELMS	Electrical Load Management System
EM	electromagnetic
EMA	electromechanical actuator
EMI	electromagnetic interference
EMP	electrical motor pump
EMR	electromagnetic radiation
EOF	end of frame
EPC	electrical power contactor
EPLD	electrically programmable logic device
EPROM	electrically programmable read only memory
ESA	European Space Agency
ESM	electronic support measures
ESS, Ess	essential
ESS	environmental stress screening
ETA	estimated time of arrival
ETOPS	extended twin operations

ETOX	erase-through-oxide
EU	electronic unit
EU	European Union
EUROCAE	European Organisation for Civil Aviation Equipment
EVS	enhanced vision system (EASA nomenclature)
EW	electronic warfare
FA	fix to an altitude
FAA	Federal Aviation Authority
FAC	Flight Augmentation Computer (Airbus)
FADEC	full authority digital engine control
FAF	final approach fix
FANS	future air navigation system
FANS1	Future Air Navigation System implemented by Boeing
FANSA	Future Air Navigation System implemented by Airbus
FAR	Federal Airworthiness Requirements
FBW	fly-by-wire
FCDC	flight control data concentrator
FCP	flight control panel
FCPC	flight control primary computer (A330/340)
FCSC	flight control secondary computer (A330/340)
FCU	flight control unit
FDAU	flight data acquisition unit
FDDI	fibre-distributed data interface
FDR	flight data recorder
FET	field effect transistor
FFT	fast Fourier transform
FGMC	Flight Management & Guidance Computer – Airbus terminology for FMS
FHA	functional hazard assessment
FIFO	first-in, first-out
FL	flight level
fL	foot-Lambert
FLIR	forward-looking infra-red
FLOTOX	floating gate tunnel oxide
FMEA	failure modes and effects analysis
FMECA	failure mode effects and criticality analysis
FMES	failure modes and effects Summary
FMGC	flight management guidance computer
FMGEC	Flight Management Guidance & Envelope Computer (A330/340)
FMGU	Flight Management Guidance Unit
FMS	flight management system
FMSP	flight mode selector panel
FOG	fibre-optic gyroscope
FoR	field of regard
FPGA	field programmable gate array

FQMS	fuel quantity management system
FRACAS	failure reporting and corrective action system
fs	sampling frequency
FSCC	flap/slat control computer
FSEU	flap/slat electronic unit
FSF	Flight Safety Foundation
FSK	frequency shift key
FTA	fault tree analysis
FTE	flight technical error
FTP	foil twisted pair
FWC	flight warning computer
G	Green Channel (hydraulics) Airbus
G4	4th generation
GA	general aviation
Galileo	European equivalent of GPS
GAMA	General Aviation Manufacturer's Association
GBAS	ground-based augmentation system
GCB	generator control breaker
GCU	generator control unit
GEOS	geostationary satellite
GHz	gigaHertz
GLC	generator line contactor
GLONASS	Russian equivalent of GPS (GLObal'naya NAvigatsionnaya Sputnikovaya Sistema)
GNSS	global navigation satellite system
GPM	general processing module
GPS	Global Positioning System
GPWS	ground proximity warning system
H	Earth's magnetic field
H_2O	water
Ha	height of aircraft
HAS	Hardware Accomplishment Summary
HDD	head-down display
HDMI	high-definition multimedia interface
HF	high frequency
HFDL	high-frequency data link
HFDS	Head-up Flight Display System (Thales)
Hg	mercury
HGS	Head-up Guidance System (Rockwell Collins)
HIRF	high-intensity radio field
HMI	human–machine interface
HOOD	hierarchical object-oriented design
HSI	horizontal situation indicator
Ht	height

HUD	head-up display
HVGS	head-up display visual guidance system
HVP	Hardware Verification Plan
H/W	hardware
HX	holding to a fix
H_X	X component of H
H_Y	Y component of H
Hz	Hertz
H_Z	Z component of H
I3, I4	INMARSAT satellites
IAP	integrated actuator package
IAS	indicated airspeed
IAWG	Industrial Avionics Working Group
IC	integrated circuit
ICAO	International Civil Aviation Organisation
ICD	interface control document
ICO	instinctive cut-out
ID	identifier
IDG	integrated drive generator
IEEE	Institution of Electrical and Electronics Engineers
IF	initial fix
IFALPA	International Federation of Air Line Pilots' Associations
IFE	in-flight entertainment
IFF	identification friend or foe
IFF/SSR	identification friend or foe/secondary surveillance radar
IFR	international flight rules
IFSD	in-flight shut down
IFU	interface unit
IFZ	independent fault zone
IGSO	inclined geostationary orbit
IIT	image intensifier
ILS	instrument landing system
IMA	integrated modular avionics
IN	inertial navigation
In Hg	inches of mercury
INMARSAT	International Maritime Satellite organisation
INS	inertial navigation system
INV	inverter
I/O	input/output
IOC	initial operational capability
IOM	input/output module
IOR	Indian Ocean Region
IP	Internet protocol
IPFD	Integrated Primary Flight Display (Honeywell SVS)
IPT	integrated product team

IR	infra-red
IRS	inertial reference system
ISIS	integrated standby instrument system
ISO	International Organization for Standardization
ITCZ	Inter-Tropical Convergence Zone
JAA	Joint Airworthiness Authority
JAR	Joint Airworthiness Requirement
JSF	Joint Strike Fighter
JTIDS	Joint Information Tactical Information Distribution System
K^1	K^1 band (10.90 to 17.25 GHz)
Ka	Ka band (36.00 to 46.00 GHz)
kbps	kilobits per second
km	kilometres
Ku	Ku band (33.00 to 36.00 GHz)
kVA	kilovolt-amps
kW	kilowatt
L	Left Channel (hydraulics) Boeing
L	Left
L	L Band (0.39 to 1.55 GHz)
LAAS	Local Area Augmentation System
LAN	local area network
LBAS	locally based augmentation system
LCC	leadless chip carrier
LCD	liquid crystal display
LCoS	liquid crystal on silicon
LED	light emitting diode
LF	low frequency
LNAV	lateral navigation
LPV	localiser performance with vertical guidance
L/R DVT	linear/rotary differential variable transformer
LRG	laser ring gyro
LRM	line-replaceable module
LROPS	long-range operations
LRU	line-replaceable unit
Ls	Ls band (0.90 to 0.95 GHz)
LSB	least significant bit
LSB	lower side-band
LSI	large-scale integration
LVDT	linear variable differential transformer
LWIR	long wave infra-red
M	Mach
M_{mo}	maximum operating Mach number

MA	Markov analysis
MAC	media access control
MAD	magnetic anomaly detector
MASPS	Minimum Aviation System Performance Standard
MAT	maintenance access terminal
MAU	modular avionics unit
Mbps	megabits per second
MCDU	multi-function control and display unit
MCU	modular concept unit
MDA	minimum descent altitude
MDH	minimum descent height
MEA	more-electric aircraft
MEL	minimum equipment list
MEOS	medium Earth orbit satellite
MF	medium frequency
MFD	multifunction display
MHRS	magnetic heading and reference system
MHz	megaHertz
MIL-STD	military standard
MIPS	million instructions per second
MISRA	Motor Industry Software Reliability Association
mK	milliKelvin
MLS	microwave landing system
MMR	multi-mode receiver
Mode A	ATC Mode signifying aircraft call sign
Mode C	ATC Mode signifying aircraft call sign and altitude
Mode S	ATC Mode signifying additional aircraft data
MOPS	minimum operational performance standards
MON/Mon	monitor
MOS	metal oxide semiconductor
MOSFET	metal oxide semiconductor field effect transistor
MPA	maritime patrol aircraft
MPCD	multipurpose control and display
mr	milli-radian
MRTT	multi-role tanker transport
MSI	medium-scale integration
MSL	mean sea-level
MTBF	mean time between failures
MTBR	mean time between removals
MTI	moving target indicator
MTOW	maximum take-off weight
MVA	mega-volt amps
MWIR	medium wave infra-red
N	north
NA	numerical aperture

NASA	National Aeronautics and Space Administration
NATS	National Air Transport System
NAV	navigation mode
NBP	no-break power
ND	navigation display
NDB	non-directional beacon
NETD	noise-equivalent temperature difference
NextGen	Next Generation Air Transport System (USA)
NIC	network interface controller
NiCd	nickel cadmium (battery)
nm	nautical mile – a unit of distance used within the maritime and aeronautical community (1 nm is equivalent to 6070 feet)
nm	nanometers – electromagnetic radiation characteristic associated with electro-optic wavelengths. (0.1 nm is equivalent to 1 angstrom unit Å). Visible light is in the region of 4000 to 7000 Å
NOTAM	Notice to Airmen
NRZ	non-return-to-zero
NVRAM	non-volatile random access memory
O_3	ozone
OAT	outside air temperature
OBOGS	on-board oxygen generation system
ODICIS	One Display for a Cockpit Interactive Solution
O-LED	organic light emitting diode
OMG	Object Management Group
OMT	object modelling technique
OOA	object-oriented analysis
OOD	object-oriented design
OOOI	OUT-OFF-ON-IN: the original simple ACARS message format
OOP	object-oriented programming
Op Amp	operational amplifier
PBN	performance based navigation
PC	personal computer
PCI	peripheral component interconnect
PCU	power control unit
PDR	preliminary design review
PED	personal electronic device
PFC	primary flight control computer
PFD	primary flight display
PGA	pin grid array
PHAC	Plan for Hardware Aspects of Certification
PIO	pilot induced oscillation
PLD	programmable logic device
PMA	permanent magnet alternator
PMAT	portable maintenance access terminal

PMC	PCI Mezzanine Card
PMG	permanent magnet generator
PoR	point of regulation
POR	Pacific Ocean Region
PowerPC	Power Optimization With Enhanced RISC – Performance Computing
PPS	Precise Positioning Service (GPS)
PRA	particular risks analysis
PRNAV	Precision Area Navigation
PROM	programmable read only memory
Ps	static pressure
PSEU	proximity switch electronic unit
PSR	primary surveillance radar
PSSA	preliminary system safety assessment
PSU	power supply unit
Pt	total pressure

q	dynamic pressure
QFE	elevation
QNH	barometric altitude
Quadrax	data bus wiring technique favoured by Airbus

ρ	air density (rho)
R	range
R	Right Channel (hydraulics) Boeing
R	right
RA	Resolution Advisory
R&D	research and development
Rad Alt	radar altimeter
RAE	Royal Aircraft Establishment
RAIM	receiver autonomous integrity monitoring
RAM	random access memory
RAT	ram air turbine
RDC	remote data concentrator
RF	constant radius to a fix
RF	radio frequency
RFI	Request for Information
RFP	Request for Proposal
RFU	radio frequency unit
RISC	reduced instruction set computer/computing
RIU	remote interface unit
RLG	ring laser gyro
RMI	radio-magnetic indicator
RMP	radio management panel
RNAV	area navigation
RNAV (GNSS)	see RNP APCH

RNAV (GPS)	see RNP APCH
RNP	required navigation performance
RNP APCH	RNP Approach
RNP AR APCH	RNP with Authorisation Required Approach
ROM	read only memory
RPC	remote power controller
RPDU	remote power distribution unit
RSS	root sum squares
RT	remote terminal
RTCA	Radio Technical Committee Association
RTL	register transfer level
RTOS	real time operating system
RTR	remote transmission request
RTZ	return-to-zero
RVDT	rotary variable differential transformer
RVR	runway visual range
RVSM	reduced vertical separation minimum
Rx	Receiver, receive
S	south
S	S band (1.55 to 5.20 GHz)
SA	selective availability
SAA	special activity airspace
SAAAR	Special Aircraft & Aircrew Authorisation Required (US equivalent of RNP APCH)
SAARU	Secondary Attitude & Air Data Reference Unit (B777)
SAE	Society of Automotive Engineers
SAHRU	secondary attitude and heading reference
SAR	synthetic aperture radar
SAS	standard altimeter setting
SAT	static air temperature
SATCOM	satellite communications
SATNAV	satellite navigation
SB	sideband
SBAS	space-based augmentation system
SC	Special Committee 213 (RTCA/MASPS)
SDR	system design review
SDU	satellite data unit
SEC	secondary elevator computer
SELCAL	selective calling
SESAR	Single European Sky ATM Research
SESAR JU	SESAR Joint Understanding
SFCC	slat/flap control computer (A330/340)
SG	symbol generator
SG	synchronisation gap
SH	sample and hold

SHF	super high frequency
SiO$_2$	silicon dioxide
SID	standard instrument departure
SIGINT	signals intelligence
SIM	serial interface module
SLR	sideways-looking radar
SMD	surface-mount device
SMT	surface-mount technology
Sp	static pressure
SPS	Standard Positioning Service (GPS)
SOF	start of frame
SOIC	small outline integrated circuit
SNMP	simple network management protocol
SPC	statistical process control
S-RAM	static random access memory
SRR	system requirements review
SSA	system safety assessment
SSB	single sideband
SSD	solid state device
SSI	small-scale integration
SSPC	solid state power controller
SSR	secondary surveillance radar
SSR	software specification review
SST	Supersonic Transport
SSTP	shielded screen twisted pair
STAR	standard terminal arrival requirements
STC	Supplementary Type Certificate
STP	screened twisted pair
S/UTP	shielded unscreened twisted pair
SV	servo-valve
SVS	synthetic vision system
S/W	software
SWIR	short wave infra-red
SysML	Systems Modelling Language
TA	Traffic Advisory
TACAN	tactical air navigation
TACCO	tactical commander
TAS	true airspeed
TAT	total air temperature
TAWS	Terrain Avoidance Warning System
TCAS	Traffic Collision Avoidance System
TCV	Terminal Configured Vehicle
TDMA	time division multiplex allocation
TDZ	touchdown zone
Terprom	terrain profile mapping

TF	track to a fix
TFTP	trivial file transfer protocol
TG	terminal gap
THS	tailplane horizontal stabiliser
TI	terminal interval
TIR	total internal reflection
TLB	translation look-aside buffer
Tp	total pressure
TPMU	tyre pressure monitoring unit
TR	transmitter/receiver
TRU	transformer rectifier unit
TSO	Technical Standards Order
TTL	transistor-transistor logic
TTP	time triggered protocol
Twinax	data bus technique favoured by Boeing
Tx	Transmit, transmitter

UART	universal asynchronous receiver transmitter
UAV	unmanned air vehicle
UDP	user datagram protocol
UHF	ultra high frequency
UK	United Kingdom
ULA	uncommitted logic array
ULD	underwater locating device
UML	Unified Modelling Language
UPS	United Parcel Services
US, USA	United States of America
USNO	United States Naval Observatory
USB	upper sideband
USMS	utility systems management system
UTP	unshielded twisted pair
UV	ultra violet

V	velocity
V_{mo}	maximum operating speed
VAC	volts AC
VDC	volts DC
VDR	VHF digital radio
VF	variable frequency
VGA	video graphics adapter
VGS	Visual Guidance System (Honeywell/BAE Systems)
VHDL	very high speed integrated hardware description language
VHF	very high frequency
VHFDL	very high frequency data link
VL	virtual link
VLF	very low frequency

VLSI	very large scale integration
VMC	visual meteorological conditions
VMS	vehicle management system
VNAV	vertical navigation
VOR	VHF omni-ranging
VORTAC	VOR TACAN
VS	vertical speed
VSCF	variable speed/constant frequency
VSI	vertical speed indicator
VSD	vertical situation display
V/UHF	very/ultra high frequency

W	watt
WAAS	Wide Area Augmentation System
WWII	World War II

X	X axis
X	X band (5.20 to 10.90 GHz)
Xb	Xb band (6.25 to 6.90 GHz)

| Y | Y axis |
| Y | Yellow Channel (hydraulics) Airbus |

Z	Z axis
ZOH	zeroth order hold
ZSA	zonal safety analysis

1

Introduction

1.1   Advances since 2003

The principles of avionics systems are unchanged but new innovations have been introduced since the first edition of *Civil Avionics Systems* was published in 2003. Many of these advances have been incorporated into modern aircraft, and research continues to improve the aircraft and the air transportation system. Notable advances include:

- The A380 and B787 aircraft have been introduced into service.
- The use of commercial off-the-shelf (COTS)-based data bus networks have significantly increased: in particular, ARINC 664 at the aircraft level, and CANbus at the intra-system level have been widely adopted.
- The introduction of advanced (3rd generation) IMA implementations on A380, B787 and emergent on A350.
- More-electric aircraft (MEA) implementations; in part on A380 and more extensively on B787.
- The rapid growth of global navigation satellite systems (GNSSs) in addition to GPS. The Russian GLONASS has been reconstituted in recent years, and COMPASS (China) and Galileo (European Union) systems are being established.
- The introduction of the electronic flight bag (EFB), most recently with iPad implementations by some organisations.
- The introduction of improved ground-based augmentation systems (GBAS)-based approaches.
- Significant improvements in flight deck displays using COTS glass in rectangular format. The trend towards larger display surfaces has continued, indeed escalated.
- Wider adoption of head-up displays and the use of enhanced vision systems (EVS) to help mitigate reduced visibility as a limiting factor in flight operations.
- The development of synthetic vision systems (SVSs) to provide an aid for location of runways and other objects.

Civil Avionics Systems, Second Edition. Ian Moir, Allan Seabridge and Malcolm Jukes.
© 2013 John Wiley & Sons, Ltd. Published 2013 by John Wiley & Sons, Ltd.

Table 1.1   Comparison of Boeing and Airbus solutions

Implementation	Boeing approach	Airbus approach
IMA implementation (Chapter 6)	B777: First generation – AIMS/ELMS B787: Third generation using cabinets and supplier-furnished RIUs	A380: CPIOMs and subsystem supplier-furnished RDCs A350: CPIOMs and generic cRDCs
Onboard maintenance (Chapter 7)	Embedded maintenance display and PMAT options	Dedicated hardware in CMC
More-electric technology (Chapter 7)	B787: 500 kVA/channel at 230 VAC No bleed air off-take from engine. Electric ECS, engine starting and anti-icing	A380: 150 kVA/channel at 115 VAC 2 'H' + 2 'E' architecture – blue hydraulics channel subsumed into electrical implementation Use of EBHAs
Data bus wiring (Chapter 7)	Twinax wiring	Quadrax wiring
Aircraft wiring (Chapter 7)	Not < 22 AWG	Not < 24 AWG
Fly-by-wire (Chapter 10)	Conventional control yoke for pitch and roll inputs Trio-triplex computing using dissimilar hardware Similar software	Sidestick controller for pitch and roll inputs Multiple dual COM/MON computing Dissimilar hardware and software
Electronic flight bag (Chapter 11)	Class I/Class II Fixed or docked	Class III Laptop/iPad/Tablet
FANS embodiment (Chapter 11)	[B737;B747;B757;B767;B777] Fixed hardware with software upgrades/increments	Additional hardware: ATSU/DCDU

1.2   Comparison of Boeing and Airbus Solutions

While the avionics technologies are applied in general to provide solutions to the same problem statement, there are a range of alternative philosophies and architectures that will provide safe and certifiable solutions for an aircraft. Boeing and Airbus adopt different approaches for a number of different system implementations; that is not to say that one solution is any better than another. The different approaches merely reflect the design cultures that exist within the respective manufacturers.

Table 1.1 lists a number of areas where the Boeing and Airbus approaches differ and which are described more fully in the body of the text.

1.3   Outline of Book Content

The contents of this book are aimed mainly at commercial transport aircraft, but the principles described are also applicable to military types. This particularly applies to large military aircraft

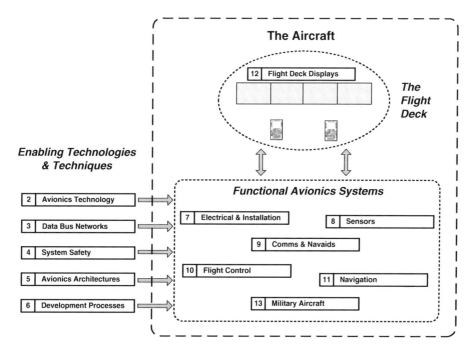

Figure 1.1 Interrelationship of enabling technologies and aircraft system

that are conversions of commercial aircraft in which the platform avionics will generally remain, with mechanisms for connecting it to military system additions. The description of avionics systems may be subdivided into three areas which together provide the total aircraft function (see Figure 1.1). The chapters that describe these functional areas are listed below under the headings:

- Enabling technologies and techniques (1.3.1).
- Functional avionics systems (1.3.2).
- The flight deck (1.3.3).

1.3.1 Enabling Technologies and Techniques

The enabling technologies described have a history that is interesting because it shows a comparison of modern implementations with their improvements in performance, mass, availability and safety. It is interesting also because it demonstrates how technology in the consumer market is being applied successfully into what was once thought to be a very specific and 'high end' market of aerospace. What is clear today is that many 'off the shelf' products can be used to advantage. Technology is still advancing and the following chapters try to point to the direction in which is it going:

- Chapter 2 – Avionics Technology
- Chapter 3 – Data Bus Networks

- Chapter 4 – System Safety
- Chapter 5 – Avionics Architectures
- Chapter 6 – Development Processes.

1.3.2   Functional Avionics Systems

The chapters in this section provide description of the functional systems of the aircraft and should be used in conjunction with companion books in the Wiley Aerospace Series to gain a complete picture of the modern aircraft and the aerospace environment. As far as possible, given the constraints of security and commercial sensitivities, there are descriptions of the latest aircraft to enter service, sufficient at least to gain an appreciation of avionic systems.

- Chapter 7 – Electrical Systems and Installation
- Chapter 8 – Sensors
- Chapter 9 – Communications and Navigation Aids
- Chapter 10 – Flight Control Systems
- Chapter 11 – Navigation Systems and PBN
- Chapter 13 – Military Aircraft Adaptations.

1.3.3   The Flight Deck

The flight deck is an amalgam of avionics technology and human–machine interface in a secure and comfortable environment to allow the flight crew to operate effectively during short haul and very long haul flights. This is an area that has seen great advances in the ability to provide information about the progress of the flight and the status of the aircraft and its systems. Advances are still being made which predict radical changes in the future, which is why this subject enjoys its own chapter.

- Chapter 12 – Flight Deck Displays.

Each of the chapters listed above contain both introductory and detailed descriptions of the respective subject matter. Given the integrated and interrelated nature of avionics technology and functions, cross-references have been made where appropriate in the main body of the text to help the reader to make the necessary links.

1.4   The Appendices

To assist the reader in understanding how some of the analytical tools such as dependency diagrams, fault tree analysis (FTA) and Markov analysis may be applied to typical systems, four appendices have included. These appendices address the following systems:

- Appendix A: Safety Analysis – Flight Control System
- Appendix B: Safety Analysis – Electronic Flight Instrument System
- Appendix C: Safety Analysis – Electrical System
- Appendix D: Safety Analysis – Engine Control System

The analyses in the Appendices are presented in a simple mathematical fashion to provide the reader with purely advisory and illustrative material: they should not be considered as definitive analyses of the standard that would be demanded during formal aircraft system design. Nevertheless, it is hoped that they will aid the reader in appreciating some of the design issues that need to be considered early on in the design process. (During formal design, engineers utilise dedicated design tools that undertake the appropriate analysis in a rigorous fashion. At the same time, these tools provide the required documentation to the standard necessary to convince the certification authorities that the design is safe.)

2

Avionics Technology

2.1 Introduction

The purpose of this chapter is to introduce the reader to the general principles and technologies employed in avionics computing. It is not intended to be a detailed and complete dissertation on the theory of computer science; the reader is directed to the references and other materials on that subject. Rather, the objective is to provide the reader with an awareness of computer science, the terminology, the principles involved and how these have been applied to avionics systems computing.

Firstly this chapter discusses the evolution of avionics systems from their inception as hardwired electromechanical (relay) logic, through analogue electronics into task-oriented digital computing with embedded application software in today's integrated modular avionics architectures.

This chapter will introduce the reader to the basic principles of digital computing and the major components used, for example, microprocessors and data storage devices such as read/write and read-only memories. It will outline the process fabrication and packaging of large-scale integrated circuits (ICs) in transistor–transistor logic (TTL) and complementary metal-oxide semiconductor (C-MOS) technologies. It will discuss the design of custom devices such as application-specific integrated circuits (ASICs) and field programmable gate arrays (FPGAs), and the special considerations required for the certification of complex hardware (RTCA-DO-254). It will discuss the interfacing of real-world analogue signals using operational amplifiers to filter, scale and condition signals prior to their transference into the digital domain by analogue-to-digital and digital-to-analogue conversion processes.

Whilst much of the technologies and components used in avionics computers are similar to those used in desktop/office computing, this chapter will highlight the different requirements and hence different implementations for real-time embedded avionics systems in terms of throughput, latency, accuracy, precision and resolution.

Civil Avionics Systems, Second Edition. Ian Moir, Allan Seabridge and Malcolm Jukes.
© 2013 John Wiley & Sons, Ltd. Published 2013 by John Wiley & Sons, Ltd.

2.2 Avionics Technology Evolution

2.2.1 Introduction

The first major impetus for use of electronics in aviation occurred during World War II. Communications were maturing and the development of airborne radar using the magnetron, thermionic valve and associated technologies occurred at a furious pace throughout the conflict. Transistors followed in the late 1950s and 1960s and supplanted thermionic valves for many applications. The improved cost-effectiveness of transistors led to the development of early avionics systems throughout the 1960s and 1970s.

For many years avionics were implemented in analogue devices and systems, with signal levels generally being related in some linear or predictive way to an analogue property, such as voltage, current, frequency, pulse-width or phase-shift. This type of analogue system is generally prone to variability due to modelling inaccuracies, intrinsic component and manufacturing tolerances, component temperature, age, drift and other non-linearities.

The principles of digital computing had been understood for a number of years before the technology was applied to aircraft. Digital computing overcomes the variability of analogue computing, providing accurate repeatable results with high precision without being subject to variation due to manufacturing tolerances and thermal effects. However, early digital computers were huge and were confined to mainframe office applications; they were quite impracticable for use in any airborne application until the integrated circuit device implementing whole logic functions on a single device became available.

The first aircraft to be developed in the US using digital techniques was the North American A-5 Vigilante, a US Navy carrier-borne bomber which became operational in the 1960s. The first aircraft to be developed in the UK intended to use digital techniques on any meaningful scale was the ill-fated TSR2 which was cancelled by the UK Government in 1965. The technology employed by the TSR2 was largely based upon transistors, then in comparative infancy. In the UK, it was not until the development of the Anglo-French Jaguar and the Hawker Siddeley Nimrod maritime patrol aircraft in the 1960s that weapon systems seriously began to embody digital computing.

Since the late 1970s and early 1980s digital technology has become increasingly used in the control of aircraft systems, as well as for mission-related systems. The availability of very powerful and low-cost microprocessors and more advanced software development tools has led to the widespread application of digital technology throughout the aircraft. No aircraft system – even the toilet system – has been left untouched

2.2.2 Technology Evolution

The evolution and increasing use of avionics technology for civil applications of engine controls and flight controls since the 1950s is shown in Figure 2.1.

Engine analogue controls were introduced by Ultra in the 1950s and comprised electrical throttle signalling used on aircraft such as the Bristol Britannia. Full authority digital engine control (FADEC) became commonly used in the 1980s.

Digital primary flight control with a mechanical backup has been used on the Airbus A320 family, A330 and A340 using side-stick controllers and on the Boeing B777 using a conventional control yoke. Aircraft such as the A380 and the B787 have adopting flight control

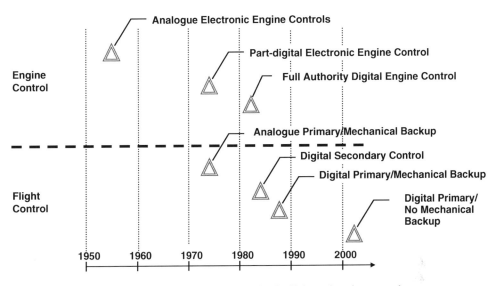

Figure 2.1   Evolution of electronics in flight and engine control

without any mechanical backup but with electrically signalled backup. Research in the military field is looking at integration of propulsion and flight control to achieve more effective ways of demanding changes of attitude and speed which may lead to more fuel-efficient operations

Avionics systems architecture evolution is summarised in Figure 2.2.

Early avionics systems can be characterised as distributed analogue computing architectures. In this era, each avionics systems function was a 'point-solution' implemented using specific hardwired analogue electronics and relays. Changes to functionality required changes to circuitry and interconnectivity.

In the mid-1970s, the first digital systems replaced analogue computers with digital computers. Each computer performed a specific 'point-solution' task, hence its name – a task-oriented computer (also known as an embedded computer system). Functionality is determined by the application software running on the target computer hardware, and changes to functionality can be effected by changing the software within the constraints of the signals available and the computer processing power.

High-speed digital data buses such as ARINC 429, MIL-STD-1553B and ARINC 629 facilitated more structured avionics architecture design in the mid-1980s, and brought with it the concept of grouping related functions into an avionics domain, with computers within the domain interconnected by a data bus. Today we call this a federated architecture. It is characterised by a number of functionally interconnected but discrete computers. Each computer has logical functional boundaries associated with the task it performs on the aircraft. Its functionality is determined by the application software running on it. Generally the computer and its embedded application software is a proprietary design of the avionics system company who supplies it. Each computer is a line-replaceable unit (LRU). Form factors are standardised to facilitate accommodation in standardised racking systems. The most universally adopted standards are the Air Transport Radio (ATR) racking system which has now largely been superseded by the modular concept unit (MCU) in civil transport aircraft. The standard fixes

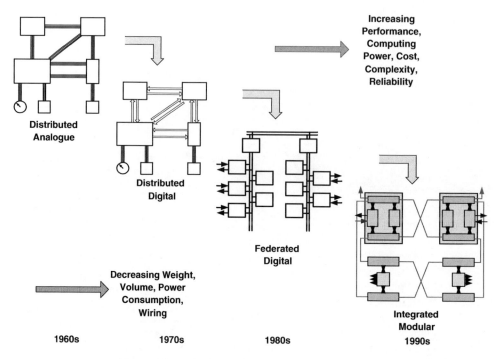

Figure 2.2 Avionics architecture evolution – summary

the LRU height, depth, connector arrangement and provisions for cooling air; the width may be varied in incremental values commensurate with the complexity of the equipment.

Integrated modular avionics (IMA) is an emerging avionics architecture and packaging technique which is being applied to current generation aircraft such as the Airbus A380, A350 and the Boeing 787. Partial implementations existed on earlier aircraft.

IMA principles introduce a common, open-architecture approach to computing hardware to provide a computationally rich resource platform on which avionics systems application software is executed. A real-time operating system manages the computational resource allocation and ensures system partitioning and segregation. Provision and certification of the hardware and software are independent. A high bandwidth communications network transports information between the computational and input/output (I/O) resources, the latter being implemented in remote data concentrators (RDCs) local to aircraft sensors and effectors. The advantages to be realised by this level of integration are:

- volume, weight and maintenance savings;
- sharing of resources, such as power supplies, across a number of functional modules;
- standard module designs yielding a more unified approach to equipment design;
- incremental certification of hardware and application software;
- management of obsolescence.

Chapter 5 provides a full discussion on avionics systems architectures.

2.3 Avionics Computing

2.3.1 The Nature of an Avionics Computer

An avionics computer is a task-oriented computer or an embedded system. It performs specific avionics functions in real time in accordance with application software stored within it and pre-loaded into its application memory on the ground.

An avionics computer may take a variety of forms. Some main processing computers such as flight management computers, flight control computers and display management computers may resemble what we expect a traditional computer to look like, a box in an avionics rack not that dissimilar to a personal computer under a desk, except that its dimensions are different. Other avionics equipment may not look like computers but in fact have similar computing hardware within them to perform the computational element of the item, such as multifunction displays, control panels, remote data concentrators, inertial reference units, and so on.

A typical avionics computer has the architecture shown in Figure 2.3 and comprises:

- **A power supply**: this converts the 115 VAC 400 Hz aircraft power to conditioned and stabilised power for the internal electronics (typically +5 V for semiconductor devices).
- **Central processing units (CPUs) plus application and data memory**: this executes the application software to perform the desired avionics function.
- **I/O interfacing**: this interfaces real-world sensors and effectors to the digital world of the CPU.
- **Data bus communications interface**: to connect the avionics computer to the avionics data bus network.

As will be seen later in Chapter 5, an integrated modular avionics architecture has the same elements but distributed differently. The Airbus A380 implements avionics computing in central processor input/output modules (CPIOMs) with standardised, common processors and a range of standardised I/O interfaces. The Boeing 787 implements I/O interfacing in separate remote data concentrators (RDCs) which communicate via the aircraft data network to centralised processors in a common computing resource (CCR) rack.

Whether a federated architecture or an integrated modular avionics architecture, the computational core has a similar architecture, although the implementation has evolved as computer architecture technology has evolved. The computational core comprises a central processing unit which executes application software held in its memory. As discussed in later paragraphs in this chapter, the central processor comprises an arithmetic logic unit which performs mathematical and logical operations in binary arithmetic; the memory comprises two elements, a read-only area of memory which contains the application software, and a read/write area of memory for computational variables.

As indicated in Figure 2.3, the core processing architecture is very similar to that of a desktop personal computer and so today it is not surprising to see many of the same components and technologies used in avionics computing. The use of commercial off-the-shelf (COTS) technology components brings with it some interesting challenges and disciplines which must be addressed in terms of both the selection and the qualification of components that will be suitable for the avionics application intended. A significant issue is the management of obsolescence. Technology turnover in the COTS world is significantly shorter than the service

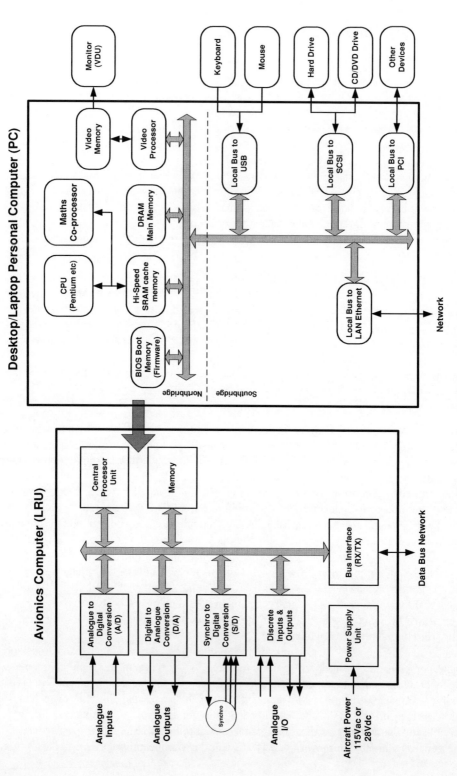

Figure 2.3 Typical computer architectures

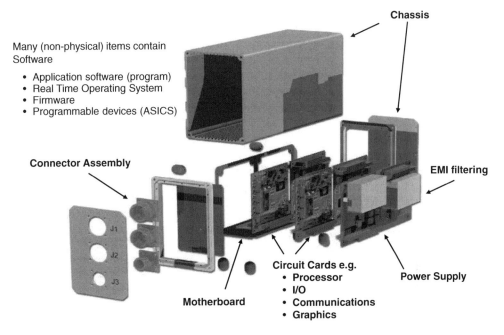

Many (non-physical) items contain
Software

- Application software (program)
- Real Time Operating System
- Firmware
- Programmable devices (ASICS)

Chassis

Connector Assembly

EMI filtering

J1

J2

J3

Motherboard

Circuit Cards e.g.
- **Processor**
- **I/O**
- **Communications**
- **Graphics**

Power Supply

Figure 2.4   Typical avionics computer – exploded view

life of an aircraft (or even the development programme), and so an obsolescence strategy must recognise that through its life, some components will need to be replaced as new parts supersede them.

Although COTS components may be used in an avionics computer, the computer itself must function reliably in the environment into which it is to be installed. This ruggedisation process must ensure that the avionics computer will operate in a severe thermal, vibration and electromagnetic environment. Figure 2.4 shows the typical construction of an avionics computer LRU.

In most applications the avionics computer is the computational element of a control system, and at a system level can be thought of as a control element with inputs, outputs and feedback. The avionics computer acquires input data from a range of aircraft sensors, computes and delivers outputs to aircraft effectors in accordance with the transfer function algorithms of its application software, and measures aircraft response in its feedback loop to provide stable and precise control of aircraft functions. Some control systems are obvious closed-loop systems, such as the autopilot and fuel management. Other control systems are open loop and the operator (pilot) closes the loop, such as the air data computer and the primary flight display. In both cases the stability criteria for a control system apply.

2.3.2   Resolution (Digitisation)

By way of an example, an air data computer (ADC) measures input quantities of pitot and static pressure (p_t and p_s) from which it derives quantities such as airspeed, barometric altitude and mach number in accordance with the simplified algorithms below, moderated by total air temperature and angle of attack.

Impact pressure (in Hg):　　$q_c = p_t - p_s$

Barometric altitude:　　$h = \dfrac{\left(29.92126^{0.190255} - p_s^{0.190255}\right)}{0.000013125214}$

Indicated airspeed:　　$IAS = 1479.1026\sqrt{\left(\dfrac{q_c}{29.126} + 1\right)^{2/7} - 1}$

Mach number:　　$M = \sqrt{5\left(\dfrac{q_c}{p_s} + 1\right)^{2/7} - 1}$

These are complex equations that seek to model the Earth's atmosphere. Prior to digital computers, these functions would have been derived by complex analogue circuit-shaping networks which would have produced a close approximation to, but not an exact evaluation of, these equations. The digital computer can repeatedly solve these equations with high precision and high accuracy. However, the difference between a digital computer and the analogue computer is the manner in which the digital computer solves these equations. In an analogue computer the equations are solved as a continuous function. In a digital computer the input signals are quantised into discrete values and the equations are solved in successive discrete time samples. The digitisation processes may introduce undesirable artefacts unless correctly understood and managed.

Figure 2.5 shows the ISO–atmosphere relationship between altitude and air pressure. If this model were to be implemented in an analogue computer, close inspection would reveal this function to be always a continuous function between input and output. In a digital computer, close inspection would reveal the function to be digitised into a series of steps. The step height is the limiting resolution or precision of the digital machine, usually determined by the number of binary digits (bits) in its computational processes and the resolution of its A to D and D to A conversion process – see Section 2.5.

Accuracy is a measure of how well the implemented function models the ideal function, in this case the Earth's atmosphere. In an analogue computer this will be the accuracy with which the analogue circuitry fits the ideal curve and will be subject to component tolerances and thermal drift. In a digital computer it will be the accuracy of the equations used to express the model of Earth's atmosphere (in itself an approximation).

2.3.3　The Sampling Frequency (Refresh Rate)

Figure 2.6 shows a simplified schematic of an aircraft pitch attitude control system in which the flight guidance computer determines and controls the required aircraft pitch attitude. The command signal from the pilot's inceptor (yoke or side-stick) is compared with the measured pitch attitude, determined via the aircraft inertial reference system, to generate an error signal which, after shaping by the control laws in the flight guidance computer, drives a servo-actuator whose output positions the control linkage to the pitch axis elevator, causing the aircraft to

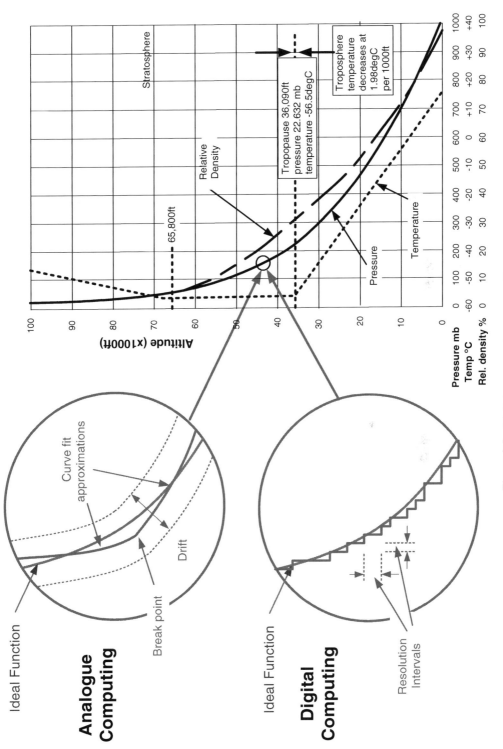

Figure 2.5 Digital versus analogue computing

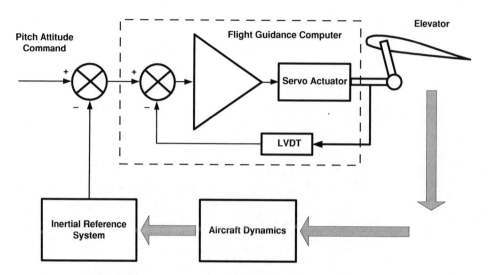

Figure 2.6   Pitch attitude control system simplified schematic

respond in pitch. The change in pitch attitude is sensed by the inertial reference system which 'backs-off' the error signal until the new commanded pitch attitude is achieved.

The desired control system performance to a step change input is a critically damped response. The open-loop and closed-loop transfer functions are tailored to the aircraft dynamic performance using classical control system design techniques including Bode diagrams, Nyquist plots and Nichols charts – see Chapter 2 of Reference [1]. Figure 2.7 shows a typical Bode diagram and illustrates the closed loop gain and phase margins with reference to the instability criteria of unity gain (0 dB) and a phase shift of 180° – a classical indication of

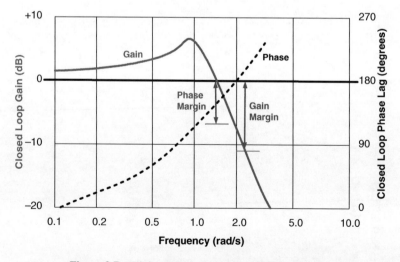

Figure 2.7   Flight control system typical Bode diagram

control loop stability. Inadequate margins will exhibit themselves as underdamped responses and control loop instability, which in the aircraft pitch axis will manifest themselves in a phugoid response as the aircraft pitches up and climbs, and then pitches down and descends, accompanied by speeding up and slowing down. With pilot-in-the-loop control this characteristic is also known as pilot induced oscillation (PIO).

An analogue computer provides a continuous function between input and output. In the example above, if the pitch attitude changes there will be a corresponding change in the elevator servo-actuator position. Analogue computing naturally tends to support reducing gain at higher frequencies due to the bandwidth characteristics of typical analogue amplifiers. However, digital computing is not so benign in this regard and must be addressed with some care and attention.

A digital computer calculates the output value based on a point-in-time sample of input quantities. If an input changes then we need to re-compute the output values. So a real-time computer must take successive samples of its inputs and re-compute new output values. Its operation is not continuous but a series of samples in time. The time interval between samples is called the refresh rate and this must be short enough to yield a pseudo-continuous response at the output. Nyquist theory for a sample data system states that to be able to faithfully reproduce a signal, the sampling rate of the system must be at least twice the highest frequency of interest within the signal. We call this the Nyquist frequency. Good design practice will set the sample rate at several multiples of the Nyquist frequency – this is known as oversampling.

A digital computer will take a finite time to work through its application software and generate an output after sampling an input. This is called latency and is somewhat equivalent to a delay or a lag in an analogue control system. Excessive delays or lags can produce instability. There are many series processing elements in a digital avionics control system end-to-end; from sensor, through A to D conversion, through computation, through data bus, through more computation, through D to A conversion to effectors. Each element has its own latency and each is running sampling processes asynchronous to the others. Variations in end-to-end system latency will produce jitter, with the possibility of introducing loop instability.

Figure 2.8 illustrates the consequences of sample rate on signal processing in the time domain. The figure illustrates the impact of a fixed sample frequency on increasing sinusoidal input signal frequencies. Figure 2.8a illustrates that at low frequencies, where there is a significant degree of oversampling, the result is a good rendition of the continuous input signal. At frequencies approaching the Nyquist frequency, shown in Figure 2.8b, distortion is present in the form of an alias component superimposed upon the desired signal. This alias signal originates from the difference or 'beat' frequency between the input and the sampling frequencies. Figure 2.8c illustrates that below the Nyquist frequency the alias component corrupts the signal to a level that renders it unusable.

As can be seen, as the ratio between the sampling frequency and the signal frequency reduces, the faithfulness by which the signal can be reconstructed progressively degrades. This consequence can be more readily explained by analysing the system in the frequency domain, and this can be seen in Figures 2.11 to 2.15 below. It is usual, therefore, to introduce a low-pass anti-aliasing filter before the sampling stage to remove from the signal any frequency components that approach the Nyquist frequency. The system designer must take care to set the sample frequency, the computational refresh rate and processing latency to ensure that the control system can faithfully process all signal frequency components of interest and remain stable under all potential operating conditions.

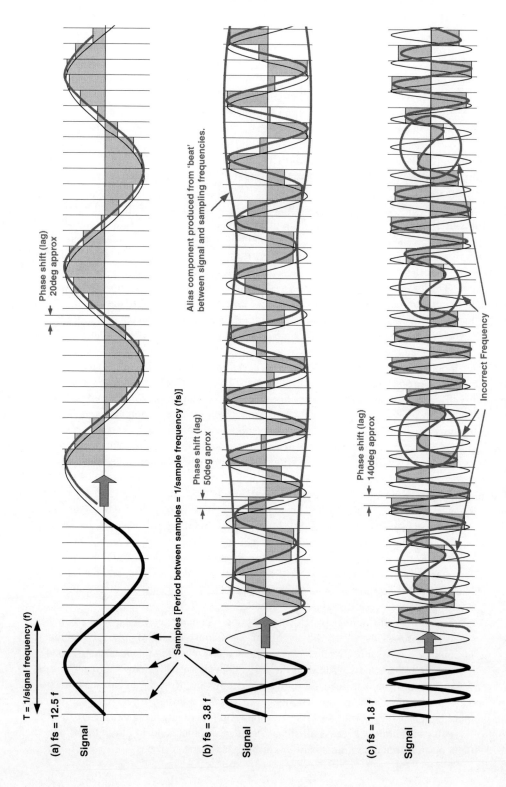

Figure 2.8 Impact of sampling frequency

Not all aircraft systems are continuous sampling control systems; some are event-driven sequences, for example extension and retraction of the landing gear. In these systems, processing latencies predominate and need careful attention to ensure the safe timing and correct sequencing of events is maintained with adequate margin. Malfunction in this regard can produce some unexpected and non-intuitive results.

2.4   Digital Systems Input and Output

2.4.1   Introduction

The foregoing describes how a digital processor operates, and the hardware that supports that operation. Digital processing may be undertaken at very high rates and with great precision. However, to perform many useful tasks the digital computer has to interface with the 'real' world, which is analogue in nature (see Figure 2.9).

In the analogue world shown on the right of Figure 2.9, parameters are represented in real units: voltages, degrees per hour, pitch rate, and so on. To permit digital processing these analogue signals have to be converted into digital form by a process called analogue to digital conversion (ADC). This converts the analogue signal into a series of digital words which the processor is able to manipulate in accordance with the instruction set that constitutes a particular algorithm or control philosophy. Once the digital processing is complete, a reverse process called digital to analogue conversion (DAC) converts the output back into an analogue

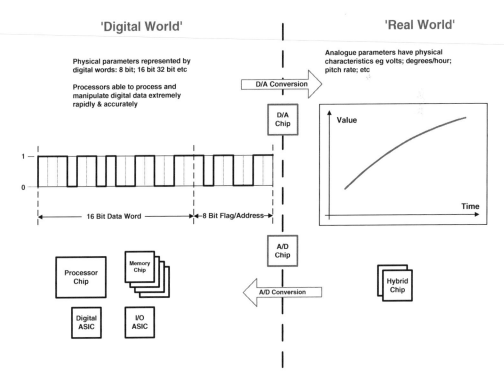

Figure 2.9   Comparison of analogue and digital domains

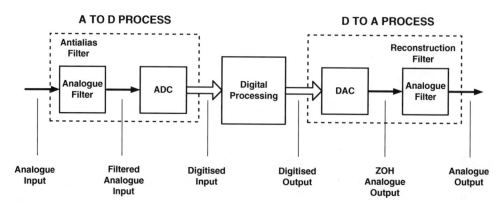

Figure 2.10 A to D and D to A conversion

parameter suitable for supplying a demand to an actuator to move: for example, a flight control actuator, or controlling a modulating valve within a fuel or environmental control system (ECS).

The A to D and D to A processes are shown as a signal flow diagram form in Figure 2.10. The main elements are as follows:

- The A to D process, which filters the analogue input using an anti-alias filter, the purpose of which will be described later. The filtered analogue signal is passed into the ADC and a stream of digital data in the desired form emerges as the digital input. The rate at which the analogue data is sampled is key to the fidelity and accuracy of the data supplied to the digital processor.
- The digital processing performs whatever tasks are demanded of it and outputs a modified digital output when processing is complete.
- The output digitised data are then passed to the D to A process which uses a DAC together with a circuit called a zeroth order hold (ZOH) to produce a raw analogue output. This output is passed in turn through a reconstruction filter and the final analogue signal is provided for use.

2.4.2 Analogue to Digital Process

In order to understand some of the key issues, the A to D process needs to be described in more detail. Figure 2.11 depicts the process as a control diagram with four main signal elements:

- The incoming analogue input signal is the desired signal which is to be subject to digital processing. This signal will vary continuously with time depending upon the behaviour of the real world system that it represents. This particular example is varying between around 7.0 V to around 2.5 V over a 60-second window. The signal is fed into a sample and hold (SH) circuit which periodically samples the analogue signal.
- The output of the SH circuit is shown as the sampled analogue signal in Figure 2.11. This closely replicates the original signal but with a staircase appearance, with the width of the

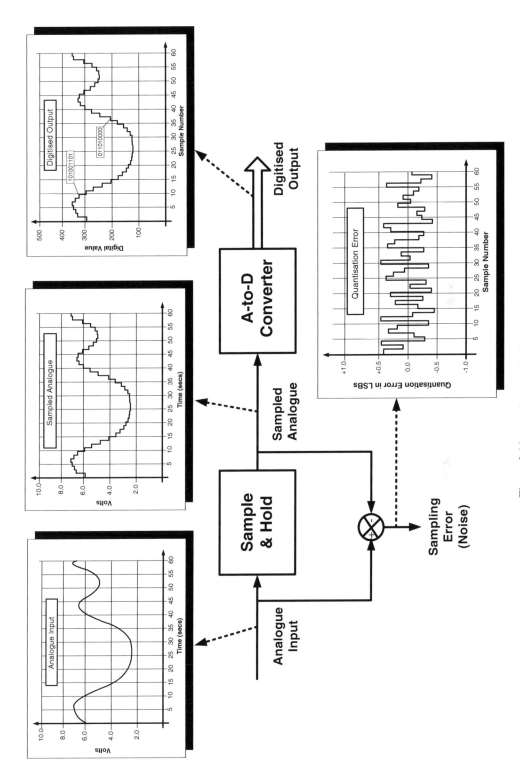

Figure 2.11 Analogue to digital conversion

'steps' depending upon the sampling rate. The higher the sampling rate, the narrower these steps will appear. In practice, sampling cannot be instantly achieved, and also in practice the sampling rate has a great effect upon the fidelity of the sampled (and therefore the digitised) signal, as has been explained in the previous section.

- The sampled data are converted into a stream of digital words which are a digital representation of each of the analogue steps. Figure 2.11 shows an 8-bit representation. This is typical of the level of resolution required for an aircraft system such as fuel or hydraulics. 12-bit or 16-bit resolution may also be used for flight data such as airspeed, altitude and attitude.
- Referring back to the original control diagram in Figure 2.11, it can be seen that there is an error signal or noise due to the fact that sampling process is not perfect. This is known as quantisation error. It varies between ± 0.5 of the least significant bit (LSB) and depends upon the scaling of the digital word. Mathematically the error is uniformly distributed between $+0.5$ LSB and -0.5 LSB. The standard deviation is $1/\sqrt{12} \times$ LSB (~ 0.29 LSB), depending upon the number of bits provided by the digitiser. For an 8-bit digitiser this is equivalent to analogue noise of $\sim 0.29/2^8$ or $\sim 0.29/256$ (1/900).

2.4.3   Sampling Rate

To fully understand the importance of the sampling rate it is necessary to gain some understanding of the frequency components of the input signal, to ensure that the sampling process will capture a sufficiently high proportion of these components and represent the signal accuracy to the desired level. To comprehend this process it is necessary to understand the interaction of both time and frequency domains (Figure 2.12).

Every time-varying analogue signal has a corresponding signature or frequency response in the frequency domain, and *vice versa*. Most engineers are more familiar with the signal performance in the time domain since that is what can be readily displayed, for example, on an oscilloscope. However, considering performance in the frequency domain can be vital to achieving the necessary system performance. Apart from a pure sinusoidal waveform which

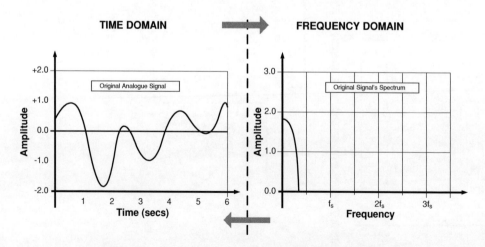

Figure 2.12   Interrelationship of time and frequency domains

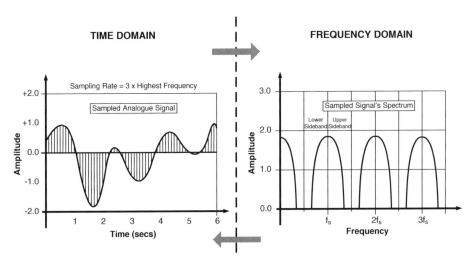

Figure 2.13   Sampling above the Nyquist rate

would have a single spectral line at the frequency of the sinusoid, most waveforms comprise multiple frequency components, all of which contribute small elements to the overall analogue signal as shown in the right-hand diagram of Figure 2.12.

If this analogue signal is sampled – as it is as part of the A to D process – then the situation becomes more complicated as shown in Figure 2.13. The importance of the Nyquist rate will be described shortly.

Figure 2.13 depicts the time and frequency domain representations for a signal sampled at three times the frequency of the highest frequency component. The time domain characteristic on the left results in a family of frequency spectra each, with an upper and lower sideband as shown on the right. Each frequency spectra element is separated from the next, so there is no possibility of adjacent spectra interfering with each other.

As the sampling rate decreases this has the effect of broadening each of the frequency spectra as shown in Figure 2.14.

Eventually there comes a point where adjacent spectra begin to touch. This occurs when the sampling rate (fs) is exactly twice the highest analogue frequency component. This is known as the Nyquist rate, and it is crucial in many sampling systems. For systems where high performance is desired, then the sampling rate should be set at least at the Nyquist rate, and preferably much higher. Figure 2.15 shows the effect of reducing the sampling rate to below the Nyqist rate. Further decreasing the sampling rate causes the adjacent spectra to interfere, an effect that is known as aliasing, shown by the dark areas. Aliasing reduces the quality of the sampled waveform and eventually the characteristics of the original analogue waveform will be lost.

2.4.4   *Digital to Analogue Process*

The D to A process is the reverse of that for A to D. However, there are subtleties that apply here.

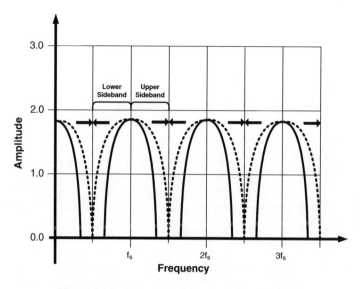

Figure 2.14   Effect of decreasing the sampling rate

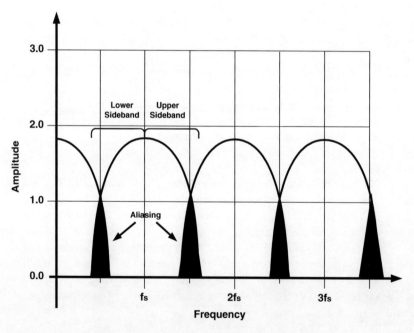

Figure 2.15   Sampling rate below Nyquist rate

Figure 2.16 portrays the D to A conversion process as a control diagram with three main signal elements:

- A series of stand-alone values – known as an impulse train – that capture a series of analogue values representing each of the digital values that the digital processor will be presenting at the end of each computation cycle. Ideally these could be passed though an analogue filter at half the sampling rate. In mathematical terms, this would faithfully reconstruct an analogue output that captured the characteristics of the digital processor output. In practice, an idealised impulse train cannot be constructed as there are practical considerations in the DAC holding the last value until a subsequent sample (impulse) is received. This process is termed a zeroth order hold (ZOH) and it represents the inverse of the sample and hold (SH) process used during A to D conversion.
- In actual terms, the ZOH circuit provides an analogue staircase which is similar to the SH function during the A to D process, except that in this case an analogue staircase is being generated from digital data.
- The staircase data are filtered to produce a time-variable analogue output as shown in Figure 2.16. This is the reconstructed analogue signal and includes correction due to the fact that an idealised pulse train cannot be implemented in reality. The true nature of the reconstruction filter that embodies this compensation is complex and is outside the scope of this book.

This description is a simplified overview: for further details in regard to a comprehensive description of digital signal processing, see Reference [2].

2.4.5   Analogue Signal Conditioning

A considerable degree of analogue computing precedes the A to D conversion process described above. The basic building block within analogue computing is the operational amplifier (Op Amp). An Op Amp may within itself comprise 30 to 40 transistors to provide the necessary signal amplification, power drives and buffering to enable it to perform its various functions. An Op Amp is classically portrayed as a triangular symbol as shown in Figure 2.17.

The arrangement of other components – resistors and capacitors – around the Op Amp itself characterise both performance and use. Op Amps may be used to provide the following functions within the analogue domain, prior to the A to D conversion process:

- signal inversion;
- signal summing;
- signal subtraction;
- signal amplification (scaling);
- signal differential amplification;
- signal filtering.

Reference [3] is useful for further details on Op Amps.

An example of a three-stage filter using three Op Amps plus associated circuitry is shown in Figure 2.18. Many filter types may be implemented and their design invariably demands a compromise of simplicity versus performance.

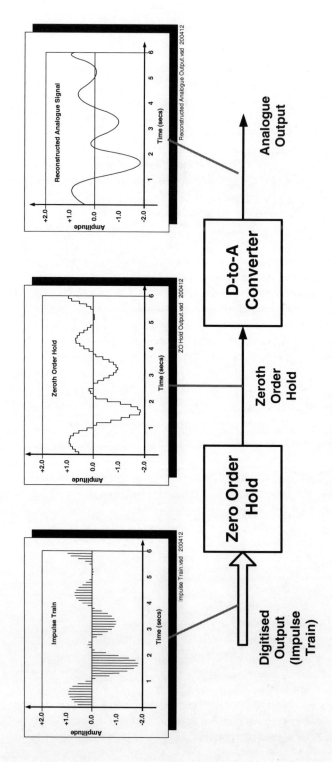

Figure 2.16 Digital to analogue conversion

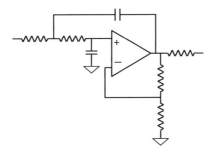

Figure 2.17 Operational amplifier

Within a typical A to D application-specific integrated circuit (ASIC), a number of functions may be integrated onto a single silicon device or chip. The ASIC example portrayed in Figure 2.19 encompasses a number of functions:

- A number of analogue input channels or 'strips' which contain three Op Amps per strip – analogous to the three-stage filter already depicted in Figure 2.18.
- A voltage level changing/test matrix probably also incorporating the ADC function.
- A 1000-gate digital memory to store present digitised data.
- Bonding pads surrounding the circumference of the chip, in this case ~70 bonding pads.

This particular ASIC is representative of one used within control and monitoring lanes of the full authority digital engine control (FADEC) unit for the Rolls-Royce BR700 series of engines used on Global Express and a number of other high-end business jets.

2.4.6   Input Signal Protection and Filtering

As well as the more pure signal processing issues already described, it has to be recognised that an aircraft represents a very hostile environment to relatively low-voltage analogue

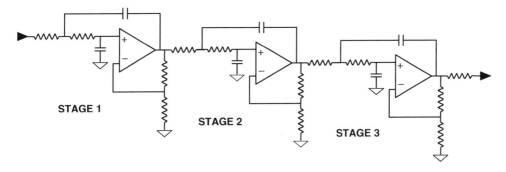

Figure 2.18 Typical three-stage analogue filter

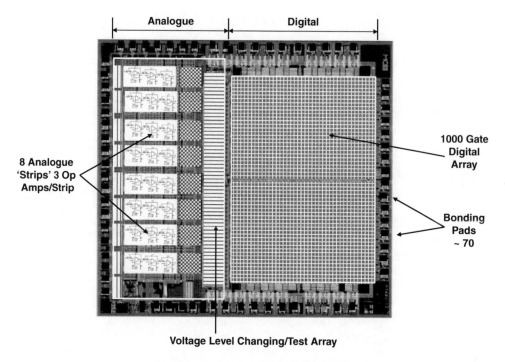

Figure 2.19   Portrayal of an input ASIC

signals. An additional screen of protection has to be applied to ensure that the foregoing (somewhat idealistic) conditions might be met. Aircraft wiring can effectively act as antennas throughout the aircraft – especially in a carbon composite structure – and specific measures need to be taken to ameliorate the effect of voltage spikes or current surges throughout the aircraft wiring due to lightning strikes, extraneous external high voltages such as high intensity-radio fields (HIRF), and other potential risks including onboard hazards due to personal electronic devices (PEDs) used by passengers. The generic signal chain described in Figure 2.20 illustrates some of these issues.The signal conditioning and ADC conversion are as portrayed in the highlighted boxes. Additional protection is included to address the following issues:

- Voltage clamping to minimize the effect of voltage 'spikes' that may be induced into the wiring.
- EMI filtering to attenuate RF noise and minimise current surges.
- Surge protection to minimise any current surges due to external effects.
- Hysteresis to avoid nuisance changes of state being recorded.

A good design will incorporate some and possibly all of these techniques, acknowledging that the aircraft is a noisy environment and preserving the accuracy and integrity of the incoming signal.

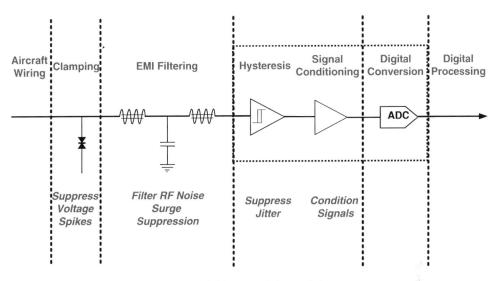

Figure 2.20   Generic input chain

2.4.7   Analogue Signal Types

An aircraft avionics system encompasses a range of different analogue signal types. Several examples are listed below:

- Potentiometer – analogue position.
- Temperature sensor – analogue temperature – fuel, hydraulics, bleed air, and so on.
- Linear/rotary variable differential transformer (L/RVDT) – analogue position – flight control surface position (nowadays used in preference to a potentiometer).
- Pressure sensor – analogue pressure – fuel, hydraulics, bleed air, and so on.
- Pulse (speed) probe – rotary analogue speed – wheel speed, engine or gearbox speed.
- Proximity switch – discrete position (proximity) – slats and flaps, doors, thrust reversers.
- Capacitive sensors to measure fuel level/contents and to measure fuel properties – density and permittivity.
- Other specialised applications: flow rates; AC and DC current flow.

2.5   Binary Arithmetic

2.5.1   Binary Notations

Calculations within a digital computer are performed using binary arithmetic, that is, numbers to base 2 (2^n), compared with the decimal arithmetic we are used to (numbers to base 10, i.e. 10^n). For example:

$$1\,0\,1\,1_{\text{base2}} = 1 \times 2^3 + 0 \times 2^2 + 1 \times 2^1 + 1 \times 2^0 = 11_{\text{base10}}$$

The range of numbers that can be held within a digital computer is limited to the physical number of storage elements (binary digits or bits) the computer can accommodate in its calculations, or its native machine resolution, usually called a word. The example shown above is a four-bit word, sometimes known as a nibble. An 8-bit word is widely known as a byte. Modern computers work in larger word sizes, usually in byte multiples; 16 bits, 32 bits, and recently 64 bits are common word sizes for desktop and avionics computers.

In a 16-bit machine the range of un-signed integer numbers that can be expressed is in the range 0 to $2^{16} - 1 = 65,535_{base10}$. To express both positive and negative numbers the most significant bit is allocated as the sign bit: 0 indicates the number is positive, 1 that the number is negative. So the range of signed integers in a 16-bit machine is $-32,767$ to $+32,767$. This is known as sign-magnitude notation.

An alternative and more usually used notation is known as 2's complement. In 2's complement notation the sign bit carries a numerical value as well as the sign. In a 16-bit integer machine it has the value -2^{15}, so the range of numbers that can be expressed is:

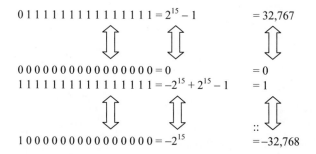

$$0\,1\,1\,1\,1\,1\,1\,1\,1\,1\,1\,1\,1\,1\,1\,1 = 2^{15} - 1 \qquad\qquad = 32,767$$

$$0\,0\,0\,0\,0\,0\,0\,0\,0\,0\,0\,0\,0\,0\,0\,0 = 0 \qquad\qquad = 0$$
$$1\,1\,1\,1\,1\,1\,1\,1\,1\,1\,1\,1\,1\,1\,1\,1 = -2^{15} + 2^{15} - 1 \qquad = 1$$

$$1\,0\,0\,0\,0\,0\,0\,0\,0\,0\,0\,0\,0\,0\,0\,0 = -2^{15} \qquad\qquad = -32,768$$

2's complement notation has the advantage that there is no discontinuity in the transition from -1 through zero to $+1$, and adding two numbers is always ADD regardless of the sign of the numbers; subtract is always SUBTRACT. There is a complication with multiplication and division but that can be resolved by a simple method called Booth's algorithm. There is also a notation called 1's complement, but it is rarely used and will not be discussed here.

The finite size of numbers that can be held in the machine gives rise to a problem if, for instance, two numbers within range are added together but their sum is greater than the number range the machine can express. This is called overflow, and is an exception that must be managed. Mathematically the associative law is invalidated; that is $(a + b) - c$ may not give the same answer as $a + (b - c)$ if $a + b$ exceeds the number range of the machine. Some machines offer the capability for double precision arithmetic which can concatenate (join together) two words to form a double-length word, for example, two 16-bit words concatenating to form a word of 32 bits. This is sometimes known as long integer notation.

Integer arithmetic does not suit all variables. Most avionics system variables are rational numbers and require a fractional part to provide the degree of precision required in computation. Fixed point and fractional notations offer alternative representations.

In fixed point notation the binary point is positioned so that the bits below the binary point afford the desired resolution of the variable concerned.

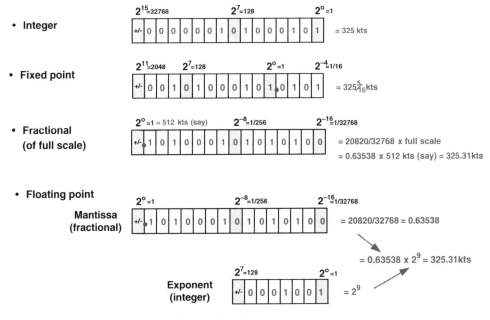

Figure 2.21   Binary notations

In fractional notation the binary number represents a fractional value of the full-scale range of the variable. This is a familiar concept for analogue system designers where, for example, the amplitude of a voltage signal represents the value of a variable quantity as a percentage of full scale. In 2's complement fractional notation the most significant bit takes the value -1, the next most significant bit then has the value of $2^{-1} = 0.5$, and so on.

Figure 2.21 shows examples of the notations discussed above to encode airspeed.

An alternative but more computing- and memory-intensive means to express rational variables is to use scientific or floating point notation, that is: $a \times 2^b$. In decimal arithmetic we are used to expressing a, the mantissa, as a decimal rational number in the range 9.99 . . . .. and b, the exponent, as an integer (or the power to which the mantissa is raised); both expressed as signed magnitude numbers, for example: $-2.35 \times 10^2 = -235$ and $7.68 \times 10^{-1} = 0.768$. In binary arithmetic it is more normal to express the mantissa as a fractional number, usually in 2's complement notation, and the exponent as a signed integer. Of course this means the quantity has to be held as two components, the mantissa and the exponent (as shown also in Figure 2.21), and mathematical functions require arithmetic operations on both the mantissa and the exponent parts. As well as being able to express very large and very small numbers, floating point arithmetic has the advantage that precision is always maintained whether the result of an arithmetic process is a large or small number.

Both fractional and fixed point notations require the programmer to maintain constant vigilance concerning the position of the binary point and/or the scaling of the variables. So these notations are rarely used in high-order languages, when the programmer usually declares a variable to be an 'integer'; a 'long integer' for double precision arithmetic; or a 'real' quantity (in some languages declared as a 'float') to designate the variable as floating point.

2.5.2 Binary Addition, Subtraction, Multiplication and Division

The rules for addition and subtraction in binary arithmetic are the same as those learned in primary school; that is, to start with the least significant digit and work to the most significant digit, using the techniques of carry and borrow. For example:

$$0\ 0\ 1\ 1\ 0_{\text{base }2} = 6_{\text{base }10}$$

$$\text{plus} \quad 0\ 1\ 1\ 1\ 1_{\text{base }2} = 15_{\text{base }10}$$

$$= \quad 1\ 0\ 1\ 1\ 1_{\text{base }2} = 21_{\text{base }10}$$

$$\text{carry} \quad 1\ 1\ 1\ 0$$

Binary multiplication is similar to decimal multiplication, again using the same rules learned in primary school, starting with the least significant digit of the multiplier and performing a set of addition and shift operations on the multiplicand working towards the most significant digit of the multiplier.

Division successively subtracts the divisor from the diminuend to derive the quotient. The diligent student is directed to references [4] and [5] to explore these principles further.

For floating point numbers the rules of exponential arithmetic apply, namely:

$$(a \times 2^p) \times (b \times 2^q) = (a \times b) \times 2^{(p+q)}$$

Two arithmetic operations are required: multiply the mantissas and add the exponents.

Addition requires the quantity with the smaller exponent to be successively shifted right (divided by 2) until the exponents of both numbers are the same, then the two mantissas can be added together; the exponent is the larger of the two.

$$(a \times 2^p) + (b \times 2^q) = (a + b \times 2^{(q-p)}) \times 2^p$$

2.5.3 The Arithmetic Logic Unit

The arithmetic logic unit (ALU), sometimes also known as an execution unit, shown in 4-bit form in Figure 2.22, is the fundamental building block of a digital processor. The ALU accumulator register is pre-loaded with operand A. Operand B is fetched from the computer memory. The processor control unit then commands the ALU to perform the desired mathematical or logical operation on the two operands (add, subtract, logic and, logic or, etc.). The result R is usually written back into the accumulator register or stored back in the computer memory. Carry-in (D) and carry-out (F) allow the result of one operation to be available for the next in long integer arithmetic.

Further information on the theory and design of digital computers can be found in references [4] and [5].

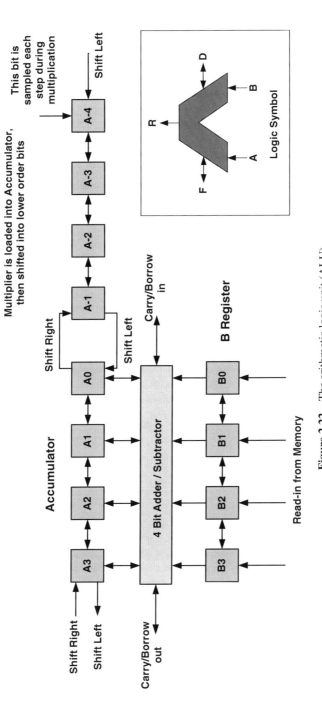

Figure 2.22 The arithmetic logic unit (ALU)

2.6 The Central Processing Unit (CPU)

The central processing unit (CPU), sometimes also more simply known as the central processor, is the functional heart of a digital computer. It fetches and executes the sequence of instructions that expresses the application software, performing arithmetical and logical operations on data and making decisions on the outcomes of those operations. The sequence of instructions (the application software) is held in the computer memory (see Section 2.9), as is the data to be operated on. So the CPU performs two functions:

- managing memory access for instructions and data (read and write);
- executing instructions on the data it accesses from memory.

A generalised and simplified CPU block diagram is shown in Figure 2.23. It has three main components, two arithmetic logic units (ALUs) and an instruction execution (or control) unit.

- **ALU(A)** generates the memory address for instructions and data. This ALU is a partial implementation, performing add and subtract operations only.
- **ALU(D)** performs a full set of arithmetic and logical operations on data (known as operands) fetched from memory.
- **The instruction execution unit** fetches instructions and data from memory and writes data to memory at the memory address provided by ALU(A). It manages the instruction sequence and controls the operations of both ALUs.

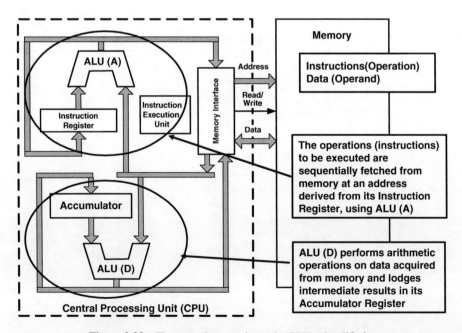

Figure 2.23   The central processing unit (CPU), simplified

Normally instructions are executed in sequence, but the sequence may be altered by the outcome of calculations or the state of discrete data. This is where the real power of a digital computer is to be found, and the real difference between it and previous technologies, in that it has the ability to make decisions based on the results of algorithms it executes, albeit the algorithms, test conditions and outcomes are pre-determined by the software designer.

2.6.1   CPU Instruction Format

The format of a typical CPU instruction is shown below:

Operation	Register	Address Mode	Operand Address

It comprises:

- the operation to be executed, or operation (op) code, for example, add, logic and, multiply, skip if greater than zero, goto, and so on;
- the internal accumulator or register to be use as the first operand (A);
- the location in memory of the second operand (B) with some address moding to extend the address field.

Table 2.1 shows a list of typical instructions the CPU can execute. It comprises:

- data transfer instructions, such as load, store, move;
- arithmetic instructions, such as add, subtract, multiply, divide;
- bit manipulation instructions, such as logic and, logic or, shift left, shift right;
- program transfer instructions, such as jump (goto), skip the next instruction if greater than zero, loop back.

2.6.2   Instruction Execution Sequence

The following example illustrates the instruction execution process. The example is the simple algorithm to evaluate the expression $y = mx + c$, where x is an input variable, m and c are constants and y is the resultant output value. Additionally the algorithm limits the value of y so that it does not exceed the value k. Although in a high-order language we would write this as a single statement, the CPU does not understand this. The high-order language statement needs to be compiled into a sequence of instructions the CPU can understand, known as machine code. Figure 2.24 shows the flowchart of the operation sequence that the CPU must execute in order to evaluate this function. The flow chart is then interpreted into a set of instructions the CPU will step through in sequence, in this example these are held in memory locations 100 to 106. Constants m, c and k are held in memory locations 200 to 203; variables x and y in memory locations 301 and 302. The reason for assigning instructions, constants and variables to different areas in memory will become clearer later.

Table 2.1   Typical CPU instructions

DATA TRANSFER		ARITHMETIC		PROGRAM TRANSFER	
General Purpose		**Addition**		**Unconditional Transfer**	
MOV	Move byte or word	ADD	Add	JMP	Jump
PUSH	Push onto stack	INC	Increment by 1	CALL	Call procedure
POP	Pop word off stack	**Subtraction**		RET	Return from procedure
Input/Output		SUB	Subtract	**Conditional Transfer**	
IN	Input byte or word	DEC	Decrement by 1	JAZ	Jump above zero
OUT	Output	NEG	Negate	JEZ	Jump equal zero
Address Object		CMP	Compare	JNEZ	Jump not equal zero
LEA	Load address	**Multiplication**		JBZ	Jump below zero
LDS	Load pointer	MUL	Multiply	JS	Jump if sign positive
		IMUL	Integer multiply	JN	Jump if sign negative
BIT MANIPULATION		**Division**		JO	Jump if overflow
Logical		DIV	Divide	JNO	Jump if not overflow
NOT	NOT byte or word	IDIV	Integer divide	**Iteration Control**	
AND	AND byte or word	**Convert**		LOOP	Loop
OR	Inclusive OR	CBW	Convert to word	LOOPZ	Loop of equal zero
XOR	Exclusive OR	CWD	Double precision	LOOPNZ	Loop if not zero
Shift				**Interrupt**	
SHL	Shift left	**STRING**		INTE	Enable interrupts
SHR	Shift right	REP	Repeat	INTD	Disable interrupts
Rotation		REPZ	Repeat while zero	IRET	Return from interrupt
ROL	Rotate left	REPNZ	Repeat non zero		
ROR	Rotate right				

Each instruction follows a stepwise sequence as shown in Figure 2.25 to Figure 2.27:

- Fetch instruction – Figure 2.25.
- Fetch data – Figure 2.26.
- Execute instruction – Figure 2.27.

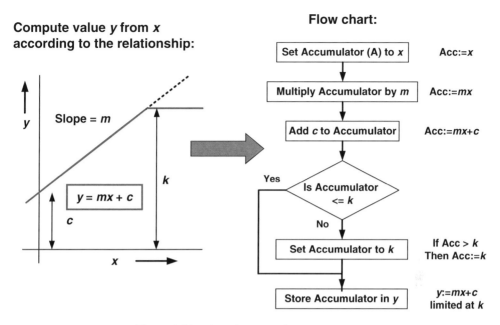

Figure 2.24   Flow chart to evaluate $y = mx + c$

Evaluation of the algorithm commences with the instruction register set to 100.

- **Fetch instruction**: ALU(A) is set to the function $R = A + 0$ and its output is routed through the memory interface to address the memory at location 100 and the instruction 'Load the Accumulator with the contents of memory location 300' is fetched from memory (memory read cycle). The operand address, in this case 300, now appears on the B port of ALU(A).
- **Fetch data**: ALU(A) is now set to the function $R = 0 + B$ and addresses the memory at location 300. Variable x is fetched from memory (since memory location 300 contains the value of variable x) and this now appears on the B port of ALU(D).
- **Execute instruction**: ALU(D) is now commanded to execute the 'Load' instruction and is set to the function $R = 0 + B$, and variable x is loaded into the accumulator register.

This completes the first instruction and the CPU steps onto the next instruction by incrementing the instruction register to the value 101, and the execution cycle commences again:

- **Fetch instruction**: Fetches the instruction at memory location 101, namely 'Multiply the Accumulator with the contents of memory location 200'. The operand address, in this case 200, now appears on the B port of ALU(A).
- **Fetch data**: Constant m at memory location 200 is fetched from memory and this now appears on the B port of ALU(D).
- **Execute instruction**: ALU(D) is now commanded to execute the 'Multiply' function $R = A \times B$ by a series of add and shift operations into the Accumulator register, resulting in it taking the value of x multiplied by m.

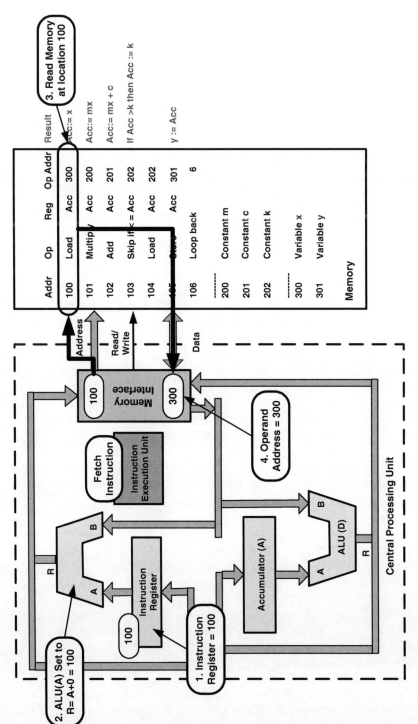

Figure 2.25 Fetch instruction

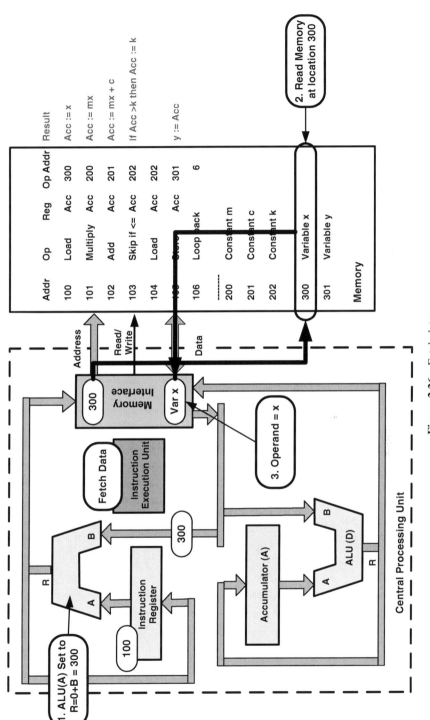

Figure 2.26 Fetch data

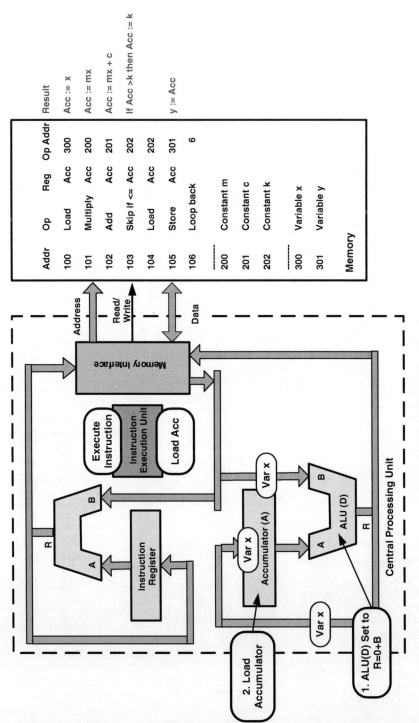

Figure 2.27 Execute instruction

The CPU steps onto the next instruction at memory location 102, and the execution cycle repeats again, this time:

- **Fetch instruction**: 'Add to the accumulator the contents of memory location 201'.
- **Fetch data**: Constant c is fetched from memory.
- **Execute instruction**: ALU(D) is now commanded to execute the 'Add' function $R = A + B$ and writes the answer back into the accumulator register resulting in it taking the value of $mx + c$.

The CPU steps onto the next instruction at memory location 103, and the execution cycle repeats again, this time:

- **Fetch instruction**: 'Skip if the accumulator contents are less the value of memory location 202'.
- **Fetch data**: Constant k is fetched from memory.
- **Execute instruction**: ALU(D) is now commanded to execute the 'subtract' function $R = A - B$, but this time it does not write the contents back into the accumulator register; instead the instruction register is incremented if the answer is negative, that is, $mx + c$ is less than constant k, then the next instruction is skipped.

The CPU steps onto the next instruction. The instruction at memory location 104 will only be executed if $mx + c$ is greater than constant k. This instruction is a 'Load' instruction like the first at location 103, and will overwrite the accumulator register with the constant k.

The instruction at memory location 105 is a little different to previous instructions.

- **Fetch instruction**: 'Write the accumulator contents to memory location 301'.
- **Store data**: ALU(A) is set to the function $R = 0 + B$ through the memory interface to address the memory at location 301 as before. ALU(D) is now commanded to execute the function $R = A + 0$, but this time it does not write the contents back into the accumulator register; instead its output is routed to the memory interface and written to memory location 301 (memory write cycle).

The final instruction at memory location 106 changes the instruction sequence:

- **Fetch instruction**: 'Loop back by 6'.
- **Load instruction register**: ALU(A) is set to the function $R = A - B$, but this time it does not address the memory; instead its output is loaded back into the instruction register which loops the instruction sequence back to begin again at memory address 100.

The above rather simple example serves to illustrate how a CPU fetches and executes instructions contained in its memory and reads and writes data to and from memory. It can be seen that the memory contains three classes of information:

- instructions;
- constants;
- variables.

The memory characteristics required for these classes differs; the implications are explored further in Section 2.9.

2.6.3   Extended Operand Addressing Modes

In general the computer memory is much larger than can be addressed by the simple mechanisms illustrated above, and a range of techniques (or modes) together with a set of registers, usually called the index registers, are used to extend the address range. The index registers are a set of address registers whose contents can be loaded and manipulated arithmetically by ALU(A) to provide various means to index or point to areas of memory. The architecture arrangement is shown in Figure 2.28.

Some of the address modes include:

- **Direct addressing**: This is used to access data (generally constants) in the same segment (page) as the instructions being executed. The resultant operand address is the current instruction page concatenated (joined) with the short operand address in the instruction.
- **Relative addressing**: This is used to access data (generally constants) in the same block of code as the instructions. The resultant operand address is the current instruction address plus the short operand address in the instruction. This mode has the benefit that the instruction code block with associated constants can be relocated in memory without revising the short operand addresses, since each address is relative to the instruction that called it.
- **Indexed addressing**: This is used to access data (generally variables), data tables, first-in first-out stacks and data arrays. The resultant operand address is the selected index register plus the short operand address in the instruction.

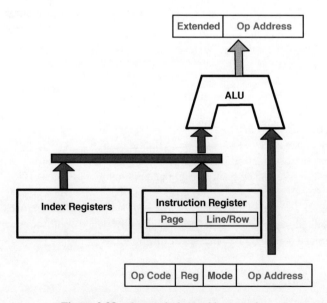

Figure 2.28   Operand (data) address modes

2.7   Software

2.7.1   Software Introduction

Software is the heart of modern digital avionics. It gives the system a flexibility that is far beyond that achievable from analogue equivalents. The software describes algorithms, logical statements, data and control processes. Designing, coding and testing software requires unique procedures to prove beyond reasonable doubt that the software is at least as reliable as its host hardware. Separate, but similar procedures have been developed for military and civil avionics systems, but with the same broad intent.

Investment in software is vast. The Airbus A320 avionics system has around 800,000 lines of code. The Boeing 777 avionics system has in excess of 4,000,000 lines of code running on 50-plus hardware platforms. The Airbus A380 and Boeing 787 will have many more lines of code. As a rough order of magnitude it has been said that the number of lines of code for an avionics system doubles every 10 years, with a commensurate increase in the man-hours/man-days required to develop and certify the software to the appropriate design assurance level (DAL).

This section is intended only as an introduction to the subject. For further information the reader is directed to other works on the subject [6].

2.7.2   Assemblers and Compilers

The software code developed in Section 2.6 is known as assembler level code. It is a set of pseudo-English language statements that have a 1:1 correspondence with the machine executable binary code that the CPU understands. Translation from the pseudo-English language assembler code to machine code is performed by a machine-specific translator known as an assembler. A linker/loader assigns specific memory addresses at the final stages of generating the machine code executable binary, to load and run the code from a specific area of memory.

Software is seldom written at assembler level. It would be totally impractical to code an avionics system this way. Possibly the only exception is device driver software which needs to specifically control and pass data between hardware components at register level. The application software will be written in a high-order language such as Ada or C++ which is not machine-specific and does not require the programmer to have intimate knowledge of the target machine architecture or its instruction set.

Good programming technique devolves the application software into a hierarchical structure of tasks, processes and objects that are then executed under the control of a real-time operating system (RTOS), which manages all the computing resources in accordance with agreed resource allocations (also known as the run-time environment). High-order source code in languages such as Ada and C++ is translated to run on the target machine by a compiler where it is combined with lower-level device drivers as necessary to produce machine loadable and executable target code. The process is illustrated in Figure 2.29.

- **The compiler**: Translates the application software (source code) into an intermediate form. It performs syntax checking, (mis)use of types and generates warnings or errors.
- **The assembler**: Translates the intermediate code to machine code (object code), which is the sequence of binary numeric values representing CPU-specific instructions plus data and references to data.

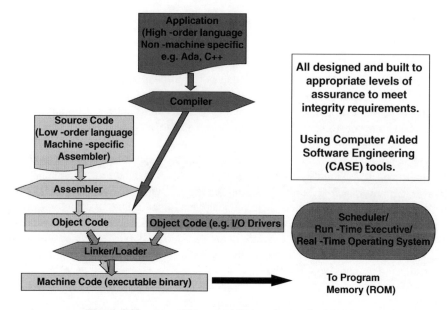

Figure 2.29   Assemblers, compilers and operating systems

- **The linker**: Combines multiple machine code entities into a single executable entity and resolves address references between entities.
- **The loader**: Writes the run-time executable code into the assigned CPU memory locations.
- **Firmware**: The Institute of Electrical and Electronics Engineers (IEEE) *Standard Glossary of Software Engineering Terminology*, Std 610.12-1990, defines firmware as "The combination of a hardware device and computer instructions and data that reside as read-only software on that device," and adds that the confusion surrounding this term has led some to suggest that it be avoided altogether.

2.7.3   Software Engineering

The development of software may turn out to be very complex. The development of most avionics applications requires teamwork. The complexity arises from the following elements:

- The size of the problem (system) domain to be dealt with.
- The difficulty in managing the development process.
- The flexibility constraints required for the final software.
- The problems caused by the compilation and analysis of the system being designed.

The aims of software engineering are to develop software which is:

- **Modifiable**: In order to modify software effectively it is necessary to assimilate all the design decisions. If not done, then the modifications are only rough indiscriminate repairs. Modifiable software must allow changes without increasing overall complexity.

- **Effective**: Uses the available resources to the best advantage. The limited resources are time (available calculation power) and space (available memory and available I/O).
- **Reliable**: Eliminating design faults and taking recovery action if dysfunctions occur (tolerant to failures).
- **Intelligible**: Readable (generally as a result of good coding style) and easily able to isolate data and operations carried out on the data.
- **Reusable**: The desired analogy is the concept of software components being as widely reusable as electronic components. In order for a library of components to be reusable, it is essential that each is fully documented and characterised.

2.7.4   Software Design Process Assurance

The software life-cycle is a subset of the system life-cycle, as illustrated in Figure 2.30. Interaction includes the exchange of data between the system and software life-cycles, categorisation of failure conditions and software design quality assurance level determination, architectural considerations and verification considerations. The system development life-cycle is described in Chapter 6.

RTCA-DO-178 [7] (now at version C) provides guidelines on the software design process for developing airborne software. The document comprises 12 interrelated sections as shown in Figure 2.31. It establishes considerations for developers, installers and users of aircraft equipment using embedded microcomputer techniques, and outlines verification, validation, documentation and configuration management and quality assurance disciplines [6]. It is a means (but not the only means) to secure approval by the certification authorities.

RTCA-DO-178 assigns software design quality assurance levels in accordance with the safety categorisations as shown in Table 2.2.

The software development process is a series of sequential phases progressively refining requirements to fully characterise the end product as illustrated in Figure 2.32. Entry and exit

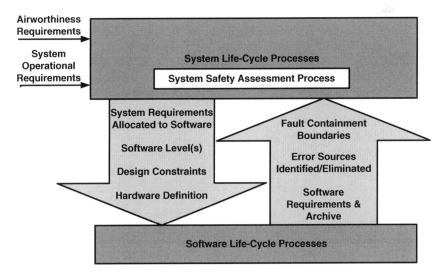

Figure 2.30   Interaction between the system and software life-cycles

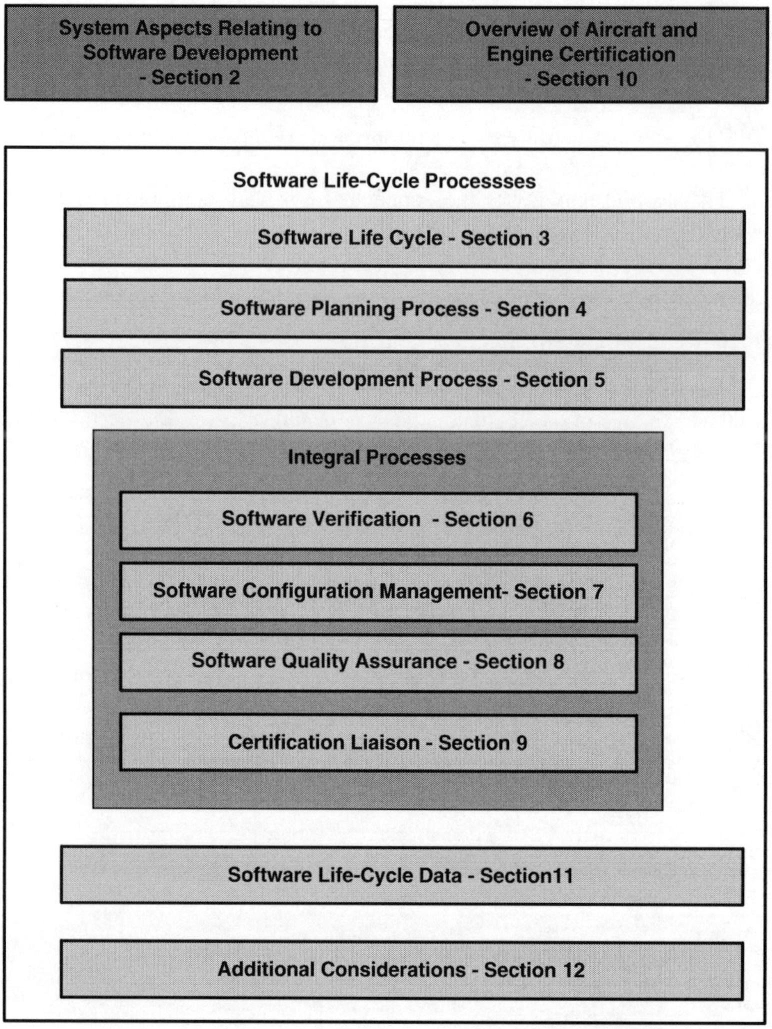

┌─────────────────────────────┐   ┌─────────────────────────────┐
│ System Aspects Relating to │   │   Overview of Aircraft and │
│   Software Development   │   │   Engine Certification   │
│   - Section 2   │   │   - Section 10   │
└─────────────────────────────┘   └─────────────────────────────┘

Software Life-Cycle Processses

Software Life Cycle - Section 3

Software Planning Process - Section 4

Software Development Process - Section 5

Integral Processes

Software Verification - Section 6

Software Configuration Management- Section 7

Software Quality Assurance - Section 8

Certification Liaison - Section 9

Software Life-Cycle Data - Section11

Additional Considerations - Section 12

Figure 2.31   RTCA-DO-178 software design process assurance

criteria are required to be specified for each phase. The philosophy is that correctness should be achieved during development rather than by subsequent elimination.

Each step in the recommended design process has a set of prescribed objectives, and activities or guidance for the process step, the rigor of which is moderated by the design quality assurance level.

- **Traceability**: Maintained through all levels of decomposition.
- **Validation**: Determine that the requirements are correct and complete and that the results have an adequate margin of safety consistent with its expected utilisation within its framework of application.

Table 2.2 RTCA-DO-178 design quality assurance levels

Design assurance level	Criteria
A	Design whose anomalous behaviour as shown by the system safety assessment would cause or contribute to a failure of system function resulting in a **catastrophic** failure condition for the aircraft
B	Design whose anomalous behaviour as shown by the system safety assessment would cause or contribute to a failure of system function resulting in a **hazardous** failure condition for the aircraft
C	Design whose anomalous behaviour as shown by the system safety assessment would cause or contribute to a failure of system function resulting in a **major** failure condition for the aircraft
D	Design whose anomalous behaviour as shown by the system safety assessment would cause or contribute to a failure of system function resulting in a **minor** failure condition for the aircraft
E	Design whose anomalous behaviour as shown by the system safety assessment would cause or contribute to a failure of system function resulting in a **no effect** failure condition for the aircraft

- **Verification**: Evaluate the results of a process for correctness and consistency to ensure that they are functionally correct and that they properly translate the choices and hypotheses made. Verification is not simply testing; it is typically a combination of reviews, analyses and tests.

2.7.5   Languages

Ada has been the preferred language of choice for embedded systems. Ada is a powerful and versatile language that includes provision for floating point arithmetic. It has been widely used; its benefits include:

- code portability and reusability;
- increased productivity;
- increased reliability (fewer coding errors);
- increased maintainability;
- good exception handling capabilities;
- supports parallel and concurrent processing;
- supports object-oriented design.

Spark Ada is a subset of Ada with particular properties that make it ideally suited to safety-critical applications. It is completely unambiguous and free from implementation dependencies. All rule violations are detectable. It was mandated for US Department of Defense programmes, but is no longer enforced due to lack of qualified programmers.

C++ is the most popular realistic alternative to Ada today. Some C++ features, if not properly controlled and verified, can lead to software code that is non-deterministic, unused,

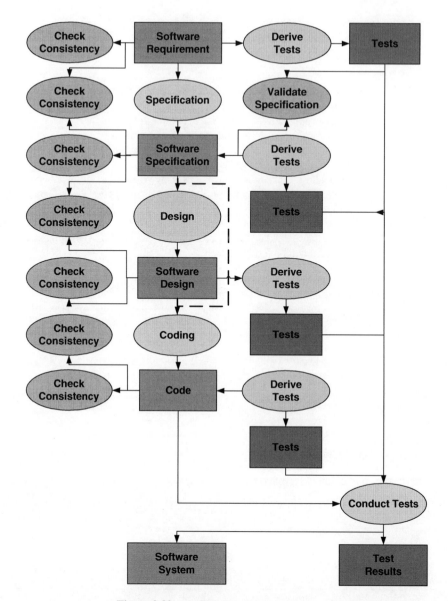

Figure 2.32   Software development life-cycle

difficult to verify, and whose configuration can change depending on the run-time state of the system. The Certification Authorities Software Team (CAST) published guidance in 2002 that identified common problems with C++ for DO-178 safety-critical applications. Many organisations have defined *ad hoc* C++ subsets to achieve DO-178 compliance. The Motor Industry Reliability Association (MISRA) specifies a 'safe' subset which defines 228 rules. Programming Research Inc. offer a qualified static code analyser tool (QA C++) which verifies source code conformance and is able to identify unsafe usage of C++.

2.7.6   Object-Oriented Design

RTCA-DO-178C has been five years in development through RTCA and EUROCAE. It was completed in November 2011 and approved/released in January 2012. The enhanced standard allows credit for modern software design technologies including:

- object-oriented programming (OOP) languages;
- formal methods;
- model-based development methods.

The hitherto 'classic' approach to software design gives priority to functions rather than data. Experience shows that the data handled by the software are more stable than the processes involved with it. Object-oriented design therefore gives priority to the data. The 'problem' to be designed is tackled by modelling a collection of objects, each of which implements a specific task. The possible states and behaviour of an object are defined by 'classes'. Each object constitutes an instance of a class. The 'solution' to the problem is provided by the interaction of the objects.

Object-oriented methods have been used for software as different from each other as mission-critical systems to commercial data processing systems. There are object-oriented methods for software requirements analysis, software design and programming. Major milestones in the development of object-oriented design and analysis techniques have been:

- object-oriented analysis (OOA) – Coad & Yourden and others (1991)
- object-oriented design and analysis (OOA & OOD) – Grady Booch and others (1997 & 1994)
- hierarchical object-oriented design (HOOD)
- object modelling technique (OMT) (1991)
- Unified Modelling Language (UML)

OOA and OOD model a system as a group of interacting objects communicating by messages, where each object is responsible for the accomplishment of tasks according to its state. Each object represents an entity of interest, characterised by its class, its state (data elements) and its behaviour. Models can be created to show the static structure, dynamic behaviour and run-time deployment of these collaborating objects.

- Object-oriented analysis (OOA) focuses on *what* the system does.
- Object-oriented design (OOD) focuses on *how* the system does it.

Object modelling technique (OMT) applies to all the software development processes, from analysis to implementation. It uses three different views to capture important aspects of the software:

- **Object model**: Represents the static aspects of the software (classes, relationships, etc.) and provides a framework into which the dynamic and functional models will be placed.
- **Dynamic model**: State transition diagrams describe the temporal aspects, the sequences of operations and events that produce changes in state within a class.

- **Functional model**: Data flow diagrams show interdependencies of input and output processes.

OMT provides a very complete static model. It uses the same graphic notation throughout the analysis and design phases and provides a powerful description of the internal behaviour of the objects

Unified Modelling Language (UML) embraces the earlier notations of object-oriented analysis (OOA) and design (OOD), the Booch method and the object modelling technique (OMT), and fuses them into a single, common, widely used modelling language. UML offers a standard way to visualise a system's architectural blueprints, including elements such as:

- actors;
- processes;
- (logical) components;
- activities;
- programming language statements;
- database schemas;
- reusable software components.

UML aims to be a standard modelling language which can model concurrent and distributed systems. It is a *de facto* industry standard, and is evolving under the auspices of the Object Management Group (OMG).

2.7.7   Auto-code Generation

There is a progressive move away from developing code from the OOA/UML formal model to automatically generating code from it:

- The advantages are:
  - reduced cost of development
  - faster timescale
  - easier to change.
- The potential issues to address are:
  - ability to certify safety-critical real-time systems
  - availability of qualified tools
  - bloated code
  - intelligibility (is it readable?).

Model-based tools model systems in very high-level, domain-specific languages. They can be used to automatically generate source code directly from the model. Both the model and the code have to be verified to DO-178. DO-178C provides objective criteria to prove that automated verification tools have been properly qualified and can be trusted. The 'Tools Supplement' explains what a tool provider has to do in order to qualify the tools. Some popular auto-code tools are:

SCADE

SCADE is a development environment used by system and software engineers to produce mission- and safety-critical applications. It is based upon the formal, synchronous and data-flow-oriented Lustre programming language, and generates production-quality C or Ada code. It is a qualified development tool for DO-178 applications up to level A and has been used in aerospace and defence applications, for example:

- flight control systems (e.g. A340 and A380);
- power management;
- reconfiguration management;
- autopilots;
- engine control systems (FADEC);
- braking systems;
- cockpit display and alarm management (e.g. NH90);
- fuel management.

Simulink©

Simulink©, developed by MathWorks, is a commercial tool for modelling, simulating and analysing multidomain dynamic systems. Its primary interface is a graphical block diagramming tool and a customisable set of block libraries. It offers tight integration with the MATLAB© environment and can either drive MATLAB© or be scripted from it. Simulink© is widely used in control theory and digital signal processing for multidomain simulation and design.

A number of MathWorks and third-party hardware and software products are available for use with Simulink©. Stateflow extends Simulink© with a design environment for developing state machines and flow charts. Real-Time Workshop extends Simulink© to generate C source code automatically for real-time implementation of systems.

VAPS

VAPS DO-178 qualified code generator from Presagis is applicable to all types of embedded display graphics ranging from digital and analogue instrumentation to menu-based displays. It offers control and flexibility in the design of dynamic, interactive, real-time human–machine interfaces (HMIs). It allows users to define both the visual appearance of a display object and the logic (state machine) that controls it in one graphical editor. Developers and human factors experts can define the look and feel of an object or group of objects, assign behaviour, and subsequently generate executable source code.

2.7.8   Real-Time Operating System (RTOS)

As discussed earlier in this chapter, avionics computations must be executed in 'real-time' on a repetitive cyclic basis. The Nyquist criteria must be obeyed. Consequences of missed deadlines are:

- control loop instability;
- data latency;

- poor frequency response;
- aliasing;
- jitter;
- missed events;
- false warnings/error messages;
- other relevant design considerations.

The application software is run on the target machine under the control of a real-time operating system (RTOS) which:

- manages task state;
- selects the task execution sequence;
- creates and destroys tasks;
- changes task priority;
- provides recovery from stuck tasks;
- manages interrupts;
- manages synchronous and asynchronous tasks;
- manages memory;
- provides memory protection and memory allocation;
- provides inter-process communication.

The operating system for integrated modular avionics systems must provide additional features in order to support the IMA objectives discussed in Chapter 5, and in particular Section 5.7. ARINC 653 specifies the requirements for an IMA RTOS. The operating system:

- Provides a partitioned environment. Periodic fixed scheduling bounds computing resource demands. Hardware memory management protects RAM/ROM utilisation. I/O resource allocation bounded in hardware configuration tables. Partitioning facilitates incremental certification.
- Supports multiple, isolated applications and guarantees that application failure effects are limited to the partition.
- Permits multitasking within a partition.
- Permits applications with different certification standards to run on same hardware platform.
- Supports application code portability and interchangeability.

ARINC 653 defines an Application Executive (APEX) and an Application Programming Interface (API) which decouples application software from the target platform. Each application software item is assigned a partition with its own memory space and dedicated time slot allocated to it by the APEX API. The APEX API provides services to manage partitions, processes and timing, as well as partition/process communication and error handling. The ARINC 653 APEX services and API calls are categorised as:

- partition management;
- process management;
- time management;

- inter-partition communication;
- intra-partition communication;
- error handling.

2.8 Microprocessors

A microprocessor is a semiconductor device that incorporates most or all of the features of a central processing unit (CPU) on a single integrated circuit (also known as a chip). Microprocessors emerged in the early 1970s and were first used in electronic calculators, point-of-sale till machines and cash dispensers. Affordable 8-bit microprocessors capable of performing control-type embedded applications appeared in the mid-1970s and spawned the personal computer (PC) industry. No longer were computers the sole domain of large corporations and government organisations. Today 32- and 64-bit multicore machines running at clock rates exceeding 1 GHz are readily available and affordable for desktop applications with processing power that far exceeds the imagination of computer systems designers in the 1970s. Then, similar processing power required a large room full of power-hungry computers, requiring a handful of computer scientists to operate and maintain them.

2.8.1 Moore's Law

In 1965, Gordon Moore, a co-founder of Intel, predicted the future of microprocessor-based computing in his now-famous Moore's Law: "Microprocessor processing power will double every two years". As illustrated in Figure 2.33, his prediction has largely proven to be correct.

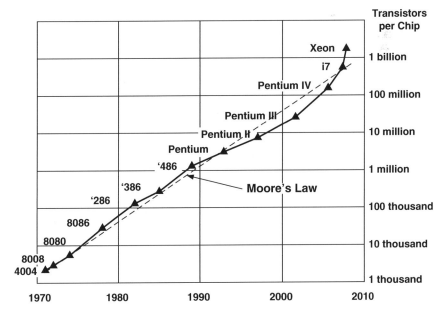

Figure 2.33   Moore's Law

2.8.2   Significant Microprocessors used in Aerospace Applications

Intel 4004

Released in 1971, the Intel 4004 was the first commercially available CPU on a single integrated circuit (IC) chip. It had a simple architecture (see Figure 2.34) with the following features:

- a 4-bit machine, with a 4-bit ALU and 8-bit instructions;
- a 12-bit memory address field accessing a maximum memory capacity of 4096 bytes of memory via a multiplexed 4-bit bus;
- a clock frequency of 740 kHz, and an instruction execution time of 10.8 μs;
- 46 instructions, and 16 registers;
- 2300 transistors, 10 μm process in a 16-pin dual in-line package (DIP).

Intel 8086

Released in 1978, this was the first 16-bit microprocessor. An earlier 8-bit version, the Intel 8080, was the basis for the first IBM PC. Intel later released a maths co-processor (Intel 8087) which supported floating point arithmetic. A simplified version of the architecture and instruction set are used in the CPU execution example in Section 2.6. It had the following features:

- 16-bit data and instructions, 20-bit address (paged);
- a clock frequency of 10 MHz and a 2 μs instruction execution time (average);
- dual ALU (data and address), 8 registers, limited pre-fetch queue (cache);
- 20,000 transistors, 3.2 μm process in a 40-pin surface mount package.

   The Intel 8086 was probably the first microprocessor to be widely used in avionics applications by several companies in the mid-1980s for embedded federated architecture avionics computers in both civil and military applications (e.g. flight management computers, display symbol generators, mission computers, autopilots, fuel gauging and management computers, etc.).

Motorola M68020

Released in 1984, the Motorola M68020 was a 32-bit processor first used in the Apple Mac II personal computer, SUN workstations and many industrial and aerospace applications. It was selected by the Eurofighter Typhoon project to be the processor of choice for all the avionics systems computers. It was also used in the French TGV. It had the following features:

- 32-bit data/instructions, and 32-bit memory address;
- a clock frequency of 33 MHz, and throughput of 10 million instructions per second (MIPS);
- 256 byte instruction cache;
- 114-pin ceramic pin-grid array package.

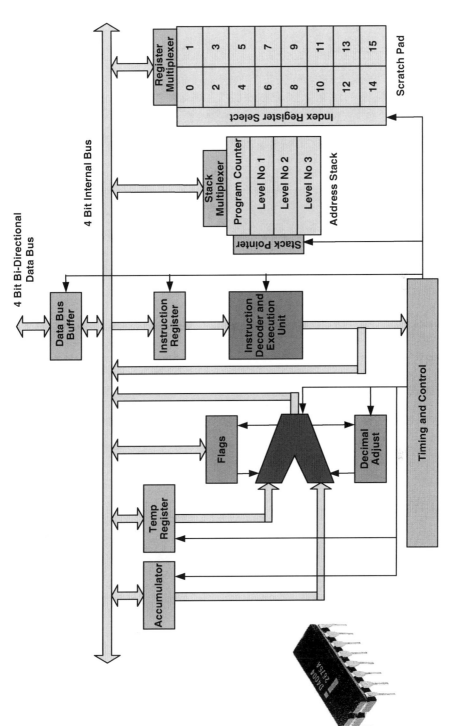

Figure 2.34 Intel 4004 microprocessor

Most significantly, the M68020 provided extensive memory management functions to protect and segregate areas of memory to facilitate robust partitioning.

Intel Pentium

Probably the most universally known microprocessor family used in desktop computing, the Intel Pentium D was released in 2008 as a dual-core, 64-bit, x86 microprocessor on the multichip module. It has the following features:

- 64-bit data/instructions, 32-bit memory address;
- a clock frequency of 3.73 GHz with a front-side bus 533 to 1066 MHz (different variants);
- x86-64 instruction set;
- 1 Mb L2 cache and 256 byte instruction cache;
- 65 nm process, dissipating 130 W in a 700-pin grid array package.

PowerPC

Performance Optimization With Enhanced RISC – Performance Computing (PowerPC), is a reduced instruction set computing (RISC) architecture created by the 1991 Apple–IBM–Motorola alliance, known as AIM. RISC allows scalable designs for low-end (8-bit) applications to high-end (64-bit and above) applications. The small size of the CPU core allows a great deal of room on each die for additional components, such as instruction and data cache or maths co-processors.

Originally intended for personal computers, PowerPC CPUs have since become popular as embedded and high-performance processors. PowerPC was the cornerstone of AIM's Common Hardware Reference Platform initiatives in the 1990s, and while the POWER1 architecture is well known for being used by Apple's Macintosh products from 1994 to 2006 (before Apple's transition to Intel), its use in video game consoles and embedded applications provided an array of uses. Most importantly, the suppliers of PowerPC components are prepared to support industrial applications by providing industrial and aerospace grade components with specified performance over a wider operating temperature range and a more severe environment than the Pentium.

In 2004, Motorola exited the chip manufacturing business by spinning off its semiconductor business as an independent company called Freescale Semiconductor. IBM exited the 32-bit embedded processor market by selling its line of PowerPC products to Applied Micro Circuits Corporation (AMCC) and focused on 64-bit chip designs. The IBM–Freescale alliance was replaced by an open standards body called Power.org. IBM continues to develop PowerPC microprocessor cores for use in their server products for large businesses, and continues to evolve POWER processors which implement the full PowerPC instruction set architecture. Freescale Semiconductor continues to evolve products for embedded systems.

The evolution and some key examples of the PowerPC range are shown below.

- POWER1 (1991) introduced in Apple-Mac machines, outperforming Intel x86.
- POWER3 ('603 and '604) full 64-bit implementation, including low cost/low power versions.

- POWER5 (2004) is a dual-core processor with support for simultaneous multithreading.
- POWER6 (2007) dual-core design operating at 5 Ghz with more than 58 million transistors produced on a 65 nm feature size process.
- POWER7 (2011) is an embedded high-end server from IBM with multicore processors available with 4, 6 or 8 cores per chip. It has approximately 1.2 billion transistors and is 567 mm^2, fabricated on a 45 nm process. POWER7 operates at 3.7 GHz and has 12 execution units per core, including two fixed point units, four double-precision floating point units and one decimal floating point unit. It is able to execute instructions out of sequence and process four simultaneous multithreads per core. It is available in a range of multichip modules with 1, 2 or 4 chips per module and 4 Mb L3 cache per core.
- MPC7447 is the fifth implementation of 4th generation (G4) microprocessors from Freescale. It implements the full PowerPC 32-bit architecture in its e600 core. The MPC447A operates at 1.4 GHz and is 106 mm^2, fabricated on a 130 nm process. It is able to process instructions out of sequence and process four instructions simultaneously in 11 execution units, which include a branch processing unit, four integer units, a floating point unit and four vector units. On-chip 32 kb L1 instruction and data cache and 512 kb L2 cache is provided, together with separate memory management for instructions and data. A simplified block diagram is shown in Figure 2.35.

The microprocessors identified above are not an exclusive list, but serve to illustrate the evolution of microprocessors used in avionics applications. A new range of processors is emerging in the tablet market based on the Advanced RISC Machines (ARM) architecture. In due course this commercial off-the-shelf (COTS) technology might well find its way into some aerospace applications.

In any application is essential to select a microprocessor which has more than adequate performance for the application with a large margin (at least 100%) for future growth in terms of:

- throughput: instructions per second;
- precision: number of bits;
- capacity: memory size.

But just as important is supportability through the expected life of the end product. To assure this, the device and associated chip set support components should be procured from a reputable source (multisourced if possible) having a good set of design support tools, and for which there is a robust obsolescence management strategy.

2.8.3   CPU Cache

A cache is used by the CPU of a computer to reduce the average time to access memory. The cache is a smaller, faster memory that stores copies of the data from the most frequently used main memory locations. As long as most memory accesses are cached memory locations, the average latency of memory accesses will be closer to the cache latency than to the latency of main memory. When the processor needs to read from or write to a location in main memory, it first checks whether a copy of that data is in the cache. If so, the processor immediately reads

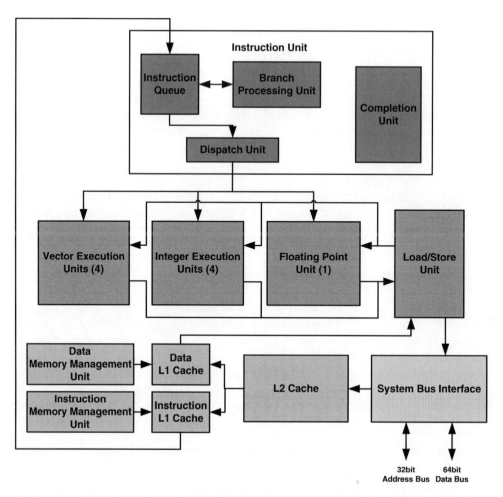

Figure 2.35   Freescale PowerPC MPC7447A simplified block diagram

from or writes to the cache, which is much faster than reading from or writing to main memory. Most modern desktop and server CPUs have at least three independent caches: an instruction cache to speed up executable instruction fetch; a data cache to speed up data fetch and store; and a translation look-aside buffer (TLB) used to speed up virtual-to-physical address translation for both executable instructions and data. The data cache is usually organised as a hierarchy of more cache levels (L1, L2, etc.).

2.8.4   Microcontrollers

Big is not always best. There are many aerospace applications where a relatively simple CPU operating at modest speed will do the job. Particularly this is true for local automated control applications such as motor controllers, sensors, effectors (servo-motors) and remote interface units. A microcontroller is a small CPU (often just eight bits) with on-chip interface components such as:

- clock;
- interval timers and event counters;
- interrupt handling;
- serial I/O ports universal asynchronous receiver transmitter (UARTS) and network interfaces (e.g. CANbus);
- software configurable signal conditioning (e.g. pulse width modulation/demodulation);
- A/D and D/A converters;
- a small amount of RAM for data;
- a moderate amount of EEPROM or Flash memory for instructions and constants.

2.8.5   Rock's Law

Current state-of-the-art technology is of the order of:

- sub-30 nm process on 300 mm wafers; 1.2 nm gate oxide thicknesses;
- tera-Hertz transistors with 'sleep' transistors to conserve power;
- strained silicon (stretching the lattice structure) to increase electron mobility.

Many have argued that at some point a technology limit will be reached and Moore's Law will no longer apply. However, Rock's Law, "the cost of a fabrication plant doubles every four years", suggests that the limit will not be technological but financial.

2.9   Memory Technologies

To be of any practical value, a digital computer must have memory as well as a CPU. The memory holds the sequence of instructions to be executed as well as providing storage for constants, input variables, intermediate variables and output results of the computational process.

A considerable amount of effort has been expended in the development of memory data storage technologies. What is required is a memory technology/device with the following characteristics:

- a discrete physical cell, written to by an external signal, into two distinct states (logic 0 or logic 1);
- the state of cell can be read without degrading its state (non-destructive readout);
- all cells should be equally accessible (random access);
- the cell must remain in this set state indefinitely (static);
- the state should require no external energy sources to maintain it (non-volatile).

To date there is no single memory storage technology that satisfies all these requirements. So the choice and mix of memory technologies is a compromise between what is desirable and what can be provided to best match the application. Not surprisingly, aerospace avionics systems applications demand a different set of technologies to those found in a desktop computer, although as we shall see, the new generation of 'instant-on' ultra-book laptops closes the gap.

2.9.1   Desired Avionics Memory Attributes

An avionics computer requires memory for the following data types with the ascribed attributes:

Application software:
- read-only (once installed);
- random access;
- instant-on (immediately available as soon as the computer is powered);
- non-volatile (must not be lost when power is removed);
- alterable on the ground by an approved procedure to erase and reload the application software.

Variables:
- read/write;
- random access;
- could be volatile (i.e. data need not be retained through power-down and power-up cycles, but must survive short-term power transients).

Maintenance data/instrumentation/configuration data:
- read/write;
- random access;
- non-volatile (needs to be readable on the ground post-flight and post power-down cycle to determine in-flight performance and any maintenance action required).

2.9.2   Available Memory Technology Attributes

Magnetic disc

Widely used in desktop computers as the 'hard drive', the main attribute is read/write, non-volatile data storage. But it is not random access. Data is extracted in sequential form as the magnetic platen rotates past the read/write head. The only reason that it has survived so long is that it is very inexpensive per bit and has high capacity. In desktop applications it holds the operating system and the application software which are uploaded into volatile RAM at power-on (the boot sequence). It holds working data in the form of files which are uploaded into volatile RAM as and when required, and must be saved to the hard drive prior to power-down, otherwise work is lost.

Surprisingly, even though it is a mechanical, rotating, electromagnetic device, it is very reliable in a benign environment, but it is unsuitable for airborne avionics systems applications.

Semiconductor random access memory (RAM)

Widely used in desktop applications, it is random access, high-capacity and fast. But it is volatile, such that data are lost if power is removed.

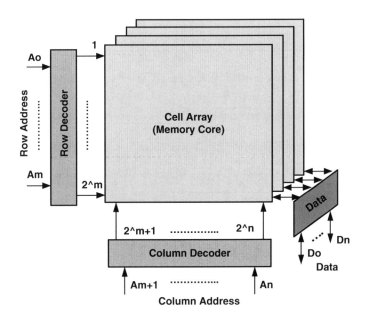

Figure 2.36   Semiconductor random access memory (RAM)

Figure 2.36 shows the general schematic for semiconductor RAM. An array of cells stores the data. The array is matrix-addressed by row and column drivers which allow random access to each cell (or bit) in the array. Data (logic '1' or logic '0') may be written to each cell (write cycle) and read from each cell (read cycle).

There are essentially three semiconductor RAM technologies:

- **Bipolar static RAM**: Uses TTL J-K flip-flops (see Section 2.11). Fast access, but has a significant silicon footprint per cell, uses more power than metal oxide semiconductor (MOS) technologies and is expensive per bit.
- **Static MOS RAM (S-RAM)**: Uses MOS field effect transistor (FET) devices to make J-K flip-flops (see Section 2.11). Takes less space per cell than bipolar TTL and therefore has higher capacity, but slower than bipolar TTL. It consumes less power and is less expensive per bit than bipolar TTL.
- **Dynamic MOS RAM (D-RAM)**: This technology uses MOS FET devices to charge a cell capacitor. However, data must to be refreshed regularly to recharge the cell capacitor since its charge will slowly leak away. Higher capacity than static MOS, it takes less power and costs less. It is the technology of choice for desktop computing, but the need for regular data refresh can be problematic in real-time avionic computing applications.

Figure 2.37 shows the circuit topologies for static and dynamic MOS RAM. It can be readily recognised that static RAM uses more silicon area per cell than dynamic RAM, and therefore is more expensive per bit, has a lower capacity per device and consumes more power. But static RAM does not need to be periodically refreshed. Data are retained provided the device remains powered. Static RAM is therefore preferred in most airborne avionics applications

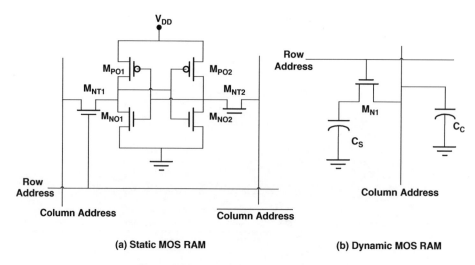

(a) Static MOS RAM   (b) Dynamic MOS RAM

Figure 2.37   MOS RAM circuit topologies

because data will survive transients, provided local power can be maintained to the device for the short period of the transient power loss.

Semiconductor read only/read mostly memories (ROM)

Avionics systems applications software requires a solid-state non-volatile technology. The technology choices available today are:

- **Mask programmed read only memory (ROM)**: A mask programmed ROM comprises an array of cells (like the RAM) with a MOS FET transistor fabricated at each intersection of the matrix. This arrangement is shown in Figure 2.38a. The transistor is connected to the matrix at the final metallisation stage of the fabrication process. A connection produces a logic '1' when the device is read; omission of the connection produces a logic '0'. The ROM is thus a device which is mask programmed at the fabrication plant. Once programmed it cannot be altered. It is an inexpensive and effective solution for a production configuration once the software is stable. For this reason, software stored this way is frequently called firmware.
- **Electrically erasable programmable read only memory (EEPROM)**: The basis for the design is a special MOS FET called a FLOating gate Tunnel OXide (FLOTOX) transistor that replaces the simple MOS FET transistor in the mask programmed ROM (see Figure 2.38b). It can be erased and programmed electrically using quantum mechanical tunnelling. Quantum mechanical tunnelling allows electrons to cross a thin barrier such as an insulator. The FLOTOX transistor has a thin region (<10 nm) over the drain that allows transport of electrons to/from the floating gate. A logic '1' is programmed by placing a positive voltage on the row and a negative voltage on the column, causing electrons to tunnel from the drain to the floating gate. To program a logic '0' the opposite biasing is used. The EEPROM requires two transistors per bit.

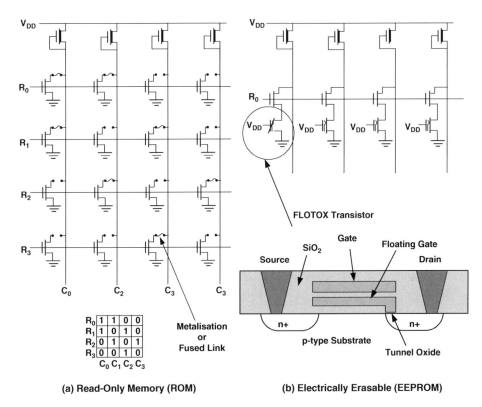

R_0	1	1	0	0
R_1	1	0	1	0
R_2	0	1	0	1
R_3	0	0	1	0

$C_0 \ C_1 \ C_2 \ C_3$

Metalisation or Fused Link

(a) Read-Only Memory (ROM)

(b) Electrically Erasable (EEPROM)

Figure 2.38   ROM schematic and EEPROM FLOTOX transistor

Flash memory

This type of memory uses Erase-Through-OXide (ETOX) devices, which use a combination of the technology used in the now obsolete UV erasable EPROM and the FLOTOX transistor of the EEPROM. The result is a device that has higher bit density and faster reprogramming time than the EEPROM, but must be bulk erased in blocks. Popular for camera memory cards and USB memory sticks, this technology is now maturing for use as a replacement for magnetic disc hard drives in the emerging Ultrabook PC and tablets (known as solid state device (SSD) technology). It is a suitable technology for the storage of application software in avionics systems computers and is used in the latest generation of IMA processor cards (see Chapter 5). However, the limited number of write cycles (typically $<10^6$) precludes its use as a replacement for RAM.

Historically, two other technologies were used, but both of these have been superseded by the technologies described above, namely:

- **Programmable read only memory (PROM)**: The device is programmed once at the equipment supplier by selectively blowing a fusible link by the controlled application of an electrical current.

- **Electrically programmable read only memory (EPROM)**: The device is programmed at the equipment supplier by removing the device from circuit, bulk erasing it with ultra-violet light, and then reprogramming it electrically. The device is slow to program. In-service modifications are cumbersome as the LRU must be removed from the aircraft to a servicing depot where the memory device can be removed from the circuit card and reprogrammed.

2.9.3   Memory Device Summary

The characteristics of the memory technologies discussed are summarised in Table 2.3.

2.9.4   Memory Hierarchy

As we have seen, it is not yet possible to find a single memory technology for all avionics systems applications. An appropriate mix of technologies is required, and the designer must match the capability of each to his application requirements (see Figure 2.39).

In this example, the application software is loaded into non-volatile EEPROM or Flash memory by a rigorously controlled ground maintenance process. The application software may be executed at run-time directly from the non-volatile memory; however, in integrated modular avionics (IMA) applications it is becoming custom and practice to upload the application software of each partition at run-time into the RAM memory area for execution in a process known as context switching. Execution from RAM is faster, and the process of context switching ensures a perfect copy of the application software is made every run-time cycle. The RAM memory area may incorporate additional error correcting codes (ECCs) to detect and correct transient data errors. Non-volatile RAM is used to retain configuration and maintenance data between power cycles and provide 'hot-start' capability in the event of short-term power loss.

2.10   Application-Specific Integrated Circuits (ASICs)

An application-specific integrated circuit (ASIC) is a general-purpose digital integrated circuit which is customised by the end user to implement his application requirement. As integrated circuit feature sizes have reduced and design tools have improved, ASICs have grown in size from 5000 gate equivalents to over 100 million gate equivalents. Moore's Law has again prevailed. Modern ASICs often include entire microprocessors, memory blocks (RAM, ROM, EEPROM and Flash), I/O and network communication interfaces (e.g. UARTS and CANbus), discrete and analogue interfaces, and A to D and D to A converters.

2.10.1   Main Types of ASICs

ASICs come in many different sizes and varieties. They can be summarised as:

- **Gate arrays or uncommitted logic arrays (ULAs)**: This device was the earliest type of ASIC. It comprises little more than a set of standard gates arranged in an orthogonal array and prefabricated onto the silicon wafer. The end-user configured the ASIC to the function

Table 2.3 Memory device summary

Category	Device	Access	Read/write	Programmable	Non-volatile	Speed	Capacity	Cost per bit
Semiconductor	D-RAM	Random	Read/write	Yes	Volatile	V fast	100s Mb	High
	S-RAM	Random	Read/write	Yes	Volatile	Fast	10s Mb	V high
	ROM	Random	Read only	No	Non-volatile	Fast	Gb	Medium
	EPROM	Random	Read only	UV erasable <100 cycles	Non-volatile	Fast	100s Mb	V high
	EEPROM	Random	Read mostly	Electrically erasable	Non-volatile	Fast	100s Mb	High
	Flash	Random	Read mostly	Electrically erasable <10^6 cycles	Non-volatile	Fast	Gb	Medium
Mechanical	Optical	Cyclic	Read/write		Non-volatile	Slow	Gb	Low
	Magnetic hard disk	Cyclic	Read/write		Non-volatile	Slow	Tb	Low

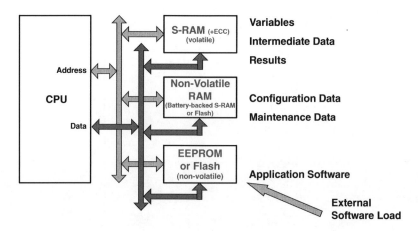

Figure 2.39   Avionics system memory configuration

desired by determining the interconnections between these gates. This interconnect is the final step metallisation layer in the fabrication process, similar to a mask programmed ROM. These devices have now largely been superseded by field programmable gate arrays.

- **Field programmable gate arrays (FPGAs)**: Also known as programmable logic devices (PLDs) and electrically programmable logic devices (EPLDs), these devices offer more complex and comprehensive logic structures than the simple gate array ULA. As with the ULA, these logic structures are prefabricated onto the silicon wafer. The interconnections between these logic structures are made by the end-user using similar technologies to those discussed in Section 2.9.2 for electrically programming ROMs. The ASIC vendor toolset includes a library of preconfigured logic functions to assist the end-user to design his custom ASIC. FPGAs are the modern-day technology for building breadboards and prototypes. For small designs and/or low production quantities, they can be more cost-effective than a custom design.
- **Semi-custom standard cell design**: The designer compiles a complete fabrication mask set for a semi-custom design using a set of standard logic cells and functional building blocks pre-designed and validated by the ASIC vendor. The designer is responsible for the logic functionality, while the ASIC vendor is responsible for the detailed circuit design of the standard logic cells and functional building blocks.

2.10.2   Field Programmable Gate Array (FPGA)

FPGAs contain programmable logic components called configurable logic blocks (CLBs). A CLB can be programmed to perform complex combinational functions. In addition to CLBs, most FPGAs include higher-order embedded functions such as adders and multipliers, which may be used to implement decoders and mathematical functions, and embedded memory elements such as S-RAM, EEPROM and Flash memory.

In addition to 'glue logic', FPGA applications include digital signal processing, cryptography, graphics processing, speech recognition, fast Fourier transforms (FFTs), and so on.

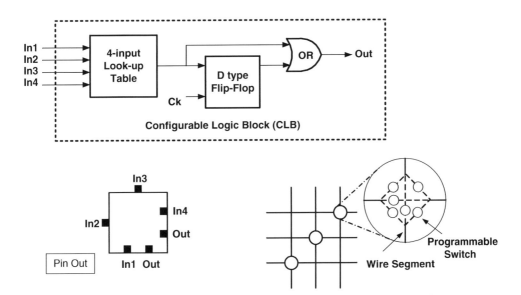

Figure 2.40   FPGA circuit topology

FPGAs are increasingly used in high-performance applications where specialised parallel computing is required, instead of using an array of microprocessors. Historically FPGAs have been slower and consume more power than standard cell designs, but the gap is closing.

A typical FPGA consists of an array of CLBs as shown in Figure 2.40. Each CLB has one output, which can be the latched or the unlatched output of a 4-input lookup table. The lookup table can be configured to implement such functions as AND, OR, Exclusive OR, and so on, and with the latch or flip-flop any logic function can be synthesised. Each CLB pin is accessible to orthogonal wiring segments in the interconnection channel adjacent to it. A programmable switch box at vertical and horizontal channel intersections routes signals to wires in adjacent channel segments. Note that in this example the clock is hardwired to all CLBs (not user configurable), so all latched functions are synchronous.

Configuring a FPGA requires the designer to determine the contents of each CLB lookup table and the topology by which that CLB is interconnected to the other CLBs on the device. The technologies available to effect the interconnections are the same as those developed for semiconductor ROM devices, namely:

- **Fused-link PROM**: one-time programmable
- **EPROM (U/V erasable)**: usually one-time programmable in production (plastic packaging has no window).
- **EEPROM**: electrically erasable, but not all EEPROM vendor devices can be reprogrammed *in situ*.
- **Flash**: can be bulk-erased, but again not all Flash vendor devices can be reprogrammed *in situ*.
- **S-RAM**: contents defined by an external boot device, loaded at power-up. Any non-volatile ROM technology is suitable, usually battery-backed S-RAM or Flash memory is chosen.

Not surprisingly the main FPGA vendors are also major vendors of semiconductor EEPROM technologies. Xilinx and Altera are the current market leaders (and long-term rivals). Together they control over 80% of the market. Other vendors include:

- Lattice Semiconductor (SRAM with integrated configuration flash);
- Actel (fusible CMOS);
- Achronix (RAM based).

2.10.3   Semi-custom Standard Cell Design ASIC

This structured approach to ASIC design allows designers to build a complex custom part from a set of standard functional blocks offered by the chosen ASIC vendor. The ASIC manufacturers' 'standard cells' have 'standard lithography' and fully characterised electrical performance. Using these functional blocks it is possible to achieve very high gate density and good electrical performance without having to design at transistor level.

The functional requirements for the device are devolved into a description of the design, somewhat analogous to writing software, called register transfer level (RTL), in a high-order language. Logic synthesis and analysis tools transform the RTL design into a collection of lower-level standard cell constructs. The collection of standard cells and their connectivity is called the gate-level netlist. The netlist is 'compiled' to create the placement of the standard cells and the connectivity between them. The output is then used to create the photolithographic fabrication masks. The benefit of 'standard cell' over full-custom is the use of proven ASIC vendor cell libraries. The design process flow is shown in Figure 2.41.

- Derive the specification using requirements capture tools.
- Develop the functional design typically expressed in functional flow diagrams through system decomposition and partitioning.
- Translate the functional description into a hardware block diagram at the register level.
- Build the logic design from standard cells provided by the ASIC vendor, simulate and develop test vectors.
- Compile the design, place the standard cells and route the interconnections.
- Verify the design with post-layout simulation.

2.10.4   Design Tools

Requirements capture and design entry for field programmable gate arrays (FPGAs) and application-specific integrated circuits (ASICs) is usually performed in very high speed integrated hardware description language (VHDL). VHDL is a general-purpose programming language that allows the behaviour of a system to be described (modelled) and verified (simulated) before synthesis tools translate the design into a real hardware implementation (logic gates and interconnections). It has a syntax which is essentially a subset of Ada. It allows the description of a concurrent system that is a system with many parts, each working together at the same time. Many ASIC vendor 'library' elements are available to be incorporated into the designer's end-system.

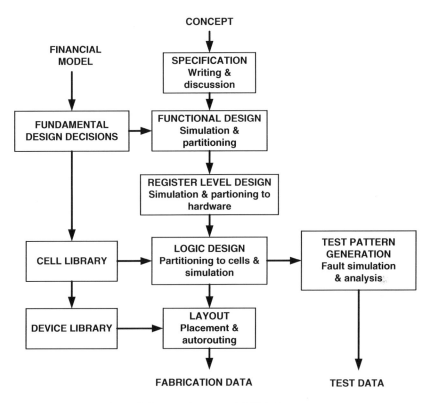

Figure 2.41   Semi-custom ASIC design process

2.10.5   RTCA-DO-254/ED 80

RTCA-DO-254, also known as ED 80 [8], provides design assurance guidelines to facilitate designers being able to demonstrate appropriate means of compliance to meet the civil airworthiness certification requirements for the provision of complex hardware embedded in airborne avionics systems. Complex electronic hardware includes devices such as FPGAs, EPLDs and ASICs. It is the counterpart to RTCA-DO-178C/EUROCAE ED-12B for airborne software and is fashioned in a similar manner.

There are five levels of compliance, A–E, in a similar manner to RTCA-DO-178, which depend on the effect a failure of the hardware will have on the operation of the aircraft. Level A is the most stringent, defined as 'catastrophic', while a failure of level E hardware will not affect the safety of the aircraft. The following items of substantiation are required to be provided:

- Plan for Hardware Aspects of Certification (PHAC);
- Hardware Verification Plan (HVP);
- Top-Level Drawing;
- Hardware Accomplishment Summary (HAS).

2.11 Integrated Circuits

Integrated circuits (chips) are the fundamental components used in a digital computer. They can be characterised by the number of logic gates on the chip as follows:

- Small-scale integration (SSI)   1–10 gates per chip
- Medium-scale integration (MSI)   10–100 gates per chip
- Large-scale integration (LSI)   100–104 gates per chip
- Very large-scale integration (VLSI) $>10^4$ gates per chip

2.11.1 Logic Functions

Figure 2.42 shows the basic logic functions, symbolic diagrams and truth tables for logic AND, logic OR and logic Exclusive OR functions. With these logic elements and a storage element like the J-K flip-flop (also shown in Figure 2.42), it is possible to develop any complex logic element including arithmetic logic units (ALUs), microprocessors and memory components.

These logic components are implemented on a silicon integrated circuit using TTL and MOS FET transistor technologies.

Figure 2.43a shows the electrical circuit of a 3-input NAND (negative AND) gate implemented in transistor-to-transistor logic (TTL) technology. A standard TTL NAND gate is constructed using multiple emitters in the input transistor. The input transistor Q_1 saturates if one or more of the inputs goes low, providing the NAND function. This provides a small voltage at the base of Q_S, therefore both Q_S and Q_O are cut off. Q_P is forward active and the output goes high.

Figure 2.43b shows the electrical circuit of a 3-input NAND gate implemented in C-MOS technology. Realisation of the NAND function in CMOS requires the series connection of n-MOSFETs (metal oxide semiconductor field effect transistors) in the pull-down branch and parallel connections of the p-MOSFET in the pull-up branch. The output of the NAND circuit will go low only if all of the n-MOSFETs are on and all of the p-MOSFETs are off. This occurs only with a logic '1' applied to all inputs. If a single input is logic '0', then the associated n-MOSFET will be cut off, whereas the associated p-MOSFET will be linear, thus bringing the output to V_{DD}.

2.11.2 The MOS Field Effect Transistor (MOSFET)

The MOSFET, shown in Figure 2.44, is one of the most important devices for digital integrated circuits today. It is a unipolar device. Electrical current is carried by the drift of one type of carrier: electrons for the n-MOS transistor, holes for the p-MOS transistor. It is a field effect or voltage controlled device. The voltage applied at the gate controls the flow of current between the drain and the source. In the n-MOS transistor (shown) a positive gate-to-source bias creates a conducting channel of electrons that move by drift from the source to the drain at a velocity proportional to the field intensity.

2.11.3 IC Fabrication

The transistor and interconnections in a digital integrated circuit are fabricated on the surface of a semiconductor wafer by a sophisticated series of steps including epitaxy, oxidation, ion implantation, and thin film deposition.

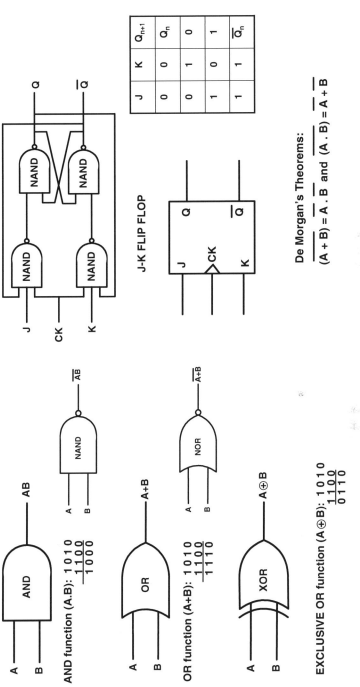

Figure 2.42 Basic logic functions

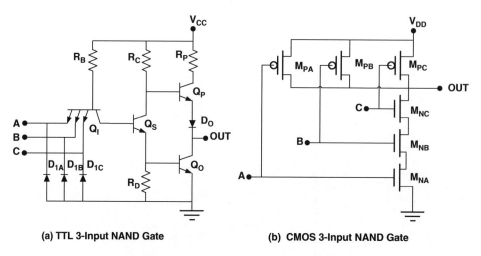

(a) TTL 3-Input NAND Gate   **(b) CMOS 3-Input NAND Gate**

Figure 2.43   TTL and C-MOS 3-input NAND gates

IC fabrication relies on the capability to transfer patterns from computer-aided designs to the physical wafers by a process called lithography. Lithographic steps are used to form semiconductor, insulator and metal regions as needed to transfer computer-generated device circuit designs to the physical wafer. There are several variations:

- photolithography;
- X-ray lithography;
- electron-beam lithography;
- ion-beam lithography;
- photoelectron lithography.

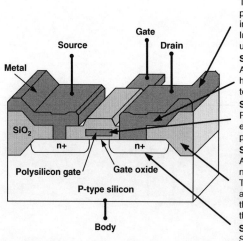

Step 5:
The first layer of metallisation (metal 1) is deposited and patterned to provide device contacts and interconnections.
In the finished circuit many more layers of metal may be used, with intermediate insulator layers.

Step 4:
After blanket deposition of an insulating layer, contact holes are defined so that metal contacts may be made to the devices.

Step 3:
Polycrystalline silicon (polysilicon) is deposited over the entire wafer. It is then etched in the required pattern to provide the gate region.

Step 2:
Active regions are defined by the deposition of silicon nitride.
Thick field oxide is produced in all areas not within the active regions using the silicon nitride as a mask. Then the nitride is stripped, and a thin gate oxide is grown over the active regions.

Step 1:
Starting with p-type substrate, an ion implant step is used to dope the source / drain regions of the MOSFET.

Figure 2.44   MOSFET fabrication

Table 2.4   Integrated circuit technology trends

	2007	2010	2015	2020
Lithography				
Stagger-contacted metal pitch (nm)	68	45	25	14
Physical gate length (nm)	25	18	10	5.6
# mask levels (up)	33	34	37	39
Wiring levels (max)	11	12	13	14
Microprocessor				
Transistors per chip (millions)	386	773	1546	6184
Chip size (sq mm)	140	140	88	111
Cost per transistor (mcents)	12.2	4.3	0.76	0.24
Clock frequency (GHz)	4.7	5.8	8.5	12.4
DRAM				
Bits	2G	4G	8G	32G
Chip size (sq mm)	93	93	59	74
Cost per bit (mcents)	0.96	0.34	0.06	0.01
General				
On-chip voltage (V)	1.1	1.0	0.8	0.65
Power dissipation – high end (W)	189	198	198	198

However, the basis for all these processes is the exposure and development of radiation-sensitive materials, called resists. Resists may be positive or negative. After a fresh layer of silicon dioxide (SiO_2) has been grown over the entire wafer, it is coated with a thin layer of the resist. The resist is spun, baked and exposed through a photo-mask. It is developed. Radiated areas degrade and are removed by solvents. The patterned photo-resist is then used as a mask to chemically etch the SiO_2. The SiO_2 assumes the same pattern as the original mask. Figure 2.44 indicates the process steps required to fabricate an n-MOS transistor. Further information on integrated circuit design and technology can be found in references [9] and [10].

Table 2.4 provides a summary of integrated circuit technology trends.

2.12   Integrated Circuit Packaging

Packaging the integrated circuit chip is the final stage of semiconductor device fabrication. The following operations are performed (see Figure 2.45):

- wafer probe and test;
- wafer separation (scribe and cut);
- die attachment (to the chosen package header);
- wire bond;
- encapsulation;
- inspect;
- test;
- burn-in and accelerated life testing.

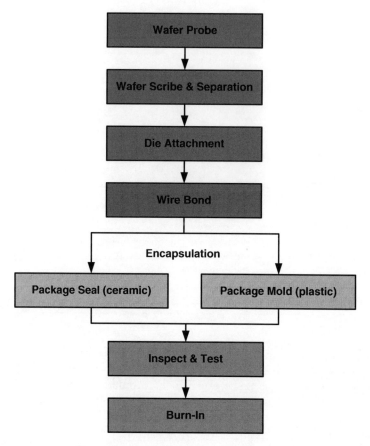

Figure 2.45   Integrated circuit packaging process

2.12.1   Wafer Probe and Test

Wafer testing is performed before proceeding to subsequent (value-added) stages. Contact is made to each circuit on the die by electrical microprobes that touch the bonding pads. Computer-generated test signals are applied to the circuit to measure and evaluate its performance. Both functional and parametric tests (noise margins, propagation delays, maximum clock frequency) are performed. Processing variations impact parametric performance. Point defects impact upon functionality – all processes are carried out in clean rooms. A dot is applied to a bad circuit (die).

2.12.2   Wafer Separation and Die Attachment

After wafer test, wafers are diced (scribed and cut) into rectangular chips (die). The die is attached to the chosen substrate prior to making electrical connections to it.

2.12.3   Wire Bonding

Interconnections to the die are made using wire bonds. 10–50 μm gold wire is drawn through a tungsten carbide wedge capillary. A gold ball is formed on the end of the wire and ultrasonically bonded to an aluminium pad on the heated die. The wire is then bonded to a metal pad on the heated ceramic substrate and pulled away to break the wire.

2.12.4   Packaging

An integrated circuit package provides the following:

- supports and protects the IC from mechanical vibration and shock;
- protects the die against the environment (moisture, contaminants, etc.);
- conducts heat away from the IC to an appropriate heat sink, to allow the die to withstand the required range of operating temperatures and repeated thermal cycling;
- provides the necessary connections to the IC, reliably and without undue impact on its electrical performance (speed, etc.).

IC packages may be classified as:

- through-hole: have metal pins that are inserted through holes in the circuit board;
- surface-mount: can be attached to a single surface of the circuit board, leaded or leadless;
- flip-chip: mount the chip upside down and make connections via solder bumps;
- bare die: for assembly onto multichip modules.

Through-hole packages

Through-hole packages have metal pins that may be inserted through the printed circuit board for soldering. The most important types are:

- dual in-line (DIPs);
- quad in-line;
- pin grid arrays (PGAs) – often installed into a board-mounted socket.

DIPs are rectangular packages with metal pins arranged along two sides. Plastic DIPs are the most popular and cost effective. Ceramic DIPs are more suitable for high-power, high-temperature applications. PGAs use pins arranged in a rectangular grid on the bottom of the package and accommodate a large number of electrical connections.

Surface-mount packages

Surface-mount device (SMD) packages are compact, lightweight and robust. Inexpensive applications use moulded plastic, but direct contact with the die means thermal mismatch is a concern. Hermetically sealed ceramic and metal SMDs are used in higher power and severe environment applications. SMDs come in a wide variety of styles and pin configurations, including:

- quad flat-pack (the original);
- small outline integrated circuit (SOIC);
- J-lead (leads bend under the package) – often installed into a board mounted socket;
- gull-wing.

Leadless surface mount packages

Ball grid array packages (BGAs) use a grid of bottom-mounted solder balls for attachment to the printed circuit card. During assembly the package is positioned over the grid of mating electrodes on the surface of the printed circuit card. The assembly is then subjected to a controlled heating process which reflows the solder balls to form an electrical connection to the printed circuit board contacts. It is not possible to inspect the contact post-assembly. SOICs and quad devices are also available in leadless chip carrier (LCC) form with solder reflow to connections around the package periphery. These are able to be inspected post-assembly.

Integrated circuit package usage

Figure 2.46 shows the current trend in integrated circuit package usage. Leadless surface-mount devices predominate since they can accommodate more interconnections in less volume than

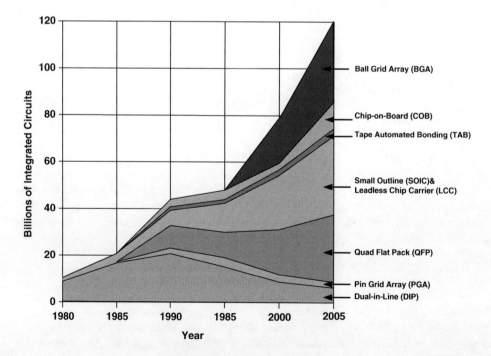

Figure 2.46 IC package usage

through-hole packages, and they are more suitable for automated assembly processes, albeit with higher investment.

References

[1] Langton, R. (2006) *Stability and Control of Aircraft Systems*. John Wiley and Sons, Ltd., Chichester.

[2] Smith, S.W. 1999, *The Scientist and Engineers Guide to Digital Signal Processing*, 2nd edn. California Technical Publishing, San Diego, CA.

[3] Mancini, R. (2002) *Op Amps for Everyone – Design Reference*. Texas Instruments/Elsevier, Burlington, MA.

[4] Bartee, T.C. (1986) *Digital Computer Fundamentals*, McGraw-Hill Inc., USA.

[5] Lewin, D. (1972) *Theory and Design of Digital Computers*, Thomas Nelson & Sons Ltd., Walton-on-Thames, UK.

[6] Rierson, L. (2013) *Developing Safety Critical Software*, CRC Press, Boca Raton, FL.

[7] RTCA-DO-178: *Software Considerations in Airborne Systems and Equipment Certification*, Radio Technical Commission for Aeronautics.

[8] RTCA-DO-254: *Design Assurance Guidance for Airborne Electronic Hardware*, Radio Technical Commission for Aeronautics.

[9] Ayers, J.E. (2010) *Digital Integrated Circuits*, CRC Press, Boca Raton, FL.

[10] Morant, M.J. (1990) *Integrated Circuit Design and Technology*, Chapman and Hall, New York.

3

Data Bus Networks

3.1 Introduction

This chapter discusses digital data buses (sometimes also foreshortened in this chapter to the bus) and communications networks used in today's civil transport aircraft.

Firstly this chapter discusses the basic principles of bus technologies including encoding techniques, attributes and topologies. Then the chapter discusses how these principles have been applied to explore the features and evolution of significant data bus networks seen in today's civil transport aircraft, namely:

- **ARINC 429**: 100 kbps; linear, unidirectional bus.
- **MIL-STD-1553B**: (also known as DEF-STAN-00-18 Part 2 and STANAG 3838), 1 Mbps, linear, bidirectional bus; centralised control, command/response protocol.
- **ARINC 629**: 2 Mbps, linear, bidirectional bus; distributed control, carrier sense multiple access/collision detection (CSMA/CD) plus dynamic time-slot allocation (DTSA) protocol.
- **ARINC 664** (AFDX Ethernet): 10/100 Mbps; star, bidirectional communications network; distributed control, with a CSMA/CD-derived protocol to provide pseudo-deterministic timing and redundancy management.
- **CANbus**: 1 Mbps; linear, bidirectional bus; prioritised collision detection/avoidance protocol.

MIL-STD-1553B is a military data bus standard used extensively in military aircraft, but is included in this chapter because it has many important features and capabilities, the lessons from which were then garnered in the evolution of subsequent civil and military avionics data bus standards. Some dual use civil and military platforms will incorporate MIL-STD-1553B equipment.

Other avionics data buses of note include:

- Avionics Standard Communications Bus (ASCB) was developed by Honeywell and is used in general aviation (GA) and business jet applications. It is a bidirectional bus with centralised control command/response protocols having some similarity to MIL-STD-1553B but using COTS technologies, operating at 670 kbps.

Civil Avionics Systems, Second Edition. Ian Moir, Allan Seabridge and Malcolm Jukes.

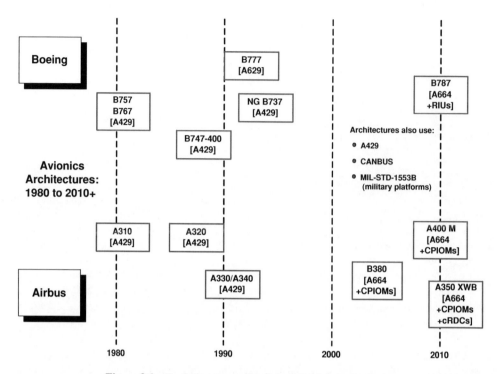

Figure 3.1 Evolutionary application of data bus technologies

- Commercial Standard Data Bus (CSDB) developed by Rockwell Collins for use in GA applications. It is an asynchronous broadcast bus, similar to ARINC 429, operating at 12.5 kbps and 50 kbps.
- RS232 and RS422 serial digital data buses are also used in some applications.
- IEEE 1394 Firewire 800 Mbps for digital video.

Figure 3.1 shows the evolutionary deployment of these data bus networks on Boeing and Airbus civil transport aircraft.

It is not the purpose of this chapter to give a detailed description of the electrical and temporal characteristics of any particular data bus type. The reader is directed to the detailed specifications for that. Rather, the purpose here is to summarise the essential features and technologies of data buses and communications networks in common use in civil transport avionics systems so the reader can appreciate the concepts involved, contrast their performance and attributes, and hence understand what makes each best suited to a particular application.

3.2 Digital Data Bus Basics

3.2.1 Data Bus Overview

Early avionics systems employed analogue computers with information transferred between sensors and effectors and computers, and between computers themselves, as an analogue signal (e.g. voltage, current, pulse width, frequency, etc.) proportional to the quantity of the measured variable. Each signal was most commonly transported over a screened copper wire, sometimes a screened twisted pair of copper wires for improved noise immunity. Discretes such as system states (e.g. undercarriage down) and commands (e.g. open cross-feed valve) were transferred by a single copper wire.

By the mid-1960s digital computers were replacing analogue computers and a means to transfer information digitally was required. The most obvious solution was to transfer the information as a series of zeros and ones (a data word) representing the binary coded equivalent of the quantity concerned (as discussed in Chapter 2). Discretes were packaged as a set of single bits into a data word. The possibility then existed of transferring more than one quantity over the same wire using the technique of time division multiplexing. With the addition of a means to identify which data word refers to which quantity (the protocol), we have a digital data bus. Information transfers need to be repeated on a continuous cyclic basis at a frequency which satisfies the Nyquist criteria (as discussed in Chapter 2).

So a digital data bus is a means to transfer information from one computer to another computer in digital form, usually over a twisted shielded pair of copper wires; or possibly a fibre-optic cable for higher bandwidth systems. Typically information will comprise messages (frames or packets of data) transmitted serially over the data bus (see Figure 3.2). Data packets comprise a series of data words preceded by a command word or header which identifies the data packet, the amount of information contained within it and other information concerning the data quality, its source and destination.

The message stream will also include a means to detect errors or corruptions within it. The simplest means to detect errors is a parity check. An extra bit (the parity bit) is added to the data stream which may take the value 0 or 1 to make the total sum of 1s in the data stream an even number (known as even parity). Parity checking can only detect a single bit error and

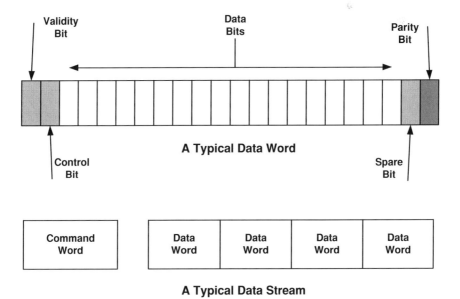

Figure 3.2   Basic data bus format

so is really only suitable for short messages. Additional bits are added to longer messages to improve the error detection capability. These are known as cyclic redundancy checks (CRCs) and with sufficient number of bits may even provide error correction as well as error detection. However, for avionics systems the general design objective is for errors to be rare events and to design alternative means to manage loss of a message.

3.2.2   Bit Encoding

Two of the most popular methods of encoding digital information into electrical signals are shown in Figure 3.3. They are the bipolar return-to-zero (RTZ) format and the Manchester biphase non-return-to-zero (NRZ) format.

The bipolar RTZ format encodes a logic 1 as a high-to-zero voltage transition during each bit time. A logic 0 is encoded by a low-to-zero voltage transition during each bit time. It is the voltage levels that signal the bit time and the bit value. The absolute voltages are unimportant, only their polarity matters. The receiver incorporates threshold detection with hysteresis to provide protection against electromagnetic interference. Each bit time comprises a portion that signifies the value of the data bit (logic 1 or logic 0) and a portion when the signal is zero which enables the receiver to synchronise itself to the transmitter. The data are self-clocking. The receiver does not need to know the frequency of the transmission, as it synchronises itself to the transmitter by sensing the period during the bit time when the signal is zero. In theory the clocking rate could be any frequency, but in practice the bit time frequency is constrained by the characteristic impedance of the cable and reflections created by impedance mismatches at cable breaks, splices and connectors.

The Manchester biphase non-return-to-zero format encodes a logic 1 as a high-to-low voltage transition during the bit time; and logic 0 as a low-to-high voltage transition during the bit time. It is the voltage transitions that signal the bit time and a bit value. Again the data are self-clocking, but a short preamble is required to synchronise the receiver to the transmitter so that both are in step with bit times.

The bipolar RTZ format has three voltage states: high, low and zero. Manchester biphase has two voltage states: high and low. Manchester biphase encoding also lends itself to fibre-optic systems as well as copper. In both arrangements the signal is transmitted in true and

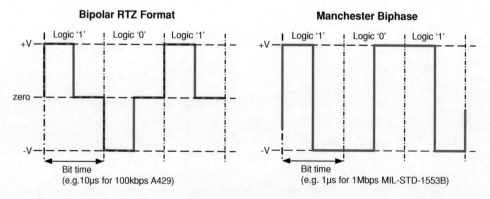

Figure 3.3   Bit encoding

complement (inverse) form over a pair of twisted copper wires, thus the electromagnetic field created by each wire is cancelled by the inverse field from the other wire to minimise interference. The twisted pair cable is shielded by a screen which is terminated at each end, extending the Faraday cage between the transmitter and receiver. The signal is received differentially (refer to the detail waveforms of ARINC 429 shown in Figure 3.3).

3.2.3   Attributes

Highly desirable attributes for avionics systems data buses include:

- **High efficiency**: how much useful data is transmitted on the bus compared with the total number of bits that could be transmitted.
- **Low latency**: a data bus is a digitally sampled system. Delays occur in many parts of the system (data latency needs careful consideration, especially if transmission is non-synchronous with the sampling rates of transmitter and/or receiver).
- **Deterministic**: time critical and safety critical systems require predictable behaviour; asynchronous and collision avoidance protocols are by nature non-deterministic.
- **High integrity**: the ability to detect and correct errors (including retransmit a message).

3.2.4   Transmission Classes

Common classes of digital data transmission are:

- **Single source – single sink (unicast)**. This classification describes a dedicated link from one equipment to another. Earliest data bus applications were of this type, replacing analogue links used in previous systems.
- **Single source – multiple sink (multicast)**. This classification describes a technique where one transmitting equipment can send data to a number of recipient equipments (sinks).
- **Multiple source – multiple sink**. In this classification, multiple transmitting sources may transmit data to multiple receivers with suitable measures to ensure only one transmitter sources data at any one time (as we shall discuss later).

3.2.5   Topologies

Common types of network topologies are shown in Figure 3.4. The distinction between a data bus and a network is a little esoteric. Linear topologies are usually referred to as a data bus, but as we add complexity with more than one link in a multiple data-link bus topology like a star topology, we tend to refer to these as networks.

- **Linear topology**
  - all line replaceable units (LRUs) listen to all transmissions
  - LRUs can be added sequentially to the bus.
- **Ring topology**
  - messages are passed sequentially from one LRU to the next round the ring
  - the ring must be broken to add a new LRU.

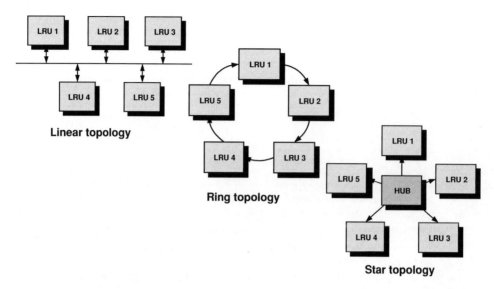

Figure 3.4   Network topologies

- **Star topology**
  - LRUs are connected to a central hub, switch or router (the term switch will be used)
  - messages pass through the hub (or switch) to all other LRUs on the network
  - LRUs can be added by adding ports at the switch.

3.2.6   Transmission Rates

Transmission rates for commonly used military and civil avionics data buses are shown in Figure 3.5:

- ARINC 429, introduced in the mid-1980s, is a single source, multiple sink, linear topology data bus operating at 100 kilo-bits per second (kbps).
- ARINC 664 P7 (AFDX Ethernet), introduced in the early 2000s, is a multiple source, multiple sink, star topology communications network typically operating at 10 and 100 Mega-bits per second (Mbps).

For further information the reader is directed to references [1] and [2].

3.3   Transmission Protocols

3.3.1   Transmission Protocols Overview

Single source, multiple sink unidirectional data buses such as ARINC 429 may be simple in command (protocol) structure, but the physical network to implement an avionics system can be complex since a data bus emanates from every transmitting equipment (see Figure 3.6).

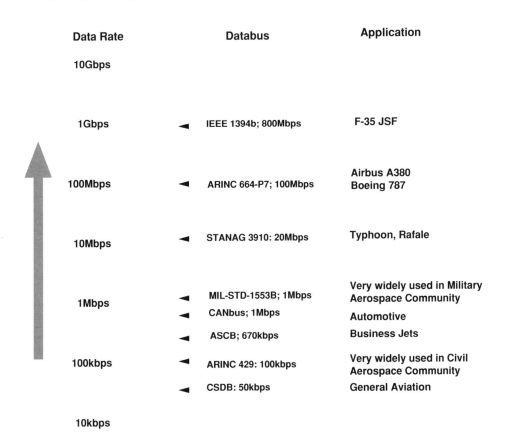

Figure 3.5   Data bus transmission rates

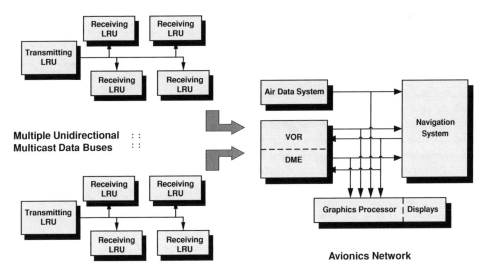

Figure 3.6   Avionics network – single source, multiple sink data bus

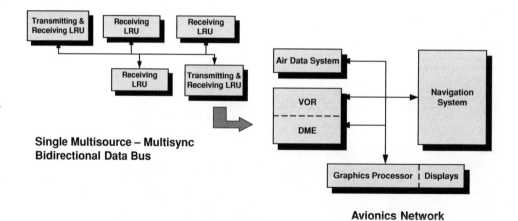

Figure 3.7 Avionics network – multiple source, multiple sink

By facilitating communication in both directions, the physical network can be simplified. A bidirectional data bus has multiple sources and multiple sinks. In this type of data bus, all terminals can transmit and receive on the same data bus; but of course not at the same time, or the messages would over-write each other. The messages need to be kept apart by a time multiplexing protocol. The physical network is simple: each terminal needs only one data bus interface (see Figure 3.7). However, the data protocol is complex. Arbitration must take place to ensure that only one terminal is able to transmit at any one time. Receivers must listen all the time. Receiver acknowledgement is possible, and error recovery strategies can be developed.

The most popular bidirectional data bus protocols are:

- **Time-slot allocation**: Each terminal is assigned a predetermined time slot on the bus. A terminal must wait (listen) for its time slot to take control of the bus and transmit.
- **Command/response:** A bus controller (BC) commands all transactions on the bus. No terminal may transmit without receiving permission from the bus controller. Failure of the prime BC may activate an alternative BC; however, only one BC can be active at any one time. The state of the bus must be replicated in all bus controllers so that a smooth transition can be effected in the event that control is handed over.
- **Token passing**: A token is passed around the bus terminals. A terminal can only take control of the bus and transmit after it receives the token. When it has completed its transactions it passes the token on to other users.
- **Contention**: Any terminal may transmit any time after the bus becomes idle (quiet). If two terminals start to transmit at once, a collision occurs. Both terminals must stop and, after a random wait, retry to transmit. Collisions are normal events. A contention bus tends to collapse under heavy loading.

3.3.2   Time-Slot Allocation Protocol

The simplest time-slot allocation protocol is time division multiplex allocation (TDMA). Each terminal is assigned a predetermined time-slot on the bus. The terminal must wait for its time-slot to take control of the bus. The time and duration of each terminal's transmission slot

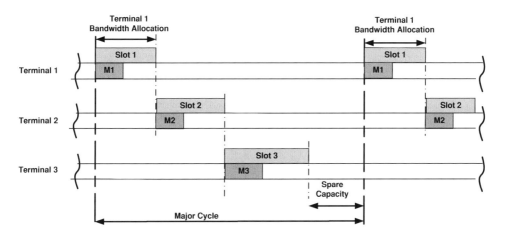

Figure 3.8 Time-slot allocation protocol

is predetermined by the bus designer. The terminal must complete its transactions within its assigned time-slot. When the time-slot expires, bus access is curtailed even if it is unfinished, and the bus is then assigned to another terminal (see Figure 3.8). Each terminal is assigned a portion of the total bus bandwidth; this is its bandwidth allocation.

In its simplest implementation, the bus designer assigns bandwidth allocation on the basis of worst-case communication requirements. Dynamic time-slot allocation (DTSA) is an enhancement that allows terminals to adjust the time of access based on bus activity. Under normal conditions one terminal will be in the transmit state, while all other terminals will be in the receive state. After a transmitter has finished its transactions, it relinquishes the bus to the next terminal. DTSA can exhibit greater throughput and shorter waiting times than baseline TDMA during periods of high bus loading.

The ARINC 629 data bus used on the Boeing 777 uses a variant of DTSA to ensure terminals access the bus in a prioritised sequence – see Section 3.6.

3.3.3 Command/Response Protocol

Command/response protocols use a bus controller to manage all bus transactions. No transmissions can be initiated without the permission of the bus controller. There may be one or more bus controllers for reasons of redundancy; however, only one may be active at any one time. Any other bus controller must be on standby and be ready to take over the bus control as and when required.

With centralised bus control, the bus controller controls all communications. Bus traffic is tightly controlled and is deterministic in nature. The network communication definition resides in the BC. With distributed bus control, control is distributed amongst several BCs. Transfer of bus control must be carefully coordinated and dynamic (non-deterministic) reconfiguration is generally not favoured for aerospace applications. The network communication definition must reside in all BCs. The BC also has knowledge of the status of every message transaction, the status of every terminal, and executes the error management strategy.

MIL-STD-1553B is the most widely used command/response protocol – see Section 3.5.

3.3.4   Token Passing Protocol

Token passing protocol avoids the single point failure of centrally controlled command/response protocols and avoids the chaos of contention protocols. A token (a special bit pattern) circulates around the terminals. When a user is in possession of the token he has exclusive use of the bus. When a terminal has completed all its transactions, it passes the token to the next user. The token is free to be passed around the bus. When a terminal wants to send a message, it must wait until it receives the token before that message may be enacted.

The bus protocol has to manage the failure case should a terminal fail to pass on the token and relinquish the bus (sometimes known as the babbling idiot). Also, a means must be provided to replicate the token in the event that the token is lost or corrupted.

DOD-STD-1773 is an example of a token passing network. In this implementation, the token is passed sequentially from the highest priority terminal to the next.

3.3.5   Contention Protocol

Carrier sense multiple access (CSMA) is the simplest contention protocol. All terminals listen to the bus. When the bus is quiet, any terminal may make a transmission. In its simplest form all terminals have equal priority. Of course, the possibility exists that more than one user may then initiate a transmission at the same time. This is a collision, and the data transmission will become corrupted.

Collision detection (CD) enhances the basic protocol (known as CSMA/CD). All terminals monitor their transmissions for collisions. If a transmission corruption is detected then the transmitting terminals must cease to transmit. The bus is now quiet again. The transmitting terminals must wait a random period of time before trying again. By nature of this randomness, one of the terminals will gain access to the bus first, thus ensuring that the retried transmission will not be corrupted.

In its basic form, CSMA/CD protocol is non-deterministic. Collisions are normal events. All transactions are equal and there is no prioritisation. As more terminals are added to the bus, the likelihood of collisions increases and bus performance degrades.

Ethernet is a CSMA/CD protocol. It is generally implemented as a star topology. This is the basis for the ARINC 664-P7 Aviation Full Duplex (AFDX) network to which a number of enhancements have been added to eliminate contention, manage bandwidth allocation and order messages in a pseudo-deterministic manner. In AFDX, network policing and fault containment is undertaken by the bus switch/router (see Section 3.7).

3.4   ARINC 429

3.4.1   ARINC 429 Overview

ARINC 429 is by far the most common data bus in use on civil transport aircraft, regional jets and executive business jets flying today. Since its introduction on the Boeing 757/767 aircraft and on Airbus aircraft in the early 1980s, hardly an aircraft has been produced which does not utilise this data bus. It is a single source, multisync, linear topology data bus. It is transmitted on a twisted pair screened cable using bipolar return-to-zero encoding. Up to 20 receiving terminals can be connected onto the bus. Connections can be made by simple splicing into the

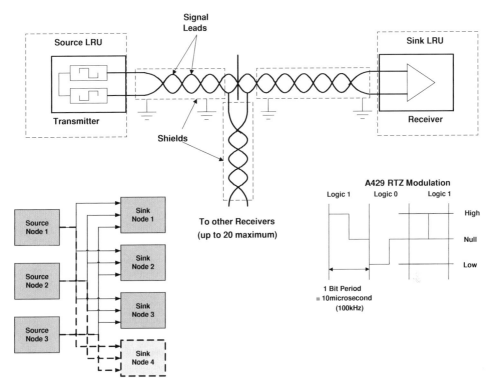

Figure 3.9   ARINC 429 bus topology

cable harness, and the bit rate is comparatively low so no matched termination is required – see Figure 3.9.

Information is transmitted one 32-bit word at a time. The word format is shown in Figure 3.10. It comprises an 18-bit data component into which information can be encoded as binary, binary coded decimal (BCD), alphanumeric (ASCII) and discretes. The data are preceded by an 8-bit label component that identifies the information content of the data component. The final bit is a parity bit. Other bits are available to identify the source and the status of the data. The ARINC 429 specification incorporates many fixed labels and data formats to facilitate an open standard. However, designers may implement alternative and additional data encryptions to suit the requirements for their system. It should be noted that information flows only in one direction from source (transmitter) to sinks (receivers). There is no means for a receiver to acknowledge receipt. The source transmits information when it is available; the receivers must take it when it comes. The transmitter does not know whether the receivers have correctly received the message. There is no handshake or error recovery mechanism for missed or corrupted messages.

The standard offers two data rates. The low-speed 12 to 14 kbps data rate is generally used for communication with sensors, effectors, control panels and annunciators with low data content. The high-speed 100 kbps data rate is generally used for communication between avionics computers. The data capacity is modest. The high-speed data rate of 100 kbps accommodates

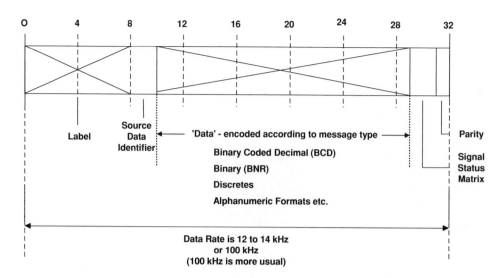

Figure 3.10   ARINC 429 word format

little more than 32 words at 50 Hz (20 ms) cycle time with a bus loading of 50%. The ARINC 429 bus can be thought of as a transport pipe for a single message (or data packet) which comprises a list of all the words transmitted over that link. This is a concept we shall return to later in the context of ARINC 664-P7 (AFDX) in Section 3.7.

3.4.2   ARINC 429 Architecture Realisation

The consequence of building an avionics system architecture using ARINC 429 is that there are a significant number of ARINC 429 data bus links. Every transmitting avionics computer (LRU) will have at least one ARINC 429 output channel, and every receiving LRU will have one ARINC 429 input channel for every LRU from which it receives data. Figure 3.11 shows a typical flight deck architecture implemented using ARINC 429 data buses. In this architecture there are over 20 ARINC 429 channels, one from every transmitting LRU. The flight deck displays are users of information from most of the avionics system LRUs, and so receive a large number of ARINC 429 input channels.

ARINC 429 has the advantage that the data protocol is extremely simple; that is, the transmitting terminal sends information when it is ready; the receiving terminal acquires data as it arrives. However, the consequential physical network is complex. For further information, see Reference [3].

Equipment-specific avionics (ARINC) specifications describe absolutely the message format that any particular equipment should transmit. While the equipment designer is constrained by this rigour, it ensures that all equipment of a particular genre will transmit identical data in an identical manner. This offers the advantage that, for example, all Air data computers from all suppliers in conformance to the ARINC specification may be considered interchangeable as far as the data bus design is concerned.

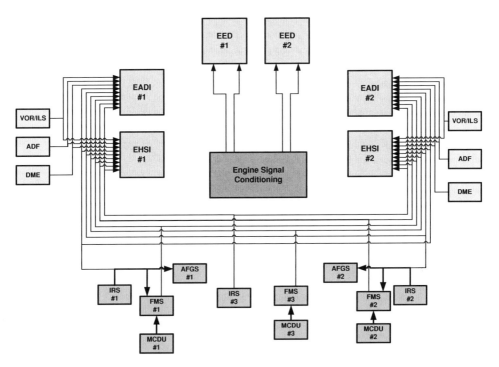

Figure 3.11 ARINC 429 architecture example – flight deck domain

3.5 MIL-STD-1553B

3.5.1 MIL-STD-1553B Overview

MIL-STD-1553B (also known as STANAG 3838) is a military standard data bus widely used on many types of military aircraft. It was originally conceived in 1973 and the 1553B standard emerged in the late 1970s. It is a bidirectional, linear topology, centralised control, command/response protocol bus. Data are transferred via a screened twisted wire pair in true and complement Manchester biphase format at 1 Mbps. Control is performed by a centralised bus controller (BC) which executes transactions with remote terminals (RTs) embedded in each of the avionics system LRUs. Each transaction takes the form of a command issued by the BC, transfer of data to/from the remote terminal, followed by a status response from the receiving remote terminal. Bus communications are highly deterministic. Comprehensive message error detection and correction capability provides high levels of data integrity.

The network is usually implemented for dual redundant operation. Up to 31 remote terminals plus the BC can be connected to the bus as shown in Figure 3.12. A typical implementation usually colocates the BC in the most compute-intensive remote terminal (usually also the terminal with highest data flow).

The standard allows two methods to connect (couple) remote terminals to the bus. If the termination length is less than 1 foot (30 cm), terminals may be directly coupled to the bus

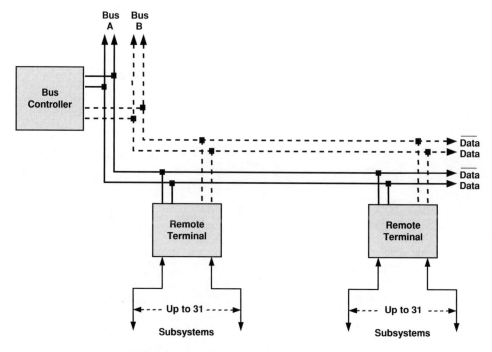

Figure 3.12 MIL-STD-1553B data bus topology

(direct coupled stub). However, this method is discouraged due to the risk of a short-circuit at the terminal causing bus failure; a transformer coupled stub is preferred (see Figure 3.13). Transformer coupled stubs can be any length, but are recommended to be less than 20 feet (6 m). Isolation resistors match the characteristic impedance of the cable, and these together with the isolation transformer provide protection against short-circuits. Transformer coupling also aids installation aspects on-aircraft as the stubs are allowed to be much longer.

3.5.2 MIL-STD-1553B Word Formats

The protocol comprises command words, data words and status words packaged into 20 bits at a data rate of 1 Mbps using self-clocked Manchester biphase encoding, as shown in Figure 3.14.

Each word commences with a synchronisation pulse to 'wake-up' the remote terminals and get them in-step:

- A *data word* comprises 16 data bits plus parity.
- A *command word* comprises five bits to identify the addressed remote terminal (maximum of 31 RTs), a transmit receive bit, five bits to identify the sub-address (usually used to identify the data packet), five bits to identify the number of data words in the transaction (maximum of 32 words per transaction) plus parity.

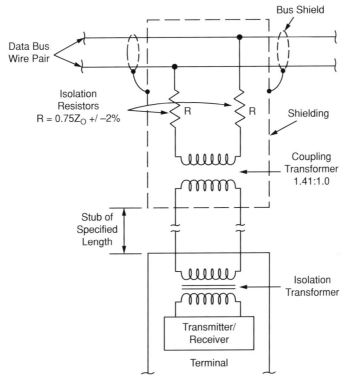

Figure 3.13   MIL-STD-1553B transformer coupled stub

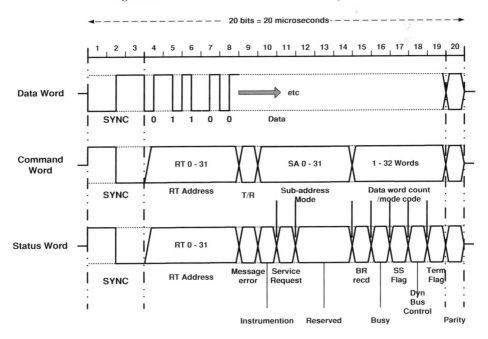

Figure 3.14   MIL-STD-1553B word formats

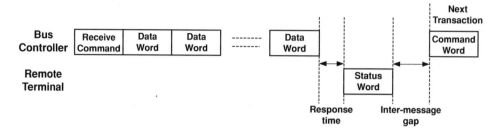

Figure 3.15   BC to RT transaction protocol

- A *status word* comprises five bits to identify the responding remote terminal, and a series of bits to indicate the remote terminal status. The reader is directed to the standard for a full explanation of these bits. For the purposes of this discussion we shall just summarise them as 'RT working correctly, message valid and correctly received', correct receipt of data transferred to it or the data it is transmitting is valid.

Typical message transactions are:

- bus controller to remote terminal;
- remote terminal to bus controller;
- remote terminal to remote terminal;
- broadcast.

3.5.3   Bus Controller to Remote Terminal (BC-RT) Protocol

The BC to RT transaction is shown in Figure 3.15. The BC issues a receive command identifying within it the bus address of the RT which is to receive data, the data packet (sub-address mode) to be received, and the number of data words in the packet. The BC then immediately appends the data packet (with no gap). After a brief time gap (4–12 µs) the receiving RT responds with its status, advising the BC that it has correctly received the message (or not). This completes the BC to RT command/response transaction handshake. After a short inter-message gap (>4 µs but typically ~8 µs) the BC can commence the next transaction.

3.5.4   Remote Terminal to Bus Controller (RT-BC) Protocol

The RT to BC transaction is shown in Figure 3.16. The BC issues a transmit command identifying within it the bus address of the RT which is to transmit data, the data packet (sub-address mode) it is to transmit, and the number of data words expected in the packet. After a brief gap, the RT responds first with its status immediately followed by the requested data packet of up to 32 words. The BC of course knows whether it has received the data packet correctly or not, so it has no need to issue a status word. This completes the BC to RT transaction handshake and after a short inter-message gap the BC can commence the next transaction.

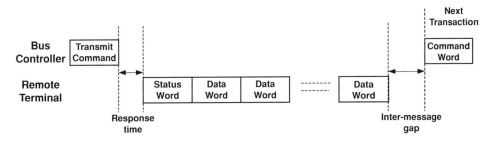

Figure 3.16   RT to BC transaction protocol

3.5.5   Remote Terminal to Remote Terminal (RT-RT) Protocol

The RT to RT transaction is shown in Figure 3.17. It is a composite of the RT-BC and BC-RT transactions. The BC first issues a receive command to make ready the RT designated to receive the data, then secondly issues a transmit command to the RT that is to transmit the data. After a short response time gap, the transmitting RT responds first with its status immediately followed by its data packet. After a short response time gap the receiving RT responds with its status. This completes the RT to RT transaction handshake, and after a further short inter-message gap the BC can then commence the next transaction.

3.5.6   Broadcast Protocol

The broadcast transaction is a command plus data issued by the BC to all RTs. The MIL-STD advises that it is to be used infrequently. Typically it is used at start-up to synchronise all RTs before bus transactions begin. There can be no status words issued in response to a broadcast command, since if there were, all RTs would respond at the same time. So the BC is unaware of any transaction errors, hence the reason that this command is to be used advisedly.

3.5.7   Error Management

There are a wide range of error management strategies possible in MIL-STD-1553B. The bus controller knows the status of every transaction message. The standard does not mandate any particular strategy; it is up to the bus designer to implement the most appropriate strategy

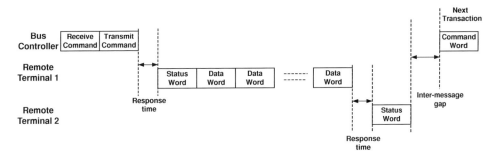

Figure 3.17   RT to RT transaction protocol

for his bus implementation to meet his availability, integrity and temporal response (latency) requirements. Typical error management strategies include:

- immediately retry the faulty transaction on the same bus;
- retry the faulty transaction on the reversionary/alternate bus;
- retry the faulty transaction when all other transactions are complete (at the end of the major cycle);
- build a fault log associated with each RT and cease communication with a faulty RT if a threshold fault rate is exceeded.

To design a particular implementation, the bus designer compiles a list of all of the messages with all the RTs on the bus. From this he compiles an execution list called a transaction table. The bus controller sequentially executes the transaction table until all transactions are complete, then after any retries the process repeats (usually at a fixed repetition rate). Thus it can be seen that the process is highly deterministic and the high degree of error detection and correction affords high integrity.

However, MIL-STD-1553B does not readily support aperiodic unscheduled transactions. A means, but admittedly rather poor means, to facilitate this is provided by the 'service request bit' which an RT can assert in its status word to signal to the bus controller that it has an aperiodic special event that requires attention. The bus controller recognises this and can then institute a series of dedicated bus transactions to service the request.

The high degree of determinism in MIL-STD-1553B carries a significant overhead burden. Figure 3.18 indicates the timing for a typical RT to RT transaction of 16 data words. The total transaction takes 466 μs total; the data element comprises 16 20-bit data words

- **Data Rate**
  Transmission rate   1 Mbps (= 1 μs per bit)
  20-bit word   20 μs per word
  Typical RT-RT transaction
  - **Receive command**   20 μs
  - **Transmit command**   20 μs
      Inter-message gap   8 μs (typical)
  - **Status word**   20 μs
  - **(1-32) Data words**   320 μs (say typically 16 words)
      Inter-message gap   8 μs (typical)
  - **Status word**   20 μs
      Inter-message gap   50 μs (typical)   Overhead = 45%
  TOTAL   466 μs (typical)

- **Capacity**
  25 messages   11.65 ms
     (16 words average each, 400 words total)
  Bus loading at 50 Hz (20 ms)   58%
                            (residual available for error handling and growth)

  Strategies to increase capacity include "rolling piano key" for <50 Hz refresh data

Figure 3.18   MIL-STD-1553 transaction timings (typical)

(320 μs); thus the overhead is 45%. 25 messages each of 16 words (400 words total) results in a bus loading of 58% at a cycle time of 50 Hz (20 ms). Strategies to increase capacity include interleaving messages requiring less than a 50 Hz update rate across several bus cycles. For example, data requiring only 25 Hz update is split across alternate frames. This technique is sometimes referred to as 'rolling piano key'.

The capacity limitation was overcome in the Eurofighter Typhoon project by using a 20 Mbps fibre-optic network for data words alongside the conventional 1 Mbps copper network for command and status response words. This variant, known as STANAG 3910, allows mixed operation of copper 1 Mbps terminals with high bandwidth 20 Mbps fibre-optic terminals. The US Department of Defense developed a 50–100 Mbps fibre-optic data bus network for the F-22. This standard, DOD-STD-1773, allows up to 128 terminals to be connected onto the bus; information is transferred in 4096 word packets.

For further information on MIL-STD-1553B the reader is directed to references [4] and [5].

3.6   ARINC 629

3.6.1   ARINC 629 Overview

ARINC 629 was developed in the late 1980s to provide civil aircraft with a multisource, multisync, linear topology network addressing the physical network issues of ARINC 429. It is built upon the experience of MIL-STD-1553, but specifically does not require a centralised bus controller; the civil aerospace community wanted to avoid the single-point failure issues surrounding that concept, and instead opted for a distributed protocol. ARINC 629 uses a 20-bit word encoded in Manchester biphase format operating at 2 Mbps, twice the bit rate of MIL-STD-1553. It also permits 128 terminals to be connected to the bus. Quadruple and triple redundancies are also supported in addition to dual redundancy (see Figure 3.19).

3.6.2   ARINC 629 Protocol

ARINC 629 uses a carrier sense multiple access/collision detection (CSMA/CD) protocol but with a form of dynamic time-slot allocation (DTSA) that resolves contention and assigns priority by the judicious adjustment of three timers within each terminal, as shown in Figure 3.20:

- A synchronisation gap (SG) timer begins when terminals sense the bus is quiet. This timer is the same for all terminals.
- A terminal gap (TG) timer starts when the synchronisation gap has elapsed. The terminal gap timer is different for each terminal. The highest priority terminal has the shortest terminal gap timer followed by the next shortest timer for the terminal with the next priority.
- A terminal interval (TI) timer sets the time that must elapse from the current transmission before the terminal can transmit again.

The timer arrangement ensures there are normally no collisions. The TG timer sequences terminal access to the bus on the basis of the shorter timer (higher priority terminal) first. The TI timer assures all terminals have an opportunity to access the bus in each bus cycle.

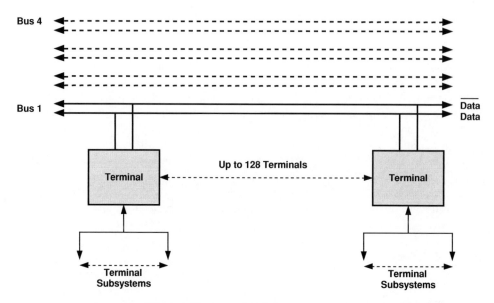

Figure 3.19   ARINC 629 databus topology

To protect against corrupted data, each transmitter monitors its own transmission and checks each packet as it is transmitted to verify its data is correct with the correct channel information and the correct number of data words. Any errors inhibit the transmitter for the rest of that message.

The protocol guarantees periodic access to the bus for each terminal and prevents any terminal from obtaining access before all the others have had chance to transmit. Periodic data are transmitted on a regular basis once every bus cycle. Aperiodic data may be transmitted occasionally. If the bus overloads it switches to aperiodic mode with no data loss.

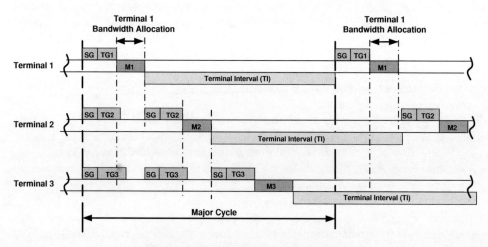

Figure 3.20   ARINC 629 message protocol

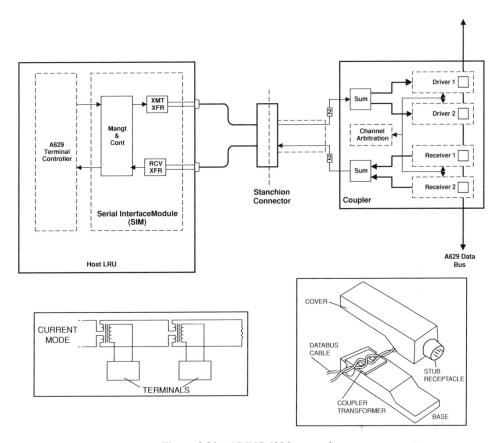

Figure 3.21   ARINC 629 bus coupler

3.6.3   ARINC 629 Bus Coupler

Each terminal connects to the ARINC 629 bus by a serial interface module (SIM) and a stanchion connector to a bus coupler. There are separate channels for transmit and receive (full duplex). Current mode transformer coupling is preferred. This means that the bus does not have to be broken to insert a terminal (see Figure 3.21). The bus coupler, arguably the greatest asset of ARINC 629, was perhaps also its downfall. The bus coupler together with the SIM is very expensive.

3.6.4   ARINC 629 Architecture Realisation

ARINC 629 was really only used on one aircraft type, the Boeing 777. The B777 architecture implements dual, triplex and quadruplex manifestations of ARINC 629 buses, as described later in Chapter 5.4 (Figure 5.12). To simplify the situation:

- Dual redundancy is used for most of the aircraft systems – left (L) and right (R) buses.

- Triple redundancy is used for the flight control ARINC 629 bus and for certain elements of the aircraft systems – left (L), centre 1 (C1) and right (R) buses.
- Exceptionally, the engine's electronic controllers (EECs) and AIMS cabinets interface in a quadruplex redundancy fashion to the left (L), centre 1 (C1), centre 2 (C2) and right (R) aircraft systems buses.

3.7 ARINC 664 Part 7

3.7.1 ARINC 664 Overview

The ARINC 664-P7 communications network is more than a data bus. It is currently used on the Airbus A380, A350, A400M, Boeing 787 Dreamliner, the COMAC ARJ21 and the Sukhoi Super-jet 100, and is likely to become the standard communications network on most future civil transport aircraft. The standard evolved from commercial communications packet switched origins, which is why terms such as 'subscribers' are often encountered.

ARINC 664 Part 7, also known as Aviation Full Duplex (AFDX, an Airbus trademark), is based on 10/100 Mbps switched Ethernet technology (IEEE 802.3) media access control (MAC) addressing, Internet protocols (IP), and user datagram protocol (UDP), but with special protocol extensions and traffic management to achieve the deterministic behaviour and appropriate degree of redundancy required for avionics applications.

The constituent elements are:

- **End-system** – a device whose applications access the network to receive or send data.
- **Switch** – a device that performs traffic policing and filtering, and routes data in the form of message packets towards their destination end-systems.
- **Virtual Links** – end-systems exchange message frames through virtual links. A virtual link defines a unidirectional logical connection from one source end-system to one or more destination end-systems (employing single source, multisync, unicast techniques in a manner similar to ARINC 429).

End-systems (or avionic computing LRUs) exchange frames (messages) based on the concept of virtual links (VL) with traffic shaping, policing, integrity checking and routing provided by ARINC 664-P7 switches using configuration tables. The configuration table embeds a 'contracted' performance guarantee in terms of bandwidth, latency, availability and integrity between the end-system and the network, and is a key element in the incremental system certification process.

Ethernet was developed in the 1970s and has dominated commercial local area networks (LANs). It is widely used; many of us now have an Ethernet network in our home and are familiar with the terminology. Ethernet holds powerful attractions for use in aviation. It is a proven technology; it offers a huge increase in bandwidth over previous avionics data buses (typically implemented at 10 and 100 Mbps in aerospace applications); and uses multisourced, relatively low-cost components.

Ethernet is a packet switched network. Its protocols are highly developed and use CSMA/CD techniques. The network is generally implemented as a star topology. Each terminal or user is connected to the network through a router or switch. Information is transferred in frames.

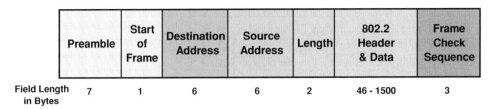

Figure 3.22   Ethernet frame format

3.7.2   Ethernet Frame Format

The frame format is shown in Figure 3.22 and comprises:

- *Preamble* – informs receiving terminals that a frame is coming.
- *Start of frame* – synchronises the frame reception at all terminals.
- *Destination and source addresses* – the source is a single terminal address. The destination can be to a single terminal, a group of terminals or broadcast to all terminals.
- *Length* – the number of data bytes that follow (1 byte = 8 bits).
- *Data* – the data portion of the frame.
- *Frame check sequence* – a cyclic redundancy check (CRC) sequence.

3.7.3   Network Topology

Ethernet was first used in civil aerospace IT ground systems, air traffic management, aircraft operations centres, and for route planning and navigation systems. On aircraft it was first used for passenger in-flight entertainment systems and now has gradually migrated into the avionics area.

A much-simplified ARINC 664-P7 avionics star topology network is shown in Figure 3.23. This shows a simplified network implemented as a star topology across two avionics domains.

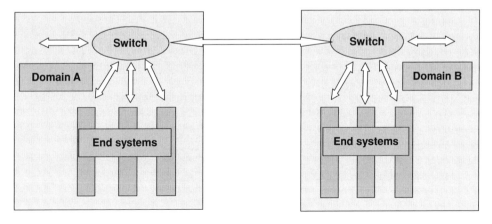

Figure 3.23   ARINC 664-P7 star topology

Each domain comprises a number of avionics systems computing elements, each of which for the purposes of an ARINC 664-P7 network hosts an end-system terminal. Each end-system terminal within a domain communicates with other end-system terminals within that same domain through the domain network switch. Communication between domains flows from domain switch to domain switch. The switch routes message frames (data packets) from the transmitter (known as the publisher) to the receiver (known as the subscriber). The switch replicates the message to more than one subscriber as requested. The physical layer conforms to Ethernet Cat5 standard for 10 Mbps or 100 Mbps implementations using components (connectors, cabling, etc.) hardened to withstand the airborne environment. On the Airbus A380 the switch is typically implemented as a 2-MCU LRU; the end system terminal is typically implemented as an ASIC plus support components. This is housed on a PCI circuit board which forms a mezzanine card on the end-system avionics processor card.

Chapter 5 will discuss a more comprehensive ARINC 664-P7 network supporting an integrated modular avionics architecture as shown in Figure 3.24. This figure shows a network comprising a set of end-systems in each connecting node and multiple network switches.

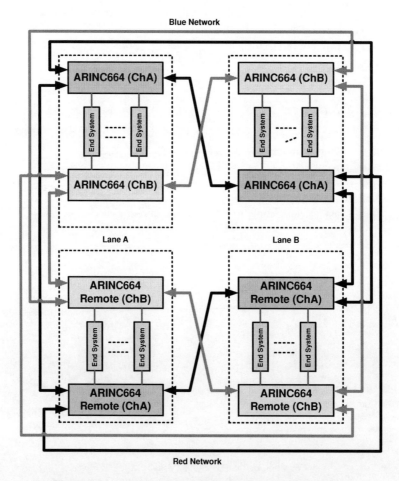

Figure 3.24   ARINC 664-P7 dual redundant network topology

The network is arranged as a dual-channel, dual-redundant switched star topology. Each end-system terminal has a point-to-point connection with two independent communication pathways (channel A and channel B). Each network switch has a direct connection to every other network switch in that channel. The A/B dual channel connection allows redundant transmission and reception of data through two independent paths. Channel redundancy is managed in the end system. This architecture ensures that loss of communication is extremely improbable.

3.7.4   Contention Avoidance

Ethernet is a contention (CSMA/CD) based protocol. In its basic form it would not be suitable for an avionics system network. As discussed earlier, avionics system data transactions must operate in real time with a high degree of confidence that messages will be transferred with high integrity and with minimum delay (latency). To accomplish this, an additional layer of confidence is provided by the ARINC 664-P7 switch. The first element of this process is to adapt the Ethernet frame format as shown in Figure 3.25.

In a traditional Ethernet switch, incoming frames (data packets) are routed to output links based on the destination address. ARINC 664-P7 replaces the source address, destination address(es) and length with a code that represents an identifier for a virtual link (known as the virtual link ID). The switch holds a record (configuration table) of all permitted virtual link IDs and therefore can verify that the data packet originates from a valid source for that ID and has the requisite message content. The switches are configured to route an incoming frame to one or more outgoing links. A frame associated with a particular virtual link ID must originate from only one end system. The switch then delivers that virtual link ID data packet to the fixed set of end systems as pre-determined by the configuration table. Thus a virtual link originates at a single end-system and delivers message frame packets to a fixed set of end-systems. The virtual link should be considered as the network transport system for data packets from one publisher to one or more subscribers, in a method that is analogous to an ARINC 429 single-source, multisink (multidrop or multicast) data bus link (see Figure 3.26).

The second element of this process removes contention. ARINC 664-P7 is a full duplex network. Each end system is connected to the switch by two twisted pairs of copper wires, one pair for transmit (Tx) and one pair for receive (Rx) – see Figure 3.27.

Rx and Tx buffers store incoming and outgoing data packets in first-in, first-out (FIFO) memories. The switch processor examines incoming data packet virtual IDs in the order received in its Rx FIFOs using its configuration tables, verifies their conformance to the expected data packets, and moves these packets to the appropriate outgoing Tx FIFO buffers.

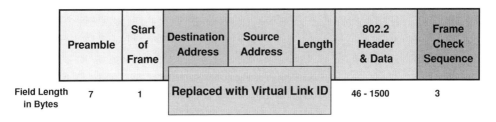

Figure 3.25   ARINC 664-P7 virtual link packet routing

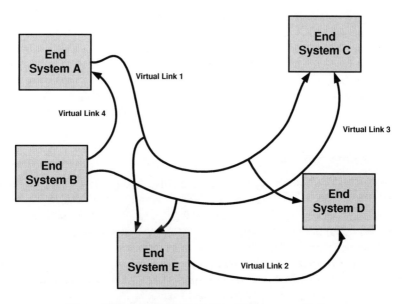

Figure 3.26 ARINC 664-P7 virtual link communications

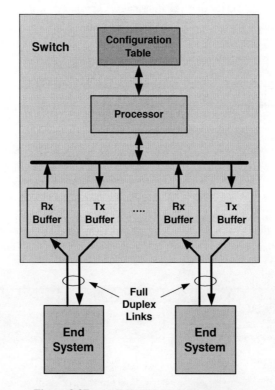

Figure 3.27 ARINC 664-P7 network switch

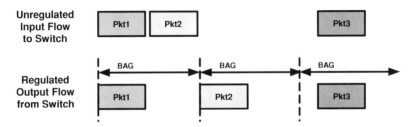

Figure 3.28   ARINC 664-P7 bandwidth allocation gap

Data are transmitted in the order received. There are no collisions, there is no loss of data, but packets may experience delays (jitter).

The network designer must ensure that traffic design avoids Rx and Tx buffer overflow. This is achieved by contract guarantees for network bandwidth (data packet size and rate) and maximum network delivery latency and jitter. The switch also checks that the packet size and transmission rate do not exceed the contracted bandwidth allocation for the specified data packet. The end-system controls the input flow for each message. Message frames will only be transmitted when they are within their contracted bandwidth allocation. This technique is known as bandwidth allocation gap (BAG), and is similar to the terminal interval (TI) discussed earlier in connection with ARINC 629. BAG is the primary means for regulating dataflow and protects the network from the babbling idiot (see Figure 3.28). BAG values are typically set as 1, 2, 4, 8, 16, 32, 64 or 128 ms intervals. A 16 ms BAG interval equates to a 62.5 Hz refresh cycle time. An end-system will only transmit data if it has new data available, irrespective of whether or not the message is BAG-eligible; in other words, if it has no new data to transmit, then no VL message will be generated.

ARINC 664-P7 is implemented as a dual redundant network. Message frames are transferred around the network by two independent data paths (channel A and channel B). Message frames are transmitted concurrently on both networks. Transmission corruption is detected in the switch using the frame check CRC sequence. Invalid frames are discarded. The end-system will select the first of the valid redundant frames to arrive and presents a single message to the end-system host processor. This process is shown in Figure 3.29. The dual redundant network assures availability; CRC checking assures integrity.

3.7.5   Virtual Links

Network data partitioning is enforced using virtual links (VLs) – see Figure 3.30. Virtual links have guaranteed allocated network bandwidth (data size and message rate) and guaranteed maximum network delivery latency and jitter. The network switch routes VLs from input port to output port(s). The switch checks that the size and transmission rate of each VL does not exceed its allocation against its network configuration file (table), as contracted between the system provider and the system integrator. VL priorities can be configured at the switch to increase performance for time-critical messages.

The switch performs real-time packet management. It manages congestion by controlling the volume of traffic sent from an end-system onto the network in a process known as traffic shaping or traffic scheduling.

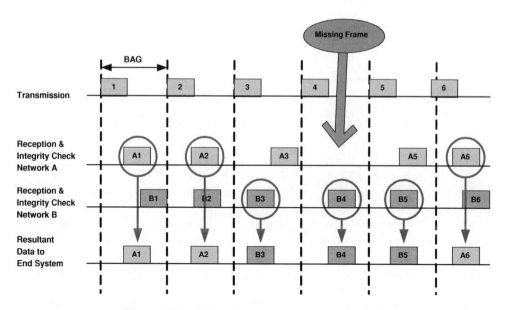

Figure 3.29   ARINC 664-P7 redundancy management

The switch buffers input data to avoid contention. Fault containment guards the network against breach of traffic contract (i.e. the babbling idiot) which might otherwise cause the input buffer to overflow, leading to loss of data.

The switch buffers data destined for the same end-system in an output buffer. Buffering resolves contention of more than one end-system sending data to the same destination.

Redundancy, dual switches and two communication paths ensure a high degree of availability. Error detection mechanisms such as CRC ensure high integrity. No data are lost, but the process of resolving contention inevitably results in messages being delayed. This delay is known as network latency. Jitter is the difference between the minimum and maximum time a receiving end-system may receive a message from a source node. The potential impact of uncontrolled jitter is discussed in Section 2.3.2. Jitter is a function of network design and multiplexing multiple VLs onto one port, as well as switch design. Maximum allowable jitter for each VL is held in the configuration table. The potential impact of uncontrolled jitter on signal quality is shown in Figure 3.31.

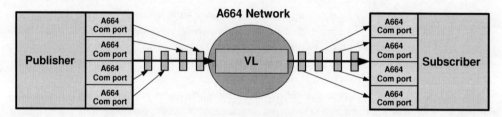

Figure 3.30   ARINC 664-P7 virtual links

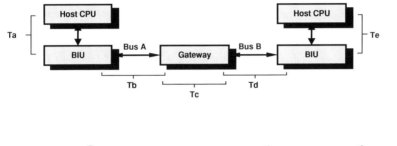

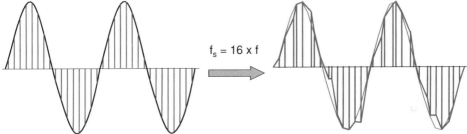

Figure 3.31   Impact of jitter on signal quality

The mechanisms to manage bus traffic are held in the switch configuration tables and are akin to the BC transaction tables in MIL-STD-1553B. The parameters to populate the tables are agreed between the subsystem designer and the network designer and form the basis of a 'contract' between them. For the subsystem designer, the 'contract' expresses his needs in terms of network performance to achieve the specified performance of his subsystem. For the network designer, the 'contract' expresses the share of network performance he must guarantee to be allocated to the subsystem. The 'contract' is the cornerstone for incremental certification of the subsystem and for the network (see also Section 5.10). Design and analysis tools are available to assist the bus designer to model bus performance and to garner evidence to support certification.

3.7.6   Protocol

Avionics applications residing at end-systems exchange messages through the ARINC 664-P7 network via the services of the user datagram protocol (UDP, layer 4 with underlying Internet protocol (IP, layer 3) – see Figure 3.32). The UDP is also the base for upper layer protocols for maintenance purposes, simple network management protocol (SNMP) and file transfer service trivial file transfer protocol (TFTP). ARINC 664-P7 switching is based on the media access control (MAC) destination address (layer 2). Deterministic provisions are implemented in layer 2.

Applications send and receive messages through communication (comm) ports which are basically mapped to UDP ports. There are two types of comm ports whose detailed characteristics are defined by ARINC 653 (Avionics Application Software Standard Interface) – see Figure 3.33:

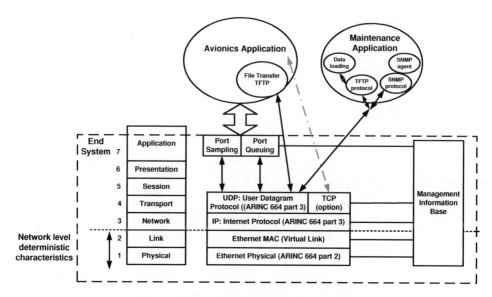

Figure 3.32   ARINC 664-P7 protocol layers

- **Sampling ports** – messages are sent over one frame, data may be lost or overwritten. The maximum amount of data is limited by the associated VL maximum frame size. Usually used for regularly scheduled but asynchronous processes that want to use latest fresh data.
- **Queuing ports** – messages may be sent over several frames (fragmentation by the IP layer is possible and is dependent on the associated VL maximum frame size). No data are lost or overwritten. Usually used for event-driven processes where it is important to maintain the messages in sequence.

The IP source is always a single source (unicast) address. The IP destination may be either a single receiver (unicast) or multiple receivers (multicast) in an end-system. A receiver is in effect an end-system application. In other words, data addressed to a unicast IP address shall only be used by one application; data addressed to a multicast IP can be used by multiple applications (or partitions) in the same end-system.

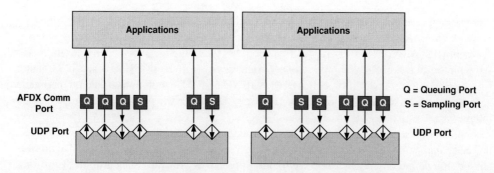

Figure 3.33   ARINC 664-P7 UDP protocol

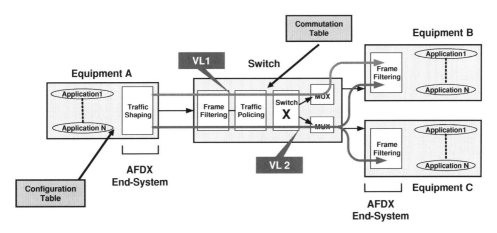

Figure 3.34   ARINC 664-P7 network communications summary

3.7.7   Summary

The means to assure network communications can be summarised as (see Figure 3.34):

- **Traffic shaping**: The end-system is configured to transmit a predetermined set of virtual links at a contracted bandwidth in accordance with a configuration table.
- **Traffic policing**: The transmitting switch statically routes VLs from prescribed sources within contracted bandwidth and latency constraints in accordance with its configuration/commutation table. It discards any traffic that exceeds its predefined VL configuration.
- **Traffic reception**: The receiving end system verifies frame and data (end-to-end CRC) and passes the first valid frame to the host application.
- **The switch protects against**:
  - the babbling idiot
  - malformed data packets
  - VL impersonation (emanating from an incorrect source).
- **Communication is pseudo-deterministic**:
  - Ethernet contentions are eliminated by the switch that buffers and orders data
  - the virtual link contract guarantees minimum bandwidth and maximum latency
  - data are received in the order sent
  - no data are lost.

3.7.8   Cables

Airbus and Boeing adopt different installation strategies for cabling and connectors, as shown in Figure 3.35. Airbus adopt the quadrax approach (two twisted pairs within a common outer shield and a single 4-way connector) using 24 AWG wire size. Boeing adopt the twinax approach (two twisted pair shielded cables and two 2-way connectors) using 22 AWG wire size.

The Boeing approach is clearly heavier, but possibly physically more robust. For further information the reader is directed to references [6], [7] and [8].

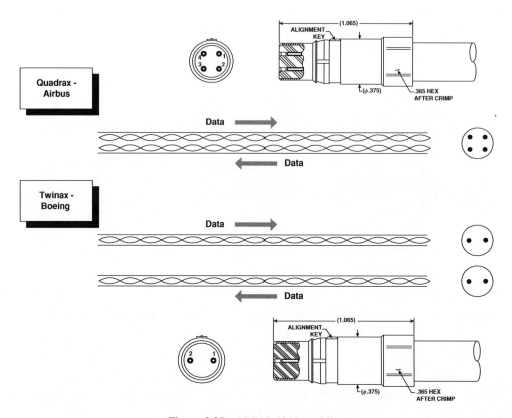

Figure 3.35   ARINC 664-P7 cabling

3.8   CANbus

3.8.1   CANbus Overview

CANbus was originally developed by Bosch in the 1980s, as a low-cost data bus for automotive applications. It has been very successful and widely adopted for ground vehicles. It is now finding its way into aerospace applications. CANbus nodes are typically sensors, actuators and other control devices with small amounts of sometimes time-critical data.

CANbus physical media is most commonly a 5 V differential (true and complement) signal transmitted over a twisted pair screened cable as shown in Figure 3.36, terminated at both ends in 120 ohms. The transmitter driver is open collector.

3.8.2   CANbus Message Formats

CANbus is a multisource, multisync, broadcast data bus. Data are transmitted at 1 Mbps, although typical realisable data rates are nearer 320 kbps. Data encoding is Manchester biphase non-return-to-zero (NRZ) format. CANbus uses an automatic arbitration version of carrier sense multiple access/collision detection (CSMA/CD) protocol. Each node is able to

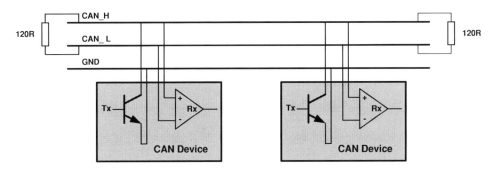

Figure 3.36 CANbus topology

send and receive messages, but not simultaneously. Messages have identifiers (ID); CAN nodes do not have addresses. The CANbus message format is shown in Figure 3.37.

The message comprises:

- SOF: start of frame, aids synchronisation (no clock is transmitted)
- ID: unique message identifier (with priority)
- RTR: remote transmission request
- DLS: specifies the number of data bytes to follow (0–8 bytes)
- DATA: data field (length as specified by DLS)
- CRC: cyclic redundancy check
- ACK: receiver acknowledge (receiver asserts dominant zero)
- EOF: end of frame.

The reader will note that the number of data bits is small (eight bytes maximum). This is because the intended application is for the transfer of information from sensors, actuators and other control devices.

If the bus is quiet, then any node can begin to transmit (CSMA protocol). If two or more nodes begin to send messages at the same time, then the node with the more dominant message ID (zeros) will overwrite the node with the less dominant message ID, due to the electrical properties of the open-collector driver. Both nodes must listen to their transmission. The node transmitting the less dominant ID must back off, giving priority to the node with the more dominant message ID. This process is illustrated in Figure 3.38.

Note that the receiving node can insert bits into the transmission. The most obvious is the receiver acknowledge bit (ACK) into which the receiver inserts a dominant zero if it has received the message correctly.

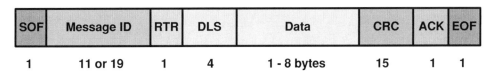

SOF	Message ID	RTR	DLS	Data	CRC	ACK	EOF
1	11 or 19	1	4	1 - 8 bytes	15	1	1

Figure 3.37 CANbus message format (simplified)

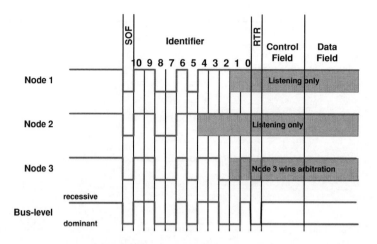

> • **At bit 5, nodes 1 and 3 send a dominant identifier bit, node 2 sends a recessive bit but reads back a dominant bit. Node 2 loses the bus.**
> • **At bit 2 node 1 loses arbitration against node 3.**
> • **Node 3 wins. Nodes 1 and 2 send their message after node 3 has finished.**

Figure 3.38   CANbus message prioritisation

CAN is a very reliable system with multiple error checks as follows:

- **Frame error**: certain bits in the frame must be set to predefined values (stuffed).
- **Stuffing error**: a receiving node will flag a stuffing error if a transmitting node fails to insert a high after five consecutive low bits (and *vice versa*).
- **Bit error**: a transmitter always reads back and verifies the message it is sending.
- **Cyclic redundancy check (CRC) error**: although the possibility of errors/corruption is relatively high, a very strong (15 bit) CRC code provides a very low probability of undetected errors.
- **Acknowledge error.**

3.8.3   CANbus Variants

There are a number of variants of CANbus, each with specific attributes to suit particular applications:

- **TJA 1054**: low-power, low-speed version used in automotive applications.
- **CANopen**: designed with control applications in mind, it is a software standard based on the full CAN-ISO-11898-2 5V bus.
- **TTCAN**: time-triggered CAN. Nodes report in predefined time windows, making the bus deterministic, and avoids overload even in worst case.

- **MILCAN**: defined primarily for use in military land vehicles where a deterministic protocol is required. It sets some rules for a software layer on top of CAN. A pseudo-hardware sync is created by one node (Sync-Master) that sends sync CAN frames with a 'sync slot number'.

3.9 Time Triggered Protocol

Time triggered protocol (TTP) was originally developed more than 20 years ago at Vienna University, and is now maintained by an international cross-industry consortium: the TTA-Group.

TTP was designed for time-critical, highly deterministic real-time applications where timely error detection, fault isolation and fault tolerance must be provided. It can be implemented in linear or star topology networks. Media access is controlled by a conflict-free time division multiplex allocation (TDMA) strategy. Every node is assigned a transmitting slot in a TDMA round. Every node contains the dispatching (transaction) table. Event-triggered and time-triggered messages can be sent. Bandwidth is reserved for aperiodic event transmissions inside the TDMA slots.

Key attributes of TTP are as follows:

- Bit rates: 25 Mbps synchronous, 5 Mbps asynchronous
- Encoding: Manchester biphase
- Message length: 240 bytes with 4–8 bit header and 24-bit CRC
- Media access: TDMA
- Topology: dual channel linear or star
- Media: copper or fibre
- No. of nodes: 64 maximum

Although of potential benefit for some time-critical avionics subsystems (e.g. flight control), TTP has yet to be applied to civil transport avionics systems.

3.10 Fibre-optic Data Communications

3.10.1 Attributes of Fibre-optic Data Transmission

Fibre-optics have been used for some time for digital data bus networks in military avionics systems, but have yet to see any extensive use in civil transport aircraft. The obvious advantages to using fibre-optics over copper are:

- immunity to interference (emission and susceptibility), especially high-intensity radiated fields (HIRF) from high-power transmitters on the ground;
- immunity to lightning strikes;
- wide bandwidth (>1 Gbps – useful for graphics and video);
- low losses (not so important for the cable lengths found on aircraft);
- low weight (especially when combined with higher bandwidth – kg/Mbps).

However, these advantages come with some disadvantages, namely:

- **Difficult to work with and maintain** – the fibre is fragile. It needs a protective casing, but can still suffer internal fracture if mishandled. Bending and routing in a tight radius can increase signal loss. Repairs are not simple (unlike copper, it cannot be crimped or soldered).
- **Connectors** – connecting two fibres requires microscopic alignment and joining so that light energy couples with low loss from one fibre to the next.

3.10.2 Physical Implementation

The physical layer of a point-to-point, duplex, fibre-optic data bus link is shown in Figure 3.39.

The transmission source is a laser diode which provides good coupling to the fibre optic cable. A laser diode is a special form of light emitting diode (LED) which stimulates laser action to produce a coherent (in-phase), monochromatic (single wavelength) and collimated (parallel) light source. Typical wavelengths are in the region of 900 nm to 1350 nm (c.f. the visible spectrum at 400–700 nm). The receiver is a photodiode detector, usually with integral preamplifier.

Digital encoding can be accomplished using offset Manchester biphase as light-on/light-off pulses. Thus a fibre-optic cable could be used in place of the copper cable in any of the Manchester biphase encoded data bus networks previously described.

Figure 3.40 illustrates that the propagation is multimode. The fibre-optic cable effectively acts as a light guide. Light is reflected at the fibre/sheath interface as it propagates along the fibre.

Fibre-optics operates on the principle of total internal reflection. If a light wave encounters a boundary between material having different indices of refraction, then the light wave is 'bent' at the interface. If the light wave is at a shallow angle, then all the light is reflected. Total internal reflection occurs when the angle of incidence exceeds the critical angle. Numerical aperture (NA) is a figure of merit for the coupling efficiency to a fibre. A point source is the worst type to couple to a fibre. It is also desirable for the light to be monochromatic.

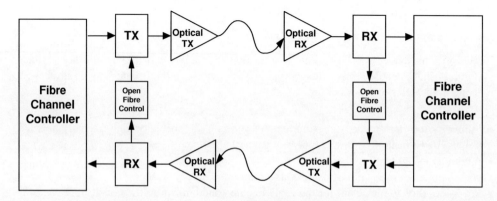

Figure 3.39   Fibre-optic transmission physical layer

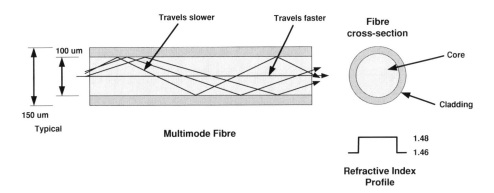

Figure 3.40 Light propagation in a fibre-optic cable

There are an infinite number of possible paths that light can take down the fibre. There is therefore a spread in time for each path to travel the length of the fibre. This phenomenon is known as modal dispersion and results in the sharp edges of the square-wave light modulation input signal becoming slightly distorted, with longer rise and fall times, similar to a low-pass filter and analogous to bandwidth limitations of a copper cable. In graded-index multimode fibres, the refractive index reduces gradually away from the core centre to the same value as the cladding at the edge of the core. The change in refractive index causes light rays to bend back towards the fibre axis instead of total internal reflection. This reduces modal dispersion.

An open fibre represents a potential for laser eye damage. Protection needs to be provided if the fibre is unterminated. A safety system needs to detect an open fibre and pulse the transmitting laser at low levels until the path is restored.

Tapping into a fibre-optic cable requires special fibre-optic couplers. Both active and passive fibre-optic couplers have been used in military avionics networks, but are expensive. Fibre-optic technology suits point-to-point star or ring topology networks best, such as ARINC 664 Part 7, but does not suit linear multicast networks well, such as ARINC 429 or CANbus.

The IEEE 802.3 Higher Speed Study Group is currently developing standards to extend the switched Ethernet 802.3 protocol to operating speeds of 40 Gbps and 100 Gbps, in order to provide a significant increase in bandwidth while maintaining maximum compatibility with the installed base of 802.3 interfaces, previous investment in research and development, and principles of network operation and management.

3.11   Data Bus Summary

3.11.1   Data Bus Overview

Table 3.1 below compares the characteristics and performance of the main data buses and communications networks in use in civil transport aircraft today. MIL-STD-1553B is included as a point of reference for information only.

All these data buses have their place (with the exception of MIL-STD-1553B which is only used in the adoption of civil platforms for military roles) in a modern civil transport

Table 3.1   Data bus network comparisons

Attribute	MIL-STD-1553B	ARINC 429	ARINC A664-P7	CANbus
Max. message length (data element)	32 × 16-bit words	18-bit word	1518 bytes (1 byte = 8 bits)	8 bytes
Max. bit rate	1 Mbps	100 kbps	10/100 Mbps	1 Mbps
Topology	Linear Single source, single sync (unicast)	Linear Single source, multisync (multicast)	Star Single source, multisync (multicast)	Linear Single source, multisync (multicast)
Communication type	Bidirectional Half-duplex	Unidirectional	Bidirectional Full-duplex	Bidirectional Half-duplex
Media access (protocol)	Command/response	Direct	CSMA/CD plus extensions	CSMA/CD
Max. bus length	100 m (6 m stub)	65 m	<100 m	40 m
Latency	Immediate	Small	Depends on network load	Depends on message priority
Jitter	Small	Small	Depends on network load	Depends on message priority
Error containment	Parity bit	Parity bit	Cyclic redundancy check	Extensive cyclic redundancy check
Error handling	Extensive retry capability	Ignored by receivers	Shut-off erroneous nodes	Immediate retry
Redundancy	Dual redundant	Simplex	Dual redundant	Simplex
Physical layer	Shielded twisted pair	Shielded twisted pair	Shielded twisted (quadrax/2× twinax) or optical fibre	Shielded twisted pair or optical fibre
Price per chipset	$1000	$30–$200	$250–$700	$1–$2

aircraft. The avionics system architect must select and apply the most appropriate data bus(es) and network technologies to meet his performance, integrity and availability requirements. Generally:

- ARINC 664-P7 is appropriate for the high bandwidth, information backbone for all avionics systems including flight controls, cockpit avionics, navigation systems and utility systems.

Table 3.2   Traffic management and policing policies

Databus/network	Features	Message policing
MIL-STD-1553B	• Prescriptive list of messages in a transaction table. • Executed in strict sequence. • Under direct command of the bus controller.	Deterministic Centralised control
ARINC 429	• Simple list of expected data words transmitted from terminal.	Deterministic by virtue of sole source transmitter
ARINC 664-P7	• Configuration table of expected message transactions, bandwidth allocation and jitter constraints. • Messages are buffered, no messages are lost. • Prioritises time-critical messages. • Traffic policing discards non-expected messages and messages that exceed contracted network resources.	Pseudo-deterministic Assured distributed control
CANbus	• Simple list of expected data words transmitted from terminal.	Non-deterministic Uncontrolled

- ARINC 429 is appropriate for lower bandwidth legacy avionics equipment (e.g. radio navigation equipment) and also sometimes used as a dissimilar means to communicate and verify safety critical information.
- CANbus is appropriate for the transfer of information to/from low data content sensors and effectors, and for intra-system data transfers.

Airbus and Boeing have different architecture philosophies that integrate CANbus and ARINC 429 with the core ARINC 664-P7 network.

The Airbus A380 core avionics computing element is the CPIOM which integrates domain-specific processing with I/O functions and communicates directly with domain sensors and effectors using ARINC 429 and CANbus interfaces and other non-digital interface types. Domain-specific remote data concentrators (RDCs) located at strategic points in the aircraft structure interface some local sensor and effector devices to the ARINC 664-P7 network.

The Boeing 787 Dreamliner makes more extensive use of distributed RDCs to interface ARINC 429, CANbus and other interface types to the ARINC 664-P7 network in a non-domain-specific manner. Chapter 5 explains this in more detail. However, the differences should be considered to be more evolutionary than fundamental.

3.11.2   Contrasting Traffic Management Techniques

Table 3.2 summarises the differences in traffic management and message policing of the aforementioned data buses/networks.

References

[1] Buckwalter, L. (2005) *Avionics Databuses*. Airline Avionics.

[2] Helfrick, A. (2004) *Principles of Avionics*, 3rd edn, Avionics Communications Inc., Leesburg, VA.

[3] Martinec, D.A. and Buckwater, S.P. (2006) ARINC Specification 429 Mark 33 Digital Information Transfer System, in *Digital Avionics Handbook*, 2nd edn (ed. C.R. Spitzer), CRC Press, Boca Raton, FL.

[4] deLong, C. (2006) AS 15531/MIL-STD-1553B Digital Time Division Command/Response Multiplex Data Bus, in *Digital Avionics Handbook*, 2nd edn (ed. C.R. Spitzer), CRC Press, Boca Raton, FL.

[5] MIL-STD-1553 Tutorial GE Intelligent Platforms Inc. Available at www.globalspec.com/reference/233495/mil-std-1553-tutorial-pdf [accessed April 2013].

[6] AFDX Training (2010) AIM GmbH. Available at www.afdx.com/pdf/AFDX_Training_October_2010_Full.pdf [accessed April 2013].

[7] AFDX/ARINC 664 Protocol Tutorial (2010) GE Intelligent Platforms Inc. Available at www.ge_ip.com/user files/file/afdx_protocol_tutorial_wp_gft604a.pdf [accessed April 2013].

[8] Walker, R. and Watkins, C. (2006) The genesis platform, in *Digital Avionics Handbook*, 2nd edn (ed. C.R. Spitzer), CRC Press, Boca Raton, FL.

4

System Safety

4.1 Introduction

This chapter discusses safety concepts in relation to the design of avionics systems. It explores the differences between reliability, available, integrity and dispatch availability. It outlines the safety assessment process, hazard analysis (the safety case) and the inverse relationship between the probability of failure and the consequent severity of the effects on the air-vehicle and its occupants. This allows the system designer to develop a means to design fault-tolerant systems and meet the systems safety objectives. It develops the mathematical basis for, and provides examples of, statistical analysis methods for analysing the consequences of random failures, namely dependency diagrams, fault tree analysis and Markov analysis. It reviews the means to assess and demonstrate component reliability. The Appendices to this book give simplified examples of the analysis methods applied to some avionics systems.

The complex nature of modern air transport aircraft systems means that special design rules need to be employed. These methodologies are described in Chapter 6 and are a crucial part of the development process. Many of the systems that are vital to flying the aircraft are required to preserve the safety and wellbeing of the flight crew and passengers. In the parlance of the aerospace community, these are flight-critical systems

During the engineering design phase the system architect devises system concepts employing various levels of redundancy that provide the necessary levels of system performance, availability and safety. These architectures are carefully crafted and reviewed using an industry-wide series of methodologies, tools and techniques that allow the provisional system design to be evaluated and to give a high degree of confidence that the eventual final design will meet its safety objectives. These tools enable a range of possible architectures to be evaluated that may be invoked to meet the system design requirements. The provision of redundant channels of control bears a burden. Additional channels cost more due to the additional provision of hardware, and are also less reliable since there are now more channels to fail. Modern technology – particularly in terms of electronics/avionics – is a help, as new technologies become more reliable and rugged in the aerospace environment through the use of improved materials and manufacturing processes. Reduced cost and development risk – sadly – does not reduce commensurately.

Civil Avionics Systems, Second Edition. Ian Moir, Allan Seabridge and Malcolm Jukes.
© 2013 John Wiley & Sons, Ltd. Published 2013 by John Wiley & Sons, Ltd.

It is not the purpose of this chapter to provide a detailed discussion of all the aspects of flight safety, but to set these principles in the context of designing avionics systems to meet the flight safety objectives of the air-vehicle. The architecture descriptions broadly outline the main candidate architectures and implementations, although in practice there may be considerable subtleties in specific implementations.

4.2 Flight Safety

4.2.1 Introduction

MIL-STD-882D defines safety as: "freedom from those conditions that can cause death, injury, occupational illness, damage to or loss of equipment or property, or damage to the environment." It is a good and absolute definition we can easily give our assent to. To attain it, our objective in designing systems must also be absolute. Conversely, any system that is not free from these undesirable outcomes is, by definition, unsafe. That means all of human endeavour is unsafe, which it is, of course, to a degree. Nothing would be achieved or attained without taking some risk.

DEF-STAN-00-55 recognises this by requiring contractors to the Ministry of Defence to apply ALARP practices: As Low as Reasonably Practical. Good though this objective is, it is an open-ended requirement. What is reasonably practical? How does a contractor demonstrate he has fulfilled his obligations? The job is never done.

SAE ARP 4764 gives what at first sight seems a rather strange definition of safety. It states that "safety is the state in which risk is lower than the boundary risk. The boundary risk is the upper limit of acceptable risk." However, this gives a clue with respect to 'as low as reasonably practical' by introducing the concept of risk. SAE ARP 4764 is not an absolute definition, but introduces the concept of a level at which we consider the risk versus outcome to be acceptable. This acceptable level of risk is a relationship between the probability of occurrence, the consequence or severity of the outcome, and the cost and timescale required to mitigate that risk.

4.2.2 Flight Safety Overview

Flying in a civil transport aircraft is the safest means of transport in human endeavour – see Table 4.1 below. But when things go wrong, they go wrong spectacularly.

The rate of fatal accidents in the worldwide commercial aircraft jet fleet (above 60,000 lbs MTOW) is shown in Figure 4.1 [1].

Table 4.1 Fatal accident rates

Mode of transport	Deaths per billion kilometres
Air	0.05
Bus	0.4
Rail	0.6
Car	3.1
Motorcycle	109

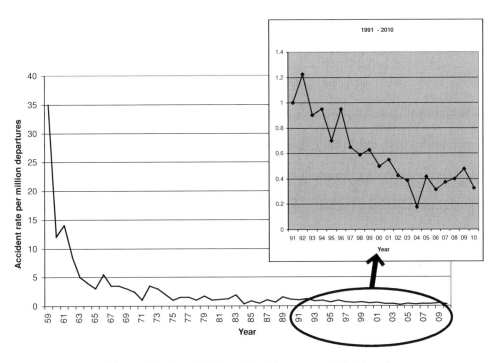

Figure 4.1   Rate of fatal accidents in commercial jet aircraft

The rate of accidents fell dramatically through the 1960s and 70s, perhaps coincidentally with the introduction of digital avionics systems, but certainly with much more rigour about air-vehicle design and manufacturing processes and a better understanding of failure mechanisms and their consequences. From 1990 onwards the failure accident trend has been progressively reducing from around 1.0 fatal accident per million departures to now just less than 0.4 fatal accidents per million departures. Given that this rate of fatal accidents is to be seen against a backdrop of the doubling of air traffic during the same period (see Figure 4.2), it indicates a progressive improvement in safety levels, so the number of fatal accidents per year has remained static. Arguably this can be said by custom and practice to represent the boundary level of acceptable risk.

In 2010 there were about 21,000 civil transport aircraft operating worldwide, making in excess of 22 million departures per year and flying in excess of 50 million flight hours per year. There were fewer than 600 on-board fatalities in 2010, a slight rise on the previous two years; the three-year running average is around 400 fatalities per annum. While each fatality is regrettable, and a personal tragedy, civil air transport is by far the safest means of transportation. From 1959 to 2010 there have been nearly 600 million civil transport jet aircraft departures worldwide and over 1000 million hours flown.

This increase in air traffic has resulted in a substantial increase in the complexity of air-traffic management procedures demanding reducing vertical and lateral separation minima and introducing new precision-based navigation techniques. Figure 4.3 shows the rate of fatal accidents by phase of flight and indicates the time of risk for each flight segment based

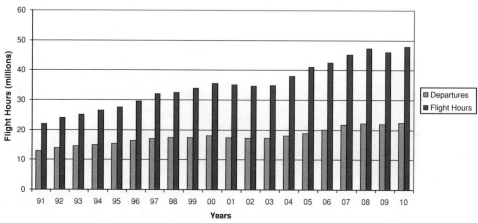

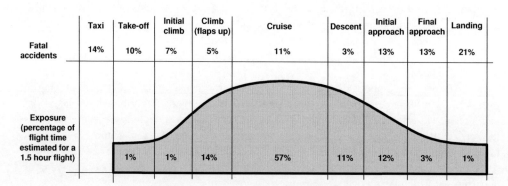

Figure 4.2   Increase in air traffic

	Taxi	Take-off	Initial climb	Climb (flaps up)	Cruise	Descent	Initial approach	Final approach	Landing
Fatal accidents	14%	10%	7%	5%	11%	3%	13%	13%	21%
Exposure (percentage of flight time estimated for a 1.5 hour flight)	1%	1%	14%		57%	11%	12%	3%	1%

Figure 4.3   Fatal accidents by flight phase

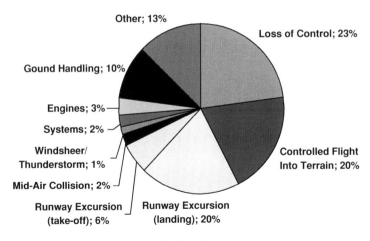

Figure 4.4   Fatal accident categories

on an estimated flight time of 1.5 hours. As might be expected, the most critical times are take-off/climb-out and approach/landing.

Figure 4.4 shows the percentage of fatal accidents by major categories for the period 2001 to 2010. The introduction of digital fly-by-wire flight control systems and precision-based navigation coupled to new approach procedures will improve these figures in the next decade.

Figure 4.5 views the causes of these failures from a different perspective [2]. It is rather sobering to recognise that more than 50% of fatal accidents are in some way due to human error. Pilot error (weather-related) represents accidents in which pilot error was the cause brought about by weather-related phenomena. Pilot error (mechanical-related) represents accidents in which pilot error was the cause brought about by some type of mechanical failure. Other human error includes air traffic controller errors, improper landing aircraft, fuel contamination and improper maintenance procedures.

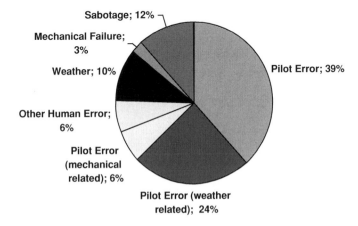

Figure 4.5   Fatal accident causes

4.2.3   Accident Causes

Most fatal accidents occur from multiple causes that conspire to produce a fatal outcome. Individually the primary causes seldom produce a fatal outcome. System-related accidents can arise from:

- single and multiple material (component) failures;
- lack of adequate performance;
- errors in manufacture and/or maintenance;
- crew mismanagement;
- environmental factors (weather, EM environment, etc.);
- behavioural factors of crew, maintenance staff and passengers.

Many of the first four categories can be related to poor design, such as poor control layout, poor information presentation, inadequate procedures and training.

Failures can be classified into the following types, with appropriate mitigation strategies:

- **Active:** deterioration in performance, such as loss of power, jamming of controls/control surfaces, loss of instruments, or loss of navigational sensors. [*Risk mitigation: designing multichannel redundant systems.*]
- **Passive:** no immediate observable effect, but fails to provide desired function when need arises, e.g. loss of reversion/monitor lane in multichannel system, loss of warning. [*Risk mitigation: periodic check for desired performance.*]
- **Cascade:** single non-hazardous failure triggers a series of other events which in total become hazardous. [*Risk mitigation: segregation, containment.*]
- **Common Mode:** same root cause affects all parts of multiredundant system, such as electromagnetic interference, temperature, altitude, ice, design/manufacturing error, maintenance error, major event such as local fire or engine disc burst. [*Risk mitigation: segregation, dissimilar redundancy, dissimilar power sources/actuation, design process assurance, quality assured parts, environmental stress screening (ESS) testing, statistical process control, etc.*]
- **Performance:** due to adverse buildup of subsystem component tolerances which impact the performance of other systems. [*Risk mitigation: functional stress testing outside the specified performance envelope.*]
- **Crew (pilot) error:** incorrect operation of controls, misinterpretation of instrumentation, controlled flight into terrain (CFIT). Murphy's Law: "If it is possible for something to go wrong, then one day it will." [*Mitigation: good human–machine interface design, simulation, training.*]

4.3   System Safety Assessment

4.3.1   Introduction

Airworthiness requirements are based on the principle that there is an inverse relationship between the probability of occurrence and the degree of hazard or consequence in its effect (see Figure 4.6).

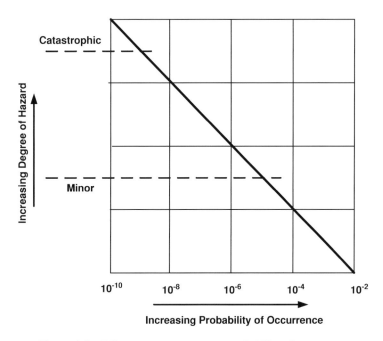

Figure 4.6   Failure consequence versus probability of occurrence

In Section 4.2 we have seen that the currently achieved boundary level of acceptable risk is less than one fatal accident in 1 million departures from all causes. For the air-vehicle itself the design objective is generally accepted to be to achieve a fatal accident (or catastrophic) rate of less than one in 10 million hours from all systems causes. The share for avionics subsystems is generally set at one in 1000 million hours or a probability of catastrophic failures of less than 1×10^{-9} per flight hour. These high levels of safety can only be achieved by multiple redundant, fail-safe systems using highly reliable components and by the application of intense scrutiny to assure targets are more than likely to be achieved.

4.3.2   Key Agencies, Documents and Guidelines

The key agencies, documents and guidelines involved in the design, development and certification of air vehicles and in particular avionics systems are identified in Section 6.2.

JAR/CS 25 paragraph 1309 is the key top-level requirement for the design and development of avionics systems. In summary this paragraph states:

"The airplane systems and associated components, considered separately and in relation to other systems, must be designed so that:

- *any catastrophic failure condition is extremely improbable and does not result from a single failure condition,*
- *any hazardous failure condition is extremely remote,*
- *any major failure is remote."*

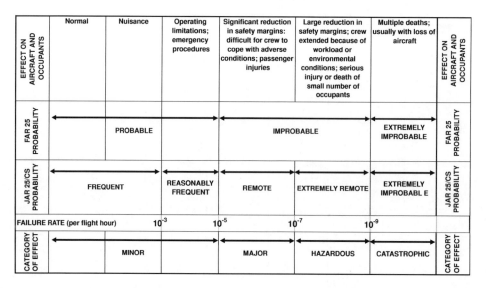

Figure 4.7   Probabilities and severity of effects

The advisory material recommends that compliance should be demonstrated by analysis augmented, where necessary, by appropriate ground, flight, or simulator tests. The analysis must consider possible modes of failure (including malfunctions and damage from external sources), the probability of multiple and undetected failures, and the resulting consequential effects on the airplane and occupants.

4.3.3   Failure Classification

Failures are classified in accordance with their degree of consequential hazard as follows, and summarised in Figure 4.7.

- **Catastrophic:** A failure condition that would prevent continued safe flight and landing. The consequence is probably a multifatal accident and/or loss of the aircraft. Probability of occurrence shall be less than 1×10^{-9} per flight hour.
- **Hazardous:** A failure condition that would reduce the capability of the aeroplane or the ability of the crew to cope with adverse operating conditions. The result is a large reduction in safety margins and high crew workload. The crew cannot be relied upon to perform tasks accurately or completely. The consequence is a serious incident with some injuries or loss of life. Probability of occurrence shall be less than 1×10^{-7} per flight hour.
- **Major:** A failure condition that results in a significant reduction in safety margins or functional capabilities and a significant increase in crew workload. The consequence is expected safe flight and landing, but with difficulties. Probability of occurrence shall be less than 1×10^{-5} per flight hour.
- **Minor:** A failure condition that would not significantly reduce aeroplane safety and which involve crew actions that are well within their capability. It is a reportable occurrence only. Probability of occurrence shall be less than 1×10^{-3} per flight hour.

The inverse relationship of failure probability to consequential hazard applies. For avionics subsystems the failure rate has generally been taken to represent the probability of failure per flight hour. However, there are some exceptions. Some systems are used only once during the flight, for instance the landing gear and braking systems. For these, the failure rate probability applies to the event. For some systems the risk is more appropriately expressed as per flight, rather than per flight hour. This is especially true when considering long haul flights and extended twin operations (ETOPS) sectors.

It is difficult to comprehend the implications of figures such as 1×10^{-9} per flight hour or one in 1000 million hours for a catastrophic event. But to put this into context, consider the following:

- A single aircraft might fly 10 hours per day for 300 days/year ($= 3000$ flight hours per year).
- The service life of the aircraft might be 20 years ($= 60,000$ flight hours through life).
- A fleet of 200 aircraft might accumulate a total of 1.2×10^7 hours over 20 years.

So, in the whole fleet the probability of a system failure causing:

- a catastrophic event, 1×10^{-9} would be extremely improbable.
- a hazardous event, 1×10^{-7} would be extremely remote, but might arise a few times.
- a major event, 1×10^{-5} would be remote, but might arise once in an aircraft lifetime and several times in the life of the whole fleet.
- a minor event, 1×10^{-3} could arise several times in the life of the aircraft.

4.3.4 In-Service Experience

The Boeing 737 is perhaps one of the most successful transport aircraft in the last 40 years. It first flew in 1967 and has been operated by more than 500 airlines, flying to 1200 destinations in 190 countries. As at July 2011, over 7000 aircraft have been delivered and over 4500 are still in service. At any given time there are on average 1250 airborne worldwide. On average, somewhere in the world, a B737 takes off or lands every five seconds. Since entering service in 1968, the 737 has carried over 12 billion passengers over 65 billion nautical miles (nm), and has accumulated more than 296 million hours in the air. The B737, at the time of the reference, represented more than 25% of the worldwide fleet of large commercial jet airliners. During that period there have been a total of 315 incidents, including 159 hull losses with fatalities from all causes. That equates to a fatal accident rate for the air-vehicle of around 0.5 per million flying hours, or about 1 per million departures.

4.3.5 Safety Assessment Processes

The safety assessment process, SAE ARP 4761, is outlined in Section 6.2.4.

The functional hazard assessment (FHA) examines the aircraft and system functions to identify potential hazards. The preliminary system safety assessment (PSSA) establishes specific safety requirements and undertakes preliminary analyses to determine that the anticipated system architecture can meet its safety objectives. These are performed early in the life-cycle of the aircraft and associated systems, and will establish the safety criteria against which each system will be designed, as discussed later in this chapter. The safety assessment must take into account the maturity, complexity and criticality of the system being developed, the technology being used, the architecture being proposed, the functionality being offered, and its interaction with other related systems.

As the product life-cycle progresses, the system safety assessment (SSA) will collect analyses and data to yield a more detailed understanding of the design in order to verify that the system meets the safety objectives identified earlier in the FHA and PSSA.

4.4 Reliability

4.4.1 Introduction

Reliability is the ability of a product to perform as intended (that is, without failure and within its performance requirements) for an expected period of time within its application environment. Aircraft avionics systems contain many components, mostly electronic in nature. Electronic hardware is a combination of electronic components (integrated circuits, transistors, resistors, capacitors, and so on), printed circuit boards and interconnects, all with various and different failure mechanisms. A failure mode is the outcome by which a combination of mechanical, electrical, chemical and thermal stresses induces failure. The effect (or consequence) of a failure is closely related to the functional performance of the product and its functionality in the whole avionics system.

Overstress failures occur when a single stress event exceeds the intrinsic strength of the component or material. Failure mechanisms due to the accumulation of incremental damage beyond the endurance of the component or material are called wear-out mechanisms. The natural process of entropy applies. It is a natural property of the universe that everything moves from a high state of order towards disorder and chaos. Decay is all around us. We all have our three score years and ten. When the accumulated damage exceeds the endurance limit of the component, a failure will occur. Unanticipated large stress events can either cause an immediate overstress failure, or shorten the life of the component by the accumulation of wear-out damage (e.g. electrostatic discharge). In a well-designed and well-manufactured product, stresses should only cause uniform accumulation of wear-out damage; the threshold at which overstress damage might occur should not be breached during the lifetime of the product, provided the product is operated within its specified environment. Further information is provided in Reference [3].

4.4.2 Failure Mechanisms

Accumulated stress in avionics components will arise from many sources. The major sources are:

- **Electrical stress:** Electrical failures in avionics equipments can be induced due to electrical overstress (voltage or current) which alters the device parameters to the point where the component causes an irreversible performance malfunction of the circuit in which it is operating. Common causes of overstress are electromagnetic interference (conducted or radiated), electrostatic discharge and particle radiation.

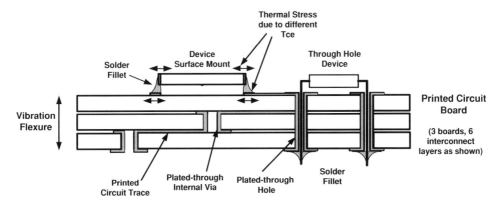

Figure 4.8   Mechanical stresses induced in mounting electronic components

- **Mechanical stress:** Mechanical failures in avionics equipments can be induced due to vibration-induced mechanical overstress which causes irreversible failure of circuit inter-connections due to fatigue. There are many vulnerabilities. Figure 4.8 shows an example of the means to mount through-hole and leadless integrated circuit components onto a printed circuit board and the potential for inducing mechanical stress.

  The printed circuit board will comprise several layers (a sandwich) of interconnect copper traces fabricated by photo-resist and chemical etching processes. Connection between the layers is made by means of drilling through the layers at intersections and electroplating the holes (plated-through holes). Components with leads are mounted through the holes and soldered on the reverse side. Leadless components are mounted by forming a solder fillet between the component termination and the printed circuit board mounting pad. Flexure of the printed circuit board will stress the solder joint and potentially induce a fatigue fracture of the joint. Solder is a non-ductile material, and careful control of the manufacturing process is required to ensure good crystalline structure and uniform shape of the solder fillet. Thermal cycling will produce mechanical stresses. The printed circuit board and the integrated circuit are fabricated from different materials which inevitably have different thermal coefficients of expansion.

- **Thermal stress:** The reliability of electrical components is inversely proportional to their operating temperature, as shown in Figure 4.9. The properties of most electrical components change with temperature due to the change in conductivity of the materials as the energy state of the atoms and molecules change. This change in electrical performance may cause a malfunction of the circuit in which the component is operating. Most integrated circuits are designed to operate with specified performance envelopes in three temperature ranges.
  - commercial grade: 0 to +70°C
  - industrial grade: −15 to +90°C
  - military grade: −55 to +125°C

Operation outside this range is normally reversible provided the component does not heat or cool beyond a critical temperature at which a phase change occurs in the properties of a material used to fabricate the component, such as the glass-transition temperature, melting point or flash point.

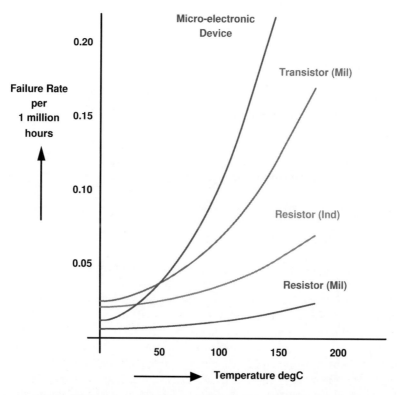

Figure 4.9 Impact of operating temperature on component reliability

- **Chemical stress:** The manufacturing processes involve the use of many corrosive chemicals, for instance, photo-resists, solder fluxes and cleaning agents. Even with the most rigorous process control, residual traces of these substances remain which then react with natural pollutants in the operational environment such as water vapour, fuel vapour, hydraulic fluid (vapour) and salt-laden atmospheres. These can be countered by providing appropriate filtration and protective barrier techniques (protective coatings) whose integrity must be preserved during repair.

4.4.3   The Relationship between Probability and Mean Time between Failures

A measure of component reliability is generally quoted as its mean time between failures (MTBF). For most practical cases we can take the probability of failure per hour (λ) as being:

$$\lambda = 1/T_m, \text{ where } T_m \text{ is the MTBF}$$

Where a component deteriorates with time due to fatigue or wear, the probability of failure clearly increases with age. If the component lifetime was relatively short, then it would be normal practice to remove the component for overhaul or replacement before the onset of this increasing risk (e.g. batteries, incandescent lamps, etc.). However, the lifetime of most avionics components is long, and they are left *in situ* until they fail. This is known as 'on-condition' maintenance. Prognostics and diagnostics techniques can assess the performance of systems and give recommendations for maintenance actions before complete failure.

Suppose that 1000 identical components, each with an MTBF of 1000 hours (that is, a probability of failure per hour of 1/1000), start to operate simultaneously. After 100 hours, a total of 100,000 hours will have been accumulated and statistically during this period 100 components will have failed. In the next 100 hours, the surviving 900 components will accumulate 90,000 hours, and statistically 90 will fail. 81 of the surviving 810 will fail in the next 100 hours, and so on. Mathematically, if the probability of failure per hour is λ, then the risk of failure before t hours is:

$$1 - e^{-\lambda t} \text{ or } 1 - e^{-t/T_m}$$

$$= \lambda t - \frac{(\lambda t)^2}{2!} + \frac{(\lambda t)^3}{3!} - \frac{(\lambda t)^4}{4!} \ldots$$

$$= \lambda t \text{ (if } \lambda t \text{ is small)}$$

The complete exponential function is shown in Figure 4.10. In an avionics system the time at risk is 10 hours or less and the probability of failure of an avionics line replaceable unit

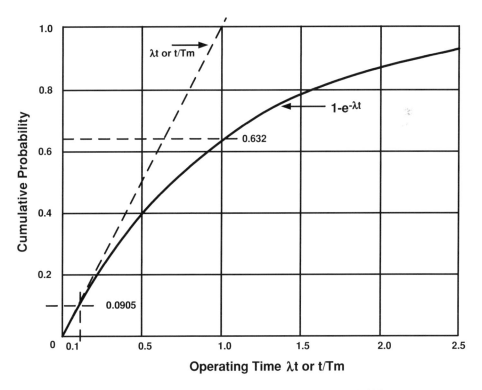

Figure 4.10   Exponential relationship of cumulative probability

(LRU) is at least better than 1/1000 hours; λt is typically less than 0.01 and so the inverse relationship quoted above between probability of failure and MTBF is a valid approximation. Further information is provided in Reference [4].

4.4.4 Assessment of Failure Probability

During development of an avionics system there are never 1000 LRUs to assess early in the life-cycle in order to establish product reliability, so another method is needed to predict mature product probability of failure. The two main methods are:

- analytical, by component count;
- historical, by means of accumulated in-service experience.

Analytical methods

MIL-STD-781 was a standard – now superseded by MIL-HDBK-781 – developed by the US military over a period of years to use an analytical bottom-up approach to predicting reliability. This method uses a component count to build up an analysis of the reliability of a complete avionics equipment. It uses type of component, environment and quality factor as major discriminators in predicting the failure rate of a particular component, module, and ultimately subsystem. Component failure rates are extracted from a database and then applied with appropriate factoring to establish the predicted value as shown in the simplified example below:

$$\text{Failure rate, } \lambda = \lambda_Q(K_1\lambda_T + K_2\lambda_E)\lambda_L$$

where λ_Q is a device quality factor, λ_T is a temperature factor, λ_E is an environmental factor, λ_L is a maturity factor, and K_1 and K_2 are constants.

There are a number of issues associated with this method:

- It is only as good as the database of components and the factors used.
- Experience has generally shown that – if anything – predicted values are generally pessimistic, thereby generating worse predicted failure rates than might be expected in real life
- It is difficult to continue to update the database, particularly with the growing levels of integration with integrated circuits (ICs), which makes device failure rates difficult to establish. The increasing number of commercial off-the-shelf (COTS) components also confuses the comparison.
- The technique has merit in comparing competing design options in a quantitative manner when using a common baseline for each design – the actual numerical values are less important than the comparison

In-service data

The technique is particularly valuable when in-service experience is available of similar technology in a similar operational environment. It does depend upon a reasonable correspondence between the components of the new design and that of the in-service database being used. Any significant variation in component usage, technology baseline or location in the air-craft/environment will undermine the comparison. Nevertheless, when used in conjunction with other methods this is a valid method. The manufacturers of civil, fighter aircraft and helicopters and their associated suppliers will generally be able to make 'industry standard' estimates using this technique.

4.4.5   Reliability Management

Achieving high reliability over the whole life-cycle of a product requires a diligent and robust methodology. This approach requires:

- A full understanding of the life-cycle environment of the product from manufacture, handling storing and shipping, in-service operations and maintenance through to disposal.
- A parts selection policy that selects components appropriate to their intended application and product life-cycle.
- A full understanding of the potential failure modes of the product and its potential impact on the operational performance of the air-vehicle.
- A programme that qualifies the product for its intended application.
- Continuous monitoring and optimisation of the manufacturing and assembly processes to assure the quality of series production products.
- Management of product reliability throughout the life-cycle through a closed-loop failure reporting and analysis programme.

Qualification and accelerated life testing

Qualification includes activities to verify that the nominal design and manufacturing spec-ifications will ensure the product will meet its performance and reliability targets over the life-cycle. This process is a combination of analysis and testing. Classic qualification testing verifies that the product will meet its performance targets throughout the environmental enve-lope – temperature, altitude, vibration, acceleration, and so on – generally in accordance with the test methods identified in RTCA-DO-160. Additional reliability growth and accelerated life testing are techniques to verify product reliability through life. Accelerated life testing is based on the concept that a product will exhibit the same failure mechanisms in a short time under high-stress conditions as it would experience in a longer time under actual life-stress conditions. Common accelerated loads include thermal stress (temperature, temperature cycling and rates of change of temperature) and mechanical loads (vibration, strain cycles and shock impulse testing). Determination of the test levels and interpretation of the results needs a careful understanding of the product and the contributions of each load to overall damage. However, the technique does highlight design and/or manufacturing weaknesses and provides useful information to improve the product design or manufacturing processes early in the product life-cycle.

Product reliability through life is frequently described as the 'bath-tub curve' – see Figure 4.11.

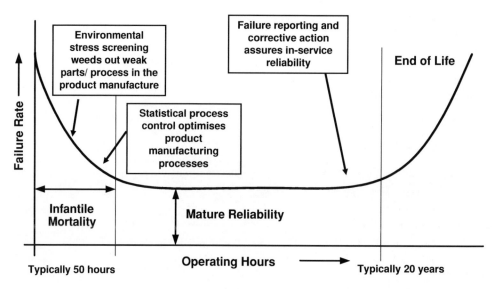

Figure 4.11   Product reliability 'bath-tub' curve

Each production article is subjected to environmental stress screening (ESS) to eliminate weak parts and/or poor manufacture. ESS screening usually involves mini-qualification testing with repeated power-on, thermal and vibration cycles to assure the product will perform as required throughout its entire operational environment. Process verification testing, including statistical process control (SPC) techniques, are used to assure the integrity of the manufacturing process.

In-service product reliability is continuously monitored through a closed loop failure reporting and corrective action system (FRACAS). Failures are analysed to establish their root cause, and corrective action is taken if a trend emerges that might otherwise compromise product reliability and safety objectives.

4.5   Availability

4.5.1   Introduction

The probability of failure of an avionics system LRU is likely to be of the order of 1×10^{-4} to 1×10^{-5} per flight hour. It is clear, then, that is not possible to design or manufacture avionics systems as single equipments which individually can meet the probability of failure safety requirements for catastrophic, hazardous or even major hazards, since:

Catastrophic: Probability of occurrence shall be $< 1 \times 10^{-9}$ per flight hour.
Hazardous: Probability of occurrence shall be $< 1 \times 10^{-7}$ per flight hour.
Major: Probability of occurrence shall be $< 1 \times 10^{-5}$ per flight hour.
Minor: Probability of occurrence shall be $< 1 \times 10^{-3}$ per flight hour.

Table 4.2 Combinations of tossing two coins, A and B

Coin A	Coin B	Probability
Heads	Heads	$\lambda = 25$
Heads	Tails	
Tails	Heads	$\lambda = \tilde{5}$
Tails	Tails	$\lambda = 25$

So it is necessary to design multiple redundant systems that are fault-tolerant and yet still able to perform the intended function in the presence of failures.

4.5.2 Classic Probability Theory

Classic probability theory provides a means to analyse the failure probability of complex systems architectures. Consider first the case of two coins, A and B. If both coins are tossed there is an equal probability that each coin will land either head-up or tail-up. Table 4.2 shows the combinations of tossing both coins.

- The probability of tossing two heads is 1 in 4, or 0.25.
- The probability of tossing a head and a tail is 2 in 4, or 0.5.

Translating that into avionics systems architectures, consider the states of two functional elements A and B as shown in the truth table of Table 4.3.

The two states of interest are 'operational' and 'failed'. They are not equally probable. The probability of failure is small (of the order of 1×10^{-4} to 1×10^{-5}). The probability of being operational is 1 minus the probability of being failed, which is approximately 1. So:

- The probability of A *and* B failing is the product of the failure rates of A and B.
- The probability of A *or* B failing is the sum of the failure rates of A and B (provided the order of failure is unimportant).

4.5.3 Simplex Architecture

Consider first one (simplex) lane of an avionics system function. It may comprise several subsystem elements as shown in Figure 4.12. Each of these elements has a probability of

Table 4.3 Combinations of avionics LRUs, A and B

LRU A	LRU B	Probability
Operational	Operational	$\lambda = (1 - \lambda_A) \times (1 - \lambda_B)$
Operational	Failed	$\lambda = (1 - \lambda_A) \times \lambda_B$
Failed	Operational	$\lambda = \lambda_A \times (1 - \lambda_B)$
Failed	Failed	$\lambda = \lambda_A \times \lambda_B$

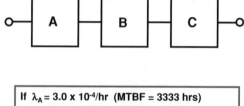

Figure 4.12   Simplex lane architecture

failure associated with it. The total lane function will be lost, or no longer available, if any of the elements fail, that is, if element A *or* B *or* C fails.

The probability of total lane failure is the sum of the probabilities of failure of the individual elements, which in the case of the architecture shown in Figure 4.12 is $\lambda = \lambda_A + \lambda_B + \lambda_C$.

If $\lambda_A = 3.0 \times 10^{-4}$ per flight hr, $\lambda_B = 5.0 \times 10^{-4}$ per flight hr and $\lambda_C = 2.0 \times 10^{-4}$ per flight hr, then $\lambda = (3.0 + 5.0 + 2.0) \times 10^{-4} = 1 \times 10^{-3}$ per flight hr.

For a flight time of longer than one hour, the probability of failure of each element needs to be computed for the time that the element is at risk. For example, for a five-hour flight the probability of failure of the architecture shown in Figure 4.12 would be 5.0×10^{-3} per flight. Clearly in this example, if failure of the system results in anything other than a minor hazard, the architecture would be unacceptable.

4.5.4   Triplex Architecture

Introducing redundancy into the architecture improves its tolerance to failure. Figure 4.13 replicates each lane of the simplex architecture of Figure 4.12 to create a triplex redundant architecture where each lane is identical.

Now the system function is lost, or no longer available, if lane 1 *and* lane 2 *and* lane 3 fail. The total system probability of failure is now the product of the probability of failure of each lane, in this example, $\lambda = \lambda_1 \times \lambda_2 \times \lambda_3 = (1 \times 10^{-3})^3$, that is, 1×10^{-9} per flight hour. This architecture would be appropriate for the autoland function of an autopilot, the failure of which could be potentially catastrophic if it occurred during the final stages of the approach (15 minutes typically).

4.5.5   Triplex Architecture plus Backup

A triplex architecture would probably not be appropriate for a fly-by-wire flight control system without a significant increase in the reliability of the individual components. Using the failure rate probabilities above, the probability of failure (in this case, potentially a catastrophic failure) would be $(5.0 \times 10^{-3})^3$, that is, 1.25×10^{-7} for a 5-hour flight, which is clearly not acceptable. A quadruplex architecture with the same failure rates of the individual elements

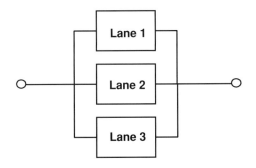

If $\lambda_1 = \lambda_2 = \lambda_3 = 1.0 \times 10^{-3}/\text{hr}$;

Then $\lambda = (1.0 \times 10^{-3})^3 = 10^{-9}/\text{hr}$

$= (5.0 \times 10^{-3})^3 = 1.25 \times 10^{-7}$ for a 5-hour flight

Figure 4.13 Triplex lane architecture

could achieve an acceptable failure probability of $(5.0 \times 10^{-3})^4 = 6.25 \times 10^{-10}$ for a 5-hour flight. Alternatively the triplex architecture plus a less reliable mechanical backup system as shown in Figure 4.14 could also achieve the required system safety objective, given that an emergency landing would be made in the event of failure of the primary system. In this example, if the backup lane has a probability of failure of 1.0×10^{-2} per flight hour, then the probability of total loss is $\lambda = \lambda_1 \times \lambda_2 \times \lambda_3 \times \lambda_{\text{backup}}$, which, for a 5-hour flight plus 1 hour emergency, is $(5.0 \times 10^{-3})^3 \times 1.0 \times 10^{-2} = 1.25 \times 10^{-9}$.

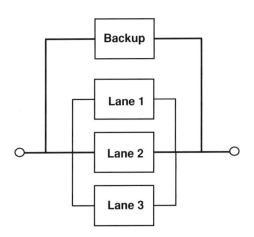

If $\lambda_1 = \lambda_2 = \lambda_3 = 1.0 \times 10^{-3}/\text{hr}$; and $\lambda_B = 1.0 \times 10^{-2}/\text{hr}$

then $\lambda = (5.0 \times 10^{-3})^3 \times (1.0 \times 10^{-2}) = 1.25 \times 10^{-9}$ for a 5-hour flight with 1-hour emergency

Figure 4.14 Triplex avionics architecture with backup

Table 4.4 Relationship between redundancy and availability

No. of lanes	Failure of a single lane (per flight hour)	Failure of all lanes (per flight hour)
Simplex	λ	λ
Duplex	2λ	λ^2
Triplex	3λ	λ^3
Quadruplex	4λ	λ^4

Table 4.4 summarises how system availability increases as the number of lanes increase by the power of the number of lanes. But note that the probability of a single failure increases linearly with the number of lanes, simply because the number of components increases.

This simplified analysis ignores any interaction between lanes. In a real system there will be a number of influences that can broadly be classed as common-mode failures which will degrade the gain in system availability as the degree of redundancy increases. However, as a 'rule-of-thumb' it is possible to make a generalised correlation between safety objectives and redundancy as follows:

- **Minor** – a simplex architecture will probably meet the safety objectives.
- **Major** – a duplex architecture is likely to meet the safety objectives.
- **Hazardous** – a triplex architecture is almost certainly required.
- **Catastrophic** – a quadruplex or triplex plus backup architecture will be necessary.

Another popular architecture is command:monitor in various degrees of redundancy, and this is discussed in the paragraph 4.7.3. As with all generalisations, the devil is in the detail. A full understanding of the failure modes and consequences is required before implementing any system architecture.

4.6 Integrity

Integrity is a further essential element to consider in designing safe systems. Not only does the system need to be available, it needs to be working correctly. It is likely that data integrity requirements are more onerous than availability. If a system is not available, then provided it is known to be unavailable and its loss is not catastrophic, alternative action can be taken. However, if the system is providing undetected erroneous outputs and the consequences can be hazardously misleading, then the outcome may lead to a situation which becomes catastrophic.

For example, loss of primary flight data (altitude, attitude, speed and heading) from the main flight deck displays is deemed to be hazardous but not catastrophic since the crew can revert to their standby instruments if information on the main displays is lost. But incorrect primary flight data on the main displays is deemed to be hazardously misleading and potentially catastrophic since it could mislead the crew into placing the aircraft into an unsafe flight trajectory. Such a situation must be extremely improbable, that is, to have a probability of occurrence of less than 1×10^{-9} per flight hour.

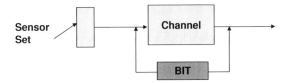

Figure 4.15   Simplex architecture with built-in-test (BIT)

There are a number of techniques and strategies that can be used to check the integrity of an avionics system.

4.6.1   Built-in-Test

Built-in-test (BIT) is a technique particularly well-suited to digital computing to check that any computing lane or channel is capable of performing correctly. It is a mechanism by which a simplex computing channel checks its hardware functionality – see Figure 4.15.

BIT is not a check of the results of a calculation but a series of tests that the computer can perform all its computing functions and that its inputs and outputs are functioning correctly. As discussed in Chapter 2, an avionics LRU is a task-oriented computer. It executes application software under the control of a real-time operating system which schedules the tasks it performs, including procedures to test itself. Three types of BIT are common:

- **Power-up or start-up BIT**: performs extensive tests on the machine (hardware) to check it can execute all its computing functions correctly.
  - Checks the machine software configuration.
  - Checks all inputs and outputs are functional (may involve user operations).
  - Checks main and backup functions.
- **Interruptive BIT**: The machine is taken off-line temporarily. Test signals (inputs) are injected and the results (outputs) compared against test results. This is most useful for automated end-to-end system checks prior to entering a critical phase.
- **Continuous BIT**: The computer checks itself as a 'background' task. Typical tests include:
  - central processing unit (CPU) check: all arithmetic and logic functions against test data patterns;
  - volatile random access memory (RAM) memory: write and read operations;
  - fixed read only memory (ROM) application memory: sum-check, cyclic redundancy check (CRC);
  - input/output (I/O): wrap around tests output to input for all I/O types, analogue and discrete;
  - backplane: write and read operations;
  - watchdog timer checks the processor is not 'stuck' and data has become frozen!

None of these tests can check the computer completely and none are instantaneous. Continuous BIT may take many computation cycles to complete. However, BIT confidence can be in excess of 95%. The BIT philosophy should be designed to give the appropriate level of confidence in accordance with the consequence of failure of the system of interest. Higher confidence levels require more extensive testing and hence take longer and require more extensive hardware features to be incorporated to facilitate testing of input and output circuits.

If the probability of failure of the computing channel is λ_C and the BIT confidence of that channel is λ_B, then:

- the probability of total loss of function (availability) is λ_C;
- the probability of undetected failure (integrity) is $\lambda_C(1 - \lambda_B)$.

If a fault is detected, then the non-exclusive possibilities available to the system integrator are to:

- Signal a warning flag to users that the function is faulty and its outputs are not to be relied upon.
- Cease to output data from the function (fail silent behaviour). A user must detect the data has ceased to be output from the function and has become stale.
- Set the function output to a safe (default) state.

4.6.2   Cross-Monitoring

BIT is a powerful tool to check the ability of a computing lane or channel to perform its intended function, but it does not check that the instantaneous output is correct. Even if BIT were 100% effective, the outputs from a single (simplex) channel rely on having correct inputs. For high levels of integrity, a further level of independence is required.

The simplest form of cross-monitoring is provided by a command:monitor architecture as shown in Figure 4.16.

Here two independent channels execute the desired functionality. One channel is the command channel and it outputs source data to onward processes. The other channel monitors the performance of the command channel to instantaneously verify the correctness of the command channel outputs. A number of possibilities exist. The monitor channel may just check that the command channel is within an expected and safe performance envelope, or it may be identical to the command channel, or it may have a degree of dissimilarity (hardware and/or software) to improve resilience to common-mode failures. The choice must depend on

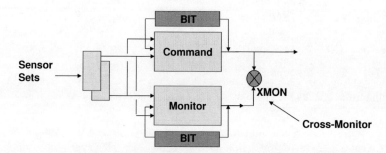

Figure 4.16   Command:monitor architecture

the safety case analysis and the precision required of the monitoring function. However, any degree in dissimilarity must be accounted for in setting the threshold of error detection, since any difference in implementation is likely to produce some differences in results in addition to any differences in input sensor data. To minimise these differences, some implementations present all sensor data to both command and monitor channels to harmonise input variables before computing outputs, and hence reduce the error budget threshold. This is useful, but it must be done with care to avoid the possibility of sneak paths (that is, faults on one side) propagating to and contaminating both command and monitor channels. Section 4.9.3 provides further discussion on dissimilar, or multiversion software.

If the probability of failure of the cross-monitor is λ_{XMON}, the probability of a command channel failure is λ_C, and the monitor channel is λ_M, and the BIT confidence of either is λ_B, then:

- The system function is lost if the either the command channel or the monitor channel fails, or if the cross-monitor generates a false alarm:

$$\text{that is: } \lambda_C + \lambda_M + \lambda_{XMON \text{ (false alarm)}}$$

- The system fails to detect an error if either command channel or monitor channel BIT fails to detect an error, and if the cross-monitor fails to detect a discrepancy:

$$\text{that is: } \{\lambda_C(1 - \lambda_B) + \lambda_M(1 - \lambda_B)\}. \lambda_{XMON \text{ (failure to detect)}}$$

As before, the possibilities available to the systems architect on detecting a fault are:

- signal a warning flag to users that the function is faulty;
- cease to output data (fail silent behaviour);
- set the output into a safe (default) state.

4.7 Redundancy

Redundancy replicates command channels. Some architectures average the output of redundant lanes so that all channels are functional all of the time; faulty lanes are disconnected in the event of failure. Some architectures nominate one of the lanes as the command lane, and in the event of its failure an alternative lane is available to take command.

Within the aerospace community the diversity of redundancy varies between a single lane implementation (simplex or ×1) through to a four-fold lane implementation (quadruplex or ×4). There is an obvious practical limit as to what levels of redundancy may be sensibly employed. In practice, quadruplex implementations have only been used in specific military applications.

The following description broadly outlines the main candidate architectures and implementations, although in practice there may be considerable subtleties between specific implementations.

The main architectures to be outlined include:

- simplex;
- duplex redundancy;

- dual/dual redundancy;
- triplex redundancy;
- quadruplex redundancy.

Examples of each of these architectures will be portrayed and the implications of various failures examined and explored. The choice of which architecture to select is subject to a rigorous examination using the design tools already mentioned. These techniques analyse risk per flight hour and the level of redundancy should then be chosen based on the failure consequence.

4.7.1   Simplex Architecture

Many control systems within an aircraft will be relatively simple and their loss will not be of great consequence. Such systems are likely to be implemented in a simplex form in terms of sensors and control, and if there is a failure then the control function will be lost. Failures will be detected by built-in-test (BIT) functions that may be continuous or interruptive in nature. BIT is not perfect, however, and typically the effectiveness of BIT is around 90 to 95%. There is a possibility that the control system may be configured to revert to a known, safe value or state, and some limited control may still be possible.

Nature of control:

First Failure	Fail Silent

4.7.2   Duplex Architecture

For systems with a higher level of safety involvement, the dual lane implementation shown in Figure 4.17 may be preferred. The sensor set and the control channel are replicated, and if a sensor or control channel fails there is an alternative available. Superficially this architecture looks very similar to the command:monitor architecture described in Section 4.6.2, but here both channels are functionally identical and have the same precision and

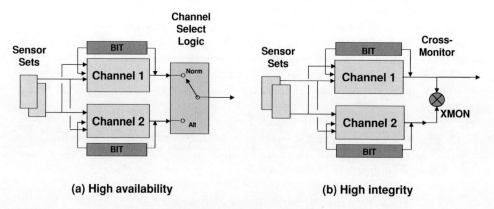

Figure 4.17   Duplex architecture

performance characteristics. This architecture offers close to 100% fault coverage and has the advantage that should one channel fail, the system may continue to operate in a simplex mode, albeit with reduced safety margins. Two implementations are shown to illustrate the range of choices available to the system architect to meet differing safety requirements.

- A high availability may be the prime driver; for example, to have high availability that the landing gear can be deployed when required. To satisfy this requirement the simple normal/alternate architecture in Figure 4.17a may be most appropriate in which control is available if either lane is operational.
- However, high integrity may dominate; for example, it may be a requirement that inadvertent deployment of the landing gear must be highly unlikely. To satisfy this requirement the architecture in Figure 4.17b would be more appropriate. in which a cross-monitor compares the output of both channels and control is only available if both lanes agree.

Analysis of the high-availability duplex architecture shown in Figure 4.17a yields:

- Degraded performance occurs with the first failure of either command channel, that is: $2\lambda_C$.
- If the faulty channel can be isolated, then the system may be permitted to continue to operate in a degraded mode with reduced safety margins. The system function is lost only if both command channels fail, that is: λ_C^2.
- The system fails to detect an error if either command channel BIT fails to detect an error, that is: $2\lambda_C(1 - \lambda_B)$.

Nature of control:

First Failure	Fail Operational
Second Failure	Fail Silent

Analysis of the high-integrity duplex architecture shown in Figure 4.17b yields:

- The system function is lost if either channel fails, or if the cross-monitor generates a false alarm, that is: $2\lambda_C + \lambda_{XMON \text{ (false alarm)}}$.
- The system fails to detect an error if either command channel BIT fails to detect an error, and the cross-monitor fails to detect a discrepancy, that is: $2\lambda_C(1 - \lambda_B).\lambda_{XMON \text{ (failure to detect)}}$.

4.7.3 Dual Command: Monitor Architecture

A more sophisticated arrangement is the dual–dual architecture which is often implemented in a command:monitor fashion as shown in Figure 4.18, and offers both high availability and high integrity.

Each channel has a command (COM) and a monitor (MON) lane within it, the command lane being in control and the monitor lane checking for correct functioning of the command element. The command and monitor lanes may be similar or dissimilar in terms of implementation, as described in Section 4.6.2. There will be a cross-monitor function

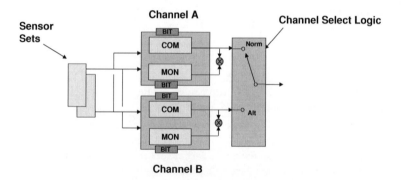

Figure 4.18   Dual COM:MON architecture

associated with each channel. A COM:MON discrepancy in a channel will disconnect that channel. The remaining COM:MON channel may continue to operate and provide full system functionality. It must be remembered that the cross-monitor function that allows the MON lane to arbitrate whether the COM lane has failed can itself be subject to a failure. Two failure modes are possible. The cross-monitor function may produce a false warning, in which case the channel will be deemed to have failed even if the COM and MON lanes are themselves fully serviceable. However, the cross-monitor may fail in such a manner that it fails to detect a malfunction of the command or monitor channels. This is an undetected malfunction and must be accounted for in the system safety analysis.

The dual COM:MON architecture is used widely within the civil aerospace community, typically for the control of major utilities systems and for full authority digital engine control (FADEC) applications.

The failure analysis is a combination of the analyses for the duplex and command:monitor architectures described above, and a simplified worked example is given in Appendix D.

- The loss of one channel, but no loss of performance occurs with the first failure, that is: $2\{\lambda_C + \lambda_M\}$.
- The system function is lost if both the command lane and the monitor lane fail in opposite channels, that is: $\{\lambda_{Ca}\,\lambda_{Cb} + \lambda_{Ca}\,\lambda_{Mb} + \lambda_{Ma}\,\lambda_{Cb} + \lambda_{Ma}\,\lambda_{Mb}\}$.

Other second-order failures of combinations of cross-monitor failures and opposite lane failures can also deny system function, and these demand a more complex Markov state diagram analysis as discussed in Section 4.8.3.

- The system fails to detect an error if any of the command lane or monitor lane BIT fails to detect an error; and if the cross-monitors fail to detect a discrepancy, that is: $2\{\lambda_C (1 - \lambda_B) + \lambda_M(1 - \lambda_B)\}$. $\lambda_{XMON\ \text{(failure to detect)}}$ (much simplified).

Design of the cross-monitors is a complex and critical task to ensure that the probability of generating a false error trip is small, but at the same time the probability of detecting a real error is high (conversely, the failure to detect a real error is small). It is simply summarised here as $\lambda_{XMON\ \text{(false alarm)}}$ and $\lambda_{XMON\ \text{(failure to detect)}}$.

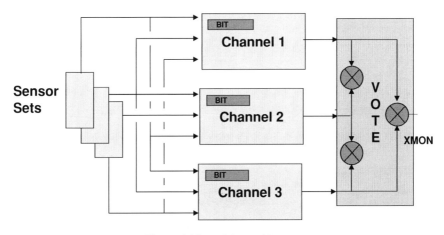

Figure 4.19   Triplex architecture

Nature of control:

First Failure	Fail Operational
Second Failure	Fail Silent (in opposite channel

4.7.4   Triplex Architecture

Higher levels of integrity demand higher levels of redundancy. Figure 4.19 shows a triplex architecture.

In this architecture all three channels have equal authority, with three independent sets of sensors and control. Arbitration in such an architecture is typified by a voter/comparator where the output of all three channels is compared, and if one channel should deviate from the others then it is discounted and 'voted out'. In its simplest form the output is the median (average) of the operational channels. If one lane differs significantly from the other two, then that lane is disconnected and the system will continue to provide full functionality in duplex mode. A second fault in either of the two remaining channels will then disconnect the system, leaving it in the (safe) state it was prior to disconnect. If the faulty channel can be isolated, then the system can continue in simplex mode with degraded safety margins and possibly constrained or limited authority.

Figure 4.20 shows a typical voter/comparator. It may be implemented in hardware or software.

It should be noted that the inputs to the voter comparator are a series of impulse functions as each digital computing lane generates a new output value. The lanes are asynchronous, so instantaneous discrepancies will occur between lanes. The output value is the mid-point average (median) of the input values. This forms one input to the three comparators, one in each lane. The amplitude threshold is set so the accumulated tolerance does not cause nuisance trips of the detector, but because transient timing effects may produce short-term amplitude

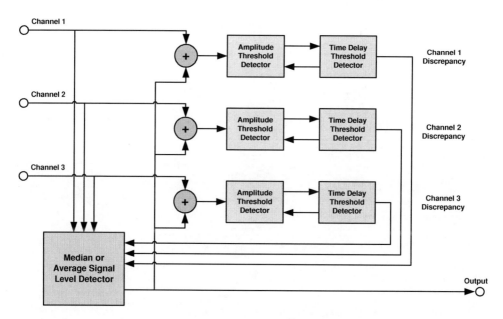

Figure 4.20   Triplex voter/comparator

trips, a timing threshold is also applied. A number of consecutive out-of-tolerance amplitude trips are necessary to declare a channel fault. Setting the amplitude and timing thresholds is critically important to provide the right balance between avoiding nuisance trips and slow response.

Nature of control:

First Failure	Fail Operational
Second Failure	Fail Safe
Third Failure	Fail Silent

4.7.5   Quadruplex Architecture

In an extreme design situation the sensors and control channels may be replicated four times, leading to a quadruplex architecture as shown in Figure 4.21. Such architectures are usually utilised for fly-by-wire flight control implementations in aircraft which are fundamentally unstable and where only the highly redundant flight control system maintains the aircraft in stable flight. Examples of this type of architecture are Eurofighter Typhoon and the Northrop B-2 Spirit stealth bomber.

The voter/comparator is significantly more complex than that described for the triplex architecture, but the same principles apply. After a first fault the system degrades to triplex. After the second fault the system degrades to dual. In operational usage a Typhoon pilot would be able to continue the mission after a first failure, whereas after the second he would be well advised to curtail the mission and land the aircraft at the first convenient opportunity.

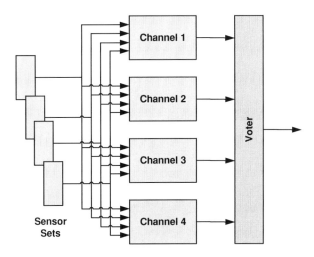

Figure 4.21   Quadruplex architecture

Nature of control:

First Failure	Fail Operational
Second Failure	Fail Operational
Third Failure	Fail Safe

4.7.6   Summary

Figure 4.22 illustrates, for simple comparative purposes only, the rough order of magnitude in terms of safety classification that can be expected from the various architectures described above in terms of their reliability, availability and integrity. As always, detailed

Architecture	Reliability (1st Failure)	Availability (Total Loss)	Integrity (Undetected Malfunction)
Simplex	5000	Minor	Major
Duplex	2500	Major	Hazardous
Dual Com:Mon	1250	Hazardous	Catastrophic
Triplex	1667	Catastrophic	Catastrophic
Quadruplex	1250	Catastrophic	Catastrophic

Figure 4.22   Comparative assessment (simplified)

analysis is required, using the techniques described in Section 4.8, to make a full and proper evaluation of any specific architectural implementation against any specific systems safety objectives.

The illustration is based on a single lane probability of failure of $\lambda = 2 \times 10^{-4}$ per flight hour, a BIT probability of 90% and a $\lambda_{XMON(\text{failure to detect})}$ of 1% of the probability of a lane failure. The analysis is much simplified. It contains gross assumptions and ignores erosion of lane independence due to the factors discussed in Section 4.9. Further information can be found in Reference [5].

4.8   Analysis Methods

The Safety Assessment Process SAE-ARP-4761 prescribes two analysis methods as being required to be performed on avionics systems: top-down methods and bottom-up methods.

4.8.1   Top-Down Methods

Top-down methods should be performed early in the development life-cycle to identify potential vulnerabilities and then design robust means to mitigate against the risk of those vulnerabilities producing an unacceptable outcome from a safety perspective. The method seeks to identify safety cases to be investigated which could produce catastrophic or hazardous outcomes, and then analyse the proposed system architecture to explore the failure mechanisms or combinations of failure mechanisms that could produce those outcomes. The probability of the safety case occurring is determined from a realistic expectation of the probability of failure of the individual architectural elements, and if the result is unacceptable, appropriate changes should be made to the system architecture and/or the technological approach being proposed until a robust system architecture is achieved. The analysis should be repeated as the design progresses and matures and as more detail is revealed, to eventually reflect the final architecture and technologies employed.

Three top-down methods have been widely used, each with their strengths and weaknesses.

- **Dependency diagram**: This method develops an architecture diagram which expresses the interdependencies between the system elements. Each element defines a failure of a part of the system and the conditions related to it connected in series or parallel arrangements, similar to the system architecture diagram.
- **Fault tree analysis**: This is a graphical method expressing the logical relationship between a particular safety concern and the failures or other causes that could lead to that particular outcome.
- **Markov analysis**: Markov analysis is useful when investigating systems where a number of states may be valid and are interrelated.

Fault tree analysis is the preferred method. There are tools available to facilitate all these analysis processes.

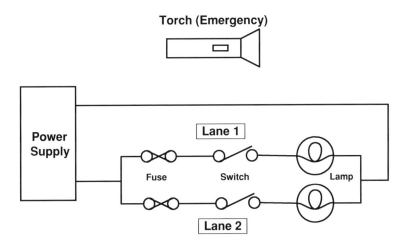

Figure 4.23 Simple electrical lighting system

4.8.2 Bottom-Up Methods

Once the design is complete, the top-down analysis should be verified by inspection of the detailed design, postulating failure mechanisms that could occur at detailed component level, and determining their effects on circuit, equipment, system and finally air-vehicle operation should they occur. The analysis is referred to as a 'failure mode effects and criticality analysis' (FMECA). It can be extensive and should include an element of practical testing to verify the theoretical analysis. With the addition of failure probability data, the FMECA should be a close correlation to the top-down method.

4.8.3 Lighting System Example

Here we use a simple electrical lighting system to illustrate the dependency diagram and the fault tree analysis methods. Some further examples are provided in the Appendices to illustrate all three analysis methods applied to real avionics systems.

The simple electrical lighting system shown in Figure 4.23 comprises a main or primary lighting system which has two (duplex) lighting circuits. Each lane has a lamp, a switch and a fuse to protect the wiring in the event that the lamp fails short-circuit. The primary system is powered from a common source (also known as the electrical system bus). A torch provides an emergency back-up system in the event that the primary system fails.

Dependency diagram

Figure 4.24 illustrates the dependency diagram for the lighting system safety case – loss of light – and illustrates the relationship in which the safety case depends on the availability of the system elements.

- Light is lost from the primary system if both primary lanes are lost (the parallel nature of these elements illustrates an AND function).

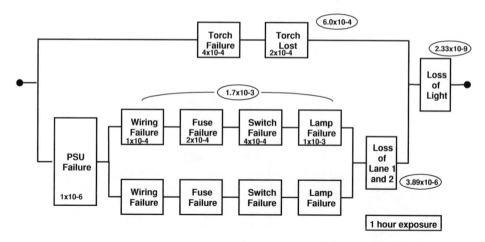

Figure 4.24   Simple electrical lighting system dependency diagram

- A primary lane is lost if any one of the lamp, the switch, the fuse or the wiring fail (the series nature of these elements illustrates an OR function).
- The power supply is a common series element to both primary lighting lanes.
- Light is lost if both the primary system AND the back-up system fails.
- The 'torch lost' dependency illustrates the point that the back-up system is not normally used, but must be available when invoked.

The total system probability of failure per flight hour can be determined by inserting the failure rate probabilities for each of the system elements and applying the probability theory developed earlier. The OR function (as determined by series elements) is computed as the sum of the individual probabilities; the AND function (as determined by the parallel elements) is computed as the product.

An example of a dependency diagram analysis is given in Appendix A which considers the likelihood of a catastrophic failure of a typical fly-by-wire flight control system. Appendix A also includes a fault tree analysis of the same system for comparison.

Fault tree analysis

Figure 4.25 illustrates the fault tree diagram for the same system and expresses the logical relationship between the failure of the subsystem elements and the safety case being examined.

- Light is lost if the primary system AND the back-up system fail.
- The primary system fails if the power supply fails OR both primary lighting lanes fail.
- Each primary lighting lane fails if the lamp, OR the switch, OR the fuse, OR the wiring fail.

The total system probability of failure per flight hour can be determined by inserting the failure rate probabilities for each of the system elements and applying the probability theory to sum the failure rate of the individual probabilities of the OR functions and compute the product of the failure rate probabilities of the AND functions.

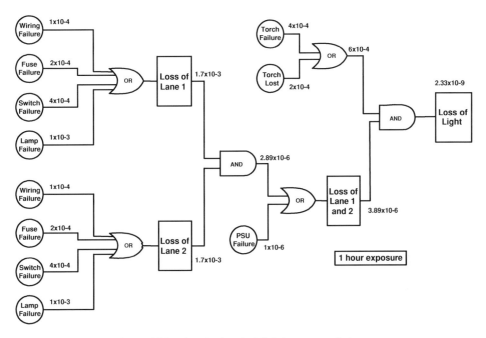

Figure 4.25   Simple electrical lighting system fault tree

Appendices 2 and 3 provide simplified fault tree analyses of an electronic flight instrument system and an electrical system of a modern civil transport aircraft.

Markov analysis

Another technique used to assist in system analysis is Markov analysis. This approach is useful when investigating systems where a number of states may be valid and also are interrelated. This could be the case in a multichannel system where certain failures may be tolerated but not in conjunction with some other failure conditions. The question of whether a system is airworthy is not a simple mathematical calculation as in previous analyses, but depends upon the relative states of parts of the system and the sequence in which they occurred. The simple methods used are insufficient in this case and another approach is required. Markov analysis is the technique to be applied in these circumstances, when it is necessary to address operation of a system in different modes and configurations.

An example of a Markov analysis is given in Appendix D, which considers the likelihood of an engine in-flight shut down (IFSD), which would be a typical analysis necessary to determine the reliability of an engine prior to seeking Extended Twin OPerationS (ETOPS) clearance.

4.9   Other Considerations

4.9.1   Exposure Time (Time at Risk)

The hazard criticality may depend on the phase of flight. For instance, failure of the autopilot during cruise is likely to be little more than major, provided it fails safe and its failure is known. The crew can take over and fly the aircraft manually. However, in the later phases of an autoland approach, especially in poor visibility conditions, failure could be hazardous, if not potentially catastrophic.

The time for which the air-vehicle is at risk will be different for different hazards. In the example above, the exposure to the risk of autoland failure during approach is short, less than 15 minutes per flight; the exposure to failure during cruise depends on the operational range, but could be more than 10 hours for a long haul aircraft.

Some systems are only used once per flight, for example, the braking system and the doors. Some systems are used throughout the flight.

Different considerations apply to operation in ETOPS sectors. The degree of hazard and the time at risk after first failure will be different than when operating in a conventional sector, since there is no possibility of an emergency landing while in the ETOPS sector.

Dormant faults may only be revealed when the other channels fail, for example, warning systems and standby or backup systems. The risk is mitigated by periodic checks on the backup or warning system. The time at risk is now the operating period between periodic checks, not the flight time. The probability of a dormant fault is thus λT_c, where T_c is the time between periodic checks.

4.9.2   Cascade and Common Mode Faults

There are various threats to the independence of channels in redundant systems which may lead to multiple failure rates higher than the simple analysis for multiple failures based on the failure rates of individual channels.

A typical triplex autopilot might have a single channel failure rate of 9.5×10^{-4} per flight. The theoretical calculated failure rate for the system would be 8.6×10^{-10} per flight; but in practice it could be several orders worse, say around 1×10^{-7} per flight as shown in Figure 4.26. The erosion in performance is due to channel failures not always being fully independent. However, the autopilot period of exposure during the critical autoland phase would be perhaps 1/10th of an hour. (An aircraft travelling at 120 kts will travel 10 nm in 5 mins.)

In multichannel systems the total systems load is often shared. In the event of one channel failing, its load will be shared between the remaining healthy channels. For example, an electrical system redistributes loads after generator failure. Even with load-shed, the load on a generator increases dramatically. A flight control system often shares the control surface hinge moment between two or more actuators (sometimes hydraulic and electric). In the event of failure the load is taken by the remaining actuator. The increase in load is likely to increase the system 'stress' and increase the probability of failure of the remaining channels.

System integrity can be undermined in other ways:

- lack of adequate electrical and mechanical segregation between channels;
- lack of segregation between comparator and main channels;
- tolerances between channels and comparator;
- use of common power supplies;
- EM interference;
- cross-coupling in cable harnesses;

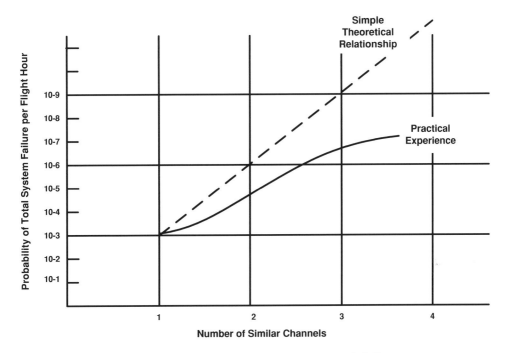

Figure 4.26   Failure rate erosion due to common mode influences

- interference between channels;
- design faults (hardware and software);
- manufacturing faults (e.g. faulty batch of components);
- maintenance faults (incorrect setup procedure);
- repair faults;
- weather (temperature, altitude, humidity, icing, lightning strike);
- corrosion;
- wear.

4.9.3   Dissimilarity

Dissimilar implementations can provide increased robustness to common-mode faults for safety critical systems by the judicial application of dissimilar hardware and or dissimilar software. For instance, the primary and secondary flight control computers of the Airbus A330/340 have different architectures and dissimilar hardware and software. The primary flight computers of the Boeing 777 have three similar lanes with dissimilar hardware but the same software.

Dissimilar, diverse or multiversion software is one of a number of techniques that may be used for design fault tolerance to minimise common-mode failure. While many design faults may be found and removed during system integration and testing, it is virtually impossible to eliminate all possible software design faults. Multiversion software is a technique in which two or more alternative versions are implemented, executed, and the results compared using some form of decision algorithm. The goal is to develop these alternative versions such that software faults in one version are not contained in the other version. Discrepancies are detected by the decision algorithm.

Software diversity maybe applied at different levels:

- design: implemented by separate development teams;
- code: in more than one language;
- compile: generate object code using more than one compiler.

However, activities before dissimilarity is introduced remain potential error sources. The specification process is the largest single source of software faults. Synchronisation and comparison accuracies between parallel components may degrade safety margins. The extent of any dissimilarity needs to be determined (Figure 4.27) in terms of top-level design, coding,

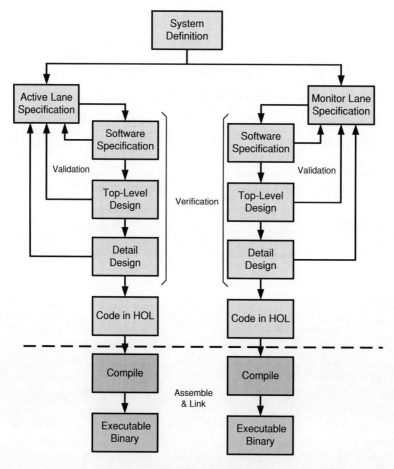

Figure 4.27   Dissimilar software

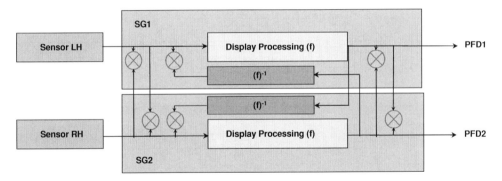

Figure 4.28   Inverse processing applied to an EFIS architecture

verification and compilation. However these are addressed, the issue still remains of validating the systems design specification.

Other software techniques include reasonableness checks and inverse computing or reversal checks. Reasonable checks are based on knowledge of what are reasonable maximum and minimum values of input data, as well as limits on rate of change and limits of authority, possibly incorporating correlation of other data (for instance, that pitch attitude, angle of attack and rates of change of speed and altitude are consistent with aircraft performance models). Inverse computing or reversal checks take the outputs from a system and calculate the inputs that should be expected to produce that output. The calculated inputs are then compared with the actual inputs to check whether there has been a processing error. Applying both techniques to a duplex electronic flight instrument system (EFIS), as shown in Figure 4.28, it is possible to demonstrate that hazardously misleading primary flight data is extremely improbable (i.e. has a probability of less than 10^{-9} per flight hour).

4.9.4   Segregation and Partitioning

Separation is a natural part of aircraft systems integration. It is usually adopted for reasons of integrity or criticality. Separation may apply to:

- hardware (segregation);
- software (partitioning);
- wiring segregation – Lane 1 routed down the right side of the aircraft whereas Lane 2 is routed down the left side;
- physical separation (e.g. ChA, ChB).

More integrated systems offer greater challenges to the principle of segregation by wishing to aggregate, rather than partition systems. This is the major challenge of integrated modular avionics (IMA) implementations – see Chapter 5. By sensibly adopting and applying the principles of partitioning and segregation, adequate integrity can be achieved within an integrated system using judicious design principles.

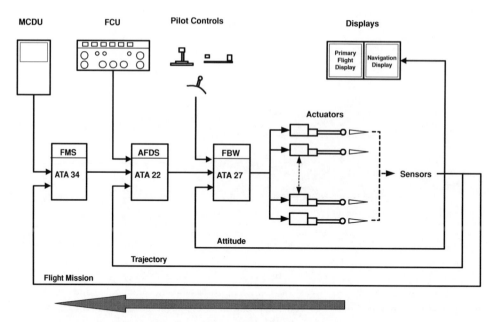

Reducing failure consequence & hence redundancy

Figure 4.29 Partitioned flight control systems

Figure 4.29 shows an example of a means to partition the flight system control functions of fly-by-wire (FBW) primary flight control, autopilot flight director system (AFDS) and flight management system (FMS). By partitioning the function in this way, the consequence of loss can be mitigated with consequent implications on degree of redundancy required for each subsystem.

4.9.5 Dispatch Availability

Dispatch availability is key to an aircraft fulfilling its mission, whether a military or civil aircraft. The ability to be able to continue to dispatch an aircraft with given faults has been given impetus by the commercial pressures of the air transport environment, where the use of multiple redundancy for integrity reasons has also been used to aid aircraft dispatch.

On the Boeing 777 the need for high rates of dispatch availability was specified in many systems, and in some systems this led to the adoption of dual redundancy for dispatch availability reasons rather than for reasons of integrity. A simplified version of the dispatch requirements is shown in Figure 4.30.

This means of specifying the dispatch requirement of part of an aircraft system leads to an operational philosophy far beyond a 'get-you-home' mode of operation. In fact, it is the first step towards a philosophy of limited unscheduled maintenance. For an aircraft flying up to 14 hours per day – a typical utilisation for a wide-bodied civil transport – this definition dictates a high level of availability for up to a 120-hour flying period. The ability to stretch

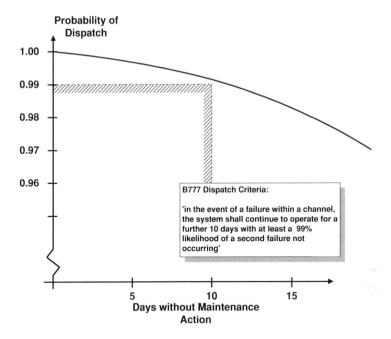

Figure 4.30   Simplified dispatch criteria

this period in the future, perhaps to 500-hour operating period including scheduled servicing checks, as more reliable systems become available, could lead to a true system of unscheduled maintenance. A 500-hour operating period roughly equates to 7 weeks of flying, at which time the aircraft will probably be entering the hangar for other specified maintenance checks.

Dispatch considerations lead to a more subtle requirement to examine the system's ability to meet integrity requirements when several failures have already occurred.

References

[1] *Statistical Summary of Commercial Jetplane Accidents 2009-2010* (2011) Boeing Airplane Company.
[2] Planecrash info Accident Statistics, available at http://www.planecrashinfo/com [accessed 2 September 2012].
[3] Vulchare, N., Ping, Z., Das, D. and Pecht, M.G. (2007) Electronic hardware reliability, in *Digital Avionics Handbook*, 2nd edn (ed. C.R. Spitzer), CRC Press, Boca Raton, FL.
[4] Lloyd, E. and Tye, W. (1982) *Systematic Safety*. Civil Aviation Authority, London.
[5] Hitt, E.F. and Mulcare,D. (2007) Fault tolerant avionics, in *Digital Avionics Handbook*, 2nd edn (ed. C.R. Spitzer), CRC Press, Boca Raton, FL.

5

Avionics Architectures

5.1 Introduction

This chapter discusses avionics architectures and their evolution from distributed analogue control systems to today's highly capable integrated modular avionics architectures. It explores the grouping of avionics functions into domains broadly in line with Air Transport Association (ATA) chapters and how data bus technologies have complemented growth in complexity of avionics systems architectures.

This chapter then reviews the key features and architecture principles employed in the distributed federated digital avionics architectures of civil transport Airbus aircraft in the mid-1980s, as realised in the Boeing 737, 757 and 767 series and in the Airbus A300, A320 and A330 series of aircraft.

The evolution of integrated modular avionics (IMA) architectures is discussed next, commencing with the proprietary, partial implementation of IMA principles in the Boeing 777 Airplane Information Management System (AIMS) through to full open-system IMA implementations on the Airbus A380 and the Boeing 787 aircraft. We shall explore key features and architecture principles of both of these implementations and review their similarities and differences.

Finally this chapter discusses the design processes to be undertaken to successfully implement and certificate an avionics system implemented as an IMA architecture. It explores the concept of a virtual (logical) systems architecture and the physical realisation of this architecture onto the IMA platform. We shall review the implementation of the architectural principles of redundancy, fault tolerance, segregation and partitioning in order to support the system safety objectives and facilitate independent and incremental certification of the hardware platform and the hosted application software.

5.2 Avionics Architecture Evolution

5.2.1 Overview of Architecture Evolution

The application of digital computing technology to civil aircraft avionics systems has occurred rapidly as aircraft performance and complexity have increased. The increase in air traffic

Civil Avionics Systems, Second Edition. Ian Moir, Allan Seabridge and Malcolm Jukes.
© 2013 John Wiley & Sons, Ltd. Published 2013 by John Wiley & Sons, Ltd.

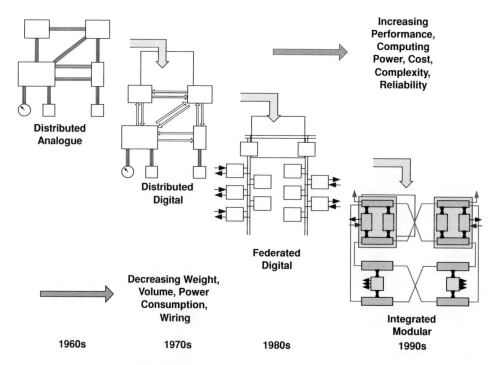

Figure 5.1 Evolution of avionics architectures

coupled with satellite navigation systems has demanded higher density route structures and improved surveillance systems with a commensurate increase in onboard computational power to plan and prosecute complex flight paths with a high degree of precision. This has been facilitated by advances in digital avionics technology in the areas of processing, software development and network-centric digital communications, enabling aircraft systems to be integrated on a much larger scale.

Aircraft avionics systems have grown rapidly in terms of capability and complexity, taking full advantage of digital processing technological enhancements. Technology has brought improvements in terms of increased performance, computing power, complexity and reliability, although with an increase in cost. Other benefits include a decrease in weight, volume, power consumption, wiring and support costs.

Figure 5.1 portrays how avionics architectures have evolved from the 1960s to the present time. The key architectural steps have been:

- Distributed analogue architectures (Section 5.2.2).
- Distributed digital architectures (Section 5.2.3).
- Federated digital architectures (Section 5.2.4).
- Integrated modular avionics architectures (Section 5.2.5).

Prior to the 1960s, civil transport aircraft had been manufactured in a similar way to their earlier electromechanical forebears. Avionics systems were disjointed 'point-solutions' to control individual and largely independent aircraft system functions, implemented in analogue (that is to say, proportional) electronics and interconnected via a considerable amount of aircraft wiring. Key advances were enabled by the advent of digital computing technology in the 1960s, and that first found application in aircraft entering service during the 1970s. The availability of digital computers that could be applied to the rugged and demanding environment of the aerospace application brought computing power and accuracies that had not been available during the analogue era. The development of serial digital data buses greatly eased the interconnection and transfer of data between the major system equipments. In the early days this was achieved by means of fairly slow unidirectional, point-to-point digital links such as ARINC 429 at a data rate of 100 kbps.

The arrival of micro-electronics technology and the first integrated circuits (ICs) enabled digital computing techniques to be applied to many more systems around the aircraft. At the same time higher capacity data buses such as MIL-STD-1553B for military aircraft and ARINC 629 for civil transport aircraft provided a bidirectional, multi-drop capability at higher data rates, of up to 2 Mbps. This enabled the federated architectures that evolved during the 1980s and early1990s (albeit only the Boeing 777 adopted ARINC 629). Avionics systems became more integrated and grouped into domain areas of associated subsystems. Multiple data bus architectures were developed to cater for increased data flow and system segregation requirements. At this time in the evolutionary process, avionics equipments were mainly proprietary designs offering dedicated solutions with few if any applications outside aerospace.

The final advance occurred when electronic components and techniques developed primarily for the demands of the commercial IT industry and the Internet yielded a far higher capability than that which the aerospace industry could develop itself. This heralded the use of commercial off-the-shelf (COTS) technology in aerospace applications to take advantage of the processing power now available. However, the use of COTS technologies brought a number of challenges, not least of which is that the faster pace of development renders devices obsolete more quickly. Integrated modular avionics architectures facilitate the promise of being able to access the computing power of COTS technologies affordably and offer a solution to obsolescence by incrementally certificating the application software independently from the hardware platform. The key attributes of each of the evolutionary stages of architectural development are described below.

5.2.2   Distributed Analogue Architecture

The distributed analogue architecture is shown in Figure 5.2. Aircraft conceived and designed throughout the 1950s and 1960s exhibit this type of architecture, some of which are still in service today.

Distributed analogue avionics architectures are characterised as a set of discrete or 'point-solutions' to individual avionics subsystems, each designed and implemented in relative isolation from one another and certificated separately.

The subsystems equipments are interconnected by hardwiring (this architecture pre-dates digital data buses) and results in a huge amount of aircraft wiring. The system function is implemented in hardwired circuitry and interconnections, and hence the system is extremely difficult to modify. This wiring is associated with power supplies, sensor excitation, sensor signal and system discrete mode selection and status signals. The most common means to signal

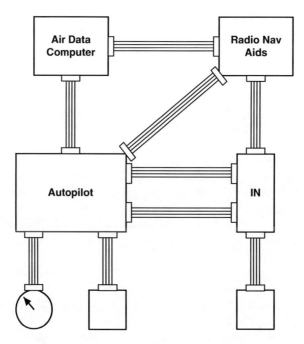

Figure 5.2   Distributed analogue architecture

angular position is by means of an electromagnetic synchro that uses a 3-wire transmission system. Aircraft of this vintage – termed classic in the industry – contain a large number of synchros and similar devices to transmit heading, attitude and other angular parameters. Each subsystem has dedicated controls and displays. The flight instruments are electromechanical and often extremely intricate in their operation, requiring instrument-maker skills for assembly and repair.

Analogue computing techniques do not provide the accuracy and stability offered by the later digital systems. The performance of analogue systems is subject to component tolerances, bias, temperature variation and long-term drift.

Typical aircraft architected in this manner include: the Boeing 707; VC10; BAC 1-11; DC-9; and early Boeing 737s. Many of these types are still flying; some such as the VC-10, the KC-135 and E-3/E-4/E-6 (Boeing 707 derivatives) fulfilling military roles. They will continue to do so for several years, but gradually their numbers are dwindling as aircraft structural problems are manifested and the increasing cost of maintaining the older systems takes its toll. Some of these aircraft have been in service for up to 50 years.

5.2.3   Distributed Digital Architecture

The maturity of digital computing devices suitable for airborne use soon saw digital computers replace their earlier analogue equivalents. The digital computers as installed on these early systems were a far cry from today, being heavy, slow in computing terms and having very limited memory capacity.

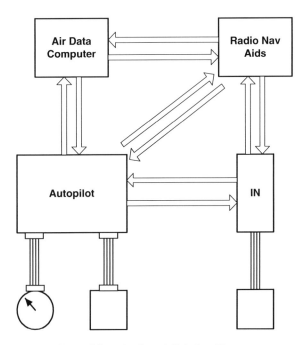

Figure 5.3   Distributed digital architecture

Initially the avionics systems architecture remained relatively unchanged, except data bus technology replaced analogue communication between computers. However, the greater speed of computation, greater accuracy and elimination of variation in performance due to component tolerance and drift problems soon led to digital computers being applied ever wider in aircraft control systems, well beyond the original narrow confines of the traditional avionics systems.

A simplified version of the distributed digital architecture is shown in Figure 5.3. Aircraft architected in this manner include the Boeing 737, 757 and 767; and the Airbus A300, 320 and 330 series and some business jets. The key characteristics of this type of architecture are described below.

Major functional units contained their own digital computer and application software stored in their internal memory. They were task-oriented, embedded computers. The functionality of a digital computer is defined by its application software, not by its detailed circuitry, so changes and modifications can be effected by changing the application software. Until electrically erasable programmable read only memory (EEPROM) became available, this could only be done by removing the equipment from the aircraft and returning it to the equipment manufacturer.

A significant development accompanying the emergence of digital processing was the adoption of serial unidirectional digital data buses – ARINC 429 being the most popular for civil transport aircraft – which allowed important system data to be passed in digital form between the major processing centres on the aircraft. Although slow by today's standards (100 kbps for ARINC 429), the introduction of these data buses represented a major step forward, bestowing major performance improvements on navigation and electronic flight instrument systems by adopting this technology.

At this stage systems were still dedicated in function, although clearly the ability to transfer data between the units had significantly improved. The adoption of data buses, particularly ARINC 429, spawned a series of ARINC standards which standardised the digital interfaces for different types of equipment. The uptake of this standardisation led manufacturers producing inertial navigation systems (INS) to prepare standard interfaces for that system. This eventually led to the standardisation between systems of different manufacturers, potentially easing the prospect of system modification or upgrade. The ARINC 429 data bus is still important and is in use today for interfacing simple equipments with low data rates to the main avionics core, and as a dissimilar means to back up critical data in safety critical systems.

Initially flight deck instruments remained as electromechanical devices dedicated to their function, as in the analogue architecture already described. However, in the mid-1980s, colour cathode ray tube (CRT) technology, and in the mid-1990s active matrix liquid crystal display (AMLCD) technology, matured sufficiently to be suitable for use in airborne environments, and the electromechanical instruments were replaced by the following multicolour display systems on the flight deck:

- Electronic Flight Instrument System (EFIS);
- Engine Indicating & Crew Alerting System (EICAS) – Boeing and others;
- Electronic Centralised Aircraft Monitor (ECAM) – Airbus.

Data buses offered a great deal of flexibility in the way that signals were transferred from unit to unit. They also allowed architectures to be constructed with a considerable reduction in inter-unit wiring and multipin connectors. This led to a reduction in weight and cost, and also eased the task of introducing large and inflexible wiring harnesses into the airframe. This in turn led to reductions in the non-recurring cost of producing harness drawings, and the recurring cost of manufacturing and installing harnesses. Data buses greatly simplified upgrades. The ARINC 429 data bus allows new equipment to be added to the data bus, up to a maximum of 20 LRUs.

Overall the adoption of even the early digital technology brought great advantages in system accuracy and performance, although the development and maintenance of these early digital systems was far from easy.

5.2.4   Federated Digital Architecture

The next development, federated digital architecture, is shown in Figure 5.4.

The federated avionics systems architecture recognises that the total aircraft avionics systems functions are interrelated and interdependent. The commonality of approach facilitates the sharing of information between previously disparate systems, enhancing the whole functionality of the avionics system. We now see related systems being grouped into domain areas (see Section 5.3) and data exchanged between equipments within the domain area on the local data bus network. Domains are interconnected at a higher level to pass global information from domain to domain.

A federated architecture generally uses dedicated task-oriented, line-replaceable computers, each with their own embedded processor and application memory. The capabilities and technologies employed are thoroughly described in Chapter 2. The significant advances made

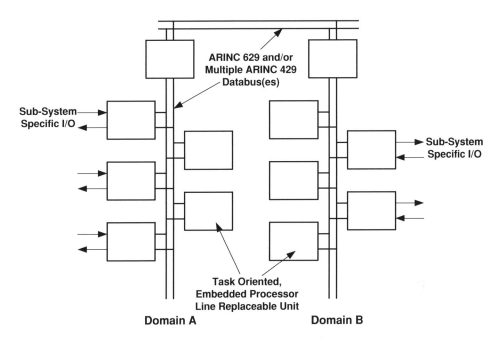

Figure 5.4   Federated avionics architecture

in computer technology meant that this architecture could be applied to other aircraft systems that were not traditionally avionics systems such as fuel systems, landing gear systems and other vehicle management systems.

The military were the first to adopt a federated architecture based around the MIL-STD-1553B 1 Mbps bidirectional data bus originally conceived by the US Air Force Wright Patterson development laboratories, as they were called at the time. It evolved through two iterations from a basic standard finally culminating in the 1553B standard, for which there are also UK DEF-STAN equivalents.

MIL-STD-1553B has been widely used across most military platforms for over two decades. It utilises a 'command:response' protocol that requires a central control entity called a bus controller (BC). The civil community voiced concerns regarding the fault tolerance of this centralised control philosophy and were less eager to adopt the federated approach, having collectively invested heavily in the ARINC 429 standard that was already widely established and proving its worth in the civil fleets. Furthermore, this group did not like some of the detailed implementation/protocol issues associated with MIL-STD-1553B, and accordingly decided to derive a new civil standard that eventually became ARINC 629. ARINC 629 supports multiple redundant operations for safety critical systems and has higher bandwidth than MIL-STD-1553B (2 Mbps).

However, in many ways ARINC 629 was overtaken by events. The only aircraft type to use it was the Boeing 777, which in architectural terms sits between a federated architecture and an integrated modular architecture. It uses multiply-redundant ARINC 629 data buses as the main avionics data bus for the flight control and utilities domains. It implements the flight deck and navigation domains in a proprietary partial IMA architecture called the Airplane

Information Management System (AIMS), to which navigational aids, air data sensors and inertial reference systems are connected using ARINC 429. It uses Ethernet for the passenger in-flight entertainment system.

Along with the developing maturity of electronic memory devices, in particular non-volatile memory, the federated architecture was able to support software re-programming of the various system LRUs via the aircraft-level data buses. This is a significant improvement in maintainability and facilitates operational improvements and updates to be speedily incorporated.

5.2.5   Integrated Modular Avionics

The advances in computer technology and local area networks and the advent of the Internet created a data explosion in the late 1990s. Powerful microprocessor technology, ever-increasing integrated circuit capability and decreasing costs have led to a revolution in avionics architectures known as integrated modular avionics (IMA).

The drivers for change were:

- data explosion:
  - network-centric air traffic management (ATM)
  - performance based navigation (PBN)
  - sensor data fusion
- commercial off-the-shelf technology being driven by commercial enterprise:
  - mobile phone networks
  - world-wide web
  - satellite communications
- obsolescence:
  - technology turns on 5-year average cycle (compared with an air-vehicle platform life cycle of 30+ years).

A typical IMA architecture is shown in Figure 5.5. Federated systems are designed to provide:

- separate processing;
- separate (and often proprietary) infrastructure;
- internal system bus;
- separate I/O routed point-to-point between sensors/effectors and processing resources.

By contrast, integrated modular systems are designed to provide:

- common processing with robustly partitioned application software;
- common infrastructure;
- distributed systems bus;
- specific I/O routed via the network between shared remote data concentrators (RDCs)/remote interface units (RIUs) and common computing resources.

Instead of dedicated, distributed (federated) task-oriented computing, IMA principles promote a general-purpose centralised computing resource comprising a set of common hardware

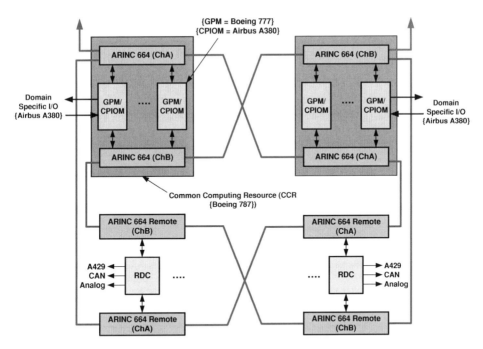

Figure 5.5 Integrated modular avionics architecture

computing modules. The avionics application software previously embedded in proprietary task-oriented computers in a federated architecture is now hosted on the general-purpose processors in the common core computing resource (see Figure 5.6).

Application software is provided and certified by the subsystem developer, independently of the IMA hardware platform. Data are exchanged between applications via a dual redundant network based on COTS Ethernet technology with enhancements to provide real-time features needed for safety-critical avionics applications. Originally developed as Aviation Full Duplex (AFDX), this network is known as ARINC 664 Part 7. It provides full duplex, bidirectional communication between network resources at 100 Mbps.

Airbus was the first airframe company to introduce an IMA architecture onto a large civil transport aircraft, the Airbus A380. It uses a central processor I/O module (CPIOM) as its common core computing element. A CPIOM is a general-purpose processor card plus domain-specific I/O interfacing. CPIOMs are grouped by domains. Each domain is connected to the ARINC 664 Part 7 network by a pair of network switches. Although collocated, CPIOMs are not housed in a common equipment rack; they are LRUs in their own right. The A380 has some domain-specific RDCs for I/O intensive subsystems, and these are usually provided by the subsystem suppliers.

Boeing adopted a subtly different approach on the B787. The common core computing element is a general processing module (GPM) similar to that in the Airbus CPIOM, but it has no domain I/O. I/O interfacing is effected in RDCs located at convenient points around the aircraft to minimise wiring. RDCs communicate digitised data to/from the centralised computing core over the ARINC 664 Part 7 network. General processing modules (GPMs) are co-located in a common computing resource (CCR) rack.

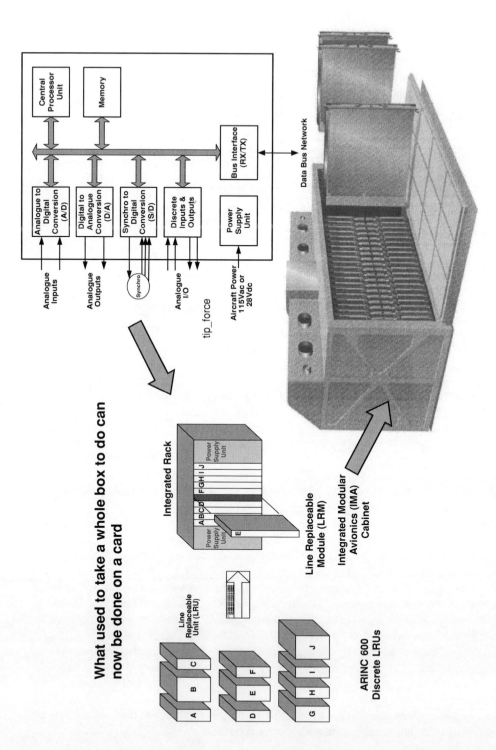

Figure 5.6 Integrated modular avionics cabinet

In an IMA architecture it is necessary to map the logical architecture of an avionics subsystem to the physical resources of the IMA platform. The logical architecture in many respects can be thought of as replicating the earlier federated architecture in terms of redundancy, fault tolerance, integrity and segregation. These features achieved by the physical realisation of a federated architecture are now being effected by mapping the subsystem logical architecture onto the physical resources of the IMA platform, and by the robust partitioning provided by the real-time operating system that runs the application software on the IMA computing hardware.

5.2.6   Open System Standards

An Open System is a system that implements open standards for interfaces, services and supporting formats to enable components to be used across a wide range of systems with minimal change. An Open System is characterised by:

- Well-defined, widely used, non-proprietary interfaces/protocols.
- Use of standards developed or adapted by recognised bodies.
- Definition of all aspects of systems interfaces, both internal and between air-vehicle and the external environment, to facilitate adding or tailoring of functions for new or modified systems capabilities.
- Explicit provision for expansion or upgrading through the incorporation of additional or higher performance elements, with minimal impact.

5.3   Avionic Systems Domains

5.3.1   The Aircraft as a System of Systems

An aircraft is a system of systems. Figure 5.7 shows some examples of aircraft systems and demonstrates that all these individual systems are interrelated. Each relies on the correct functioning of others in order to perform its system function in an optimal manner, so that the aircraft may achieve its overall operational performance effectively and efficiently.

A holistic approach is required to optimise aircraft performance. Often individual system requirements conflict, and it is necessary to trade off and balance subsystem implementations against the overall aircraft system objectives. This simple example shows:

- **Engines**: The aircraft engines are the prime source of power. Today each engine is controlled by an engine-mounted avionics style computer in the form of a full authority digital engine control (FADEC) system, usually implemented in a dual command:monitor architecture to meet its safety objectives in terms of availability and integrity.
- **Electrical system**: Engine-mounted generators provide electrical power for all the aircraft systems including the avionics systems. Electrical power is regulated, distributed and monitored by the electrical power management system, itself controlled by avionics-style computers, which ensure available power is matched to power demand, shedding loads in a controlled manner under conditions of primary power reduction.
- **Hydraulics**: The hydraulics system is possibly the least avionics-rich area of the aircraft systems. It provides the muscle to move actuators, raise and lower landing gear, operate

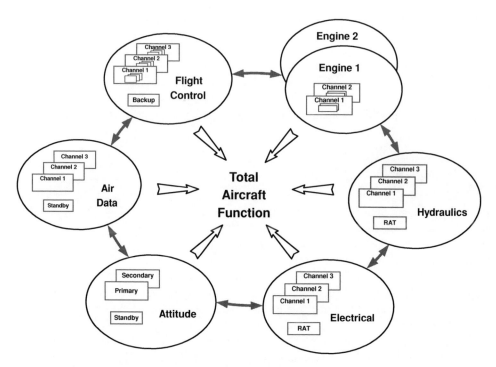

Figure 5.7   System of systems

doors and slides, and so on. But as aircraft systems migrate towards more electrical systems, centralised hydraulic systems are being replaced by distributed hydraulics systems in which hydraulic power is generated and maintained by smart electric motor pumps. The A380 flight control actuation arrangements use a combination of traditional hydraulic actuators and electro-hydrostatic actuators (EHAs). EHAs use hydraulic pressure to move the actuator generated from an internal mini-hydraulic system powered by an electric motor pump controlled by a smart electric motor controller.

- **Attitude**: Aircraft attitude is determined by a redundant set of sensors, including solid-state gyroscopes and inertial sensors (laser-ring gyro) augmented by satellite navigation systems, to compute an extensive set of parameters that determine the aircraft present position, ground velocity and attitude.
- **Air data**: Accurate knowledge of airspeed and barometric altitude are required to ensure the aircraft operates within its safe flight envelope. Typically three sets of air data sensors are used to meet the safety objectives in terms of availability and integrity, plus total air temperature and angle of incidence.
- **Flight control**: Modern civil transport aircraft incorporate fly-by-wire flight control technologies. All the other systems discussed are required to control the aircraft flight path with a high degree of precision and ensure the aircraft is always operated within its safe flight envelope from take-off to touch-down at the destination airfield. Since failure of the flight control system would be catastrophic, the flight control system is implemented in a multiple redundant architecture to meet safety, availability and integrity objectives.

5.3.2   ATA Classification

Historically civil aircraft systems have been categorised into Air Transport Association (ATA) chapters. Figure 5.8 indicates hierarchically those chapters of primary interest for avionics systems. Segment 9.2 identifies the traditional avionics segment of:

- autoflight (Chapter 22);
- communications (Chapter 23);
- recording and indication (Chapter 31) – today this is known as the flight deck;
- navigation (Chapter 34).

But today avionics-style technology is being applied in other areas, including:

- flight control (Chapter 27);
- power (engine) control (Chapter 26);
- landing gear (Chapter 32);
- fuel system (Chapter 28);
- electrical power (Chapter 24);
- ice and rain protection (Chapter 30).

The latter four are frequently combined as constituent elements of the aircraft vehicle management system (VMS) domain.

It is useful to devolve the aircraft total avionics system into manageable parts of associated functions, or domains. This can be done in a number of ways, but probably the easiest to comprehend is to adopt something close to the ATA chapter classification.

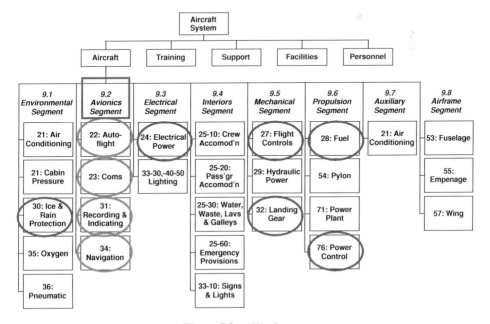

Figure 5.8   ATA chapters

Considering then the major avionics systems functions by domain areas, the top-level functionality of each domain area can be summarised as follows:

- Autoflight (ATA 22)
  - functional: flight control, auto-stabilisation, autopilot, auto-throttle, envelope protection;
  - sensors: body rates, air data, attitude, heading, inertial reference;
  - effectors: actuators, servomotors.
- Navigation (ATA 34)
  - functional: flight management, performance-based navigation (PBN), RNP/ANP, RVSM, area navigation (RNAV), LNAV and VNAV guidance, GPS/GNSS approaches, surveillance, GPWS/TAWS, TCAS, ADS-B, navigation & terrain databases;
  - sensors: ADF, VOR, DME, ILS, MLS, and so on.
- Flight deck (ATA 31)
  - functional: primary flight display (PFD), navigation display (ND), ECAM/EICAS, standby instruments;
  - sensors: weather radar, infra-red cameras;
  - effectors/devices: flight deck displays, head-up display, synthetic and enhanced vision.
- Vehicle management systems (utilities)
  - functional: electrical system: primary and secondary power management and distribution. Fuel system: fuel gauging, refuel, defuel and jettison, centre of gravity control, load alleviation. Landing gear: extension and retraction, braking, nose wheel steering. Wing anti-ice/de-ice, environmental control system, engine control – full authority digital engine control (FADEC).

For a full description and examination of all of these functional elements, the reader is advised to read the appropriate Chapter.

5.4   Avionics Architecture Examples

This section provides some aircraft type-specific examples of avionics systems as they have evolved from federated architectures to today's integrated modular avionics architectures.

Federated avionics systems are designed to provide separate processing, separate (and often proprietary) infrastructure, internal system bus and separate I/O routed point-to-point between sensors/effectors and processing resources.

By contrast, integrated modular avionics (IMA) systems are designed to provide common processing with robustly partitioned application software, common infrastructure, distributed systems bus and specific I/O routed via the network between shared remote data concentrators (RDCs) and common computing resources.

5.4.1   The Manifestations of IMA

IMA architectures have evolved through three distinct phases which can be described as:

- **First-generation IMA**: Implementations of an avionics subsystem by a single supplier using own proprietary standards, modules and parallel backplane, e.g. Boeing 777.

- **Second-generation IMA**: Open architectures with multiple supplier modules/applications hosted in a standard backplane. Application supplier certificates the application/hardware combination and supplies the module(s) on which his application is hosted, e.g. Hawker Horizon.
- **Third-generation IMA**: Independent provision of open architecture modules and applications. Applications run under an open-standard Application Executive (APEX) real-time operating system (RTOS). Multiple applications may reside on a single module. Hardware/software integration and certification is the responsibility of the system integrator, e.g. Airbus for the A380/A350 and Boeing for the 787.

5.4.2 The Airbus A320 Avionics Architecture

The A320 avionics architecture is a late example of the typical single source/multiple sink A429 arrangement used by virtually all civil airliners from around 1980 onwards. The A429 data bus is described in Section 3.4. A simplified version of the A320 family architecture is shown in Figure 5.9. Airbus utilise a Side 1/Side 2 dual architecture; for simplicity only, Side 1 is shown. Side 2 will replicate the Side 1 architecture and, as can be seen from the diagram, there is considerable cross-talk and interchange of data between Side 1 and Side 2. For some aircraft a third ADIRS (air data and inertial reference system) may be fitted.

At the centre of the architecture are the Flight Augmentation Computer (FAC) and the Flight Management & Guidance Computer (FMGC). The FAC provides the following functions

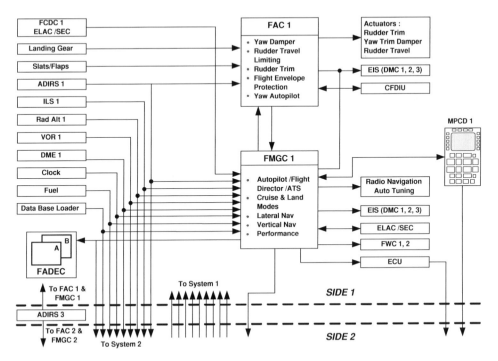

Figure 5.9   Simplified A320 avionics architecture

(mainly associated with the yaw channel):

- yaw damper;
- rudder travel limiting;
- rudder trim;
- flight envelope protection;
- yaw autopilot.

The FAC receives inputs from airframe systems such as the landing gear and slats/flaps in order to take account of changes in aircraft configuration, especially during take-off and landing. ADIRS 1 supplies air data and attitude data so that changes in flight conditions may be accommodated. The FAC drives the appropriate actuators to exercise the various yaw control functions. It also outputs display data to the electronic instrument system (EIS), specifically to Display Management Computers 1, 2 and 3.

The FMGC provides higher-level autopilot and navigation functions, namely:

- autopilot/flight director/air traffic services (ATS);
- cruise and land modes;
- lateral navigation (LNAV);
- vertical navigation (VNAV);
- performance monitoring.

The FMGC receives inputs from the fly-by-wire system: flight control data concentrators (FCDCs), elevator/aileron computers (ELACs) and secondary elevator computers (SECs). The radio navigation aids (ILS, Rad Alt, VOR and DME) provide Earth-based navigation data. Timing and fuel data provide information for the flight management functions. A database loader is provided in order that specific route and performance data may be uploaded to customise the flight management database by airline and route. The FMGC outputs data to the multipurpose control and display (MPCD) which provides the flight crew interface. Among other outputs, data are provided to the electronic instrument system (EIS) and flight warning computers (FWCs) for display and warning purposes; it also provides data to the engine FADEC in order that some of the more advanced performance optimisation functions may be executed. The FMGC also outputs commands to the radio navigation auto-tuning function which will automatically tune to the radio navigation aids mentioned above, as well as selecting the necessary communications frequencies as the aircraft executes the pre-loaded flight plan.

The architecture of the flight deck display suite is shown in Figure 5.10. The images for the six EFIS and ECAM displays are sourced from three display management computers that provide a triplex redundant set. Two flight warning computers source warning information to the display suite and the central warning panel. Display data not available digitally from the aircraft data bus network is derived from analogue signals via two signal data acquisition computers.

5.4.3 The Boeing 777 Avionics Architecture

In contrast to the earlier point-to-point data bus architectures previously used on the Boeing family, the Boeing 777 employed a new data bus – ARINC 629 – which operated at 2 Mbps.

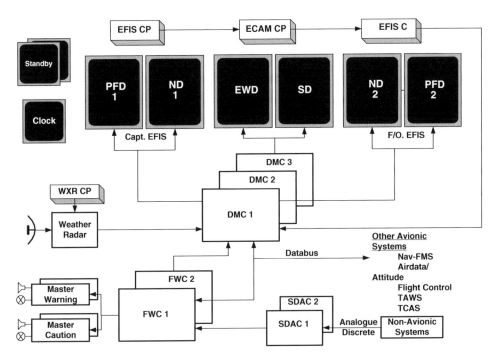

Figure 5.10 Airbus A320 flight deck display suite

In some ways similar to MIL-STD-1553B, it uses current coupling rather than transformer coupling. The bus protocol is totally different, as described in Section 3.6. At the heart of the architecture is the Honeywell Airplane Information Management System (AIMS) where the central interface with the main system data buses and displays is undertaken by two AIMS cabinets. Refer to Figure 5.11 for a simplified top-level portrayal of the Boeing 777 avionics architecture.

The Boeing 777 avionics system has a mixture of data buses and architectures. The backplane bus of the two AIMS cabinets is ARINC 659 SAFEbus. Many external systems follow the traditional 'federated' architecture using ARINC 429. Two ARINC 629 triplex data buses carry fly-by-wire and system functions. Two LANs are used for the electronic library, the cabin management system and in-flight entertainment (IFE).

Three and occasionally four A629 buses provide the data bus interface between the functional elements of the flight control and aircraft systems functions, as shown in Figure 5.12.

The flight control data buses and the systems data buses are key elements in this architecture. The flight control A629 buses incorporate Left (L), Centre (C), and Right (R) buses to offer a triplex redundant means of connecting the functional subsystems of the flight control system and also to the autopilot function. The major systems connecting to the flight control data buses are as follows:

- Total pressure (Tp) and static pressure (Sp) air data modules (ADMs) to provide basic air data information in triplex form: left, centre, right.
- A primary Air Data and Inertial Reference Unit (ADIRU) to provide a primary integrated source of air data and inertial data.

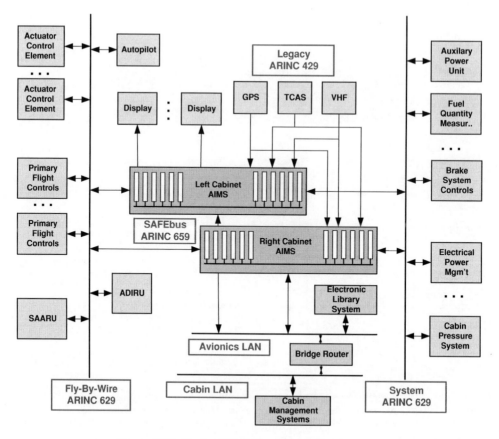

Figure 5.11 Boeing 777 top-level avionics architecture

- An additional Secondary Attitude & Air-data Reference Unit (SAARU) to provide a backup source of air data and inertial information.
- Control inputs from three primary flight computers.
- Outputs to the AIMS cabinets for computational and display purposes.
- Interface to the crew control & display units (CDUs).
- Interfaces to the three A629 data buses which interconnect the major aircraft utilities and the engine electronic controllers.

The integration of these functions is described in more detail in Chapter 10 and specifically in Figure 10.22.

The systems in A629 data buses incorporate a Left (L), Centre 1 (C1), Centre 2 (C2) and Right (R). The subtlety of the C2 bus is described below. The systems buses interconnect the major engine, electrical and other subsystems as described in brief below. It will be noted that some of these systems only have two A629 connections, indicating an underlying dual controller philosophy:

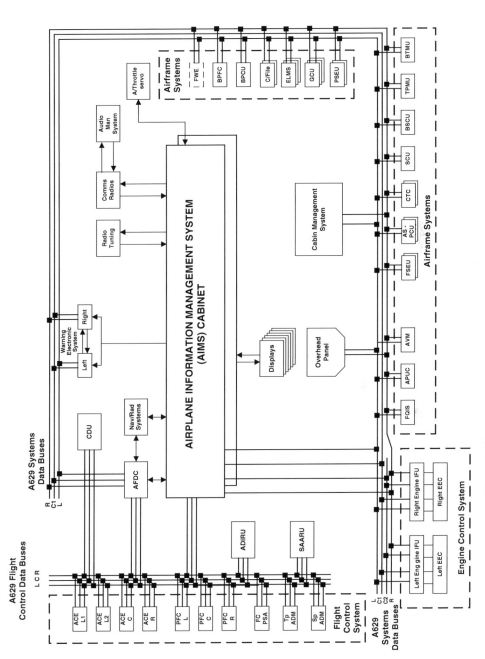

Figure 5.12 B777 flight control and aircraft systems architecture

- The engines employ a dual/dual control philosophy and therefore need four independent interfaces between engines and displays. Hence the need for the C1 and C2 channels. The need for the left and right engine interface units (IFUs) was dictated by the fact that the electronic engine controllers were originally designed for an A429 data bus system. This design was fixed long before the aircraft-level A629 bus structure was finalised.
- Electrical system control and distribution: GCUs, BPCU, ELMS and auxiliary power unit (APU) generator
- Elements supporting the flight control system: proximity switch electronic units (PSEUs) and flap/slat electronic units (FSEUs).
- Cabin conditioning: air supply and pressure control unit (ASPCU) and cabin temperature controller (CTC).
- Landing gear and braking system: brake system control unit (BSCU), tyre pressure monitoring unit (TPMU) and brake temperature monitoring unit (BTMU).
- Other major elements are the overhead panel and the cabin management system.

The B777 incorporated two subsystems that may be considered to be first-generation IMA implementations: the dual-channel AIMS cabinet provided by Honeywell, and the three-channel Electrical Load Management System (ELMS) provided by Smiths (now GE Aviation).

The AIMS cabinet is a partial but significant integration of hardware and software in a proprietary integrated modular architecture. The AIMS functions performed in both cabinets include the previously traditional federated architecture functions of the flight management system (FMS), the electronic flight instrument system (EFIS), the Engine Indicating & Crew Alerting System (EICAS), displays management, central maintenance, airplane condition monitoring and communications management, and provides a data communications conversion and gateway to legacy ARINC 429 systems. Each cabinet has common design hardware line replaceable modules (LRMs) consisting of power supplies, processors, memory and shared input/output. Each cabinet also has common software for the operating system, I/O device handling and built-in-test. Figure 5.13 shows the AIMS cabinet internal structure. Each cabinet has a capacity for 13 LRMs, typically assigned as four processor modules, four input/output modules (IOMs) and two power supply modules with spare capacity for growth. Applications resident on the processor modules include (the figure in brackets indicates the number of redundant copies):

- displays (4);
- flight/thrust management (2);
- central maintenance (2);
- data communications management (2);
- flight deck communications (2);
- airplane condition monitoring (1);
- digital flight data acquisition (2);
- data conversion gateway (4).

The ARINC 659 SAFEbus AIMS backplane uses a deterministic control in time and space. All bus timing information and message locations are contained in table memories within each terminal. Messages are transmitted simultaneously on two self-clocking buses.

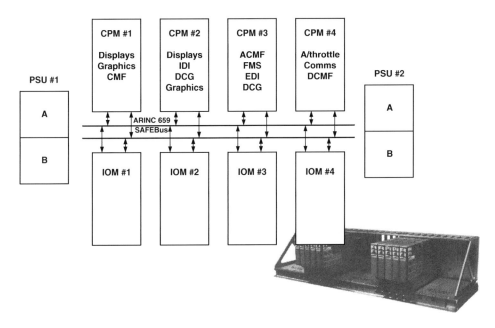

Figure 5.13   Boeing 777 AIMS cabinet

5.4.4   Honeywell EPIC Architecture

The business jet community is often at the forefront in adopting new technology – in this market, equipment or system cost is not necessarily a prime driver, whereas performance is. In this sense, the business jet is set apart from normal commercial airline solutions.

A good example of a business jet avionics system is the Honeywell EPIC system which is used on a wide range of feeder liners and business jets: the Embraer 170 family; the Dassault family and the Gulfstream family, as well as other platforms. The EPIC system utilises an avionics architecture taking the generic form shown in Figure 5.14. The basic core of the system is represented by dual 10 Mbps data buses. The data bus is known as Avionics System Communications Bus – Variant D (ASCB-D). These avionics data buses are targeted at lower-performance aircraft and have been developed by the General Aviation Manufacturer's Association (GAMA) to suit the requirement of general aviation aircraft. In fact, ASCB-D is a deterministic and ruggedised version of 10BaseT Ethernet – a 10 Mbps network, using twisted wire pairs.

At the core of the system are up to four modular avionics units (MAUs) which host a range of line-replaceable modules (LRMs). Pilot inputs into the system are generated by two cursor control units where a 'point and click' philosophy is adopted. In the example, five display surfaces are depicted – usually 8" × 10" displays in portrait format as shown. As the architecture has evolved into more sophisticated forms, fewer and larger displays in 10" × 13" landscape format have been adopted.

The functions provided by the EPIC system include:

- computing (MAU);
- sensors (ADM, IMU);

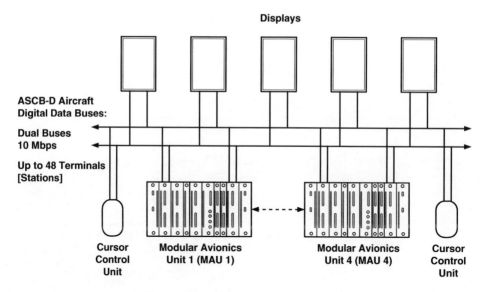

Figure 5.14   Honeywell EPIC architecture

- displays (DU-1080, and others);
- controls (CCD, voice);
- functions (navigation, system utilities, etc.);
- data (electronic charts, maps, manuals, video).

The MAUs effectively embody a second-generation implementation of IMA. The backplane is proprietary to Honeywell and many of the 'core' modules are provided by that supplier. However, third-party subsystem suppliers may also design their own modules to be hosted in the cabinet; these modules incorporate an interface with the proprietary backplane to fulfil the functionality of dedicated aircraft systems such as fuel. Reference [1] describes how the fuel system was integrated into the MAU arrangement for the Hawker Horizon aircraft. Typical modules hosted within the MAUs are listed below:

- processor;
- memory;
- power supply;
- network interface controller (NIC) to interface the MAU proprietary backplane to the aircraft level ASCB-D bus;
- interface (I/O) modules – provided by specialist suppliers for fuel and other dedicated systems.

5.4.5   The Airbus A380 and A350

Airbus introduced a different concept of avionics systems integration with the A380. The basis of this architecture was the adoption of a packet-switching technology using fast-switched

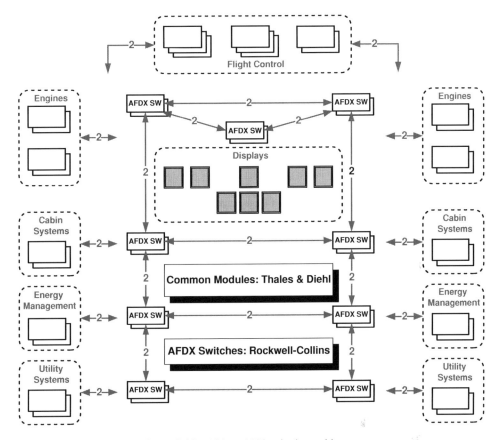

Figure 5.15   Airbus A380 avionics architecture

Ethernet (100BaseT) twin copper wire buses. At the centre of the architecture is a dual redundant AFDX switched network, comprising AFDX switches operating at 100 Mbps (refer to Section 3.7). In Figure 5.15 the AFDX switching network is shown centrally and consists of eight pairs of switches arranged according to their avionics domain of interest. The dual redundant switch network is distributed longitudinally throughout the aircraft and also comprises left/right elements.

The avionics function in each domain area is implemented in a set of computer processor input/output modules (CPIOMs). The general architecture of the CPIOM, for illustrative purposes, is shown in Figure 5.16. It comprises a general-purpose processing function and a set of I/O interface functions. The reader should note that the CPIOM architecture reflects the generic avionics computer architecture discussed in Section 2.3.1. The implementation uses current generation PowerPC processor architectures and memory technologies discussed in Sections 2.8 and 2.9. The operating system and application software are loaded on aircraft into the onboard non-volatile Flash memory and uploaded into the RAM memory. Partitioning is assured by context switching the application software at run-time. Longer-term storage of configuration and maintenance data is provided non-volatile RAM (NVRAM). A peripheral

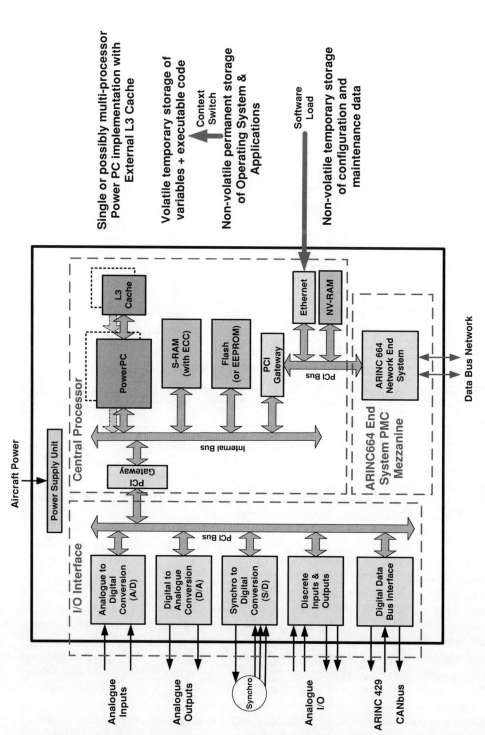

Figure 5.16 Airbus A380 CPIOM architecture (illustrative)

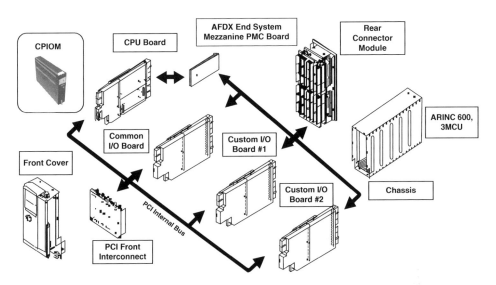

Figure 5.17   Airbus A380 CPIOM physical arrangement

component interconnect (PCI) bus interfaces the I/O devices to the CPU main system bus. The network interface is provided by an ARINC 664-P7 end-system PCI Mezzanine Card (PMC) which mounts onto a second (PCI) socket on the CPU board.

The CPIOM physical arrangement is shown in Figure 5.17. It comprises four circuit boards contained in an ARINC 600 3MCU enclosure. The CPU circuit board and I/O common board are common to all CPIOMs; the other I/O boards are domain-specific. Interconnections between the CPU board and the I/O boards are via the internal PCI bus. The AFDX end-system is housed on a PMC mezzanine board mounted on the CPU board.

In the A380 architecture there are a total of 22 CPIOMs of seven different types in which the central computing core is common and the personality of each CPIOM variant is determined by the input/output (I/O) and henceforth by the intended system function. These CPIOMs are utilised across a number of functional domains featuring the flight deck, cabin, energy and utilities. In the utilities domain, four CPIOMs–Fuel (CPIOM–F) and four CPIOMs–Gear (CPIOMs–G) provide the computing core for the fuel and landing gear functions respectively. A major advantage of this concept is that common development tools and software languages may be used across all CPIOM variants. The CPIOM responsibilities within the Airbus consortium are shown in Figure 5.18.

In the A380 architecture, system-specific remote data concentrators (RDCs) are provided by the major subsystem supplier. For example, two fuel quantity management system (FQMS) RDCs provide the fuel system-specific interface, while three landing gear RDCs provide the interface to the landing gear and braking systems. A similar system concept was adopted for the A400M central avionics core.

In the case of the A350, which followed several years later, a similar concept has been used. The central AFDX switching network is unchanged, but dedicated subsystem RDCs have been replaced by ∼29 multipurpose common remote data concentrators (cRDCs) of two types. While the number of CPIOMs variants has been substantially reduced from seven to

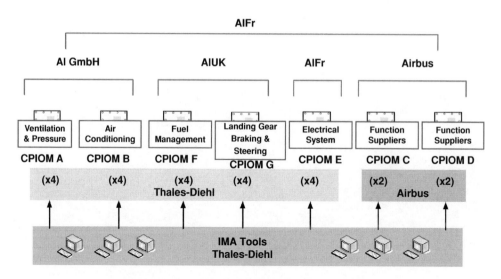

Figure 5.18   Airbus A380 CPIOM responsibilities

two, the total number of CPIOMs remains virtually the same (see Figure 5.19). The outcome of this further hardware consolidation gives rise to the following issues:

- reduced hardware variability;
- increased hardware content for the Airbus partners at the expense of the subsystem supplier;
- an increase in the burden of aircraft-level configuration control, as overall system configuration is largely dictated by the cRDC configuration.

5.4.6   The Boeing 787

The B787 also utilises AFDX 100 Mbps technology as defined by ARINC 664 – see Section 2.7. The architecture is, however, quite different to the Airbus approach, using two closely coupled common computing resource (CCR) cabinets as shown in Figure 5.20, and distributed I/O interfacing through the services of RDCs. The CCR represents the common computing core of the system, as opposed to the Airbus philosophy of a distributed system using a network of 16 AFDX switches.

In many respects the Boeing 787 GPM + RDC IMA architecture for illustrative purposes can be viewed as having the same constituent elements as the Airbus CPIOM, with the processing function housed in the CCS and the I/O function housed in RDCs, with communication between these elements via the ARINC664 network. This is illustrated in Figure 5.21.

Each CCR comprises a number of general processing modules (GPMs) and two ARINC 664 Part 7 switches (Channels A and B). The GPM is an independent computing platform that hosts core software and provides the hosted applications with a robust partitioned environment and infrastructure, services including I/O, health monitor and non-volatile file storage and retrieval based on the ARINC 653 standard. Computational resources in terms of time window, cycle

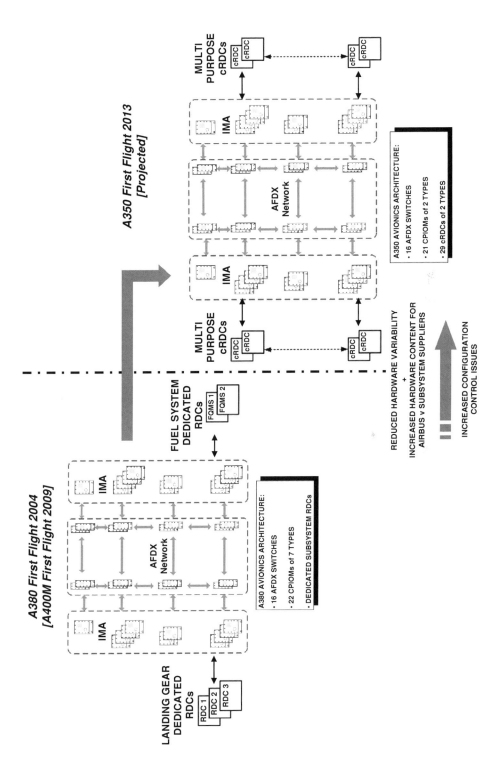

Figure 5.19 Top-level comparison of A380 and A350 architectures

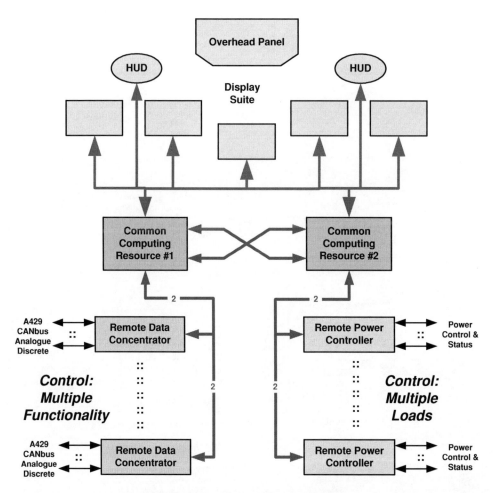

Figure 5.20 B787 avionics architecture

time, memory allocation and I/O needs are conveyed to the core software through configuration files. All these allocations are enforced by partitioning mechanisms.

The CCR cabinets are believed to use 100Base F fibre-optic star hubs to interconnect the high-speed data bus elements. The CCRs are interconnected to one another for data exchange. They also communicate with a number of RDCs and remote power controllers (RPCs) across the aircraft: in the region of 25 to 30 of each, depending upon the configuration:

- The RDCs have local interfaces including A429, CANbus, analogue and discrete signals to interface with avionic and aircraft systems. Their function may be considered analogous to the cRDCs used in the A350 architecture.
- The RPCs serve a similar function but are concerned with switching power to the various aircraft loads and monitoring status. The RPCs effectively provide a distributed power management system.

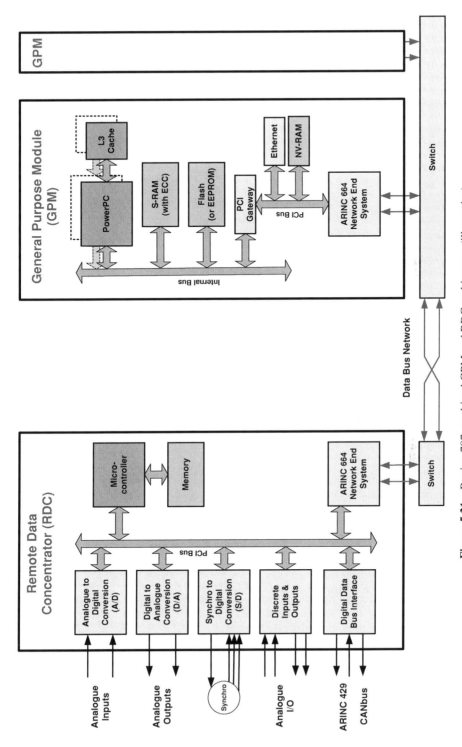

Figure 5.21 Boeing 787 combined GPM and RDC architecture (illustrative)

The CCR and RDC hardware are provided by GE Aviation who act as avionics system integrators [2]. The RPCs are provided by Hamilton Sundstrand, who also provide the electrical power generation and primary power distribution panels.

Another key feature of the B787 aircraft is that it is a 'more-electric' aircraft where the bleed air function has been removed except for engine intake cowling anti-icing. Many functions previous powered by bleed air such as the aircraft pressurisation, cabin temperature control and wing anti-icing are powered electrically. This introduces a huge electrical demand on the aircraft, and the aircraft has two 250 kVA, 230 VAC generators per channel or 1 MVA on the aircraft in total for primary power generation. Detailed discussion of these issues are outside the scope of this book, but are fully addressed in a companion volume [3].

5.5 IMA Design Principles

An IMA architecture is a configurable resource platform. Functions are allocated the resources they require on the platform in the terms of:

- processing time;
- memory;
- network I/O communications;
- interface resources (analogue signals, discretes and other digital bus types).

These resource allocations are mechanised through configuration tables. The resource guarantee (or contract) together with the platform's system partitioning characteristics are the cornerstone of hosted system independence. These properties allow individual functions to change without collateral impact on other functions. This fundamental philosophy facilitates acceptance at the individual hosted function level, and is the foundation for incremental certification on a hosted function-by-function basis.

The process for architecting a system onto the platform is as follows:

1. Define the functions that comprise the system.
   - System elements emerge as processing applications.
   - I/O needs and non-platform elements emerge as sensors, effectors and devices.
2. Define the system elements required to implement each function.
   - Develop the logical system architecture to achieve the availability, integrity and performance needs.
3. Define the resource usage demands for each system element.
4. Define the data exchange between system elements and functions.
5. Allocate the logical system elements to the platform to define the physical architecture.

Hosted functions are allocated to the platform resources to form a functional or 'logical' virtual system architecture to meet the availability, operational, safety and topological requirements of each subsystem. Hosted functions can own unique sensors, effectors, devices and non-platform line replaceable units (LRUs). The virtual system partitioning environment guarantees that hosted functions are isolated from each other and cannot interfere with each other, regardless of faults. Each hosted function is allocated its share of computing, network and

I/O resources. These predetermined resource allocations are communicated to the platform through configuration files loaded during the installation process. These configuration files enforce run-time execution of hosted function guarantees.

The IMA architecture represents a 'virtual system' concept replacing the 'physical' systems of a federated architecture. The 'virtual system' consists of the same logical groupings of components as a 'physical' system mapped onto the physical resources of the IMA platform of:

- application software;
- infrastructure/operating system;
- processor;
- system bus;
- I/O.

The 'virtual system' extends beyond the IMA computing core to the ARINC 664-P7 (AFDX) network and to the remote data concentrators and interface units (RDCs/RIUs).

5.6   The Virtual System

5.6.1   Introduction to Virtual Mapping

The key difference between an integrated modular avionics (IMA) architecture and a federated architecture is the logical system. In a federated architecture the logical system is the same as the physical system. In an IMA architecture the logical system is not the same as the physical system; the architecture is a virtual system mapped onto the physical resources of the IMA platform.

In a federated architecture a simplex avionics system function is implemented by a single dedicated task-oriented LRU complete with processor (CPU and memory) whose I/O resources interface directly with the sensors and effectors of the subject function. The application software that turns this general-purpose computing resource into the specific avionics function task-oriented processor is resident in the CPU memory and runs under the control of a real-time operating system (RTOS).

Figure 5.22 shows the mapping of this simplex system architecture onto the physical resources of an IMA platform. The figure shows a dual rack IMA platform with just two processor cards in each rack; a real implementation would of course have more cards in each rack. The simplex function is mapped onto one card in one rack. The function application software is one of several applications that can run on this card (limited by its throughput) under the control of the RTOS. The application software gains access to its systems sensors and effectors via the ARINC 664-P7 network to I/O resources resident in RDCs/RIUs that interface directly with the sensors and effectors.

In a similar manner, Figure 5.23 shows the mapping of a duplex system architecture onto the resources of the same two-rack IMA platform. In this example the federated architecture would have been implemented in two identical LRUs; Lane A and Lane B. Consolidation of the lane outputs is generally performed in the I/O elements. Mapping this logical architecture onto the physical resources of the IMA platform replicates the application software onto two processor cards: one in rack one (Lane A) and the other in rack two (Lane B). In this generic

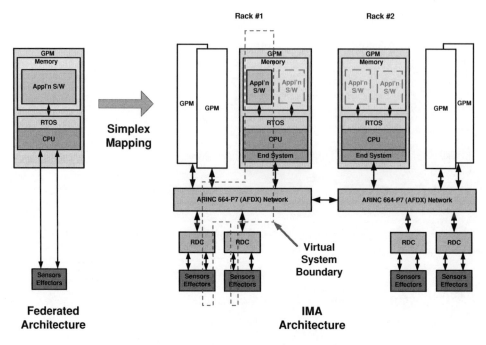

Figure 5.22 Simplex architecture mapping

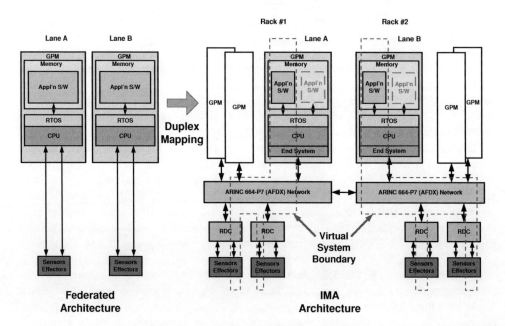

Figure 5.23 Duplex architecture mapping

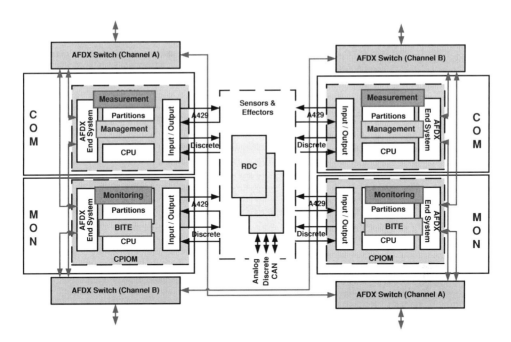

Figure 5.24   A380 landing gear IMA implementation

diagram, communication with sensors and effectors is through the ARINC 664-P7 network to I/O resources in the RDCs (although as will be seen shortly, the Airbus A380 architecture routes I/O directly to sensors and effectors). Consolidation of the lane outputs is performed in the RDCs.

5.6.2   Implementation Example: Airbus A380

Airbus mapped the A380 landing gear functions as shown in Figure 5.24. Logically the arrangement is a dual command:monitor architecture. It is implemented in four identical CPIOMs (Type G).

As discussed earlier in Section 5.4.4, the A380 CPIOM structure implements a common core processor board with an ARINC 664-P7 end system mezzanine card and a set of I/O cards in a single 3MCU line-replaceable unit. Different domain optimisations of the I/O modules result in seven variants of the CPIOM, each with a common processor/ARINC 664-P7 mezzanine card. The A380 CPIOM I/O cards interface directly with function (domain)-specific sensors and effectors. In some domains (including the landing gear domain) the I/O capability is augmented with local domain-specific RDCs that aggregate sensor/effector signals close to their source in order to minimise aircraft wiring, improve signal quality and reduce EM interference. Some domain-specific RDCs (e.g. the fuel system RDCs) include local processing to reduce network bandwidth requirements and IMA core processor load.

The Airbus A380 IMA architecture does not implement the concept of an IMA rack. However, we can consider the architecture as two 'open' racks. In Figure 5.24 the left-hand

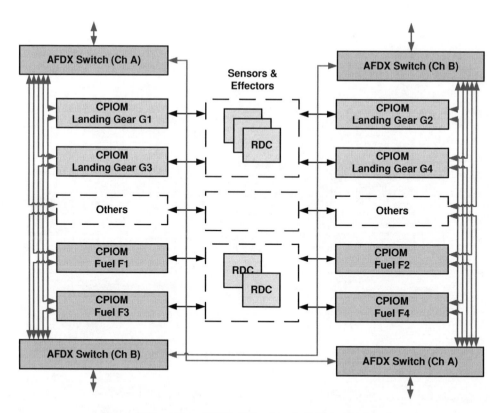

Figure 5.25   Airbus A380 utilities domain IMA implementation

pair of CPIOMs implement the command:monitor function of Side 1, while the right-hand
pair of CPIOMs implement the command:monitor function of Side 2. The reader should note
that Airbus tend use the terms Side 1 and Side 2 rather than Lane A or Lane B when describing
a duplex architecture. All four CPIOMs are identical; the architectural function implemented
in each CPIOM is determined by the application software loaded into it. The command pair in
Lane A and Lane B execute the measurement and management applications of the command
channel; the monitor pair in Lane A and Lane B execute the monitoring and built-in-test
applications. Consolidation of the command and monitoring lanes is implemented in the I/O
sections of the CPIOMs.

The Airbus A380 IMA implements a common modular avionics platform resource architecture (hardware, I/O, network and RTOS); however, the IMA platform resources are strongly
assigned on a system-by-system domain basis more closely aligned to a more traditional federated architecture approach. A strong physical segregation exists between domains. This is
illustrated in Figure 5.25 which shows the high-level mapping of the fuel system (CPIOM
Type F) and the landing gear system (CPIOM Type G) functions within the utilities domain.
Each function (fuels and landing gear) have their own set of CPIOMs and dedicated RDCs.
The four fuel system CPIOMs are identical to each other but are different to the four landing
gear CPIOMs by virtue of a different set of interface cards optimised to the fuel system sensor
and effector signals requirements.

5.6.3 Implementation Example: Boeing 787

The Boeing approach on the B787 implements the IMA architecture as two common core processing racks, each with a set of common core processing modules, but with no I/O resources. Communication with domain I/O sensors and effectors is via the ARINC 664-P7 network to a set of RDCs distributed throughout the aircraft. Mapping a duplex command:monitor architecture similar to, say, the A380 landing gear functions would result in an implementation typically like that shown in Figure 5.26.

As with the Airbus A380, two processor cards are assigned to implement the landing gear function in each rack. The pair of cards in the left-hand rack implement Lane A, and the pair of cards in the right-hand rack implement Lane B. The COM card in each lane implements the management and measurement functions of the command channel; the MON card in each lane implements the monitor function. However, communication with the landing gear sensors and effectiveness is now by network shared resources, rather than directly via dedicated I/O interfaces as in the Airbus A380.

The key difference between these two architectures is the way in which they communicate with I/O. We should not see these as two rival implementations, but rather as incremental steps in the evolutionary process of IMA architecture development. As with any architecture, their strengths and weaknesses should be assessed on the basis of:

- effective use of resources;
- fault containment (segregation and partitioning);
- certification (and incremental certification);
- maintainability;
- adaptability to change;
- obsolescence management;
- risk.

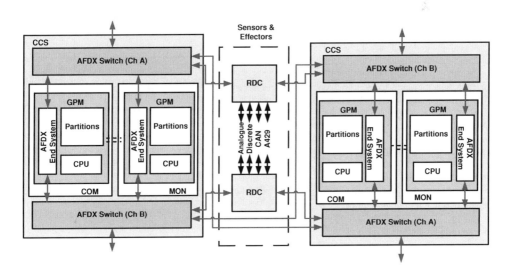

Figure 5.26 Boeing 787 possible COM:MON implementation

In both, the objective is to provide a flexible, safe, cost-effective avionics systems architecture which is supportable throughout the life of the air-vehicle.

The fundamental difference between a federated architecture and an IMA architecture is that in a federated architecture the target computer and the application software are packaged and certificated as a single physical system. In an IMA architecture the application software is integrated with the platform to form a virtual system. The platform hardware, the configuration table that expresses the share of the platform resources assigned to the application and the real-time operating system become the equivalent of the target computer for the purposes of RTCA-DO-178.

Multiple systems can be hosted on a single processor module in partitions segregated from each other by the real-time operating system. The number of systems that can be accommodated is constrained only by the module capacity and their guaranteed share of the computing resources. The examples above show related applications hosted on a single processor, but *in extremis* this need not be so. However, if otherwise unrelated systems share a common resource, then the systems architect must account for the common mode failure of that single resource on all the functions hosted upon it.

5.7   Partitioning

Partitioning is perhaps the toughest challenge in an IMA architecture. Its equivalent, segregation, in a federated architecture is easier to demonstrate by physical separation of system resources. However, in an IMA architecture multiple applications are hosted on shared platform resources, and this places significant demands on the real-time operating system to ensure that one application cannot interfere with or disturb another.

Robust partitioning demands that no application function in one partition can under any circumstances:

- impact the temporal processing resource allocation of another partition;
- access the memory assigned to another partition;
- adversely affect the I/O resources of another partition.

To achieve this, the real-time operating system must build a protective barrier around each application, as shown in Figure 5.27.

The features to be provided by the real-time operating system are defined in ARINC 653. Its key properties are summarised here as:

- The application software is independent of the hardware platform.
- The operating system provides a partitioned environment.
  - Periodic fixed-scheduling bounds computing resource demand.
  - Hardware memory management protects RAM/ROM utilisation.
  - I/O resource allocation is bounded in hardware configuration tables.
- Partitioning facilitates incremental certification.

In principle, applications can be written in different high-order languages and then compiled to run on the target platform. Applications communicate with the platform hardware resources

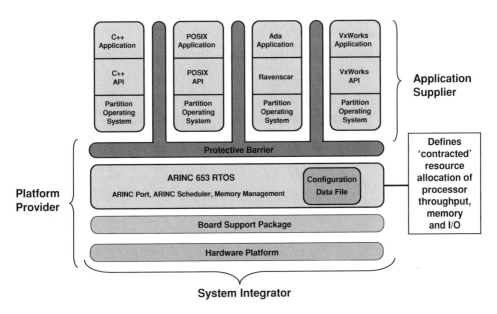

Figure 5.27   ARINC 653 real-time operating system (RTOS)

through open-system Application Programmable Interfaces (APIs) which are independent of the detailed hardware implementation. The board support package defines the implementation of a specific physical hardware platform. The architecture is future-proofed. Modifications, technology refresh, obsolescence, upgrades and alternatives can be accommodated with commensurate changes to the board support package, transparent to the application.

Functions are allocated the platform resources they require to perform their task. These resource allocations are mechanised through configuration tables. They allow individual functions to change without collateral impact on other functions.

5.8   IMA Fault Tolerance

5.8.1   *Fault Tolerance Principles*

Fault tolerance is designed into the IMA virtual system by allocating sufficient redundant elements to support the required system availability and integrity in accordance with the principles described in Sections 4.6–4.8; that is, multiple copies for processing elements and multiple channels (independent fault zones) for I/O elements. As with a federated architecture, it is important to ensure there are no common-mode failures that can simultaneously impact the independence of multiple lanes in a multi-redundant architecture.

The platform is an asynchronous system. Scheduling of processing tasks must be independent of one another in order to prevent individual function behaviour propagating through the system and disrupting the operation of other hosted functions. This means that, for instance, in a duplex architecture the computational output from one channel may be time-skewed with respect to the other channel and therefore slightly different. This has to be accounted for in

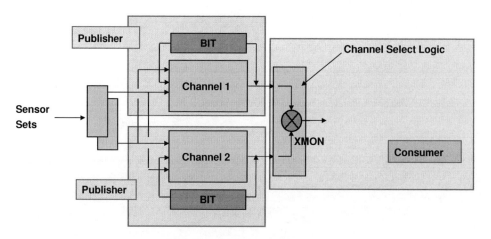

Figure 5.28 Data integrity

any cross-channel monitoring process. As discussed in Chapter 2, a digital system is a set of discrete time samples. In a complex asynchronous system there may be many sampling stages which, when taken in series, can create significant jitter in the calculations. Applications must account for this environment in their design, avoiding implementations that depend on synchronous behaviour and ensuring that appropriate bandwidth and latency requirements are built into the contracted platform performance in order for the application to meet its functional operational requirements.

5.8.2   Data Integrity

A 'fail-silent' philosophy is fundamental to systems integrity and fault containment. Each computational element continuously monitors its own health using the built-in-test techniques discussed in Section 4.6.1. An element should cease to output (publish) valid data in response to any uncorrectable faults. The user (consumer) should select source data on the basis of its view of received data validity. The data integrity arrangement for a duplex architecture is shown in Figure 5.28 and is analysed in Section 4.7.

Since failure to publish will result in data becoming stale (frozen), a user needs to maintain an awareness of data freshness and should correlate data from multiple sources wherever possible. It is the responsibility of each user to determine the appropriate response to input data failure.

Platform integrity is achieved by redundancy, fault isolation mechanisms, 'virtual system' partitioning and fault containment zones.

- Processor modules are designed to be 'fail-passive'. Undetected fault sequences are extremely improbable.
- The A664-P7 network is designed to detect and contain all faults. Any data corrupted during transportation do not propagate to the host system. Receive-time tagging allows the consumer to determine data freshness and decide when a source has become 'invalid'.

The network reports status to hosted functions as header information associated with each message.

- Remote data concentrators employ physical separation between I/O resources and temporal partitioning mechanisms for computational resources to segregate functionality into multiple independent fault zones (IFZs). Each IFZ is generally an order of magnitude better than sensors/effectors. Downstream consolidation of multiple lower-integrity devices achieves high integrity. RDCs continuously monitor their health and return a status message which allows hosted functions to monitor I/O and sensor/effector status.

5.8.3   Platform Health Management

The IMA platform has a built-in health manager function that aggregates and reports platform health. The platform health manager provides the following services:

- platform fault isolation;
- monitors connectivity of all platform and non-platform LRUs;
- communicates configuration status;
- controls 'return to service' tests;
- determines when data-load is possible.

Each platform component periodically reports health to the platform health manager (using internal BIT techniques). The platform health data are correlated to provide definitive fault isolation for operational maintenance action. They also indicates dispatch-readiness based on the minimum equipment list (MEL) and indicate any platform failures that would prevent safe dispatch.

5.9   Network Definition

Once the logical architecture has been formulated, it is then possible to develop the data exchanges between the logical system elements. The communication links implement the data exchanges of the logical system, but will move data between the physical resources of the IMA platform. For this reason they can be thought of as 'virtual links' (VLs) in a similar way that the logical architecture is a virtual architecture implemented on the physical resources of the IMA platform.

The task of structuring the communications framework is a top-down process. The first step is to identify the primary data exchanges between publishers and subscribers. A top-level data flow diagram is shown in Figure 5.29.

It is useful to commence from the publisher's perspective, identifying the class of information for a communication link and the publisher's view of possible subscribers or users of that data. This then needs to be correlated with the subscriber's view of the data needed in order to perform their intended avionics system function. Further devolution explores the detail of each message, the data format, resolution, update rates, size, and so on, to develop an interface control document (ICD) (see Section 6.9 and specifically Figure 6.18). This process is much like defining ARINC 429 single-source, multi-drop data links between LRUs in a traditional federated avionics architecture.

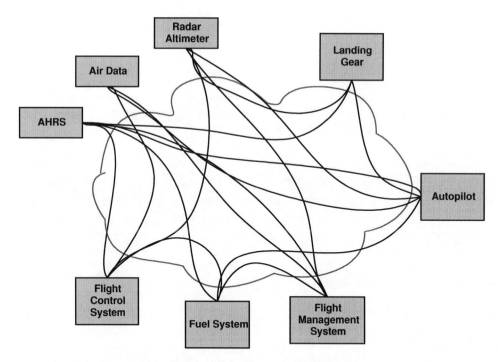

Figure 5.29   Network data flow diagram example

Now the detailed design can commence to map the virtual links onto the physical ARINC 664-P7 network resources, and separate aperiodic from periodic messages, define message formats, group periodic data according to refresh rate (bandwidth), and agree the detailed network performance guarantees of bandwidth, latency and jitter with hosted function providers.

5.10   Certification

5.10.1   IMA Certification Philosophy

The certification philosophy for avionics systems implemented in an IMA architecture is significantly different to the traditional certification approach for a federated avionics system. In a federated avionics architecture, each system is certificated by the system provider as a functional systems entity, comprising hardware and software. In an IMA architecture the platform and the system functions hosted on the platform are certificated separately. IMA is certificated on an incremental basis: first the platform, then the hosted avionics systems functions, finally building to the certification of the whole. Each constituent element, component or module, is incrementally accepted for certification and requires its own set of design assurance and certification evidence. The whole system certification argument is based upon the arguments for the certification of the individual components [4].

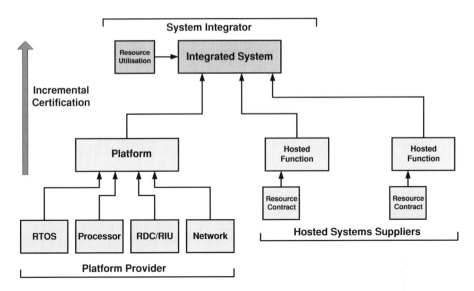

Figure 5.30   Incremental certification process

The certification process for incremental acceptance can be characterised as having the following distinct phases:

Phase 1: Platform acceptance (Section 5.10.2).

Phase 2: Hosted function acceptance (Section 5.10.3).

Phase 3: Integrated system acceptance.

Phase 4: Aircraft integration and certification.

Phase 5: Incremental change of platform and/or hosted functions.

The hierarchical structure for incremental certification is shown in Figure 5.30.

5.10.2   Platform Acceptance

The platform is independently accepted for certification prior to acceptance of hosted functions. This acceptance itself can be devolved into the sub-elements of the platform resources:

- processing module(s);
- real-time operating system;
- ARINC 664-P7 network; and
- I/O elements (RDCs/RIUs).

The acceptance of the real-time operating system follows the established practice for certification of airborne software in accordance with RTCA-DO-178. The acceptance of platform hardware follows the established practice for certification of airborne hardware elements including complex hardware in accordance with RTCA-DO-254. Both should include sufficient data to characterise fully the performance, interface, physical and environmental aspects (for hardware elements) of each constituent component, including usage limitations and guidance in order to facilitate future upgrades, manage obsolescence and maximise the opportunity for reuse.

A portion of the platform certification evidence will be required by hosted function developers to support acceptance of their system function when installed onto the platform.

5.10.3   Hosted Function Acceptance

Hosted functions should be certificated according to their logical systems architecture, not their physical architecture. This is achieved and justified through a 'contract-based' approach for guaranteeing platform resources. The hosted functions argument for certification acceptance is based on a guaranteed set of resources and hence guaranteed performance of the platform. These guarantees describe the logical architecture of the hosted function. It would be possible that the physical architecture could be altered later without impacting certification of the hosted function, provided the platform guarantees are maintained.

The performance of the hosted function relies upon the provision of the guaranteed resources by the platform. The full benefits of IMA and the hardware/software independence it affords can only be realised if the guaranteed resources are fully characterised and assured by the platform. This is accomplished by a 'contract-based' approach. The contract establishes a formal means of communication between the platform and the hosted function developer to describe the resources and services required and to be provided by the platform in terms of processor time allocation and throughput, memory allocation, I/O resources and network communication bandwidth, latency and jitter. The parametric details of the contract are held in loadable configuration tables on the platform processor module(s) and in the network switch(es), and are rigorously enforced at run-time. The 'contract' forms a vital piece of certification evidence which the hosted function developer references in his certification arguments and safety analyses.

The roles and responsibilities for successful integration are split between the platform provider, the hosted function developer and the system integrator. The system integrator guarantees the platform resource allocations upon which the hosted function supplier bases certification evidence for the performance of his system. The platform provider verifies the claim guarantees from all function suppliers and aggregates this evidence to the certification authorities to demonstrate that the total platform resources are appropriately deployed with adequate spare capacity.

5.10.4   Cost of Change

Managing the cost of change is critically important due to the high degree of integration between the platform and hosted functions. The 'contract-based' modular incremental certification approach is an essential strategy to minimise the cost of change. Without this approach, changes in a hosted function could cause dramatic and unnecessary costs for suppliers sharing platform resources but not otherwise involved in the change. Similarly, without this approach, updates to platform components due to technology refresh or obsolescence could cause significant impact on hosted functions. Therefore it is essential that the platform implementation including the real-time operating system must support:

- robust partitioning;
- separation/independence of software applications from hardware;
- network partitioning;
- I/O segregation and fault containment.

This can only be achieved if the certification arguments at every step in the document hierarchy support the 'contract-based' incremental certification process and are based on the hosted functions' logical system architectures, not their physical architectures.

5.10.5   Configuration Management

It is evident, then, that changes will be made to the platform and its hosted functions in an incremental manner. Maintaining tight control over the system configuration is essential. A platform configuration manager must provide the following services:

- Validate software and hardware compatibility.
- Provide consistency check for hosted function software.
- Communicate configuration status for maintenance purposes.

The primary means to confirm the platform configuration is a platform manifest. The manifest is validated against the hardware installation and then broadcast to all functional units to confirm that their loadable hosted software applications are compatible. The configuration manager should also provide a consistency check for hosted function software, ensuring that each redundant copy is consistent across the platform.

5.11   IMA Standards

The following documents or standards are useful points of reference in developing and certificated integrated modular avionics systems:

- **RTCA-DO-297/EUROCAE ED-124** – Integrated Modular Avionics (IMA) Design Guidance and Certification Considerations [5].
- **FAA TSO-C153** – Integrated Modular Avionics (IMA) Hardware Elements [6].
- **FAA Advisory Circular AC 20-170** – Guidance for Integrated Modular Avionics (IMA) that implement TSO-C153 [7].
- **FAA Advisory Circular AC20-148** – Reusable software components [8].
- **ARINC 653** – Avionics Application Software Interface (API) Standard [9].

RTCA-DO-297 provides the following useful definitions of terms used in integrated modular avionics:

- **Application software**: The part of an application implemented in software.
- **Approval**: Act of giving formal acknowledgement of compliance with regulations.

- **Component**: A self-contained, configuration controlled hardware or software item.
- **Module**: A component or collection of components. It may be software, hardware or both.
- **Partition**: An allocation of resources whose properties are guaranteed and protected from adverse interaction from outside the partition.
- **Partitioning**: An architectural technique to provide separation and independence of functions.

RTCA-DO-297 defines IMA as a shared set of flexible, reusable and interoperable hardware and software resources designed and verified to a defined set of safety and performance requirements that form an integrated platform and services to host applications performing aircraft functions.

- It provides guidance for IMA developers, integrators and airworthiness approval authorities.
- It sets out objectives, processes and activities related to the development and integration of IMA modules, applications and systems; provides guidance on the incremental accumulation of design assurance data; and describes the process to install and approve an IMA system on an aviation product.
- It supports incremental acceptance (certification) of individual items (hardware and software) and facilitates the reduction of follow-on certification efforts without compromising system safety.

RTCA-DO-297 recognises that IMA requirement analysis is a significant challenge. It recommends that a long-term through-life perspective should be taken. A strategy for obsolescence management and for incremental certification needs to be established. It recommends that installed IMA resources should be less than 50% utilised at initial entry into service in terms of:

- processing power (throughput)
- input/output
- memory capacity

(plus growth slots for spare modules).

RTCA-DO-297 has six interrelated chapters, as shown in Figure 5.31:

1. Introduction.
2. IMA concepts, system, hardware and software characteristics.
3. IMA general development and integration guidelines.
4. Guidelines for acceptance and relationship to approval of the installed system.
5. Describes the integral (umbrella) development processes.
6. Guidelines for continued airworthiness of IMA systems.

Annex A to RTCA-DO-297 has six tasks that describe the incremental acceptance certification process:

Task 1: Module acceptance.

Task 2: Application software or hardware acceptance.

Task 3: IMA systems acceptance.

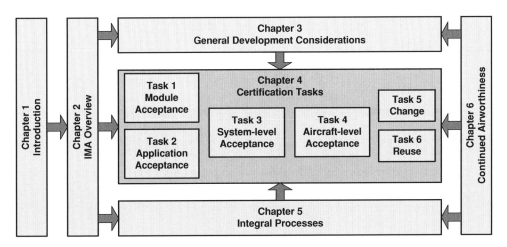

Figure 5.31 RTCA-DO-297 structure

Task 4: Aircraft integration, including validation and verification.

Task 5: Change of modules or applications.

Task 6: Reuse of modules or applications.

Each task has multiple objectives, set out in tables in Appendix A to Annex A.

RTCA-DO-297 also identifies life-cycle data to be submitted to the Certification Authority.

References

[1] Tully, T. (1998) *Fuel Systems as an Aircraft Utility.* Civil Aerospace Technologies Conference, FITEC 98, London.

[2] Walker, R. and Watkins, C. (2006) The genesis platform, in *Digital Avionics Handbook*, 2nd edn (ed. C.R. Spitzer), CRC Press, Boca Raton, FL.

[3] Moir, I. and Seabridge, A.G. (2008) *Aircraft Systems, Mechanical, Electrical and Avionics Subsystem Integration*, 3rd edn. John Wiley & Sons, Ltd., Chichester.

[4] Spitzer, C.R. (2006) RTCA-DO-297 Integrated Modular Avionics (IMA) design guidance and certification requirements, in *Digital Avionics Handbook*, 2nd edn (ed. C.R. Spitzer), CRC Press, Boca Raton, FL.

[5] RTCA-DO-297/EUROCAE ED-124 Integrated Modular Avionics (IMA) Design Guidance and Certification Considerations. Available at http://boutique.eurocae.net/catalog/product_info.php?products_id=276&osCsid=zucmsoeoqlujf [accessed April 2013].

[6] FAA TSO-C153 Integrated Modular Avionics (IMA) Hardware Elements. Available at www.airweb.faa.gov/Regulatory_and_Guidance_Library/rgTSO.nsf/0/C07999442E496ADA86256DC700717DB5?OpenDocument [accessed April 2013].

[7] FAA Advisory Circular AC 20-170 Guidance for Integrated Modular Avionics (IMA) that implement TSO-C153. Available at www.faa.gov/documentLibrary/media/Advisory_Circular/AC%2020-170.pdf [accessed April 2013].

[8] FAA Advisory Circular AC20-148 Reusable software components. Available at www.faa.gov/regulations_policies/advisory_circulars/index.cfm/go/document.information/documentID/22207 [accessed April 2013].

[9] ARINC 653 Avionics Application Software Interface (API) Standard. Available at www.arinc.com/cf/store/index.cfm

6

Systems Development

6.1 Introduction

As the reader will judge from the contents of this book, aircraft systems are becoming more complex and more sophisticated for a number of technology and performance reasons. In addition, avionics technology, while bringing the benefits of improved control by using digital computing and greatly increased integration by the adoption of digital data buses, is also bringing greater levels of complexity to the development process. The disciplines of avionics system development – including hardware and software integration – are now being applied to virtually every aircraft system.

The increasing level of system sophistication and the increased interrelations of systems is also making the development process more difficult. The ability to capture all of the system requirements and interdependencies between systems has to be established at an early stage in the programme. Safety and integrity analyses have to be undertaken to ensure that the system meets the necessary safety goals, and a variety of other trades studies and analytical activities have to be carried out.

These increasing strictures need to be met by following a set of rules, and this chapter gives a brief overview of the regulations, development processes and analyses that are employed in the development of modern aircraft systems, particularly where avionics technology is also extensively employed.

The design of an aircraft system is subject to many rigours and has to satisfy a multitude of requirements derived from specifications and regulations. There are also many development processes to be embraced. The purpose of this chapter is not to document these *ad naseum*, but to give the reader an appreciation of the depth and breadth of the issues that need to be addressed.

6.1.1 Systems Design

There are references to some of the better known specifications and requirements, but also attempts to act as a tutorial in terms of giving examples of how the various design techniques and methods are applied. As the complexity and increasing interrelationship and reliance

---

Civil Avionics Systems, Second Edition. Ian Moir, Allan Seabridge and Malcolm Jukes.
© 2013 John Wiley & Sons, Ltd. Published 2013 by John Wiley & Sons, Ltd.

between aircraft systems has progressed, it has become necessary to provide a framework of documents for the designer of complex aircraft systems.

6.1.2   Development Processes

An overview of a typical life-cycle for an aircraft or equipment is given and the various activities described. Furthermore, some of the programme management disciplines are briefly visited.

6.2   System Design Guidelines

Key documentation is applied under the auspices of a number of agencies. A list of the major documents that apply is included in the reference section of this chapter, and it is not intended to dwell on chapter and verse of those documents in this brief overview. There are several agencies who provide material in the form of regulations, advisory information and design guidelines whereby aircraft and system designers may satisfy mandatory requirements.

6.2.1   Key Agencies and Documentation

These agencies include:

- Society of Automotive Engineers (SAE):

  ARP 4754 [1]

  ARP 4761 [2]

- Federal Aviation Authority (FAA):

  AC 25.1309-1A [3]

- European Aviation Safety Authority (EASA)

  AMC 25.1309 [4]

- Air Transport Association (ATA):

  ATA-100 [5]

- Radio Technical Committee Association (RTCA):

  RTCA DO-160G [6]

  RTCA DO-178C [7]

  RTCA DO-254 [8]

This list should not be regarded as exhaustive but merely indicative of the range of documentation that exists. In addition, it is worth mentioning that for military aircraft both US MIL standards and UK Def standards apply.

6.2.2   Design Guidelines and Certification Techniques

References [1] and [2] offer a useful starting point in understanding the interrelationships of the design and development process:

- ARP 4761: Safety Assessment Process Guidelines and Methods [1]
- ARP 4754: System Development Processes [2]
- Def Stan 00-970 for military aircraft.

Figure 6.1 shows the interplay between the major techniques and processes associated with the design and development process. This figure, which is presented as part of the SAE ARP 4761 document, gives an overview of the interplay between some of the major references/working documents that apply to the design and development process. In summary:

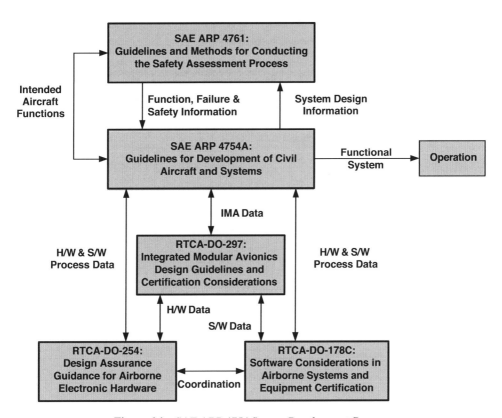

Figure 6.1   SAE ARP 4754 System Development Process

- SAE ARP 4754 is a set of system development processes.
- SAE ARP 4761 represents a safety assessment process including a set of tools and techniques.
- RTCA DO-178B offers guidance for the design and certification of software.
- RTCA DO-254 offers guidance for hardware design and development.
- More recently, RTCA DO-297 has been introduced to address IMA design and certification.

Serious students or potential users of this process are advised to procure an updated set of these documents from the appropriate authorities. The main subject headings are summarised below.

6.2.3   Guidelines for Development of Civil Aircraft and Systems – SAE ARP 4754A

Key areas are:

- system development;
- certification process and coordination;
- requirements determination and assignment of development assurance level;
- safety assessment process;
- validation of requirements;
- implementation verification;
- configuration management;
- process assurance;
- modified aircraft.

6.2.4   Guidelines and Methods for Conducting the Safety Assessment – SAE ARP 4761

Major elements include:

- functional hazard assessment (FHA);
- preliminary system safety assessment (PSSA);
- system safety assessment (SSA);
- fault tree analysis (FTA);
- dependence diagrams;
- Markov analysis (MA);
- failure modes and effects analysis (FMEA);
- failure modes and effects summary (FMES);
- zonal safety analysis (ZSA);
- particular risks analysis (PRA);
- common mode analysis (CMA);
- contiguous safety assessment process example.

6.2.5 Software Considerations – RTCA DO-178B

Major development areas include:

- introduction;
- system software development;
- software life-cycle;
- software planning process;
- software development process;
- software verification process;
- software configuration management process;
- software quality assurance process;
- certification liaison process;
- overview of aircraft and engine certification;
- software life-cycle data;
- additional considerations.

6.2.6 Hardware Development – RTCA DO-254

Key points in this document map almost directly on to the RTCA DO-178B software equivalent. They are:

- introduction;
- system aspects of hardware design assurance;
- hardware design life-cycle;
- hardware planning process;
- hardware design process;
- validation and verification;
- configuration management;
- process assurance;
- certification liaison;
- design life-cycle data;
- additional considerations.

6.2.7 Integrated Modular Avionics – RTCA DO-297

Key points are:

- overview;
- IMA concepts, systems, hardware and software characteristics;
- IMA-specific development and integration guidelines;
- guidelines for acceptance and relationship to approval of the installed system;
- describes the integral (umbrella) development processes;
- guidelines for continued airworthiness of IMA systems;
- Annex A: Objectives to be satisfied (based on the six tasks in Chapter 4). For further details refer to Chapter 5.11, IMA Standards.

Table 6.1   Equivalence of US and European specifications

Specification topic	US specification	European EUROCAE specification
Systems development process	SAE ARP 4754	ED-79
Safety assessment process guidelines and methods	SAE ARP 4761	
Software design	RTCA DO-178B	ED-12B
Hardware design	RTCA DO-254	ED-80
Environmental test	RTCA DO-160C	ED-14B
Integrated modular avionics	RTCA DO-297	ED-124

[1]European Organisation for Civil Aviation Equipment.

6.2.8   Equivalence of US and European Specifications

To the uninitiated, the abundance of US and European development specifications can be most confusing. To alleviate this problem, Table 6.1 establishes equivalence between these specifications.

6.3   Interrelationship of Design Processes

There are a number of interrelated processes that are applied during the safety assessment of an aircraft system. These are:

- functional hazard assessment (FHA);
- preliminary system safety assessment (PSSA);
- system safety assessment (SSA);
- common cause analysis (CCA).

Figure 6.2 shows a portrayal of the interplay between these processes as the system design evolves and eventually the system achieves certification.

The diagram effectively splits into two sections: design activities on the left and analysis on the right. As the system evolves from aircraft-level requirements, aircraft functions are evolved. These lead in turn to system architectures, which in turn define software requirements and the eventual system implementation. At corresponding stages of the design, various analyses are conducted that examine the design in the light of the mandated and recommended practices. At every stage the analyses and the design interact in an evolutionary manner as the design converges upon a solution which is both cost-effective and able to be certificated, meeting all the safety requirements.

6.3.1   Functional Hazard Assessment (FHA)

A FHA is carried out at both aircraft and system levels; one flows down from the other. The FHA identifies potential system failures and identifies the effects of these failures. Failures are

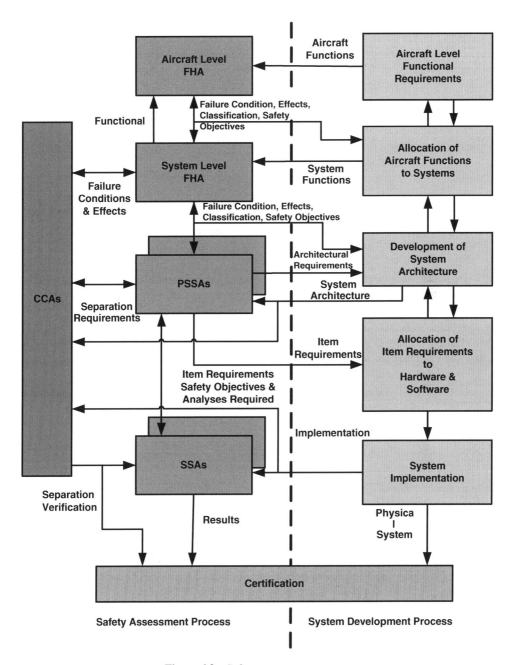

Figure 6.2 Safety assessment process

Table 6.2   Overview of failure classification and safety objectives

Failure condition classification	Development assurance level	Safety objectives	Safety objectives quantitative requirement (probability per flight hour)
Catastrophic	A	Required	$<1 \times 10^{-9}$
Hazardous/severe major	B	May be required	$<1 \times 10^{-7}$
Major	C	May be required	$<1 \times 10^{-5}$
Minor	D	Not required	$<1 \times 10^{-3}$
No safety effect	E	Not required	None

tabulated and classified according to the effects that failure may cause and the safety objectives assigned, according to the criteria briefly listed in Table 6.2.

FHA identifies the data in first two columns of the table: the failure condition classification and the development assurance level. These allow the safety objectives to be assigned for that particular condition and a quantitative probability requirement assigned.

For a failure that is identified as having a catastrophic effect, the highest assurance level A will be assigned. The system designer will be required to implement fail-safe features in his design and will have to demonstrate by appropriate analysis that the design is capable of meeting the criterion of a probability of failure of less than 1×10^{-9} per flight hour. In other words, the particular failure should occur less than once per 1000 million flight hours. This category of failure is assigned to systems such as flight controls or structure, where a failure could lead to the loss of the aircraft, death or serious injury to crew, passengers or the overflown population. The vast majority of aircraft systems are categorised at much lower levels where little or no safety concerns apply.

A more user-friendly definition quoted in words as used by the Civil Airworthiness Authority (CAA) may be:

- Catastrophic: less than 1×10^{-9} – extremely improbable.
- Hazardous: between 1×10^{-9} and 1×10^{-7} – extremely remote.
- Major: between 1×10^{-7} and 1×10^{-5} – remote.
- Minor: between 1×10^{-5} and 1×10^{-3} – reasonably probable.
- Greater than 1×10^{-3} – frequent.

6.3.2   Preliminary System Safety Assessment (PSSA)

The PSSA examines the failure conditions established by the FHA(s) on a system-by-system basis and demonstrates how the system design will meet the specified requirements. Various techniques such as fault tree analysis (FTA) and Markov diagrams may be used to identify how the design counters the effects of various failures and may point toward design strategies that need to be incorporated into the system design to meet the safety requirements. Typical analyses may include the identification of system redundancy requirements, for example, how many channels, what control strategies could be employed, and the need for dissimilarity of control (e.g. dissimilar hardware and/or dissimilar software implementation). The PSSA is therefore part of an iterative process that scrutinises the system design and assists the system designers in ascribing and meeting risk budgets across one or a number of systems.

Increasingly, given the high degree of integration and interrelationship between major aircraft systems, this is likely to be a multisystem, multidisciplinary exercise coordinating the input of many systems specialists.

6.3.3   System Safety Assessment (SSA)

The SSA is a systematic and comprehensive evaluation of the system design using similar techniques to those employed during the PSSA activities. However, whereas the PSSA identifies the requirements, the SSA is intended to verify that the proposed design does in fact meet the specified requirements as identified during the FHA and PSSA analyses conducted previously. As may be seen in Figure 6.2, the SSA occurs at the point in the design cycle where the system implementation is concluded or finalised and prior to system certification.

6.3.4   Common Cause Analysis (CCA)

The CCA begins concurrently with the system FHA and is interactive with this activity and subsequent PSSA and SSA analyses. The purpose of the CCA is – as the name suggests – to identify common cause or common mode failures in the proposed design, and to assist in directing the designers toward strategies that will obviate the possibility of such failures. Such common cause failures may include:

- failure to correctly identify the requirement;
- failure to correctly specify the system;
- hardware design errors;
- component failures;
- software design and implementation errors;
- software tool deficiencies;
- maintenance errors;
- operational errors.

The CCA is therefore intended to scrutinise a far wider range of issues than the system hardware or software process. Rather, it is meant to embrace the whole process of developing, certifying, operating and maintaining the system throughout the life-cycle.

6.4   Requirements Capture and Analysis

It can be seen from the foregoing that requirements capture is a key activity in identifying and quantifying all the necessary strands of information that contribute to a complete and coherent system design. It is vital that the system requirements are established in a complete, consistent and clear manner, or significant problems will be experienced during development and into the service life. There are a number of ways in which the requirements capture may be addressed. Two main methods are commonly used:

- top-down approach;
- bottom-up approach.

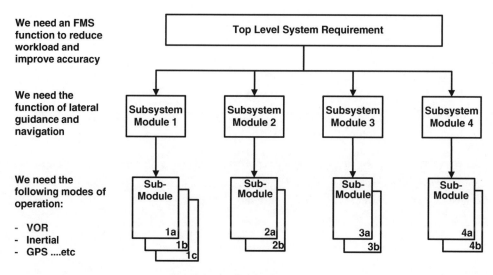

Figure 6.3   Top-down approach

6.4.1   Top-Down Approach

The top-down approach is shown in Figure 6.3. This represents a classical way of tackling the requirements capture by decomposing the system requirements into smaller functional modules. These functional modules may be further decomposed into functional submodules. This approach tends to be suited to the decomposition of large software tasks where overall requirements may flow down into smaller functional software tasks or modules. This would apply to a task where the hardware boundaries are fairly well understood or inferred by the overall system requirement. An example might be the definition of the requirements for an avionics system such as a flight management system (FMS). In such a basic system requirement – the need to improve the navigation function is well understood – the means by which the various navigation modes are implemented are well defined: the inertial navigation system (INS), Global Positioning System (GPS), and VHF omni-ranging (VOR).

6.4.2   Bottom-Up Approach

The bottom-up method is shown in Figure 6.4. This approach is best applied to systems where some of the lower-level functions may be well understood and documented and represented by a number of submodules. An example of this is adding a new functional element to an established system design. However, the process of integrating these modules into a higher subset presents difficulties as the interaction between the individual subsystems is not fully understood. In this case, building up the top-level requirements from the bottom may well enable the requirements to be fully captured. An example of this type might be the integration of aircraft systems into an integrated utilities management system. In this case the individual requirements of the fuel system, hydraulic system, environmental control system, and so on, may be well understood. However, the interrelationships between the candidate systems and the implications of adopting

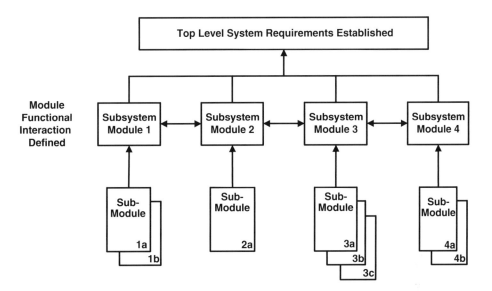

Figure 6.4   Bottom-up approach

integration may better understood and documented by working bottom-up. In fact, most development projects may use a combination of both of these approaches to best capture the requirements.

6.4.3   Requirements Capture Example

The example given in Figure 6.5 shows a functional mapping process that identifies the elements or threads necessary to implement a fuel jettison function. Two main functional subsystems are involved: the fuel quantity measurement function and the fuel management function. Note that this technique merely identifies the data threads that are necessary to perform the system function. No attempt is made at this stage to ascribe particular functions to particular hardware or software entities. Neither is any attempt made to determine whether signals are hardwired or whether they may be transmitted as multiplexed data as part of an aircraft system data bus network. The system requirements from the flight crew perspective are:

- The flight crew need to jettison excess fuel in an emergency situation in order that the aircraft may land under the maximum landing weight.
- The flight crew wish to be able to jettison down to a pre-selected fuel quantity.
- The crew wish to be given indications that fuel jettison is under way.

The information threads associated with the flight crew requirements are shown in the upper centre portion of the diagram. It may be seen that although the system requirements

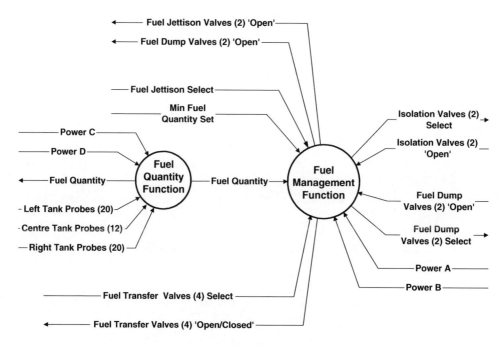

Figure 6.5   System requirements capture example

are relatively simple when stated from the flight crew viewpoint, many other subsystem information strands have to be considered to achieve a cogent system design:

- **Fuel quantity function:** The fuel quantity function measures the aircraft fuel quantity by sensing fuel in the aircraft fuel tanks; in the example given in Figure 6.5, a total of 52 probes are required to sense the fuel held in three tanks. The fuel quantity calculations measure the amount of fuel which the aircraft has onboard taking account of fuel density and temperature. It is usual in this system, as in many others, to have dual power supply inputs to the fuel quantity function to assure availability in the event of an aircraft electrical system busbar failure. Finally, when the calculations have been completed they are passed to the flight deck where the aircraft fuel quantity is available for display to the flight crew. Fuel quantity is also relayed to the fuel management function so that in the event of fuel jettison, the amount of fuel onboard may be compared with the pre-set jettison value. The fuel quantity function interfaces to:
  - the fuel quantity system measurement probes and sensors
  - the flight deck multifunction displays
  - the fuel management system
  - the aircraft electrical system.
- **Fuel management function:** The fuel management function accepts information regarding the aircraft fuel state from the fuel quantity function. The flight crew inputs a 'Fuel Jettison Select' command and the minimum fuel quantity that the crew wishes to have available at the end of fuel jettison. The fuel management function accepts flight crew commands for

the fuel transfer valves (4), fuel dump (jettison) valves (2), and fuel isolation valves (2). It also provides 'Open'/'Closed' status information on the fuel system valves to the flight crew. As before, two separate power inputs are received from the aircraft electrical system. The fuel management function interfaces with:

- the fuel system valves
- the flight deck displays multifunction displays and overhead panel
- the fuel quantity function
- the aircraft electrical system.

This example shows how a relatively simple function interfaces to various aircraft systems and underlines some of the difficulties that exist in correctly capturing system requirements in a modern integrated aircraft system. This simple example illustrates how this system – like most systems – can operate in different configurations or 'modes' of operation and that the desired behaviour of a system needs to be specified for all eventualities.

6.5 Development Processes

6.5.1 The Product Life-Cycle

Figure 6.6 shows a typical aircraft product life-cycle from concept through to disposal at the end of the product's useful life.

Individual products or equipment may vary from this model, but it is a sufficiently good portrayal to illustrate the role of systems engineering and the equipment life-cycle. The major phases of this model are:

- concept phase (Section 6.5.2)
- definition phase (Section 6.5.3)
- design phase (Section 6.5.4)
- build phase (Section 6.5.5)
- test phase (Section 6.5.6)
- operate phase (Section 6.5.7)
- refurbish or retire (Section 6.5.8).

This model closely approximates to the Downey cycle used by the UK Ministry of Defence, for the competitive procurement of defence systems The model is equally applicable for systems

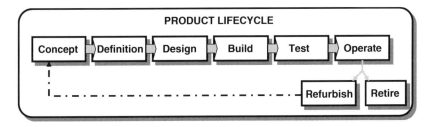

Figure 6.6 Typical aircraft product life-cycle

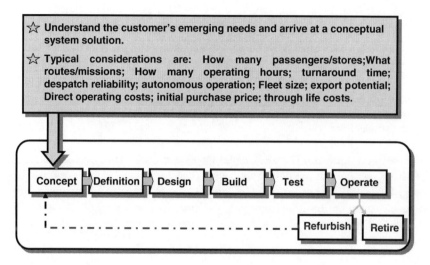

Figure 6.7   Concept phase

used in commercial aircraft as it is for military applications. It is used to describe the role of systems engineering in each phase of the product life-cycle.

6.5.2   Concept Phase

The concept phase is about understanding the customer's emerging needs and arriving at a conceptual model of a solution to address those needs. The customer continuously assesses his current assets and determines their effectiveness to meet future requirements. The need for a new military system can arise from a change in the local or world political scene that requires a change in defence policy. The need for a new commercial system may be driven by changing national and global travel patterns resulting from business or leisure traveller demands.

The customer's requirement will be made available to industry so that solutions can be developed specifically for that purpose, or can be adapted from the current research and development (R&D) base. This is an ideal opportunity for industry to discuss and understand the requirements to the mutual benefit of the customer and his industrial suppliers, to understand the implications of providing a fully compliant solution or one which is aggressive and sympathetic to marketplace requirements (see Figure 6.7).

Typical considerations at this phase are:

- Establishing and understanding the primary role and functions of the required system.
- Establishing and understanding desired performance and market drivers such as:
    - range
    - endurance
    - routes or missions
    - technology baseline
    - operational roles

- number of passengers
- mass, number and type of weapons
- availability and dispatch reliability
- fleet size to perform the role or satisfy the routes
- purchase budget available
- operating or through-life costs
- commonality or model range
- market size and export potential
- customer preference.

This phase is focused on establishing confidence that the requirement can be met within acceptable commercial or technological risk. The establishment of a baseline of mature technologies may be first solicited by means of a Request for Information (RFI). This process allows possible vendors to establish their technical and other capabilities and represents an opportunity for the platform integrator to assess and quantify the relative strengths of competing vendors and also to capture mature technology of which he was previously unaware for the benefit of the programme.

It is in this phase that important decisions are made that determine whether a type is modified or extended (e.g. Boeing 737-400, -500, -600) or whether a new type emerges (e.g. Airbus A380).

6.5.3 Definition Phase

As shown in Figure 6.8, the customer will usually consolidate all the information gathered during the concept phase to firm up his requirement. A common feature used more frequently by platform integrators is to establish engineering joint concept teams to establish the major

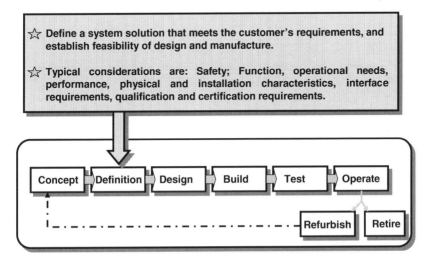

Figure 6.8 Definition phase

system requirements. These teams are sometimes called integrated product teams (IPTs). They may develop a cardinal points specification, and perhaps even undertake a preliminary system or baseline design against which all vendors might bid. This results in the issue of a specification or a Request for Proposal (RFP). This allows industry to develop their concepts into a firm definition, to evaluate the technical, technological and commercial risks, and to examine the feasibility of completing the design and moving to a series production solution. Typical considerations at this stage are:

- Developing the concept into a firm definition of a solution.
- Developing system architectures and system configurations.
- Re-evaluating the supplier base to establish what equipment, components and materials are available or may be needed to support the emerging design.
- Defining physical and installation characteristics and interface requirements.
- Developing operational and initial safety models of the individual systems.
- Quantifying key systems performance such as:
  - mass;
  - volume;
  - growth capability;
  - range/endurance.

The output from this phase is usually in the form of feasibility study reports, performance estimates, sets of mathematical models of individual system's behaviour and an operational performance model. This may be complemented by breadboard or experimental models or laboratory or technology demonstrators. A preliminary design is also likely to examine installation issues with mock-ups in three-dimensional computer model form (using tools such as CATIA) which replaces in the main the former wooden and metal models.

6.5.4   Design Phase

If the outcome of the definition phase is successful and a decision is made to proceed further, then industry embarks upon the design phase within the programme constraints as described later in the chapter. Design takes the definition phase architectures and schemes and refines them to a standard that can be manufactured (see Figure 6.9).

Detailed design of the airframe ensures that the structure is aerodynamically sound, is of appropriate strength, and is able to carry the crew, passengers, fuel and systems that are required to turn it into a useful product. As part of the detailed design, attention needs to be paid to the mandated rules and regulations that apply to the design of an aircraft or airborne equipment. Three-dimensional solid modelling tools are used to produce the design drawings, in a format that can be used to drive machine tools to manufacture parts for assembly.

Systems are developed beyond the block diagram architectural drawings into detailed wiring diagrams. Suppliers of bought-in equipment are selected and they become an inherent part of the process to start to design equipment that can be used in the aircraft or systems. Indeed, in order to achieve a fully certifiable design for many of the complex and integrated systems found on aircraft today, an integrated design team comprising platform integrators and supplier(s) is

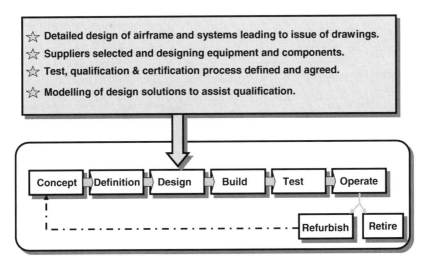

Figure 6.9 Design phase

essential. In fact, many of these processes are iterative, extending into and even beyond the build and test phases.

6.5.5 Build Phase

The aircraft is manufactured to the drawings and data issued by design, as shown in Figure 6.10. During the early stages of the programme, a delivery schedule would have been established. Some long lead-time items – those which take a long time to build – may need to be ordered well ahead of aircraft build commencing. In the case of some of the more complex, software-driven

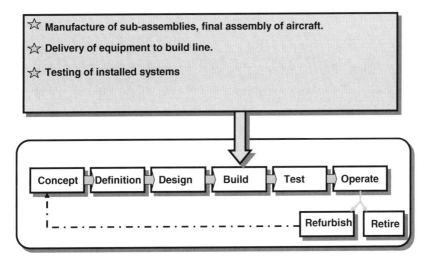

Figure 6.10 Build phase

equipment, design will be overlapping well into the test phase. This is usually accommodated by a phased equipment delivery embracing the following:

- **Electrical models** – equipment electrically equivalent to the final product but not physically representative.
- **Red label hardware** – equipment that is physically representative but not cleared for flight.
- **Black label hardware** – equipment that is physically representative and is cleared for flight either by virtue of the flight-worthy testing carried out and/or the software load incorporated.

These standards are usually accompanied by a staged software release which enables a software load that progressively becomes more representative of the final functionality.

6.5.6 Test Phase

The aircraft and its components are subjected to a rigorous test programme to verify its fitness for purpose, as shown in Figure 6.11. This phase includes testing and integration of equipment, components, sub-assemblies, and eventually the complete aircraft. Functional testing of equipment and systems on the ground and flight trials verify that the performance and the operation of the equipment is as specified. Conclusion of the test programme and the associated design, analysis and documentation process leads to certification of the aircraft or equipment

In the event of a new aircraft, responsibility for the certification of the aircraft lies with the aircraft manufacturer. However, where an equipment is to be improved or modified in the civil arena, equipment suppliers or other agencies can certify the equipment by means of the Supplementary Type Certificate (STC) in a process defined by the certification authorities.

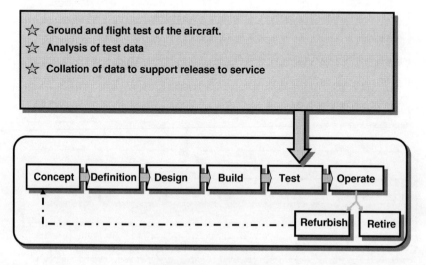

Figure 6.11   Test phase

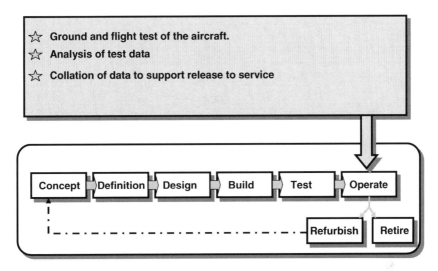

Figure 6.12   Operate phase

This permits discrete equipment – for example, a more accurate fuel quantity gauging in a particular aircraft model – to be changed without affecting other equipment.

6.5.7   Operate Phase

During this phase, shown in Figure 6.12, the customer is operating the aircraft on a routine basis. Its performance will be monitored by means of a formal defect-reporting process, so that any defects or faults that arise are analysed by the manufacturer. It is possible to attribute causes to faults such as random component failures, operator mishandling, or design errors. The aircraft manufacturer and their suppliers are expected to participate in the attribution and rectification of problems arising during aircraft operations, as determined by the contract.

6.5.8   Disposal or Refurbish Phase

At the end of the useful or predicted life of the aircraft, decisions have to be made about its future, as depicted in Figure 6.13. The end of life may be determined by unacceptably high operating costs, unacceptable environmental considerations – noise, pollution, and so on – or by predicted failure of mechanical or structural components determined by the supplier's test rigs. If it is not possible to continue to operate the aircraft, then it may be disposed of – sold for scrap, or an alternative use, such as aircraft enthusiast or gate guardian.

If the aircraft still has some residual and commercially viable life, then it may be refurbished. This latter activity is often known as a mid-life update, or even a conversion to a different role, for example, a VC10 passenger aircraft converted to in-flight re-fuelling use, as has happened with the Royal Air Force. Similarly, in the civil arena, many former passenger aircraft are being converted to a cargo role.

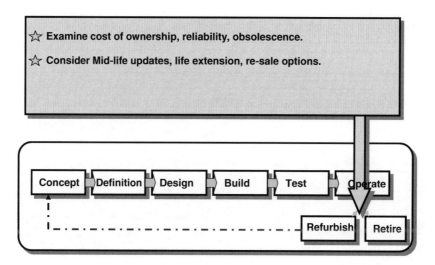

Figure 6.13   Disposal or refurbishment

6.6   Development Programme

6.6.1   Typical Development Programme

So far, the processes, methods and techniques used during aircraft system design have been described. However, these need to be applied and controlled within an overall programme management framework. Figure 6.14 shows the major milestones associated with the aircraft systems development process. It is assumed – as is the case for the majority of aircraft systems developed today – that the system has electronics associated with the control function, and that the electronics has a software development content.

The main characteristic of the development is the bifurcation of hardware and software development processes into two separate paths, although it can be seen that there is considerable interaction between the two. The key steps in the avionics development programme that are primarily designed to contain and mitigate against risk are:

- **System requirements review (SRR)**: The SRR is the first top-level, multidisciplinary review of the perceived system requirements. It is effectively a sanity check upon what the system is required to achieve: a top-level overview of requirements and review against the original objectives. Successful attainment of this milestone leads to a preliminary system design, leading in turn to the parallel development of the hardware and software requirements analysis, albeit with significant coordination between the two.
- **System design review (SDR)**: The hardware SDR immediately follows the preliminary design phase and will encompass a top-level review of the system hardware characteristics such that preliminary design may proceed with confidence. Key hardware characteristics will be reviewed at this stage to ensure that there are no major mismatches between the system requirements and what the hardware is capable of supporting.
- **Software specification review (SSR)**: The SSR is essentially a similar process to the hardware SDR, but applying to the software when a better appreciation of the software

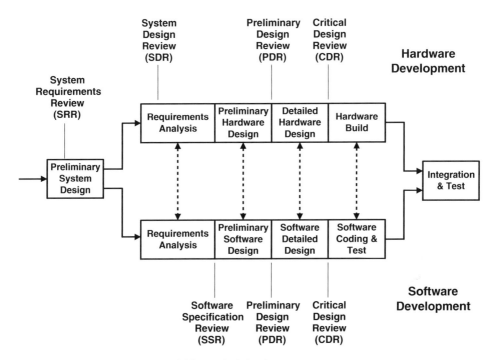

Figure 6.14 Typical development programme

requirements has become apparent, and possibly embracing any limitations such as throughput, timing or memory which the adopted hardware solution may impose. Both the SDR and SSR allow the preliminary design to be developed up to the preliminary design review (PDR).

- **Preliminary design review (PDR)**: The preliminary design review process is the first detailed review of the initial design (both hardware and software) versus the derived requirements. This is usually the last review before committing major design resource to the detailed design process. This stage in the design process is the last before major commitment to providing the necessary programme resources and investment.
- **Critical design review (CDR)**: By the time of the CDR, major effort will have been committed to the programme design effort. The CDR offers the possibility of identifying final design flaws, or more likely, trading the risks of one implementation path versus another. The CDR represents the last opportunity to review and alter the direction of the design before very large commitments and final design decisions are taken. Major changes in system design – both hardware and software – after the CDR will be very costly in terms of cost and schedule loss, to the total detriment of the programme.

The final stages following CDR will realise the hardware build and software coding and test processes, which bring together the hardware and software into the eventual product realisation. Even following system validation and equipment certification, it is unusual for there to be a period free of modification either at this stage or later in service when airlines may demand equipment changes for performance, reliability or maintainability reasons.

Table 6.3 Software levels

Level	Contribution to resultant failure condition for the aircraft
A	Catastrophic
B	Hazardous
C	Major
D	Minor
E	No effect

6.6.2 'V' Diagram

The rigours of software development are particularly strict and are dictated by Reference [6]. For obvious reasons, the level of criticality of software used in avionics systems determines the rigour applied to the development process. The levels of software are generally defined according to the operational impact of a software failure. These are defined in Table 6.3.

The software development process is generally of the form shown in Figure 6.15, which shows the development activities evolving down the left of the diagram and the verification activities down the right. This shows how the activities eventually converge in the software validation test at the foot of the diagram that is the confluence of hardware and software design and development activities. Down the centre of the diagram the various development software stages are shown. It can be seen that there is considerable interaction between all the processes that represent the validation of the requirements and of the hardware and software design at each level. Any problems or issues discovered during the software validation tests are fed back up the chain, if necessary back into the top level. Therefore any minor deviations are reflected back into all the requirements stages to maintain a consistent documentation set and a consistent hardware and software design.

Whereas the earlier stages of software development and test might be hosted in a synthetic software environment, it is increasingly important as testing proceeds to undertake testing in a representative hardware environment. This testing represents the culmination of functional testing for the LRU or equipment, short of flight test.

6.7 Extended Operations Requirements

6.7.1 ETOPS Requirements

Extended Twin OPerationS (ETOPS), allowing for twin-engine aircraft to operate up to 180 minutes from a suitable diversion airfield, have been in force for a number of years. A number of different definitions have emerged in recent years with subtle variations between the following aviation authorities, with reference to Figure 6.16.

- International Civil Aviation Organisation (ICAO).
- European Aviation Safety Authority (EASA).
- Federal Aviation Authority (FAA).

ETOPS operations are addressed by Reference [9]. The increasing long-range capability of some new transport aircraft and the ability to navigate accurately over the entire surface of

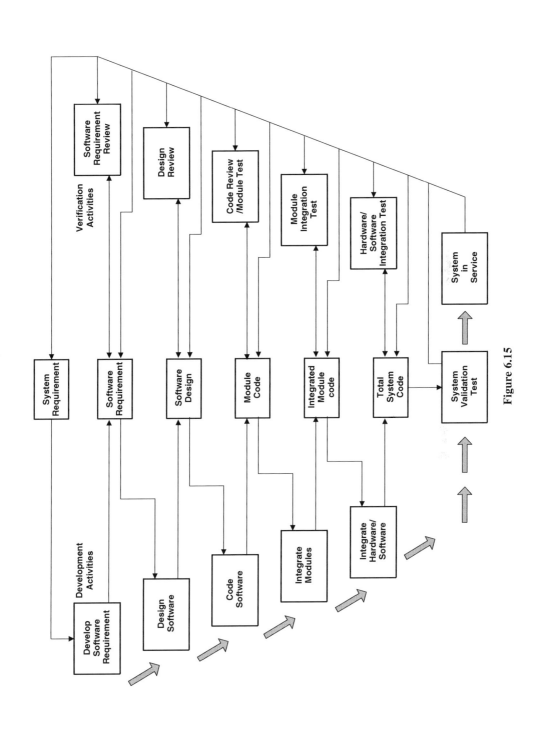

Figure 6.15

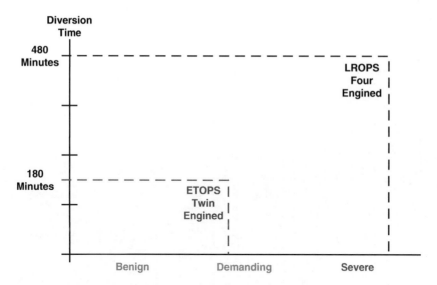

Figure 6.16   Extended operations definitions

the world is leading to the opening up of long-range transpolar oceanic and wilderness routes which require the drafting of new regulations. Perhaps the easiest approach to understand is that taken by the FAA. Further information on the present situation may be found in Reference [10] which represents a recent Airbus view. Present regulations permit operation between 78°N and 60°S.

Key issues and definitions that appear to be emerging may be summarised as follows:

- ETOPS begins at 60 minutes diversion time.
- ETOPS up to 180 minutes will be controlled using the existing ETOPS Advisory Circular – Reference [9]
- Long-Range OPerationS (LROPS) will apply to all two-, three-, and four-engined aircraft beyond 180 minutes (realistically only 180 minutes will apply to twin-engined aircraft with the possible exception of 207 minutes (180 minutes plus 15%).
- In some cases in Pacific areas the FAA may permit 240 minutes ETOPS flights, subject to specific criteria.

It is outside the scope of this book to delve in detail into all the subtle details relating to the alternate approaches outlined above; rather, readers should conduct their own research depending upon the conditions that prevail.

6.7.2   Equipment Requirements

Key constraints apply to the flight planning process and the aircraft communications fit. Particular attention also needs to be given to the following aircraft systems issues:

- Engine reliability – in-flight shut down (IFSD) of one or more engines.
- Conditioned air supply/bleed air system integrity and reliability.
- Cargo hold fire suppression capacity and integrity of smoke/fire alerting systems.
- Brake accumulator and emergency braking system capacity/integrity.
- Adequate capacity of time-dependent functions.
- Pressurisation/oxygen system integrity/reliability/capacity.
- Integrity/reliability/capacity of backup systems (electrical, hydraulic).
- Fuel system integrity and fuel accessibility. Fuel consumption with engine failure and other system failures.
- Fuel quantity and fuel-used indications and alerts.

6.8   ARINC Specifications and Design Rigour

The list of ARINC specifications produced by Air Radio Inc. (ARINC) in the US at the time of writing was as follows:

- ARINC 400 Series – 60 items.
- ARINC 500 Series – 38 items.
- ARINC 600 Series – 179 items.
- ARINC 700 Series – 148 items.
- ARINC 800 Series – 50 items.
- ARINC 900 Series – none yet agreed.

The overall interrelationship of the ARINC specifications is portrayed in Figure 6.17.

6.8.1   ARINC 400 Series

The 400 Series ARINC specifications and reports provide a design foundation for equipment specified per the ARINC 500 and 700 Series. They include guidelines for installation, wiring, data buses, databases, and general guidance. Examples: A404A line-replaceable unit (LRU) racking; A429 data bus.

6.8.2   ARINC 500 Series

The 500 Series ARINC characteristics define analogue avionics equipment still used widely on the B727, DC-9, and DC-10, as well as on early models of B737, B747 and A300 aircraft. Examples: A562 Terrain Avoidance Warning System (TAWS); A566A-9 very high frequency (VHF) transceiver.

6.8.3   ARINC 600 Series

The 600 Series ARINC specifications and reports define enabling technologies that provide a design foundation for equipment specified by the ARINC 700 Series of digital avionics

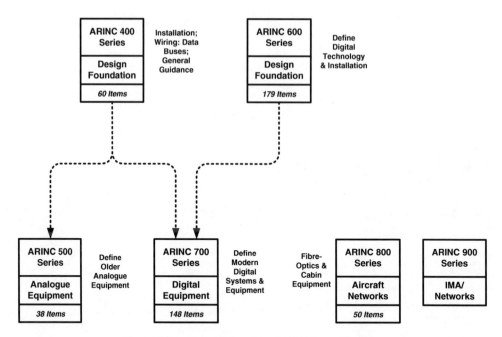

Figure 6.17   Interrelationship of ARINC specifications

systems. Among the topics covered by the Specifications are data link protocols. Examples: A600-16 equipment interfaces; A629 B777 data bus.

6.8.4   ARINC 700 Series

The 700 Series ARINC characteristics define digital systems and equipment installed on current-model production aircraft. They include definitions of form, fit, function, and interface for line-replaceable units (LRUs) in a federated architecture. Examples: A708-6 weather radar; A738-8 Air Data and Inertial Reference System (ADIRS).

6.8.5   ARINC 800 Series

The 800 Series ARINC specifications and reports define enabling technologies supporting the networked aircraft environment. Among the topics covered in this series is fibre-optics used in high-speed data buses. Examples: A802-1 fibre-optic cables; A808-1 cabin distribution system.

6.8.6   ARINC 900 Series

The 900 Series ARINC characteristics define avionics systems in an integrated modular and/or networked architecture. They will include detailed functional and interface definitions.

6.9   Interface Control

6.9.1   Introduction

The increasing use of modern commercial off-the-shelf (COTS) technology offers greater system functionality and performance, but with an accompanying increase in complexity. All of the aircraft system interfaces have to been defined and bounded. Every aircraft system interacts with others, so the aircraft is truly a system-of-systems.

To define and control the system interfaces, an interface control document (ICD) is used that defines all of the electrical interfaces. To illustrate the point a notional system is portrayed consisting of four units – refer to Figure 6.18. The example chosen could be typical of a fuel gauging and management system on a large transport aircraft. Units A and B represent fuel gauging and management computers, whereas units C and D represent remote data concentrators interfacing directly with the components in the aircraft fuel tanks, such as fuel measurement probes, temperature and densitometer sensors, and fuel pumps and valves.

6.9.2   Interface Control Document

In the simplified system example portrayed, there are four major types of system interfaces:

- **Aircraft level data buses**: For an aircraft such as the Boeing 787 or the Airbus A380/A350, the aircraft-level data buses will be implemented in a form of ARINC 664 data bus. These aircraft-level data buses will be transmitting data typically at 100 Mbps using COTS technology which originated within the telecommunications industry using either conventional twisted wire pairs or fibre-optic technology.
- **System internal data buses**: Within the system, digital data will need to be exchanged between units at a lower bandwidth. A COTS data bus called CANbus, developed by Bosch and originally intended for automobile automatic braking systems (ABS), is commonly used, albeit in a deterministic and ruggedised form. Typical data rates are of the order of 1 Mbps.
- **Internal system input/output signals between the system units**: Hardwired signals between units to compare data synchronise the operation of the system computers and establish which computer/channel is in control.
- **System internal interfaces** in which the remote data concentrators interface with the components within the aircraft fuel tanks. Key issues include the provision of intrinsically safe electrical interfaces where the power allowed into the tank to feed a fuel probe is constrained to miniscule levels to ensure that the system is safe.

6.9.3   Aircraft-Level Data-Bus Data

Aircraft-level data will include top-level aircraft data which is useful to the flight crew in terms of operating the aircraft. In many cases this will be data needed for other aircraft systems, or which needs to be displayed to the flight crew. In the example given, typical data presented by the system could include the total Fuel On-Board (FOB) or the contents of individual fuel tanks. Warning and advisory data would also be provided.

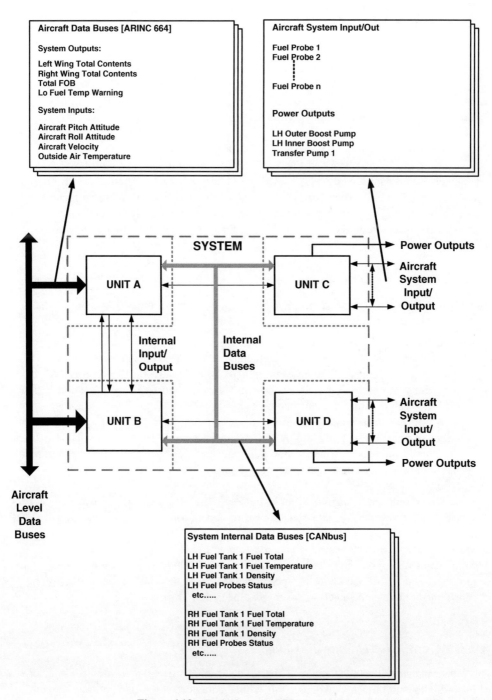

Figure 6.18 Typical system ICD components

Inputs to the fuel system would include aircraft attitude information in order that the fuel contents may be accurately calculated, aircraft velocity, and outside air temperature (OAT), which is of particular interest in understanding cold fuel issues during prolonged cold soak at altitude.

6.9.4 System Internal Data-Bus Data

Many systems use an internal system data bus to exchange system-specific data. In the example shown, fuel probe and other sensor data are exchanged. System built-in-test (BIT) and other sensor health-related data will be included. The system will also have in-built monitors to ensure that hazardous events do not occur, or ensure that the flight crew are kept fully informed of any failures and advised of what remedial action to take.

6.9.5 Internal System Input/Output Data

There will be a number of hardwired interfaces between the system units that are not appropriate to be passed over the internal data buses. These will be the physical input/output signals that connect the control elements to the real world of sensors, valves, pumps and actuators.

6.9.6 Fuel Component Interfaces

The ICD defines and controls all of the parameters defining:

- electrical signal parameters;
- wire sizes and types;
- bonding and screening;
- termination and matching;
- data resolution and accuracy;
- data rates and refresh rates;
- power levels;
- EMI categorisation.

References

[1] *SAE ARP 4761 Guidelines and Methods for Conducting the Safety Assessment Process on Civil Airborne Systems.*
[2] *SAE ARP 4754 Guidelines for Development of Civil Aircraft and Systems.*
[3] Advisory Circular, AC 25.1309-1A, 21 June 1988, System Design and Analysis.
[4] *AMC 25.1309 System Design and Analysis, Advisory Material Joint.*
[5] *ATA-100 ATA Specification for Manufacturer's Technical Data.*
[6] *RTCA D0-160G Environmental Conditions and Test Procedures for Airborne Equipment.*
[7] *RTCA DO-178C Software Considerations in Airborne Systems and Equipment Certification.*
[8] *RTCA DO-254 Design Assurance Guidance for Airborne Electronic Hardware.*
[9] Federal Aviation Authority (2008) *Advisory Circular AC 120-42B, Extended Operations (ETOPS and Polar),* Federal Aviation Authority.
[10] Fortunato, E. (2011) Overview of new ETOPS/LROPS/EDTO rules, ICAO Fourth Meeting of the Regional Aviation Safety Group – Pan American (RASG-PA), Miami, 19–21 October 2011.

7

Electrical Systems

7.1 Electrical Systems Overview

7.1.1 Introduction

Electrical systems have made significant advances over the years as aircraft have become more dependent upon electrically powered services. A typical electrical power system of the 1940s and 1950s was the twin 28 VDC system. This system was used a great deal on twin-engined aircraft, each engine powered a 28 VDC generator which could employ load-sharing with its counterpart if required. One or two DC batteries were also fitted, and an inverter was provided to supply 115 VAC and then 26 VAC to the flight instruments.

The advent of the V-bombers in the UK, and similar developments in the US, altered this situation radically due to the much greater power requirements imposed by the wide range of systems in these aircraft. In the UK, one aircraft, the Vickers Valiant, incorporated electrically actuated landing gear. These aircraft were fitted with four 115 VAC generators, one being driven by each engine. To provide the advantages of no-break power these generators were paralleled, which increased the amount of control and protection circuitry. The V-bombers had to run high-power military mission loads such as radar and electronic warfare jamming equipment.

Over the years the generated power levels on civil aircraft have increased considerably, and that trend is continuing, even accelerating today. Figure 7.1 shows this trend from 1950 onwards. To illustrate the point the Boeing 747-100 (200 kVA), Boeing 747-400 (360 kVA), Airbus A380 (600 kVA) and Boeing 787 (1000 kVA or 1 MVA) (more-electric) aircraft are highlighted.

Most aircraft utilisation equipment in use today is accustomed to a constant frequency supply of 115 VAC. In order to generate constant-frequency 115 VAC at 400 Hz, a constant speed drive or CSD is required to negate the aircraft engine speed variation typically over an approximately 2:1 speed range (full power speed to flight idle speed). These are complex hydromechanical devices that by their very nature are not highly reliable without significant maintenance penalties. Therefore the introduction of constant-frequency AC generation systems was not without accompanying reliability problems, particularly on fighter aircraft where engine throttle settings are changed very frequently throughout the mission. In modern aircraft the generator and CSD are a combined unit called an integrated drive generator (IDG).

Civil Avionics Systems, Second Edition. Ian Moir, Allan Seabridge and Malcolm Jukes.
© 2013 John Wiley & Sons, Ltd. Published 2013 by John Wiley & Sons, Ltd.

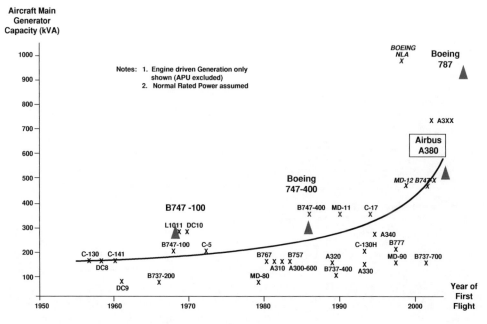

Figure 7.1   Trends in aircraft power levels

7.1.2   Wider Development Trends

The total picture in terms of system development of aircraft electrical systems is more involved than this, as portrayed in Figure 7.2 which depicts the use of differing generation and load management systems from 1950.

During the 1950s the main types of electrical power generation were narrow and wideband variable frequency (VF) alternating current (AC) or twin 28 VDC direct current (DC) systems that satisfied the modest electrical power requirements of the time. Some aircraft that were derivatives of World War II (WWII) aircraft used 130 VDC, as did the first of the V Bombers – the Vickers Valiant and the Vulcan B1. These generation methods were used for the following reasons:

- Twin 28 VDC systems were used on twin-engine aircraft where the necessary power levels could be attained by using DC generation – sensibly limited to around 12 kW per channel for technical reasons. Examples in the military were the English Electric Canberra and the Gloster Javelin; civil aircraft used a similar arrangement, but in the frugal post-WWII years civil aircraft were not in great abundance.
- VF systems were used on aircraft that required higher levels of power of around 40 kVA per channel either for aircraft systems use or for aircraft anti-icing purposes. Particular examples were the Vickers Viscount and Vickers Vanguard. A proportion of the VF AC power was converted to 28 VDC using transformer rectifier units (TRUs).

The use of DC power offered another advantage – it is a relatively simple task to parallel two DC machines to provide a source of no-break power (NBP). In older architectures, supplies

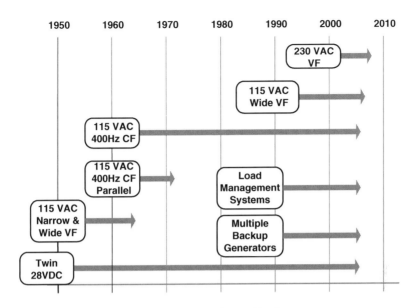

Figure 7.2   Trends in aircraft generation and load management systems

of 115 VAC power were required to supply the aircraft flight instruments. To convert this power from 28 VDC to 115 VAC, rotary machines called inverters were provided. In more modern systems inverters are still used, although the implementation today is solid state – static inverters.

As power generation technology developed, it became possible in the late 1950s/early 1960s to generate constant frequency (CF) 400 Hz 115 VAC, as described above. Also schemes were developed that allowed two AC machines to be run in parallel to provide a no-break power (NBP) capability. In the UK this technique was adapted for the later V Bombers and the Vickers VC10. The V Bombers have long retired from active service, but the VC-10 is still being used today by the Royal Air Force as an air-to-air refuelling tanker.

The continued increase in the power levels and in the complexity of aircraft electrical loads resulted in the adoption of load management systems and multiple backup generators. The further reliance of high integrity systems upon the aircraft electric systems was also a significant driver.

The advances in high-power solid-state switching technology, together with enhancements in the necessary control electronics, have made variable speed/constant frequency (VSCF) systems a viable proposition in the last decade. The VSCF system removed the unreliable CSD portion, the variable frequency or frequency wild power from the AC generator being converted to 400 Hz constant frequency 115 VAC power by means of a solid-state VSCF converter. VSCF systems are now becoming more commonplace; the F-18 fighter uses such a system and some versions of the Boeing 737-500 did use such a system. In addition, the Boeing 777 airliner utilises a VSCF system for backup primary AC power generation which uses IDGs.

In US military circles, great emphasis is been placed by the US Air Force and the US Army on the development of 270 VDC systems. In these systems, high-power generators derive 270 VDC power, some of which is then converted into 115 VAC 400 Hz or 28 VDC required to power legacy AC and DC equipments and loads. Certain specific airborne platforms in the US military inventory with huge power requirements have adopted the use of 230 VAC power generation. The use of higher voltages than 115 VAC for civil aircraft has not been generally favoured in the past; however, the Boeing 787 has adopted this technology as it has implemented 'more-electric aircraft' technology. A historical overview of the trends of electrical power development [1] and the more-electric aircraft (MEA) issues are described more fully in a companion volume [2].

7.1.3 Typical Civil Electrical System

Most civil aircraft today are twin-engine aircraft and the basic electrical system architecture varies little between different aircraft types. This simple architecture based upon 115 VAC CF power generation is shown in Figure 7.3.

The system comprises the following main elements, which form two major electrical power channels as follows:

- A generator located on the accessory gearbox of each engine; in most cases this will be a 3-phase, 115 VAC, CF 400 Hz generator using an IDG. These generators feed power respectively to the left and right main AC buses. The power levels will vary from aircraft to aircraft, but 90 kVA is a typical rated value for a primary generator in this configuration.

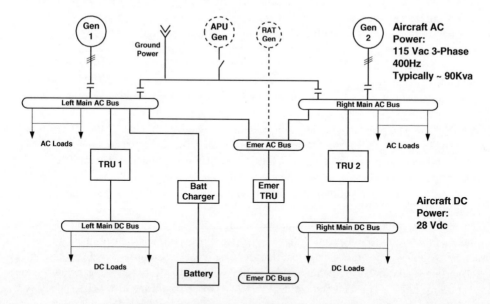

Figure 7.3 Typical electrical system architecture

- An APU generator is used to provide a source of 3-phase, 115 VAC, 400 Hz electrical power independent of the main engines. This may be used on the ground for servicing purposes, but it can also be used in the air on certain installations for use in failure conditions and to support Extended Twin OPerationS (ETOPS).
- There is also provision for the connection of an external 115 VAC power source – either a stand-alone ground power cart or an electric–electric supply connected to the National Grid.
- For emergency use – a ram air turbine is included to provide a short-term source of electrical power following multiple system failures.
- A number of power conversion units are also provided: TRUs and battery chargers that supply a number of 28 VDC buses and maintain battery charge.

In this typical electrical system arrangement, most high-power loads are fed from the AC buses. Numerically, most electrical loads are small in terms of power drawn and in general are supplied by the DC buses.

A more detailed explanation of this architecture is described later in the chapter using the Airbus A320 system as an example.

7.2 Electrical Power Generation

7.2.1 Generator Control Function

The primary functions associated with power generation control are:

- **DC generation systems:**
  - voltage regulation
  - parallel operation
  - protection functions.
- **AC generation systems:**
  - voltage regulation
  - parallel operation
  - supervisory functions.

Prior to discussing the particular attributes of AC and DC generation systems, it is worth examining the issues associated with generator control, particularly the practical aspects. This is necessary as key parts of the generation system are not collocated, but for practical reasons are positioned at different locations within the airframe. Figure 7.4 shows an AC generator control loop. As can be seen, the generator is located on the engine; 3-phase power feeders, sensing and control wiring have to be routed through the aircraft, and the control, power contactors and point of regulation (PoR) are located in the aircraft electrical equipment (EE) bay.

In the example, the generator output voltage and current are monitored by the generator control unit (GCU) which controls the excitation to ensure that the voltage derived at the generator terminals meets the specified voltage at the PoR. This has to allow for voltage drops in the feeders. When the power delivered to the PoR meets all the necessary voltage and protection conditions described below, then power is enabled to the aircraft primary AC busbars.

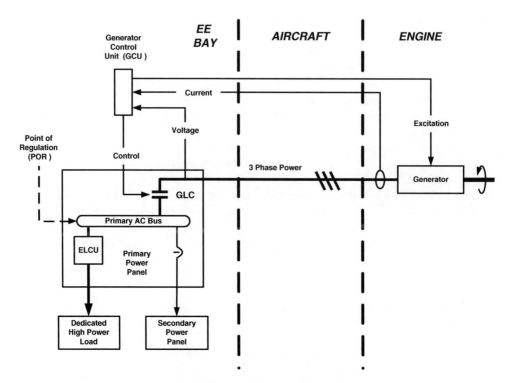

Figure 7.4   AC generator control loop

7.2.2   DC System Generation Control

Voltage Regulation

DC generation is by means of shunt-wound self-exciting machines. The voltage regulation needs to be in accordance with the standard used to specify aircraft power generation systems. The standards specify the voltage at the point of regulation and the nature of the acceptable voltage drops throughout the aircraft distribution, protection and wiring system. Typical standards for commercial or military use are presented in the electrical power quality section below. DC systems are limited to around 400 A or 12 kW per channel maximum for two reasons:

- The size of conductors and switchgear to carry the necessary current becomes prohibitive.
- The brush wear on brushed DC generators becomes excessive with resulting maintenance costs if these levels are exceeded.

Parallel Operation

In multi-engined aircraft each engine will be driving its own generator, and in this situation it is desirable that 'no-break' or uninterrupted power is provided in cases of engine or generator

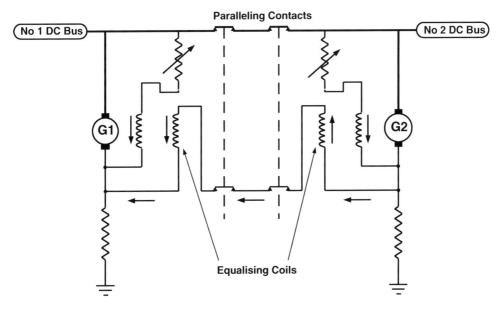

Figure 7.5   DC generator parallel operation

failure. A number of sensitive aircraft instruments and navigation devices that comprise some of the electrical loads may be disturbed and may need to be restarted or re-initialised following a power interruption. In order to satisfy this requirement, generators are paralleled to carry an equal proportion of the electrical load between them. Individual generators are controlled by means of voltage regulators that automatically compensate for variations. In the case of parallel generator operation, there is a need to interlink the voltage regulators such that any unequal loading of the generators can be adjusted by means of corresponding alterations in field current. This paralleling feature is more often known as an equalising circuit and thereby provides 'no-break' power in the event of a major system failure. A simplified diagram showing the main elements of DC parallel operation being supplied by generator 1 (G1) and generator 2 (G2) is shown in Figure 7.5.

Protection Functions

The primary conditions for which protection needs to be considered in a DC system are as follows:

- **Reverse current**. In a DC system it is evident that the current should flow from the generator to the busbars and distribution systems. In a fault situation it is possible for current to flow in the reverse direction and the primary system components need to be protected from this eventuality. This is usually achieved by means of reverse current circuit breakers or relays. These devices effectively sense reverse current and switch the generator out of circuit, thus preventing any ensuing damage.

- **Over-voltage protection**. Faults in the field excitation circuit can cause the generator to over-excite and thereby regulate the supply voltage to an erroneous over-voltage condition. This could then result in the electrical loads being subject to conditions that could cause permanent damage. Over-voltage protection senses these failure conditions and opens the line contactor, taking the generator offline.
- **Under-voltage protection**. In a single generator system, under-voltage is a similar fault condition to the reverse current situation already described. However, in a multigenerator configuration with paralleling by means of an equalising circuit, the situation is different. Here an under-voltage protection capability is essential as the equalising circuit is always trying to raise the output of a lagging generator; in this situation the under-voltage protection is an integral part of the parallel load-sharing function.

7.2.3   AC Power Generation Control

Voltage Regulation

AC generators differ from DC machines in that they require a separate source of DC excitation for the field windings, although the system described earlier does allow the generator to bootstrap the generation circuits. The subject of AC generator excitation is a complex topic for which the technical solutions vary according to whether the generator is frequency-wild or constant frequency. Some of these solutions comprise sophisticated control loops with error detectors, pre-amplifiers and power amplifiers.

Parallel Operation

In the same way that DC generators are operated in parallel to provide 'no-break' power, AC generators may also be controlled in a similar fashion. This technique only applies to constant-frequency AC generation as it is impossible to parallel frequency-wild or variable-frequency (VF) AC generators. To parallel AC machines the control task is more complex as both real and reactive (imaginary) load vectors have to be synchronised for effective load-sharing.

The sharing of real load depends upon the relative rotational speeds and hence the relative phasing of the generator voltages. Constant speed or constant frequency AC generation depends upon the tracking accuracy of the constant speed drives of the generators involved. In practice, real load-sharing is achieved by control laws that measure the degree of load imbalance by using current transformers and error detection circuitry, thereby trimming the constant speed drives such that the torques applied by all generators are equal.

The sharing of reactive load between the generators is a function of the voltage generated by each generator, as for the DC parallel operation case. The generator output voltages depend upon the relevant performance of the voltage regulators and field excitation circuitry. To accomplish reactive load-sharing requires the use of special transformers called mutual reactors, error detection circuitry and pre-amplifiers/power amplifiers to adjust the field excitation current. Therefore by a using a combination of trimming the speed of the constant speed drives (CSDs) and balancing the field excitation to the generators, real and reactive load components may be shared equally between the generators, as shown in Figure 7.6. This has the effect of providing a powerful single-vector AC power supply to the aircraft AC system, providing a

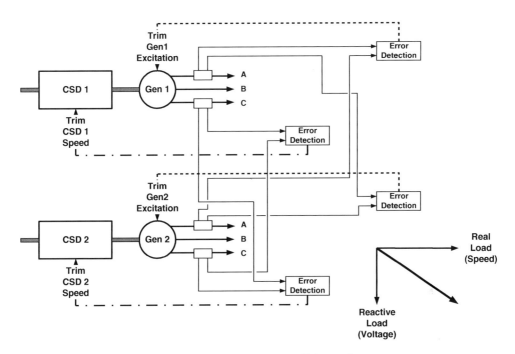

Figure 7.6   AC generator parallel operation

very 'stiff' supply in periods of high power demand. Perhaps the biggest single advantage of paralleled operation is that all the generators are operating in phase synchronism, therefore in the event of a failure there are no change-over transients.

Supervisory and Protection Functions

Typical supervisory or protection functions undertaken by a typical AC generator control unit (GCU) are listed below:

- over-voltage;
- under-voltage;
- under/over-excitation;
- under/over frequency;
- differential current protection.

The over-voltage, under-voltage and under/over-excitation functions are similar to the corresponding functions described for DC generation control. Under or over frequency protection is effectively executed by the real load-sharing function already described above for AC parallel operation. Differential current protection is designed to detect a short-circuit busbar or feeder line fault which could impose a very high current demand on the short-circuited phase. Differential current transformers sense the individual phase currents at different parts of the system. These are connected so that detection circuitry will sense any gross difference in phase current (say, in excess of 30 A per phase) resulting from a phase imbalance, and disconnect the generator from the busbar by tripping the generator control breaker (GCB).

Modern Electrical Power Generation Types

So far the basic principles of DC and AC power-generating systems have been described. The DC system is limited by currents greater than 400 A, and the constant frequency AC method using an integrated drive generator (IDG) has been mentioned. In fact, there are many more power generation types in use today. A number of recent papers have identified the issues and projected the growth in aircraft electrical power requirements in a civil aircraft setting, even without the advent of more-electric systems. However, not only are aircraft electrical system power levels growing, but the diversity of primary power generation types is increasing.

The different types of electrical power generation currently being widely used in civil applications are shown in Figure 7.7. The constant frequency (CF) 115 VAC, 3-phase, 400 Hz options are typified by the integrated drive generator (IDG), variable speed/constant frequency (VSCF) cycloconverter and DC link options. Variable frequency (VF) 115 VAC, 3-phase power generation – sometimes termed 'frequency-wild' – is also a more recent contender, and although a relatively inexpensive form of power generation, it has the disadvantage that some motor loads may require motor controllers. Military aircraft in the US are inclining toward 270 VDC systems. Permanent magnet generators (PMGs) are used to generate 28 VDC emergency electrical power for high-integrity systems.

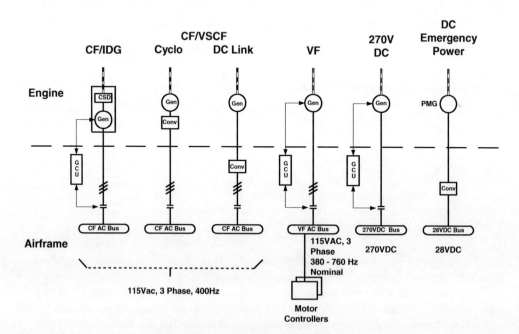

Figure 7.7 Electrical power generation types

Figure 7.7 is also interesting in that it shows the disposition between generation system components located on the engine and those within the airframe. Without being drawn into the partisan arguments regarding the pros and cons of the major types of power generation in use or being introduced today, it is worth examining the main contenders with reference to Figure 7.8.

- constant frequency using an IDG (CF/IDG);
- variable frequency;
- variable speed/constant frequency (VSCF) options.

Constant Frequency/IDG Generation

The main features of CF/IDG power are shown in Figure 7.8. In common with all the other power generation types, this has to cater for a 2:1 ratio in engine speed between maximum power and ground idle. The constant speed drive (CSD) in effect acts as an automatic gearbox, maintaining a constant generator shaft speed which results in a constant frequency output of 400 Hz, usually within ~10 Hz or less. The drawback of the hydromechanical CSD is that it needs to be correctly maintained in terms of oil charge level and oil cleanliness. Also, to maintain high reliability, frequent overhauls may be necessary. That said, the IDG is used to power the majority of civil transport aircraft today, as shown in Table 7.1.

Variable Frequency

Variable-frequency (VF) power generation as shown in Figure 7.8 is the simplest and most reliable form of power generation. In this technique no attempt is made to nullify the effects of the 2:1 engine speed ratio, and the power output, although regulated to 115 VAC, suffers a frequency variation typically from 380 to 720 Hz. This wide band VF power has an effect on frequency-sensitive aircraft loads, the most obvious being the effect on AC electric motors that are used in many aircraft systems. There can therefore be a penalty to be paid in the performance of other aircraft systems such as fuel, ECS and hydraulics. In many cases variations in motor/pump performance may be accommodated, but in the worst cases a motor controller may be needed to restore an easier control situation.

VF is being widely adopted in the business jet community as their power requirements take them above the 28 VDC/12 kW limit of twin 28 VDC systems. Aircraft such as Global Express had VF designed in from the beginning. VF power was adopted for the Airbus A380, Boeing 787 and Airbus A350 aircraft.

VSCF

Figure 7.8 also shows the concept of the VSCF converter. In this technique the variable frequency power produced by the generator is electronically converted by solid-state power-switching devices to constant frequency 400 Hz, 115 VAC power. Two options exist, but the method used in civil applications is the DC link. Here the raw power is converted to an intermediate DC power stage at 270 VDC – the DC link – before being electronically converted

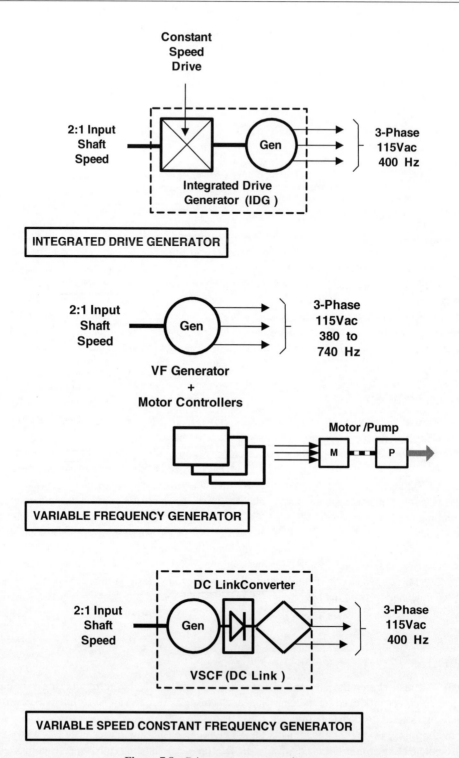

Figure 7.8 Primary power generation types

Table 7.1   Recent electrical power generation examples

Generation type	Civil application	Rated power	Remarks
IDG/CF (115 VAC, 400 Hz)	Boeing 717	2 × 40 kVA	
	Boeing 737 NG	2 × 90 kVA	
	Airbus A320 family	2 × 90 kVA	
	Boeing B777	2 × 120 kVA	
	Boeing B747-400	4 × 90 kVA	
	Airbus A340	4 × 90 kVA	
VSCF direct link (115 VAC, 400 Hz)	Boeing B777 (backup)	2 × 20 kVA	Backup to main and APU generators
	McDonnell Douglas MD-90	2 × 75 kVA	
VF (115 VAC, typical ~380–760 Hz)	Hawker Horizon	2 × 20/25 kVA	
	Global Express	4 × 50 kVA	2 generators/engine
	Airbus A350	2 × 90 kVA	
	Airbus A380	4 × 120 kVA	
VF (230 VAC, typical ~360–760 Hz)	Boeing B787 (more-electric)	4 × 250 kVA	2 generators/engine or 500 kVA per channel

to 3-phase AC power. DC link technology has been used on the B737, MD-90 and B777, but has yet to rival the reliability of CF or VF power generation.

As suggested earlier in Figure 7.8, each of these techniques may locate the power conversion section on the engine or in the airframe.

Table 7.1 lists the power generation types developed and proposed for civil aircraft platforms throughout the 1990s. Not only are the electrical power levels increasing in this generation of aircraft, but the diversity of electrical power generation methods introduce new aircraft system issues to be addressed. For example the B777 standby VSCF and the MD-90 VSCF converters, being located in the airframe, increase the environmental control system (ECS) requirements since waste heat is dissipated in the airframe, whereas the previous IDG solution rejected heat into the engine oil system. Similarly the adoption of variable frequency (VF) can complicate motor load and power conversion requirements. The adoption of 270 VDC systems by the US military has necessitated the development of a family of 270 VDC protection devices, since conventional circuit breakers cannot be used at such high voltages.

Electrical Power Quality

The quality of the electrical power onboard an aircraft is carefully controlled in a number of regards. Power quality will be defined by the nature of the electrical power generation system, distribution system and the nature of loads – particularly the high-power loads that are connected. Power quality is defined by a number of specifications that are similar in many regards, which are:

- **RTCA DO-160C**: The generally recognised US civil specification [3].
- **ADB-0100**: The Airbus specification [4].

- **MIL-STD-704E:** The US military specification [5]. Although apparently disassociated from either of the above, it becomes highly relevant when fitting military equipment onboard an aircraft which was predominantly civilian in origin – civil aircraft such as Boeing 767 or A330 – and which has been designed using a civil electrical power standard. Topical examples are the UK and US military programmes in which these aircraft platforms are being seen as ideal candidates for adoption as military tanker aircraft. For a discussion of some of these issues, see Chapter 13, Military Aircraft Adaptations.

These electrical power system references typically refer to specification of the following defined, although not exclusive, parameters:

- Voltage transients – both normal and abnormal – for AC and DC networks.
- Normal voltage excursions – 115 VAC and 28 VDC networks.
- Normal frequency excursions – CF systems.
- Voltage spikes – 115 VAC and 28 VDC networks.
- Power quality – harmonic distortion – AC systems.
- Power factor limits – AC systems.
- Emergency power requirements – both AC and DC.

These system requirements demand a rigorous assessment of the aircraft power electrical generation system, distribution system and aircraft electrical loads, particularly load utilisation. Load assessments have to be conducted to ensure that the system will meet or exceed the supply of electrical power at the necessary voltage and power levels, whilst also meeting the integrity of certain flight-critical recipient equipment such as fly-by-wire (FBW) and others.

Many of the requirements are similar in concept, although not necessarily identical in approach. For a detailed review of the differing requirements, readers are recommended to refer to the appropriate electrical power specifications for the aircraft type in mind.

7.3 Power Distribution and Protection

7.3.1 Electrical Power System Layers

Figure 7.9 shows how the electrical system is configured using a number of primary power elements and controllers.

- Power generation (section 7.2).
- Primary power distribution (section 7.3.2).
- Power conversion (section 7.3.4).
- Secondary power distribution.

7.3.2 Electrical System Configuration

The aircraft electrical system has to be configured at a top level depending upon what electrical power sources are available and how that power needs to be made available throughout the aircraft, depending upon the operating requirements. Figure 7.10 depicts a typical top-level configuration.

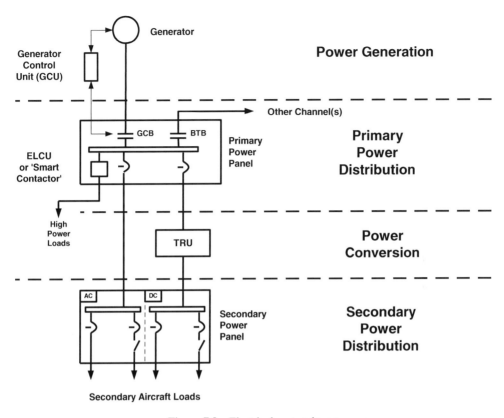

Figure 7.9 Electrical system layers

The main power sources on a typical civil aircraft are switched using the following contactors:

- Generator control breakers (GCBs) apply power to the left and right AC buses under the control of the respective generator control unit.
- The auxiliary power unit (APU) applies power under the control of the APU GCB.
- The external power source is applied via the external power contactor under the control of the bus power control unit (BPCU).
- The BPCU also controls the operation of the two bus tie breakers (BTBs). The BTBs allow the left and right buses to be connected together in the event of a left or right main generator failure. The BTBs also allow power from the APU or ground power source to be switched to the main AC bus bars once the appropriate APU GCB or ECB has been closed.
- Feeding from the main AC busbars electrical load control units (ELCUs) or 'smart contactors' feed specific high-power AC loads directly. These ELCUs may operate autonomously or may be part of a load shed scheme.
- Sub-bus feeders feed power to lower-level AC buses or to a TRU to produce a source of DC power.

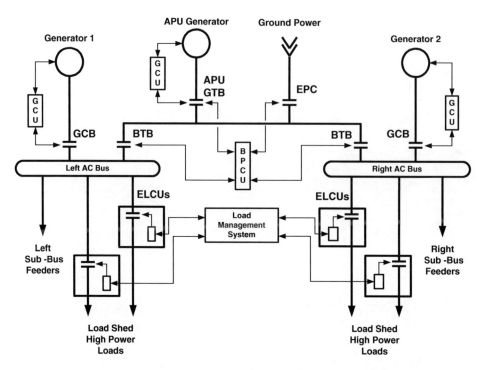

Figure 7.10   Top-level configuration

'Contactors' is the name given to power-switching devices when the power exceeds 20 amps per line or per phase in a 3-phase AC system. Above these current levels, care has to be taken with a positive action when the contacts are opened or closed in order that they are not damaged. During contact opening arcing can occur; during closure contact bounce may be experienced. Both these events can damage the contacts causing excessive voltage drop and local heating.

7.3.3   Electrical Load Protection

Higher-power aircraft loads are increasingly switched from the primary aircraft bus bars by using electrical load control units (ELCUs) or 'smart contactors' as outlined above for load protection. Like primary contactors, these are used where normal rated currents are greater than 20 A per phase, that is, for loads of around 7 kVA or greater. Figure 7.11a shows the comparison of a line contactor such as a GCB with an ELCU or 'smart contactor' in Figure 7.11b. The ELCU has inbuilt current sensing coils that enable the current of all three phases to be measured. Associated electronics allow the device trip characteristics to be more closely matched to those of the load. Typical protection characteristics embodied within the electronics are I^2t, modified I^2t, phase imbalance and earth leakage. For a paper explaining more about 'smart contactors', see Reference [6]. A brief outline of these schemes is shown briefly in Figure 7.12.

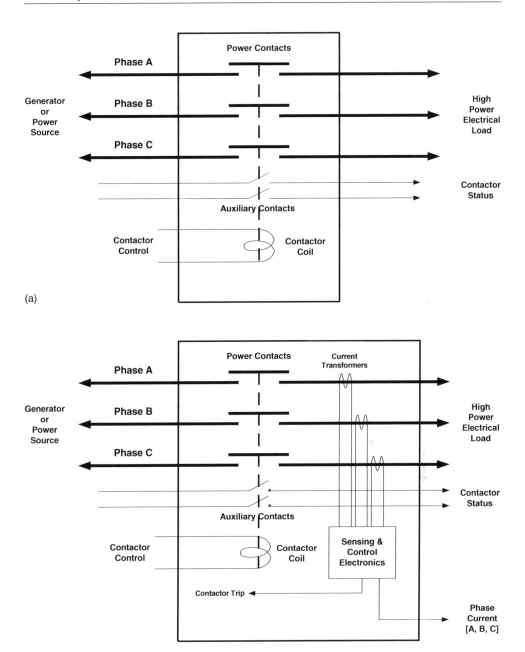

Figure 7.11   (a) Power contactor; (b) ELCU or 'smart contactor'

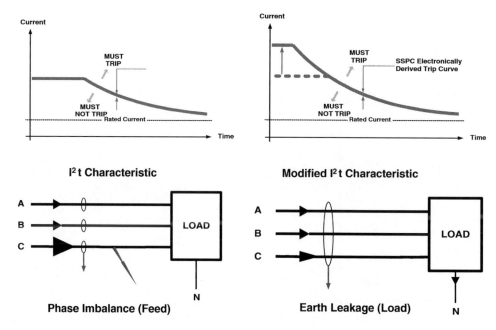

Figure 7.12   Typical ELCU trip schemes

Lower power loads are protected by conventional circuit breakers or by solid state power controllers (SSPCs). In actual fact the protection is provided to protect the aircraft wiring rather than the load itself. As will be seen later in this chapter, aircraft wiring is a complex system in its own right. Wiring bundles are largely inaccessible after aircraft build and cannot easily be replaced. In many cases where an aircraft suffers wiring damage – usually during maintenance – repair of the aircraft is uneconomic.

Circuit breakers rely upon the use of a bimetallic strip that trips the breaker during overload conditions. SSPCs are more precise in their operation and nuisance trips are less likely to occur. SSPCs can also be used to 'tailor' the trip characteristic to give an I^2t or modified I^2t where high in-rush currents are expected, as illustrated in Figure 7.13. More information on SSPCs may be found in Reference [7].

More recently the industry has been developing methods of arc-fault detection. Arc faults can occur in two ways, especially on older aircraft where the wire or cable insulation has degraded over time:

- **Series arcing**. Series arcing can occur when a wire conductor is broken, resulting in a gap in the wire where arcing can occur. This can be the result of a break in the line or a loose pin or connector. The associated energy is limited to the current that the conductor can carry and is therefore relatively low. In most cases this does not generate enough heat to cause a fire.
- **Parallel arcing**. Parallel arcing is experienced when there is a breakdown of the insulation between two wires within a wire bundle. This is manifested as a short-circuit or an earth fault. A conductor-to-conductor circuit arc decreases the dielectric strength of insulation

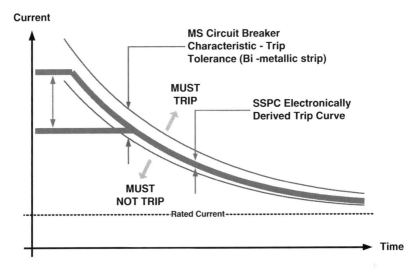

Figure 7.13   Comparison of circuit breaker and SSPC trip characteristics

separating the conductors, allowing a high-impedance, low-current arc fault to develop that carbonises the conductor's insulation. This situation escalates, further reducing the insulation dielectric and leading to an increase in fault current, finally resulting in a rise in temperature that may lead to a fire.

Arc fault protection devices are able to determine and isolate arc faults far quicker than conventional circuit breaker or SSPC designs. Electrical arcs are caused by worn, damaged or contaminated insulation in wiring or connectors, and improved detection methods are demanded [8].

7.3.4   Power Conversion

Power conversion is used at various places in the system and for a variety of reasons. Several typical applications commonly used in a civil aircraft are described as follows:

- A transformer rectifier unit (TRU) converts 3-phase 115 VAC power to 28 VDC as shown in Figure 7.14, and there are usually three or four TRUs in a typical aircraft system. In a civil application TRUs are usually unregulated, meaning that the voltage characteristic 'droops' as power output is increased. In other words, the voltage output reduces as DC current supply increases, possibly up to 4 or 6 VDC from the nominal 28 VDC output. Some sensitive equipment used on military aircraft may require the provision of a regulated supply whereby the voltage is maintained at 28 VDC irrespective of the load being supplied from the TRU.
- Static inverters are solid-state power converters that convert a 28 VDC power source into a 3-phase 115 VAC, 400 Hz power source for emergency conditions.

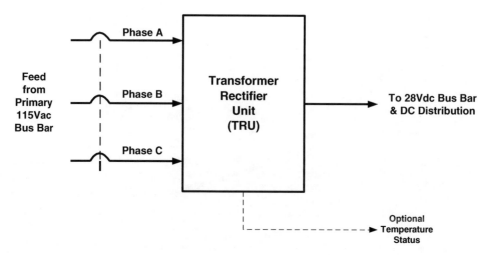

Figure 7.14 Transformer rectifier unit

- Motor controllers are used to control high power loads, especially in VF AC systems where the motor/pump combination may be very sensitive to variations in generator frequency. They may also be used to 'soft start' certain high-power loads with heavy load characteristics – thereby minimising the effect of 'inrush' currents during start-up. Historically, a motor controller would have taken the form shown in Figure 7.15a where the control is exercised over the inverter section of the controller – almost equivalent to a distributed form of a DC link VSCF system. The advent of more robust power switching devices has led to the more modern arrangement shown in Figure 7.15b where the control is exercised using a matrix converter.
- Line-replaceable units (LRUs) require internal power supply units (PSUs) to provide a regulated supply of low-level DC power for the electronics residing within the unit. In some LRUs there may be a need for several PSUs to be provided for integrity and availability reasons. Typical supply voltages supplied by a PSU are ±15 V, $+5$ V and $+3.3$ VDC.

7.4 Emergency Power

Modern aircraft electrical systems have been developed over many years as power requirements have increased and more high-integrity systems evolved: all depending upon the provision of high-quality electrical power. Electrical systems have multiple generators and can be reconfigured to maintain power following a range of failures. Nevertheless, there is still a need to provide emergency power for the aircraft in the direst of failure conditions. Typical emergency sources used are:

- ram air turbine (RAT);
- permanent magnet generators;
- backup systems;
- batteries.

(a)

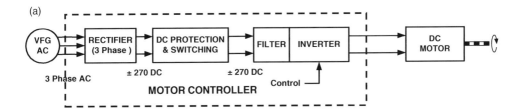

The need for motor controllers:

- In variable frequency (VF) AC systems, high inductance loads such as motors are adversely affected by the frequency variation as the reactive portion of the load is frequency ($\dot{\omega}$) dependent

- Motor controllers alleviate this problem and also allow a 'soft start' for certain high -power motor loads – e.g. fuel pump motors

- Existing solutions accept AC 115 VAC 3 -phase power and output DC to the motor as shown.

(b)

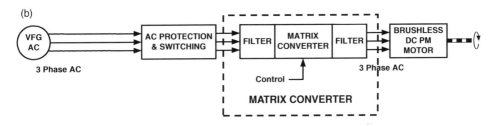

- Recent advances in power switching electronics technology have enabled the development of reliable power conversion using matrix converters

- Matrix converters are analogous to VSCF direct link systems and convert 3-phase VF AC power input to a constant frequency AC output

- The motor is a brushless DC permanent magnet (PM) motor to deliver shaft horsepower to the appropriate aircraft mechanical load

Figure 7.15   (a) Typical motor controller; (b) motor control using a matrix converter

7.4.1   Ram Air Turbine

The ram air turbine or RAT is deployed when most of the conventional power generation system has failed or is unavailable for some reason, such as total engine flame-out or the loss of all generators. The RAT is an air-driven turbine, normally stowed in the aircraft ventral or nose section, that is extended either automatically or manually when the emergency commences. The passage of air over the turbine is used to power a small emergency generator of limited capacity, usually enough to power the crew's essential flight instruments and a few other critical services (see Figure 7.16). Typical RAT generator sizing may vary from 5 to 15 kVA

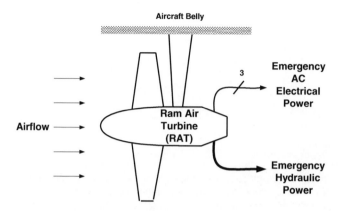

Figure 7.16   Typical ram air turbine

depending upon the aircraft, although more recent aircraft such as the Airbus A380 and Boeing B787 generate much higher levels. The RAT also powers a small hydraulic power generator for similar hydraulic system emergency power provision. Once deployed the RAT remains extended for the duration of the flight and cannot be re-stowed without maintenance action on the ground. The RAT is intended to furnish the crew with sufficient power to fly the aircraft while attempting to restore the primary generators, to reach the engine relight envelope, or to divert to the nearest airfield. It is not intended to provide significant amounts of power for a lengthy period of operation.

Due to the importance of the supply of emergency power, the RAT will normally be expected to deploy and come online in around 5 seconds.

7.4.2   Permanent Magnet Generators

The use of PMGs to provide emergency power has become prominent over the last decade or so. A typical 3-phase PMG arrangement is shown in Figure 7.17. A PMG produces rough

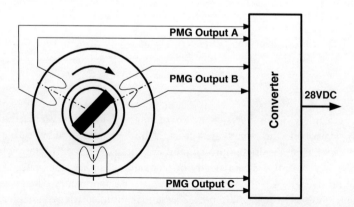

Figure 7.17   Three-phase permanent magnet generator

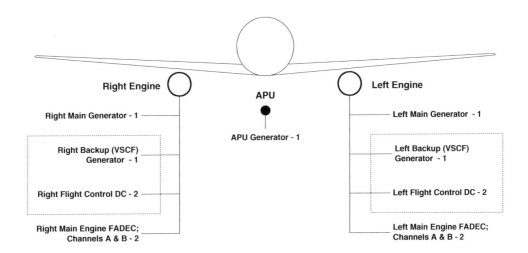

Total PMGs used on B777: 13

Figure 7.18   Boeing 777 PMG installation

and ready power where both the output voltage and frequency vary directly with the speed of the input shaft. Additionally, as the PMG may have to provide DC power over a wide range of input shaft speeds, then a power converter is required. In some cases a PMG/converter combination may have to deliver several hundred watts of power.

All AC generators include a PMG to bootstrap the excitation system. Additionally, PMGs – also called permanent magnet alternators (PMAs) – are used to provide dual independent on-engine supplies to each lane of the FADEC. As an indication of future trends on an aircraft such as the B777, there are a total of 13 PMGs/PMAs across the aircraft critical control systems: flight control, engine control and electrical systems (see Figure 7.18).

Reference [9] is an early paper describing the use of a PMG, and Reference [10] describes some of the work being undertaken in looking at higher levels of PMG power generation.

7.4.3   Backup Systems

The requirements for ETOPS have led to the need for an additional method of backup power supply, short of deploying the RAT that should occur only in the direst emergency. The use of a backup converter satisfies this requirement and is used on the B777. Backup generators are driven by the same engine accessory gearbox but are physically independent of the main IDGs, as shown in Figure 7.19.

The backup generators are VF and therefore experience significant frequency variation as engine speed varies. The VF supply is fed into a backup converter which, using the DC link technique, first converts the AC power to DC by means of rectification. The converter then synthesises 3-phase 115 VAC, 400 Hz power by means of sophisticated solid-state power-switching techniques. The outcome is an alternative means of AC power generation which

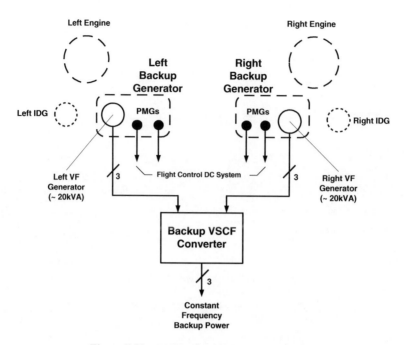

Figure 7.19   Boeing 777 backup VSCF system

may power some of the aircraft AC busbars: typically the 115 VAC transfer buses in the case of the Boeing 777. In this way substantial portions of the aircraft electrical system may remain powered even though some of the more sizeable loads such as the galleys and other non-essential loads may need to be shed by the Electrical Load Management System (ELMS).

7.4.4   Batteries

The majority of this section has described active means of providing backup power. However, it neglects an omnipresent element: the battery. This effectively provides an electrical storage medium independent of the primary generation sources. Its main purposes are:

- To assist in damping transient loads in the DC system.
- To provide power in system startup modes when no other power source is available.
- To provide a short-term high-integrity source during emergency conditions while alternative/backup sources of power are being brought online.

The capacity of the aircraft battery is limited and is measured in terms of ampere-hours. This parameter effectively describes a current/time capability or storage capacity. Thus a 40 ampere-hour battery when fully charged would have the theoretical capacity of feeding a 1 A load for 40 hours or a 40 A load for 1 hour. In fact the capacity of the battery depends upon the charge sustained at the beginning of the discharge, and this is a notoriously difficult parameter to quantify. Most modern aircraft systems utilise battery chargers to maintain the battery charge at moderately high levels during normal system operation, thereby assuring a reasonable state of charge should solo battery usage be required.

The battery most commonly used is the nickel-cadmium (Ni-Cd) type which depends upon the reaction between nickel oxides for the anode, cadmium for the cathode, and operates in a potassium hydroxide electrolyte. Lead-acid batteries are not favoured in modern applications due to the corrosive property of the electrolyte. To preserve battery health it is usual to monitor its temperature, which gives a useful indication of overcharging and whether thermal runaway is likely to occur.

Batteries are susceptible to cold temperatures, and lose capacity rapidly. This can be a problem on autonomous operations in cold climates, after a cold soak or a prolonged period of operation at low temperature where starting APU or engine starter motors can be troublesome.

7.5   Power System Architectures

The development of aircraft electrical systems has evolved as the systems themselves have adapted to meet more challenging problem statements. It is impossible to generalise; although all systems will possess the generic attributes already described, each will have its own particular characteristics depending upon the configuration of the systems of that particular aircraft. In order to draw some lessons out of these different architectures, four examples are described below: all are fly-by-wire (FBW) systems, and as may be seen below, each system has broken new ground using the availability of new technology to meet new and evolving system requirements. The particular architectures are as follows:

- **Airbus A320 family**. The first civil airliner to embrace a FBW flight control system architecture. (Entry into service 1988)
- **Boeing 777**. The first Boeing aircraft to adopt FBW and a novel Electrical Load Management System (ELMS). (Entry into service 1995)
- **Airbus A380**. The first large civil aircraft in recent times to adopt VF power generation (4 × 150 kVA) and the first civil aircraft to provide a FBW system without a mechanical backup. (Entry into service 2007)
- **Boeing 787**. The first civil aircraft to adopt a 'more-electric' architecture and dispense with the use of engine bleed air on the aircraft and rely instead upon electrical power to drive these systems. This has resulted in huge increases in power generation (2 × 2 250 kVA generators or 500 kVA per channel). Also the first civil aircraft to adopt a 3-phase VF 230 VAC power generation system. (Entry into service 2011)

7.5.1   Airbus A320 Electrical System

The A320 electrical system architecture is very conventional and similar to the generic architectures already described. The A320 power system architecture is outlined in Figure 7.20, although variations within the family may occur. The electrical system architecture is shown as in flight in normal operating mode, with a key explaining the symbols and abbreviations.

- Left generator running and feeding 115 VAC, 400 Hz power to AC Bus 1 via the generator line contactor (GLC) which is analogous to the GCB. AC Bus 1 feeds AC power to the

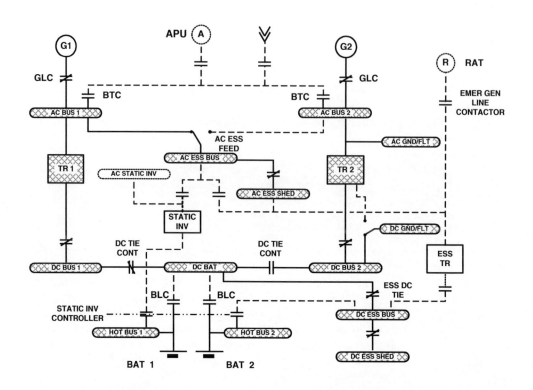

SYMBOLS

(G1)	Generator On-Line	
(A)	Generator Off-Line	
⋎	Ground Power Input	
——	Line Powered	
- - - -	Line Unpowered	
=	Contactor Open	
≠	Contactor Closed	
AC BUS 1	Bus Powered	
AC STATIC INV	Bus Unpowered	

ABBREVIATIONS

APU	Auxiliary Power Unit
GLC	Generator Line Contactor
RAT	Ram Air Turbine
BTC	Bus Tie Contactor
TR	Transformer/Rectifier
INV	Inverter
ESS	Essential
BLC	Battery Line Contactor

Figure 7.20   A320 electrical power system architecture

AC essential bus and to the AC essential shed bus, which may be switched off if the need arises.

- Right generator running and feeding 115 VAC, 400 Hz to AC Bus 2. Power is also supplied to the AC ground/flight bus.
- If necessary, both bus tie contactors (analogous to BTBs) may be closed to feed power across from AC Bus 1 to AC Bus 2 and *vice versa*. With both BTCs closed, it is possible to apply APU or external ground power to both AC buses for ground servicing purposes.
- AC Bus feeds TR 1 (transformer rectifier) which supplies 28 VDC to DC Bus 1 and thence to the DC battery bus and DC essential and DC essential shed buses. As for the AC part of the architecture, the DC essential shed bus may be switched off if necessary.
- AC Bus 1 feeds DC Bus 2 via TR 2. DC Bus 2 may supply the DC ground/flight bus with 28 VDC power or alternatively it may be powered directly from TR 2.
- In the configuration shown, Battery 1 and Battery 2 are connected to their respective hot buses but not to any other services as the battery line contactors (BLCs) remain open.
- A number of reversionary paths are shown that allow power from the RAT, essential TR and the static inverter to be fed to various buses when the system is operating in an abnormal mode. For reasons of brevity these modes are not discussed.

7.5.2   Boeing 777 Electrical System

The Boeing 777 electrical system plus the main power distribution panels are shown in Figure 7.21. The main power sources are:

- Left, right and APU generator sources each rated at 120 kVA and producing 115 VAC, 400 Hz power.
- A backup VSCF system powered by two 20 kVA VF generators. This system has already been described.
- Two external power connections.
- RAT.

An extensive description of the B777 electrical system is given in Reference [11].

Figure 7.21 shows the general layout of the Electrical Load Management System which was a novel feature on the aircraft and set new standards for the industry in terms of electrical load management. The system represents the first integrated electrical power distribution and load management system for a civil aircraft. It comprises seven power panels, three of which are associated with primary power distribution:

- P100 – Left primary power panel distributes and protects the left primary loads.
- P200 – Right primary power panel distributes and protects the right primary loads.
- P300 – Auxiliary power panel distributes and protects the auxiliary primary loads.

The secondary power distribution function is undertaken by four secondary power panels:

- P110 – Left power management panel distributes and protects power, and controls loads associated with the left channel.

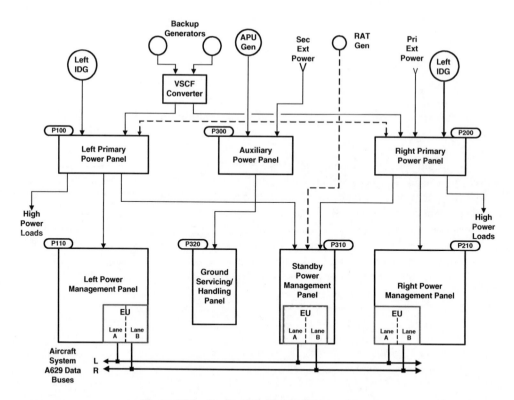

Figure 7.21   Boeing 777 electrical power system

- P210 – Right power management panel distributes and protects power, and controls loads associated with the right channel.
- P310 – Standby power management panel distributes and protects power, and controls loads associated with the standby channel.
- P320 – Ground servicing/handling panel distributes and protects power associated with ground handling.

Load management and utilities systems control is exercised by means of electronic units (EUs) mounted within the P110, P210 and P310 power management panels. Each of these EUs interfaces with the left and right aircraft systems ARINC 629 digital data buses, and contains a dual redundant architecture for reasons of dispatch availability. The EUs contain a modular suite of line-replaceable modules (LRMs) that can readily be replaced when the door is open. A total of six module types are utilised to build a system comprising an overall complement of 44 modules across the three EUs. This highly modular construction with multiple use of common modules reduced development risk and resulted in highly accelerated module maturity at a very early stage of airline service. LRMs typically have mature in-service 'mean time between failures' (MTBF): ~200,000 hours as reported by Reference [12].

The load management and utilities control features provided by ELMS are far in advance of any equivalent system in airline service today. Approximately 17 to 19 electrical load control units (ELCUs) – depending upon aircraft configuration – supply and control loads directly from the aircraft main AC buses. These loads can be controlled by the intelligence embedded within the ELMS EUs. A major advance is the sophisticated load shed/load optimisation function which closely controls the availability of functions, should a major electrical power source fail or become unavailable. The system is able to reconfigure the loads to give the optimum distribution of the available power. In the event that electrical power is restored, the system is able to reinstate loads according to a number of different schedules. The system is therefore able to make the optimum use of power at all times rather than merely shed loads in an emergency.

The benefits conferred by ELMS have proved to be important, with significant reduction in volume, wiring and connectors, weight, relays and circuit breakers. Due to the inbuilt intelligence, use of digital data buses, maintainability features and extensive system built-in-test (BIT), the system build and on-aircraft test time turned out to ~30% of that experienced by contemporary systems.

In order to give the reader some idea of the range and scope of the system, Figure 7.22 shows all of the units associated with the control and distribution of electrical power on the B777 aircraft.

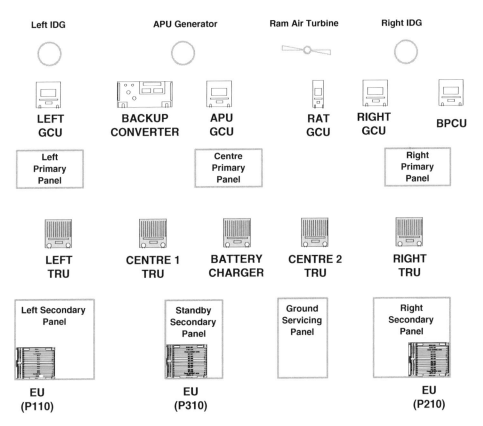

Figure 7.22

7.5.3   Airbus A380 Electrical System

As the world's largest aircraft, the Airbus A380 introduced large-scale VF generation with a four-channel VF 115 VAC system, each generator being rated at 150 kVA. The 150 kVA per primary power channel represented an increase over previous civil aircraft. Hitherto the most powerful had been the Boeing 777 with 120 kVA (CF) plus 20 kVA (VSCF backup), representing 140 kVA per channel.

The key characteristics of the A380 electrical power generation systems are listed as follows:

- 4 × 150 kVA VF generators (370–770 Hz). VF generators are reliable, but do not provide no-break power capability.
- 2 × 120 kVA CF APU generators (nominal 400 Hz).
- 4 × external power connections (400 Hz) for ground power.
- 70 kVA ram air turbine for emergency use.

These key attributes are shown in Figure 7.23. Each of the main 150 kVA AC generators is driven by the associated engine. The two APU generators are driven by the respective auxiliary power units (APUs). Each main generator supplies power to the appropriate AC bus under the control of the GCU. Each main AC bus can also accept a ground power input for servicing and support activities on the ground. Because the aircraft generators are variable frequency (VF) and the frequency of the AC power depends upon the speed of the appropriate engine, the primary AC buses cannot be used in parallel operation. An overview of the A380 electrical system is found in Reference [13].

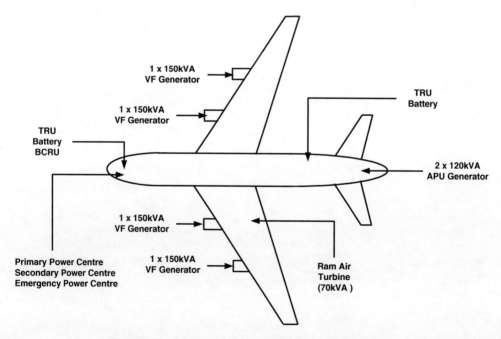

Figure 7.23   Airbus A380 key electrical system components

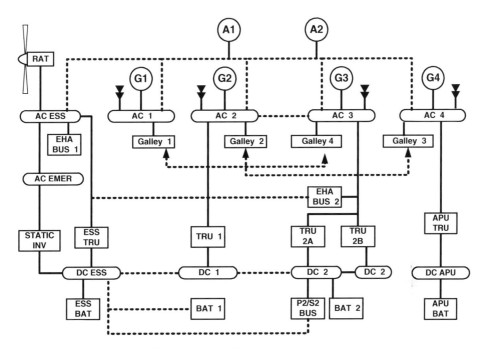

Figure 7.24   Top-level A380 electrical system architecture

The A380 power systems is a '2H' plus '2E' architecture, which means the power architecture relies upon two hydraulic channels '2H' plus two electrical channels '2E'. This is an unusual architecture, having previously been used in a simplified form on the Vickers VC-10 aircraft 50 years earlier. The left generators (1 and 2) and right generators (3 and 4) represent two independent channels of AC power generation feeding their respective buses as shown in Figure 7.24.

To provide no-break power (NBP), the TRUs are regulated and operated in parallel to provide uninterrupted DC power. A more detailed explanation and a description of the electrical system segregation on the A380 is given in Reference [2].

7.5.4   Boeing 787 Electrical System

The B787 is the first civil aircraft in modern times that can claim to be a 'more-electric' aircraft (MEA). The aircraft incorporates many novel more-electric aircraft features and is a big step toward the all-electric airplane – one in which all systems are run by electricity. Bleed air from the engines has essentially been eliminated, and while hydraulic actuators are still used, the majority of their power comes from electricity. In breaking with five decades of practice, Boeing claims that electric compressors are better suited for the cabin than engine bleed and have many savings.

The B787 electrical power system is portrayed at a top level in Figure 7.25. A key feature is the adoption of 3-phase 230 VAC electrical power, compared with the conventional 3-phase

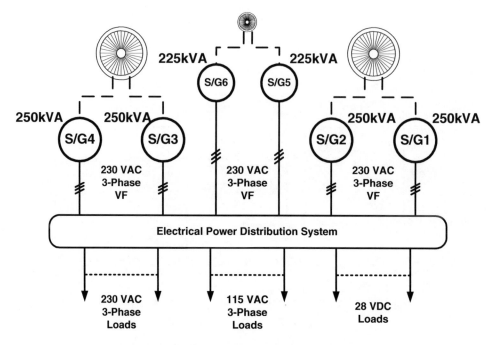

Figure 7.25 Boeing 787 top-level electrical system

115 VAC arrangement usually used. The increase in voltage by a factor of 2:1 decreases feeder losses in the electrical distribution system and allows significant wiring reduction. The use of higher 230 VAC phase voltage, or 400 VAC line-to-line, does require considerable care during design to avoid the possible effects of partial discharge, otherwise known as 'corona'. The salient features of the B787 electrical power system are:

- 2 × 250 kVA starter/generators per engine, resulting in 500 kVA of generated power per channel. The generators are variable frequency (VF), reflecting a recent industry trend away from constant frequency (CF) 400 Hz power.
- 2 × 225 kVA APU starter/generators, each starter/generator driven by the APU.

Each main generator feeds its own 230 VAC main bus before being fed into the power distribution system. As well as powering 230 VAC loads, electrical power is converted into 115 VAC and 28 VDC power to feed many of the legacy subsystems that require these more conventional supplies.

A summary of the B787 electrical loads is given in Figure 7.26. As bleed air is no longer used within the airframe, there are no air feeds to the environmental control system, cabin pressurisation system, wing anti-icing system or other air-powered subsystems. The only bleed air taken from the engine is low-pressure fan air used to anti-ice the engine cowl. Tapping bleed air off the engine compressor is extremely wasteful, especially as engine pressure ratios and bypass ratios increase on modern engines such as the General Electric GeNex and Rolls-Royce Trent 1000. An additional saving is removal of the overhead of providing large ducts

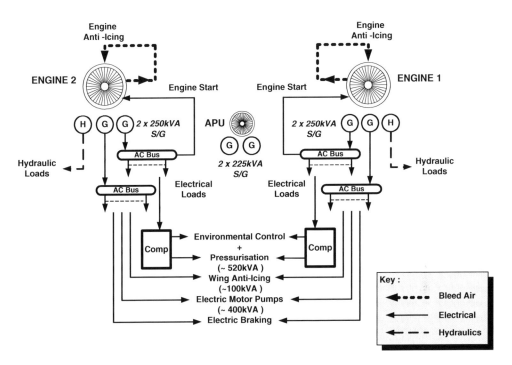

Figure 7.26   Boeing 787 electrical loads

throughout the airframe to transport the air; typically 8" diameter ducts are required between engine and airframe, and 7" ducts between APU and airframe and in the air-driven pump (ADP) feed. In some parts of the airframe, overheat detection systems are required to warn the flight crew of hot gas leaks. The main more-electric loads in the B787 system are:

- **Environmental control system (ECS) and pressurisation.** The removal of bleed air means that air for the ECS and pressurisation systems needs to be pressurised by electrical means; on the B787 four large electrically driven compressors are required, drawing in the region of 500 kVA.
- **Wing anti-icing.** Non-availability of bleed air means that wing anti-icing has to be provided by electrical heating mats embedded in the wing leading edge. Wing anti-icing requires in the order of 100 kVA of electrical power.
- **Electrical motor pumps.** Some of the aircraft hydraulic engine-driven pumps (EDPs) are replaced by electrically driven pumps. The four new electrical motor pumps require ∼100 kVA each, giving a total load requirement of 400 kVA.

A further outcome of the adoption of the 'bleedless engine' is that the aircraft engines cannot be started by the conventional means: high-pressure air. The engines use the inbuilt starter/generators for this purpose and require ∼180 kVA to start the engine.

The introduction of such high-powered electrical machines has a significant impact upon the aircraft electrical distribution system, as shown in Figure 7.27. Primary electrical power

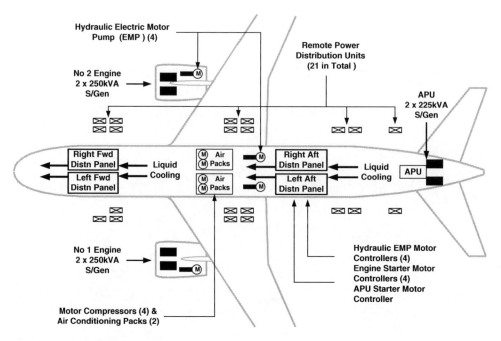

Figure 7.27 Boeing 787 electrical power distribution system

distribution is undertaken by four main distribution panels, two in the forward electrical equipment bay and two others in the aft electrical equipment bay. The aft power distribution panels also contain the motor controllers for the four electrical motor pumps (EMPs); two of the associated pumps are located in the engine pylons and two in the aircraft centre section. Also located within the aft distribution panels are the engine starter motor controllers (4) and APU starter motor controller (1). The high levels of power involved and associated power dissipation generate a lot of heat, and the primary power distribution panels are liquid cooled. The electrically powered air conditioning packs are located in the aircraft centre section.

Secondary power distribution is achieved by using remote power distribution units (RPDUs) located at convenient places around the aircraft. In all there are a total of 21 RPDUs located in the positions indicated in Figure 7.27. The operation of the Boeing 787 systems and more detailed discussion on more-electric systems may be found in Reference [2].

7.6 Aircraft Wiring

Of the many invisible attributes of an aircraft, the electrical wiring is not the easiest to understand, and its all-pervasive extent within the aircraft is most difficult to comprehend. The authors thought that this section on aircraft wiring would be useful to those studying aircraft systems. Flight control runs, fuel pipes, hydraulic lines and air conditioning ducts are easy to visualise and identify within the airframe, whereas the electrical wiring is less easily identifiable. This is compounded by the variety of different connectors used for differing

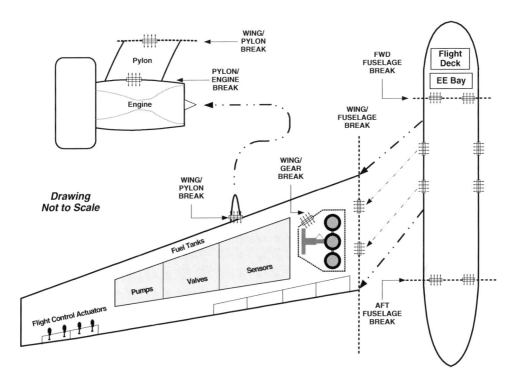

Figure 7.28 Typical aircraft wiring breaks

electrical technical reasons. A good starting point is to understand the basic aircraft structure and how aircraft wiring relates to these basic structural building blocks.

7.6.1 Aircraft Breaks

The key structural/wiring breaks within an aircraft are determined by aircraft structural breaks, and in an increasing worldwide aerospace industry there is a tendency to outsource or subcontract these major areas of work. Significant structural elements are increasingly likely to be distributed to investment risk-sharing partners around the world who will accept responsibility for that element of the aircraft structure. These risk-sharing partners may also take responsibility for the installation of systems' components and wiring that lie within that part of the airframe, whilst the prime contractor takes responsibility for final assembly. To facilitate this, individual sections will be completed with wiring up to structural break connectors.

A typical example is shown in Figure 7.28. Typical aircraft breaks include:

- **Forward and aft fuselage breaks**. These separate the forward, centre and aft fuselage sections.
- **Wing/fuselage breaks**. These breaks define the boundary between the relatively benign pressurised cabin environment and the more challenging area of the wing, in which flight

control actuators, fuel system components such as fuel pumps, valves and gauging and temperature sensors reside. The aircraft wing area presents severe challenges to aircraft wiring and electronics. Typical withstanding voltages for aircraft wing wiring are twice those that would be specified for wiring residing within the fuselage compartment.

- **Landing gear/wing compartment breaks**. The landing gear bays present another severe environmental area. Because of the exposure to a hostile environment during take-off, landing and approach, the wiring looms on the landing gear are usually armoured to survive within the environment as well as withstand the effect of flying foreign objects such as discarded aircraft wheel tyre treads or foreign objects picked up from the runway, as well as spray and contaminants.
- **Wing/pylon and pylon to engine breaks**. These breaks are important as the associated wiring carries essential information between the flight crew and the engine. The pilot transmits throttle commands and other control information to the engine.
- **Engine wiring**. The engine without doubt presents the most aggressive environment on the aircraft in terms of temperature and vibration. In a similar manner to the landing gear harnesses, wiring harnesses on the engine are typically armoured to provide protection against the severe operating conditions.

With exception of a few key systems, most of the aircraft systems are constrained by the boundaries described above. Particular exclusions may include:

- The routing of high-power generator feeder cables, due to the possibility of high resistance or high-power dissipation contacts.
- The routing of high-integrity wiring such as fire warnings and hydraulic shut-off valve selection lines, due to the consequence of a connector failing.
- Specific wiring associated with flight control.

7.6.2 Wiring Bundle Definition

Within specific wiring zones the aircraft wiring may vary from single wires connecting two items together electrically, to wiring bundles or harnesses in which a number of wires need to be routed to/from specific points within the aircraft structure. The definitions associated with individual wires/bundles/harnesses are broadly as follows, with reference to Figure 7.29:

- **Open wiring**: Any wire, wire group, or wire bundle not enclosed in a covering.
- **Wire group**: Two or more wires tied together to retain identity of the group.
- **Wire bundle**: Two or more wire groups tied together because they are going in the same direction at the point where the tie is located.
- **Wire harness**: Wire group or bundle tied together as a compact unit (open harness) or contained in an outer jacket (enclosed harness). Wire harnesses are usually prefabricated and installed on the aircraft as a single assembly.
- **Electrically protected wiring**: Those wires that have protection against overloading through fuses, circuit breakers or other current-limiting devices. Most of the aircraft electrical wiring is protected in this way. The purpose of the protection is to protect the aircraft wiring, not the load.

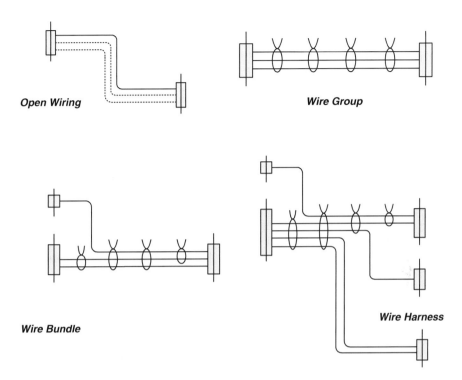

Figure 7.29   Definition of aircraft wiring grouping

- **Electrically unprotected wiring**: Those wires (generally from generator to main bus distribution points) that do not have protection from fuses, circuit breakers or other current-limiting devices. However, protection against electrical fault conditions will be inherently provided as part of the generator control loop including current and voltage fault conditions.

These definitions were extracted in the main from AC21-99, Advisory Circular from the Australian Civil Certification Authority.

7.6.3   Wiring Routing

Given the foregoing constraints and the need to transit the various structural and electrical breaks as already described, the wiring is subject to very practical considerations during installation, namely:

- taking care not to exceed the bend radius of the wire type, especially coaxial types such as antenna cables;
- prevention of chafing between wire bundles and aircraft structure;
- securing bundles through bulkheads and structure;
- fastening wires in junction boxes, panels and bundles for correct routing and grouping;

- prevention of mechanical strain that may break conductors and connections;
- prevention of the possibility of arcing or overheating wires causing damage to mechanical control cables;
- facilitation of re-assembly following repair;
- prevention of interference between wires and other equipment;
- permitting replacement of individual wires without removing the entire bundle;
- prevention of excessive movement in areas of high vibration (armoured cables in landing gear and engine zones.

7.6.4   Wiring Sizing

Aircraft wiring is generally categorised by reference to the American Wire Gauge (AWG) convention. Within the AWG convention, the higher the number, the smaller the size of the wire. Typically AWG 24 (Boeing) and AWG 26 (Airbus) are usually the smallest wires used within the aircraft for reasons of robustness. Smaller gauge – higher AWG numbers – may exist within individual equipments, as these are protected from general wear and tear within the aircraft.

The lower AWG categories are used for high-power feeders, usually from the aircraft electrical power generators or for major electrical power feeders within the aircraft electrical power distribution system. A key consideration in the selection of the wire size is the voltage drop associated with the wiring run and the power dissipation associated with feeder losses. The nature and duration of the anticipated electrical faults in association with the capability and reaction time of the wiring protection devices available is an important consideration.

The following tables give an indication of typical aircraft wiring parameters and an indication of the amount of wire in a typical large aircraft:

- Table 7.2 provides indicative information for copper (Cu) wiring in a typical large civil aircraft. Aluminium (Al) wiring is lighter but has higher resistivity, so there is scope for a selection of Cu versus Al power feeders to save installation weight at the expense of greater voltage drop/feeder losses.
- Table 7.3 gives a typical wiring weight budget for a large transport aircraft of around 20 years ago. This represents basic aircraft wiring and does not include the in-flight entertainment burden. Of particular interest is the extent of the wiring – almost 700,000 feet of wiring weighing in the region of 6500 pounds. This example is dated, and it could be expected that an aircraft of the current generation would have a much higher wiring content.

7.6.5   Aircraft Electrical Signal Types

Aircraft wiring, as has already been described, is complex and is often installed in a hostile environment. In many cases aircraft wiring cannot be accessed following aircraft initial build. Wiring types are varied, as the following examples testify:

- RF/coaxial wiring for radios and radars; subminiature coaxial wiring is used in places.
- Power feeders for primary electrical power; conventional wiring for lower power electrical supplies.

Table 7.2 Wire sizes – American Wire Gauge (AWG)

AWG	Diameter (in)	Ohms/1000 feet (Cu)	Max. current (typical)	Typical application
0000	0.46	0.049	260	
000	0.41	0.062	225	
00	0.36	0.078	195	
0	0.32	0.098	170	Primarypower feeders
1	0.29	0.124	150	
2	0.26	0.156	130	
4	0.20	0.248	95	
6	0.16	0.395	75	
8	0.13	0.628	55	
10	0.10	0.998	40	
12	0.08	1.588	30	Secondary feeders and high power loads
14	0.06	2.525	25	
16	0.05	4.016		Medium-sized loads
18	0.04	6.385		
20	0.03	10.150		Normal use
22	0.26	16.140		
24	0.02	84.22		

Table 7.3 Typical large aircraft wiring composition

AMG	Length (feet)	Weight (lbs)[1]
24	162445	887.0
22	148239	594.2
20	237713	1859.4
18	82211	732.6
16	26663	276.8
14	4998	65.4
12	9872	256.2
10	4681	146.0
8	3981	231.9
6	2048	115.3
4	2622	240.9
2	1140	170.2
1	444	50.2
*1	719	196.1
*2	2447	418.4
*3	55	12.5
Special	5574	219.0
Total	695852	6472.1

*Includes connectors but excludes IFE.

- Signal wiring for aircraft sensors; often twisted/screened pairs, triads and quads.
- Twisted copper pairs and quads for data buses.
- Fibre-optic wiring for data buses and in-flight entertainment (IFE) system.
- Specialised wiring is also required in the area of fuel gauging (tank wiring harnesses), landing gear and engine (armoured conduits). Wiring in fuel tanks has to be especially protected to limit the amount of energy associated with a fuel contents sensing probe, and also to limit the fault conditions associated with an electrically powered fuel pump.

Associated with the varied wiring configurations are a huge range of connector types.

7.6.6 Electrical Segregation

Due to the widely diverse nature of the aircraft electrical signalling, types require segregation as certain types may interfere disproportionately with others, causing detrimental performance of vital aircraft systems.

The wide diversity of aircraft signal types may be summarised by using an Airbus example. On Airbus aircraft systems the aircraft wiring system is generally divided into two main systems and further subdivided into routes. This ensures that damage is limited and any EM interference (EMI) is reduced to a minimum.

In the Airbus system, different circuits have specific identifiers (EMI classifications) as listed below:

- G Generation
- P Power Supply
- M Miscellaneous
- S Sensitive
- R Audio
- C Coaxial

Similar conventions are used by other aircraft manufacturers.

7.6.7 The Nature of Aircraft Wiring and Connectors

The discussion so far has centred on the point-to-point electrical wiring as it connects components together throughout the airframe. The means by which the various controllers and sensors are connected also deserves mention.

There are three main ways in which these components are electrically connected:

- rack-mounted;
- structure-mounted;
- bulkhead or wiring breaks.

All of these connector types will employ the separation of differing signal types as described above and illustrated in Figure 7.30.

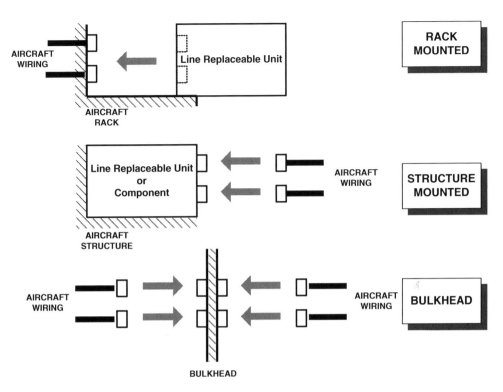

Figure 7.30   Portrayal of equipment connection schemes

7.6.8   Used of Twisted Pairs and Quads

The usual way of reducing susceptibility to EMI within the aircraft wiring is to resort to shielding and screening. This is usually employed for sensitive sensor signals and digital data buses (which can also be significant emitters). Practising aerospace engineers often have problems envisaging what the data buses in an aircraft actually look like, so it is worth giving a brief explanation.

The twisted screened wire pairs are categorised as follows and are portrayed in Figure 7.31:

- Unshielded: the twisted pair has no metallic shroud (UTP).
- Shielding is where the unshielded twisted pair is contained within a metallic shroud (S/UTP); also known as FTP (Foil TP).
- Screening is where a twisted pair is contained within a metallic shroud (STP); also known as STP-A.
- Shielding and screening together for a twisted pair (SSTP).
- Similarly twisted triple or quad wire arrangements may be used.

The shield or screen may be bonded or grounded depending upon the installation requirements.

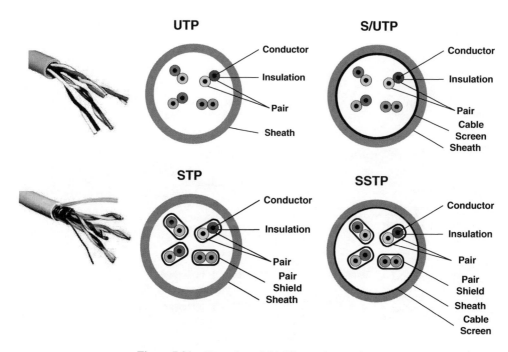

Figure 7.31   Examples of shielding and screening

There are different ways of installing data buses – Boeing and Airbus adopt different schemes as shown in Figure 7.32, showing a circular MIL-DTL-38999 connector and civil ARINC 600 type rack connectors. The different methods are described below:

- Airbus use the quadrax arrangement shown in the upper part of the diagram. A full duplex data bus – a data bus passing data in both directions simultaneously – is implemented in one self-contained cable including two twisted wire pairs. This arrangement allows a higher packing density but may arguably be less robust than a twinax arrangement.
- Boeing favour the twinax scheme shown in the lower part of the figure. This implements a half duplex arrangement where each bus only passes data in one direction and is separate from the other half of the digital link.

In an aircraft that makes extensive use of aircraft data buses, which probably represents most of the aircraft in production today, then multiconnector arrangements are used.

7.7   Electrical Installation

The installation of avionic equipment is generally rack-mounted in the aircraft electrical equipment (EE) bay. The issues surrounding aircraft wiring have already been outlined in Section 7.6. The EE bay is conveniently located just below the flight deck. Equipments located

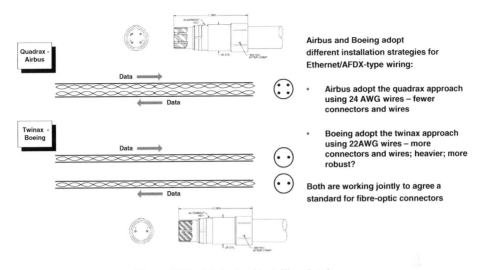

Figure 7.32 Methods of installing data buses

in the EE bay will be rack-mounted using LRUs configured to use the standard connectors specified as follows:

- ARINC 404 connectors for older legacy equipment.
- ARINC 600 connectors for modern digital equipment.

These specifications also dictate the form factor of the LRUs as shown in Figure 7.33.

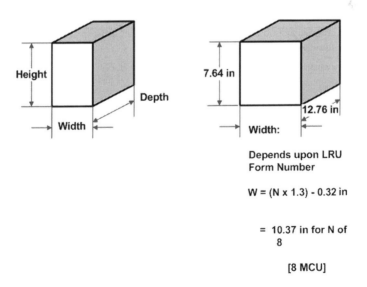

Figure 7.33 Civil aircraft LRU form factor

LRUs are specified as modular concept units (MCUs) where the height and depth of the unit are fixed to facilitate the installation of the units in standard racks within the EE bay. The width can be varied depending upon how much electronics the unit is expected to contain. The MCU number increases as the box width increases and is defined by the formula in the figure. In general, as a rough guide 1 MCU equates to ~1.25" therefore a 4 MCU box would be approximately 5" wide. The MCU format is therefore a useful guide to the volume that a set of system controllers needs within the EE bay. Key issues that need to be considered when designing an installation are:

- temperature and power dissipation;
- electromagnetic interference (EMI);
- lightning strikes.

7.7.1   Temperature and Power Dissipation

Although the electronics contained within a LRU are ruggedised and protected, there is still the question of the operating temperature. Electronics do not survive much above 90°C, and to obtain high reliability it is best to assure an environment below 40°C. The avionics equipment – like the passengers – needs a supply of cool air to moderate the operating temperature. A detailed description of cooling methods is not included here – see Reference [2].

The amount of heat an LRU dissipates depends upon size, technology and the purpose for which it is being used. Power converters in the electrical system will not be 100% efficient, and as they are delivering significant levels of power then the waste (or rejected) heat can be considerable. A 300 A regulated TRU producing 28 VDC may dissipate ~850 W, a considerable amount of heat. When the heat loads of all the units within the EE bay are summed, the total heat load is significant; fortunately most LRUs can be cooled by the air conditioning system. Exceptionally, liquid cooling may required, as is the case for the Boeing 787 primary distribution panels because of the high power levels involved.

When the total heat load comprising avionics equipment, in-flight entertainment, passengers and crew is summed, then there is a considerable amount of heat that needs to be rejected from the aircraft. An assumption of ~100 W per seat for IFE equipment and ~100 W per passenger would not be widely amiss, therefore the heat load for an A380 bearing 500 passengers would potentially be a maximum of around 200 × 500 W or 100 kW for the passengers and IFE alone. Of course, a utilisation factor needs to be applied: not all passengers will have IFE switched on, and not every passenger will radiate 100 W. However, the principle is clear. Figure 7.34 gives examples of how the heat exchange is achieved. Some of the issues relating to the environmental control system (ECS) and heat transfer are discussed in Reference [2].

7.7.2   Electromagnetic Interference

Electromagnetic interference (EMI) is omnipresent within the aircraft, whether from internal or external sources. The effects of EMI can be mitigated by careful screening throughout the aircraft wiring system and by careful LRU design. Generally the industry adopts designs similar to that shown in Figure 7.35, where the unit is split into two areas: a 'clean' area and a 'dirty' area. The clean area houses the sensitive electronics within a Faraday cage, with all

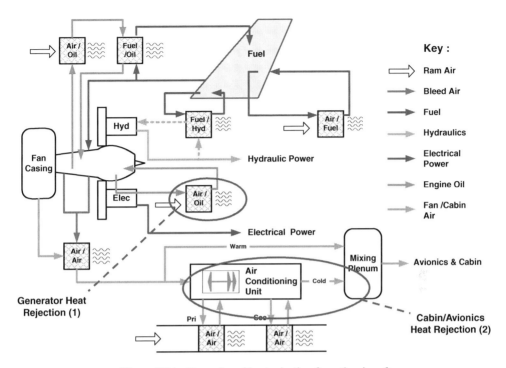

Figure 7.34   Examples of heat rejection from the aircraft

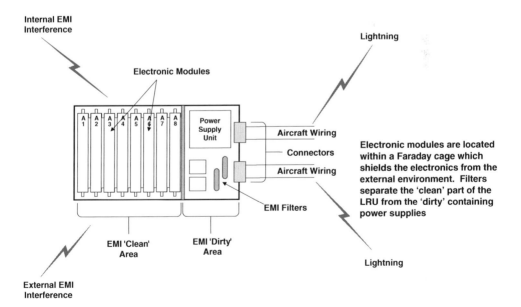

Figure 7.35   Example of LRU design to minimise EMI effects

Table 7.4   Lightning strike risk categorisation

Aircraft area	Zone
Wing; horizontal tailplane; vertical tailplane; propulsion system; landing gear; tail cone	Exposed
Radome: belly fairing	Radome/belly
Flight deck, interfloor zone	Pressurised
Protected avionics racks; main avionics compartment	Electronics bay

input/output signals being filtered. The dirty area contains some of the EMI filters and the power supply unit (PSU), which are themselves likely to be significant noise emitters.

Consideration of EMI and the design of aircraft wiring and LRUs is a complex subject and has been made more difficult by the adoption of composite airframes. Further discussion is outside of the scope of this book, but see [14].

7.7.3   Lightning Strikes

Another condition that occurs naturally and in a random and unpredictable manner is that of lightning strikes. In a similar fashion to EMI, the equipment design and installation needs to take this into account. The usual way to approach this issue is to categorise different areas of the aircraft according to risk and invoke the necessary design rules to minimise the effects. A typical categorisation scheme is shown in Table 7.4 in decreasing order of severity.

7.8   Bonding and Earthing

There are important considerations to be borne in mind relating to the aircraft electrical wiring system: these are bonding, grounding and earthing. These techniques reduce the voltage potential between adjacent items of hardware, provide a stable reference point for the aircraft electrical systems, and provide a means by which static is dissipated during ground servicing operations. Sometimes these terms are very loosely used – even by aerospace professional engineers – so it is necessary to use precise definitions. Commonly used definitions are as follows:

- **Bonding**: the electrical connecting of two or more conducting objects not otherwise adequately connected to minimise potential difference.
- **Grounding**: the electrical connecting of a conducting object to the primary structure or earth electrode, for return of current.
- **Earthing**: a specific case of bonding to an earth reference to dissipate static while the aircraft is being serviced – particularly during refuelling and/or when an external electrical power source is connected. In this sense two further definitions may apply:
  - Static ground: an approved ground point with an impedance of less than 10,000 ohms when referenced to earth.
  - Power ground: an approved ground point with an impedance of less than 10 ohms with respect to the aircraft power system neutral.

This information was extracted from AC21-99, Advisory Circular from the Australian Civil Certification Authority.

While the aircraft is in flight the aircraft structure represents 'ground'. Therefore the aircraft structure provides the return path for current flowing through an aircraft load and back to the electrical power source.

The aircraft is only 'earthed' during specific ground servicing operations such as maintenance, refuelling and arming. There are two ways in which an aircraft can be earthed during servicing operations:

- Using a dedicated earth lead to connect an earth stud on the aircraft structure to a specified earthing point on the airfield. This situation prevails if the aircraft is connected to a standalone external ground power cart. Earthing points are conveniently situated around the airfield, particularly close to where the aircraft is to be serviced (or armed). They are specifically designed and maintained to ensure the necessary high quality of earth connection required for the task in hand.
- If the aircraft is connected to a mains-generated power source (or electric/electric power source), the aircraft will be automatically connected to the external power source (including earth) via the aircraft external power connector. In this situation the aircraft is effectively connected to the National Grid earth of the country in question.

It is also necessary to make a distinction between the various types of grounding connection used for different power and signal types. This may be understood by reference to Figure 7.36,

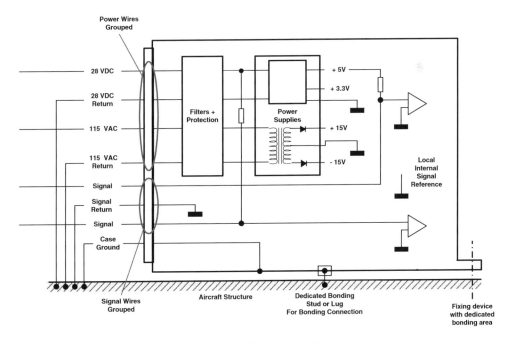

Figure 7.36 Typical bonding and grounding arrangement

which shows a typical electronic controller as installed in the aircraft – whether a rack-mounted or stand-alone unit.

The aircraft structure to which the unit is mounted represents 'ground' and provides the current return path as already described:

- The fixing or mounting device has its own dedicated bonding area.
- The unit case is directly bonded to the aircraft structure using a dedicated stud or lug.
- In this example, four other bonding connections are shown:
  - 28 VDC return connection
  - 115 VAC return connection
    Both these returns are from the internal unit power supply unit (PSU) and are separately grouped together. These are likely to be EMI emitters.
  - Signal return connection.
  - Case ground connection.
- The signal connections are separately grouped together as they are more sensitive and are EMI-susceptible. Within the unit there will also be local internal signal references for power and signal types.

Bonding, grounding and earth points will all have specified low resistance values, which may depend to some degree on the aircraft type and nature of the avionics fit. The aircraft and equipment level designers need to adhere closely to these requirements if the aircraft is to possess a satisfactory EMI performance.

7.9   Signal Conditioning

7.9.1   Signal Types

The input/output (I/O) wiring is considerable, as all the aircraft sensors for each system have to be connected to their respective controllers. There is a wide range of different sensors onboard an aircraft, each suited to the particular parameter being measured, range and accuracy. Correspondingly, there will be a range of outputs to switch power and drive sensors and effectors. Typical examples are listed below:

- **Inputs:**
  - Discrete inputs that signify the state of a component: ON or OFF; OPEN or CLOSED.
  - Potentiometer: to signify sensor position: linear or rotary.
  - Pulse probe: to signify angular speed or rate.
  - Platinum resistance bulb: to measure temperature.
  - Thermistor: to measure level or temperature.
  - Pressure transducer: to signify pressure (oil, fuel, hydraulic fluid).
  - AC resolver/synchro: to signify angular position.
  - Frequency modulation: to represent an analogue measurement as a digital pulse stream.
  - Linear/rotary differential variable transformers (L/R DVT): to signify linear or rotary position.

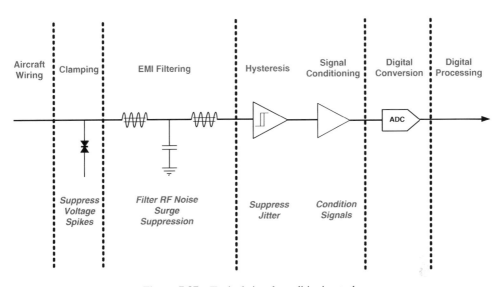

Figure 7.37   Typical signal conditioning tasks

- **Outputs:**
  - DC discrete: to signify status or command a change of state; ON or OFF; OPEN or CLOSED.
  - DC power drive (up to 1 A inductive load): for example, to command a solenoid valve to OPEN or CLOSE.
  - Low voltage analogue: to output an analogue signal.
  - Low-current servo drive: to drive a low-current servo-valve.
  - High-current servo drive: to direct drive a servo-valve.

These are typical examples – the list is not exhaustive since there are many other signal types besides this abbreviated list.

The means of converting 'real world' analogue signals from analogue to digital has been described in Chapter 2. The need to sample signals at greater than the Nyquist rate has been outlined, and the use of Op Amps described in terms of filtering, summing and scaling the signals in analogue form prior to the A-to-D conversion process.

7.9.2   Signal Conditioning

As the aircraft signals are being collected far and wide across the airframe, so they are subject to many spurious and troublesome effects such as noise, voltage spikes and EMI. These effects are addressed at the front-end of the signal chain as shown in Figure 7.37. The generic techniques that the signal conditioning function utilises to improve noise immunity and mitigate these factors include the following:

- voltage spike protection using clamping diodes and transorbs;
- EMI filtering using baluns (coils) and capacitors;

- bandwidth limiting using simple resistor–capacitor filtering;
- use of high logic thresholds;
- employing broad hysteresis bands (Schmitt buffers).

The nature of the technology used in both aircraft system sensors and the associated digital interfacing techniques is highly specialised and jealously guarded by the companies involved. As such, detailed information is proprietary in nature. Generic principles rather than specific detailed examples are outlined below; each supplier will utilise their own sophisticated diagnostic digital techniques that enable a huge range of health-monitoring and self-test capabilities for sensors and controllers alike.

Figure 7.37 depicts the generic functions that the signal conditioning performs. The precise techniques will vary depending upon system sensitivity and the susceptibility of a particular sensor type, its location within the airframe, the nature of the aircraft wiring (and structure) and the likely EMI threats that are to be presented, as well as the integrity level that the signal demands. Typical categories of protection are:

- Clamping using Zener diodes to limit any voltage spikes that may enter into the unit and potentially cause damage due to, say, a lightning strike.
- Use of EMI filtering to filter out RF noise and to suppress any current surges that external influences may present. A range of filters exists with a variety of characteristics that may be employed depending upon the frequency of the noise that needs to be rejected and the input/output characteristics of the surrounding circuitry.
- For both analogue and digital signals, steps need to be taken to introduce hysteresis or counteract pulse jitter. These techniques need to be applied sympathetically so that adequate protection is given while maintaining the fidelity of the signal that the system is trying to capture.

7.10 Central Maintenance Systems

The need for maintenance data onboard the aircraft carrying complex integrated avionics systems of today is paramount. One of the keystones of onboard avionics systems is the built-in-test (BIT) function that most systems embrace. BIT embraces a test capability within the equipment that enables the correct functioning of the equipment to be assured. BIT is performed continuously (CBIT) or can be initiated by the crew during pre-flight checks (IBIT). Early manifestations of BIT were unreliable and not always able to ascertain the correct functioning of the equipment, let alone make accurate diagnostic assessments of equipment failures. Many years of hard-learned lessons have resulted in aircraft and avionics manufacturers fielding much more capable and potent equipment in the last decade or so. Whilst never perfect, the maintenance systems of today offer much more help than hindrance in the battle to keep the aircraft airborne and earning revenue.

The availability of improved equipment BIT capabilities, together with the increased use of aircraft-level data buses such as ARINC 429 and ARINC 629, have enabled the various avionics equipment to share copious quantities of diagnostic data. The ready availability of improved systems-level multifunction displays such as EICAS and ECAM on most modern aircraft has resulted in much more information relating to system function and status being displayed at the system level on the flight deck displays. However, it is the control and display unit (CDU) – normally associated with the flight management system (FMS) – which has made the greatest contribution to onboard maintenance test and diagnosis.

It is by its very nature – text display (normally in colour), hard alpha-numeric and menu-driven soft keys – that the CDU offers a multipurpose man–machine interface for the technician on the ground in the same way that it provides similar operational functions for the flight crew while airborne. These features, coupled with the highly integrated nature of data transfer in the modern avionics system, enable the CDU to be used as a very powerful test and diagnostic tool.

Two distinct but separate implementations are examined, both of which offer considerable improvements over earlier generation systems. Many modern regional, business jet and large, medium and small transport aircraft will have similar functionality embedded: what will be described is indicative of the industry norm rather than one or two specific examples. The examples to be described are:

- Airbus A330/340 Central Maintenance System (CMS);
- Boeing 777 Central Maintenance Computing System (CMCS).

7.10.1   Airbus A330/340 Central Maintenance System

The implementation of the A330/340 Central Maintenance System (CMS), also adopted for later models of the A320 family, is described herein. The Airbus aircraft predominantly use ARINC 429 data buses throughout, and this system is based upon the integration of data based upon this level data fusion. This contrasts with the Boeing ARINC 629 centralised approach that will be described later. Both represent extremely capable systems; the fact that the approach is different is of interest. The key display elements as shown in Figure 7.38 are as follows:

- Display of aircraft system synoptic and status displays available to the flight crew on the ECAM displays.
- Use of the three CDUs as a man–machine interface for system test and diagnostic purposes.

The ECAM displays typically portray system synoptics relating to the following (and many other systems) provided by the display management computers:

- engines;
- electrical system;
- APU;
- hydraulic system;
- fuel system;
- air conditioning;
- landing gear.

The CDUs are provided with data, in this instance from one of the two dual-redundant central maintenance computers (CMC), as shown in the top-level architecture in Figure 7.39. The

Engines

IDG

APU

Hydraulic

Fuel

Landing Gear

Air Conditioning

etc

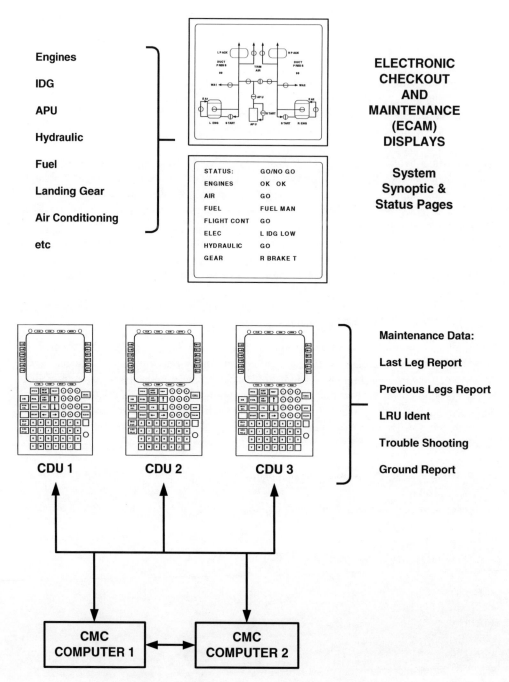

ELECTRONIC
CHECKOUT
AND
MAINTENANCE
(ECAM)
DISPLAYS

System
Synoptic &
Status Pages

Maintenance Data:

Last Leg Report

Previous Legs Report

LRU Ident

Trouble Shooting

Ground Report

Figure 7.38   Airbus CMC computation and display

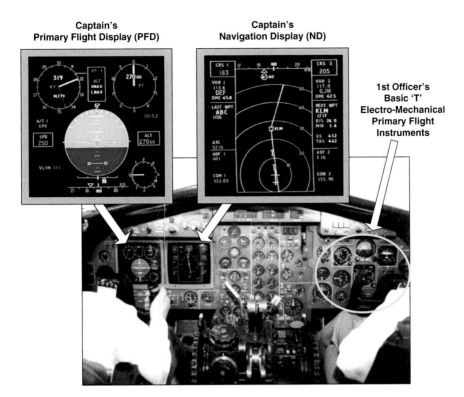

Figure 12.7 BAC 1-11 Advanced Civil Flight Deck demonstrator (April 1981). Reproduced with permission from GE Aviation (formerly Smiths Aerospace)

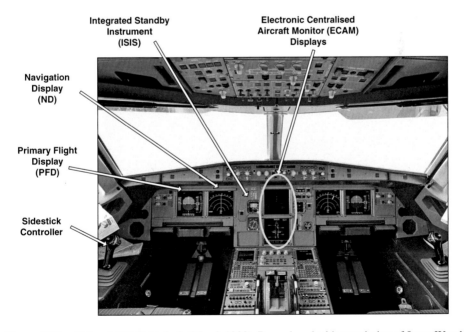

Figure 12.9 Airbus A320 flight deck (March 1988). Reproduced with permission of Jason Wood

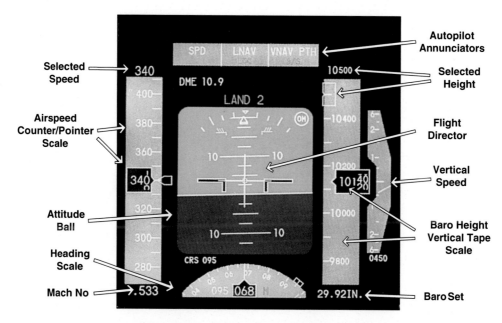

Figure 12.10   Airbus A320 primary flight display

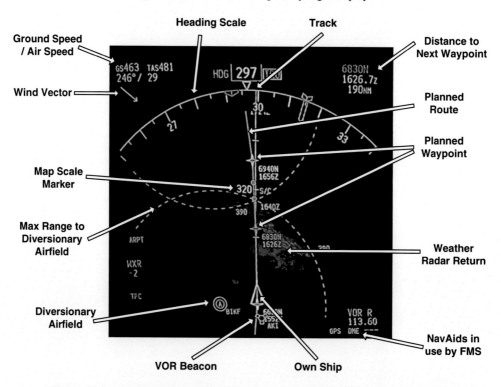

Figure 12.11   Boeing 777 navigation display with weather radar. Copyright TriplET

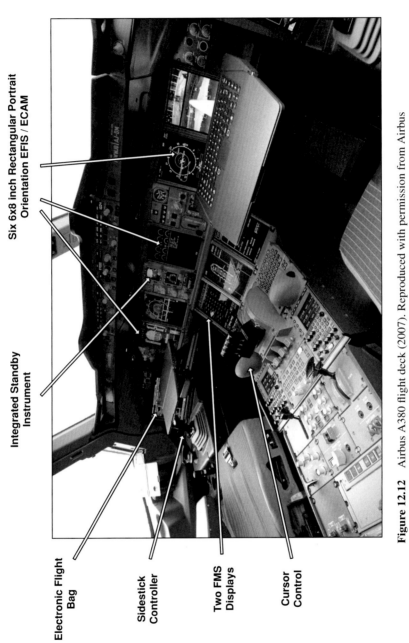

Six 6x8 inch Rectangular Portrait Orientation EFIS / ECAM

Integrated Standby Instrument

Electronic Flight Bag

Sidestick Controller

Two FMS Displays

Cursor Control

Figure 12.12 Airbus A380 flight deck (2007). Reproduced with permission from Airbus

Figure 12.13 Boeing 787 flight deck (2011). Copyright Alex Beltyukov

Head-Up
Display

Five Large Area 15.1 inch Diagonal
Multifunction Split-Screen Displays

Electronic Flight
Bag

Conventional Yoke
Control Inceptor

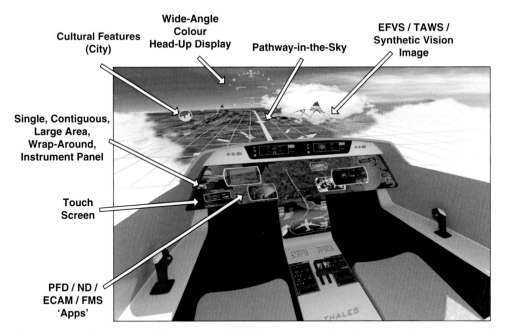

Figure 12.14 Generation 3.0 flight deck – ODICIS conceptual view. Reproduced by permission of Thales Avionics. Copyright Felix and Associés

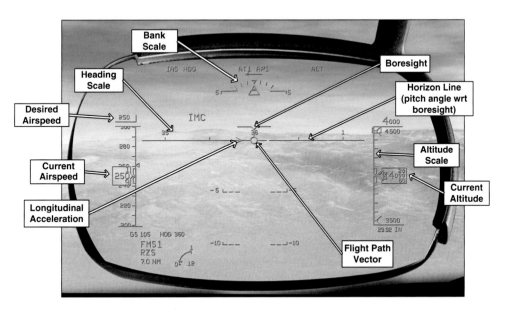

Figure 12.19 Typical HVGS symbology. Reproduced with permission from Rockwell-Collins

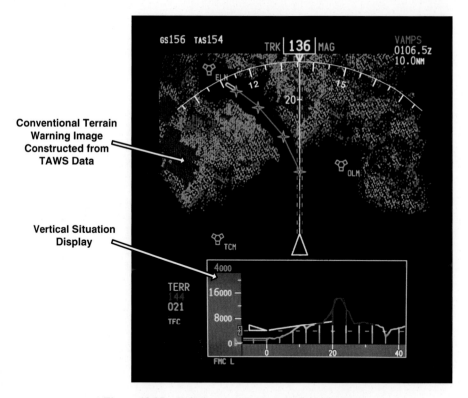

Figure 12.25 TAWS and VSD on a navigation display

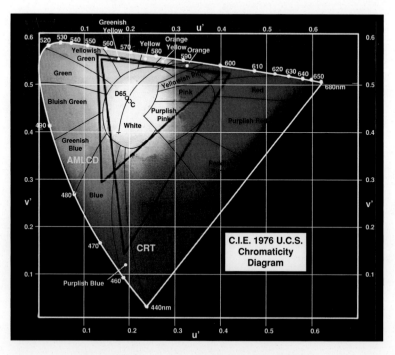

Figure 12.34 Colour perception and chromaticity

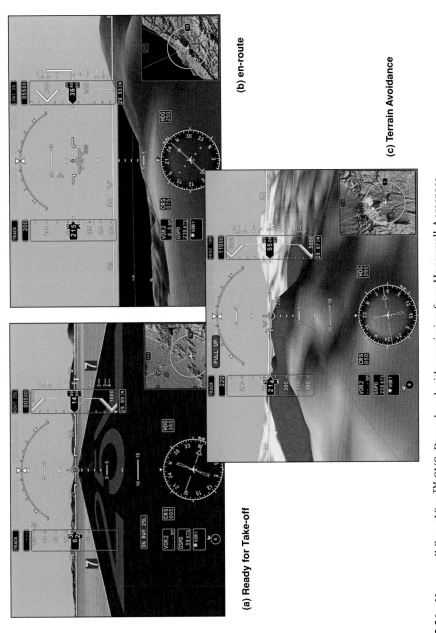

(a) Ready for Take-off

(b) en-route

(c) Terrain Avoidance

Figure 12.26 Honeywell SmartView™ SVS. Reproduced with permission from Honeywell Aerospace

HUD Symbology

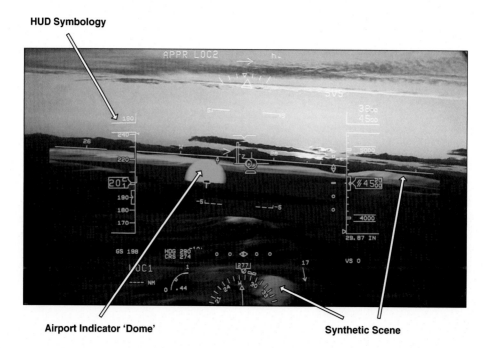

Airport Indicator 'Dome'   **Synthetic Scene**

Figure 12.27 Rockwell Collins HUD with synthetic vision. Reproduced with permission from Rockwell-Collins

PFD Symbology
(HUD style)

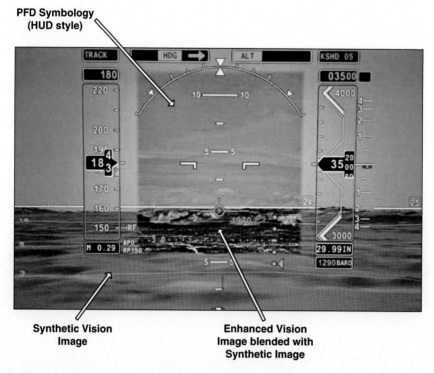

Synthetic Vision   **Enhanced Vision**
Image   **Image blended with**
                              Synthetic Image

Figure 12.28 Honeywell primary flight display with SVS and EVS overlay. Reproduced with permission from Honeywell Aerospace

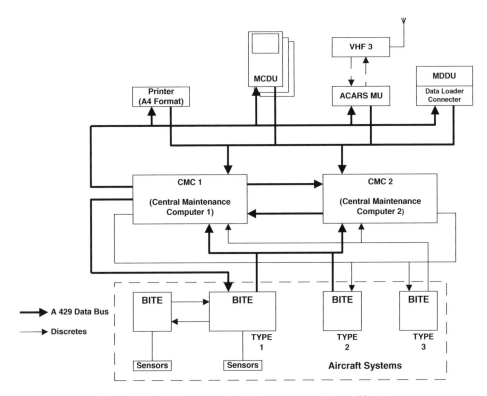

Figure 7.39   Airbus central maintenance computing architecture

data are menu-driven according to the maintenance page selected on the CDU, with further down-selection using the soft keys at the side of the display. The CMCs provide data to the CDUs for display, to the flight deck printer and to VHF system 3 for ACARS transmission as required. Reference [15] describes the operation of the system in significant detail.

In the Airbus approach, avionics systems are defined by three types:

- **Type 1** – These are characterised by ARINC 429 digital data bus inputs and outputs. A total of 34 basic and 9 optional systems comprising a total of 75 units are in this category.
- **Type 2** – These systems comprise a discrete and an ARINC 429 data bus input from the CMC. There are a total of 10 basic systems comprising a total of 19 units.
- **Type 3** – These systems typically are characterised by discrete inputs and outputs: there are a total of 4 basic and 1 optional system comprising a total of 8 units. Therefore the system as described can interface to a total of 102 LRUs if all options are fitted (91 basic-fit LRUs).

The most powerful aspect of the system relates to the menu-driven function provided by the CDU(s). On the display shown in Figure 7.40, several of the soft keys are highlighted and enable selection of sub-menus as stated on the CDU screen.

The Airbus system, in common with many other similar systems, uses the total capabilities of the CDU, albeit focused in the maintenance sense. In the example shown, various soft keys

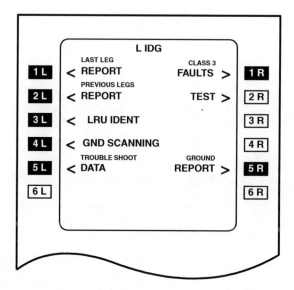

Figure 7.40   CDU maintenance menu-driven display

are highlighted and the associated maintenance/diagnostic function illustrated alongside the appropriate key. Typical systems allow the following capabilities at this level:

- last leg reporting;
- previous leg reporting;
- LRU identification;
- trouble-shooting options.

The precise options available will depend upon the implementations in a specific system; suffice to say that the menu-driven option allows the interrogation of system functions at an individual sensor level complete with status and scaling (accuracy information). Modern systems are usually also capable of interrogating data down to the module/sensor level and may also include a shop-level (second/third level) fault history for every module within the system. In this way modules with previous fault histories may be identified and correlated with the reported fault before removal from the aircraft. Rogue modules with persistent and repeatable fault histories may also be identified and investigated.

The system outlined provides a multilevel diagnostic capability based upon an ATA (functional) chapter basis. The original system did permit the download of diagnostic data via a 1.44 Mb disk if required, although this is probably somewhat outdated by the IT standards that prevail today.

7.10.2   Boeing 777 Central Maintenance Computing System

The Boeing Central Maintenance Computing System (CMCS) also uses an integrated approach to gathering maintenance data and displaying them to the technician. In the Boeing 777, the

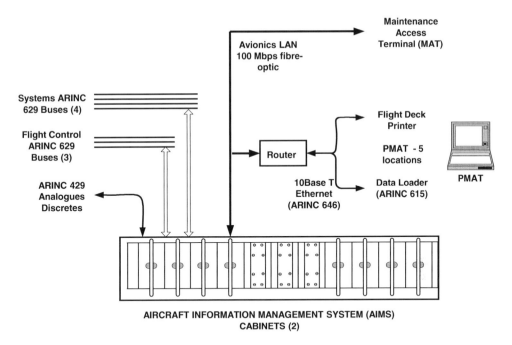

Figure 7.41   Boeing 777 Central Maintenance Computing System

Aircraft Information Management System (AIMS) is used to assemble the necessary data for the CMCS function as depicted in Figure 7.41. The AIMS comprises two integrated cabinets that interface to the aircraft systems as follows:

- to the aircraft systems and engines via 4 aircraft systems ARINC 629 data buses;
- to the flight control system via 3 flight control ARINC 629 data buses;
- to other aircraft equipment by means of ARINC 429 data buses, analogue and discrete signals.

The B777 avionics system is described in more detail elsewhere in this volume.

Either cabinet has the ability to pass data via a fibre-optic avionics LAN to the maintenance interface. This is a high data-rate fibre-distributed data interface (FDDI) bus operating at 100 Mbps that connects directly to the maintenance access terminal (MAT) and keyboard. Data are also passed via a router that performs a network interface function, enabling several devices to be connected via 10 BaseT, 10 Mbps Ethernet (ARINC 645) as follows:

- Flight deck printer.
- Portable maintenance access terminal (PMAT) at five receptacles around the aircraft.
- Data loader operating in accordance with ARINC 615.

The Boeing philosophy is therefore to use dedicated display equipment, whereas Airbus use the FMS CDU and dedicated computers. However, Boeing uses the existing AIMS cabinets for the data-gathering and processing function, whereas Airbus use dedicated computing resources – the two CMCs. The Boeing system has the following capabilities:

- LRU software loading;
- input monitoring;
- configuration reporting;
- access to shop fault history data (contained within local non-volatile memory on every LRU/module);
- onboard engine balancing;
- PSEU and air/ground rigging;
- reporting capabilities.

References

[1] Moir, I. (2006) TEOS Forum, *Fifty Years of Aerospace Power – Full Circle*, 28 June 2006, Paris.
[2] Moir, I. and Seabridge, A. (2008) *Aircraft Systems*, 3rd edn. John Wiley & Sons., Ltd., Chichester.
[3] RTCA DO-160C: *Environmental Conditions and Test Procedures for Airborne Equipment*, RTCA Inc., 1989.
[4] ADB-0100: Airbus Directives (ABD) and Procedures, ADB0100, Electrical and Installation Requirements.
[5] MIL-STD-704E: *Aircraft Electric Power Characteristics*, 1 May 1991.
[6] Boyce, J.W. *An Introduction to Smart Relays*, Paper presented at the SAE AE-4 Symposium.
[7] Layton, S.G. (1999) *Solid State Power Control*, 1999 Avionics Conference and Exhibition: Civil and Military Convergence, London Heathrow 17–18 November 1999.
[8] Khan, I. and Critchley, M. *Arc Fault Detector*. Leach International. Available at www.esterline.com/Portals/3/Products/Arc_Fault_Detection_paper.pdf. [accessed April 2013].
[9] Rinaldi, M.R. (1988) *A Highly Reliable DC Power Source for Avionics Subsystems*, SAE Conference Paper 881408, October 1988.
[10] Mitcham, A.J. and Grum, N. (1998) *An Integrated LP Shaft Generator for the More-Electric Aircraft*, IEE Colloquium on All-Electric Aircraft, London, January 1998, pp. 1–9.
[11] Andrade, L. and Tenning, C. (1992) *Design of the Boeing 777 Electric System*. Proceedings of the IEEE 1992 National Aerospace and Electronics Conference, May 18–22, pp. 1281–90, and also *IEEE Aerospace and Electronic Systems Magazine* 7, (7): 4–11.
[12] Haller, J.P., Weale, D.V. and Loveday, R.G. (1998) *Integrated Utilities Control for Civil Aircraft*, FITEC'98, London.
[13] Dodds, G. (2005) *A380 Electrical Architecture and Distribution*, RAeS Conference, London.
[14] MacDiarmid, I. (2010) Electromagnetic integration of aircraft systems, in *Encyclopedia of Aerospace Engineering*, (eds R.H. Blockley and W. Shyy), John Wiley & Sons, Ltd., Vol. 8: 5045–57.
[15] Airbus FAST magazine (1994) *A330/A340 Central Maintenance System,* April 1994.

Further Reading

[1] Wall, M.B. *Electrical Power System of the Boeing 767 Airplane*.
[2] Thom, J. and Flick, J. (1990) *New Power System Architecture for the 747-400,* Aerospace Engineering, May 1990.
[3] Thom, J. and Flick, J. (1989) *Design Features of the 747-400 Electric Power System.* SAE Technical Paper 892227, September 1989.
[4] Barton, A. (1992) *The A340 Electrical Power Generation System.* Aircraft Power Generation and Management AeroTech '92 Congress Seminar Papers, Seminar 36.

8

Sensors

8.1   Introduction

The primary autonomous sensing methods available to the modern civil airliner are:

- Air data derived from sensors that sense the pressure and temperature of the surrounding air as the aircraft flies through the atmosphere.
- Magnetic sensing of the Earth's field to derive a means of determining aircraft heading for navigation purposes.
- Inertial sensing using gyroscopes and accelerometers to derive aircraft attitude, body rates and velocities, to enable the flight crew to fly the correct attitudes and flight paths and to navigate the aircraft.
- Integrated systems combining both air data and inertial data – Air Data & Inertial Reference Systems (ADIRS).
- Radar sensors using onboard radar equipment to navigate, to avoid storm clouds and turbulent weather, and to determine aircraft altitude by radar means.

These sensors are carried on the aircraft as opposed to radio and navigation aids which are described in Chapter 9, Communications and Navigation Aids.

Each of the onboard sensor families possesses its own attributes and has associated strengths and shortcomings. A modern system will take account of these attributes, matching the benefits of one particular sensor type against the deficiencies of another. Onboard autonomous sensors are also used in conjunction with external navigation aids and systems to achieve the optimum performance for the navigation system. As will be seen, the capabilities of modern integrated navigation systems blend the inputs of multisensor types to attain high levels of accuracy and integrity that permit new navigation and approach procedures to be used. Indeed, in the crowded skies that prevail today, such highly integrated systems are becoming essential to assure safe and smooth traffic flow.

Civil Avionics Systems, Second Edition. Ian Moir, Allan Seabridge and Malcolm Jukes.
© 2013 John Wiley & Sons, Ltd. Published 2013 by John Wiley & Sons, Ltd.

8.2   Air Data Sensors

8.2.1   Air Data Parameters

Air data, as the name suggests, involves the sensing of the medium through which the aircraft is flying. Typical sensed parameters are dynamic pressure, static pressure, rate of change of pressure and temperature. Derived data includes barometric altitude (ALT), indicated airspeed (IAS), vertical speed (VS), Mach (M), static air temperature (SAT), total air temperature (TAT) and true airspeed (TAS). The simplest system provides ALT and IAS as a minimum; modern jet aircraft require Mach, VS, maximum operating speed (V_{mo}), maximum operating Mach (M_{mo}), SAT, TAT and TAS to satisfy the aircraft requirements. The evolution of the high-performance commercial and business jet aircraft of today, together with an increase in traffic on congested routes, has significantly influenced the design of the air data system in the following ways:

- By extending the dynamic range of the sensors involved with higher altitudes, higher airspeeds and greater temperatures.
- By increasing the use of air data onboard the aircraft, not just for navigation but for engine control, flight control and a whole range of other aircraft subsystems.
- The adoption of higher cruise altitudes has introduced a more severe environment for equipment located outside the pressurised cabin. Increasing complexity, density and functional requirements have also led to more complexity within the cabin.
- Demands for reduced vertical separation minima require greater accuracy of height sensing and methods of maintaining height within strict limits.

8.2.2   Pressure Sensing

Air data pressure parameters are sensed by means of pitot static probes and static sensors as shown in Figure 8.1.

Pitot pressure is the head of pressure created by the forward movement of the aircraft during flight. The dynamic pressure varies according to the square of the forward velocity of the aircraft. Static pressure is the local pressure surrounding the aircraft at a given altitude and may be used to determine the aircraft altitude. Static pressure may be measured by means of ports in the side of the pitot static probe or by means of static ports in the side of the aircraft skin. The exact relationship between these parameters will be described.

There are several types of error that affect these static pressure sensors. To avoid errors when the aircraft yaws and due to changes in the aircraft angle of attack, static ports are located on both sides of the aircraft. There is inevitably some error associated with the less than ideal positioning of the static ports – this is known as 'static source error' and will need correction as described later. A typical pitot probe used on a civil airliner is described in Reference [1].

8.2.3   Temperature Sensing

Temperature sensing involves positioning a probe in the airflow and sensing the change in resistance associated with the temperature rise at the stagnation point (see Figure 8.2).

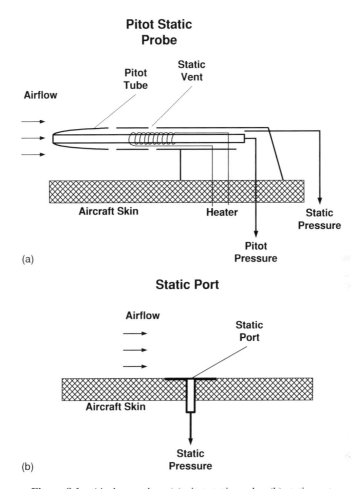

Figure 8.1 Air data probes: (a) pitot static probe: (b) static port

Outside air temperature (OAT) affects aircraft performance in a number of ways. During take-off it directly affects the thrust available from the engines and lift due to air density, both of which can significantly affect the aircraft take-off distance and operational margins. In the cruise, engine performance and fuel consumption are affected, and total air temperature (TAT) needs to be calculated. In adverse weather, static air temperature (SAT) indicates the potential for icing, while total air temperature (TAT) is closer to the leading edge temperature and is more indicative of icing accretion. The outside air temperature (OAT) has to be corrected to represent SAT or TAT. A technical paper describing the detailed working of a TAT sensor may be found in Reference [2].

Both pitot and total temperature probes are susceptible to icing and so are equipped with heating elements to prevent this from happening without affecting the accuracy of the sensing elements. Since the blockage of a pneumatic pipe will lead to erroneous air data, the correct operation of the heating element needs to be continuously monitored.

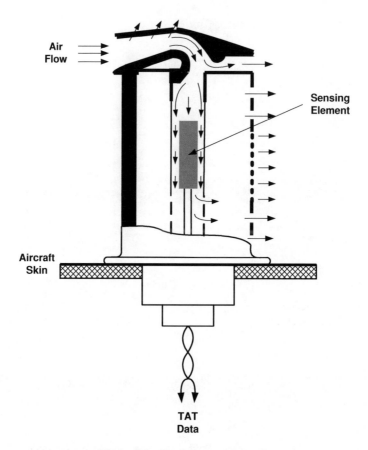

Figure 8.2   Total temperature probe

8.2.4   Use of Pressure Data

In a basic air data instrumentation system, the combination of sensed pitot and static pressure may be used to derive aircraft flight data as shown in Figure 8.3.

By using the capsule arrangement shown on the left, pitot pressure is fed into the capsule while static pressure is fed into the case surrounding the capsule. The difference between these two parameters, represented by the deflection of the capsule, represents the aircraft airspeed. This permits airspeed to be measured. The equation is:

$$q = \tfrac{1}{2}\rho V^2 = (\mathrm{Pt} - \mathrm{Ps})$$

where q is the dynamic pressure, ρ is the air density and V is the velocity.

In the centre capsule configuration, static pressure is fed into the case of the instrument while the capsule itself is sealed. In this case, capsule deflection is proportional to changes in static pressure and therefore aircraft altitude. This allows aircraft barometric altitude to be measured.

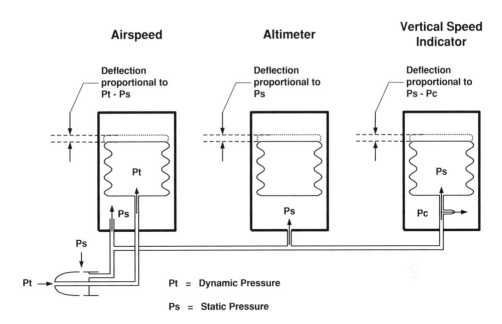

Figure 8.3   Use of pitot and static pressure to derive simple indications

In the arrangement shown on the right of Figure 8.3, static pressure is fed into the capsule. It is also fed via a calibrated orifice into the sealed case surrounding the capsule. In this situation the capsule defection is proportional to the rate of change of altitude. This permits the aircraft rate of ascent or descent to be measured.

Originally, the portrayal of aircraft airspeed, altitude and rate of change of altitude was accomplished using discrete instruments: airspeed indicator, altimeter and vertical speed indicator (VSI). In these simple instruments any scaling required was accomplished by means of the mechanical linkages.

8.2.5   Pressure Datum Settings

Determination of altitude from pressure measurements is based upon a standard atmosphere in which pressure, density and temperature are functions of altitude. The altitude resulting from these calculations is called pressure altitude and represents the altitude above sea-level under these standard atmospheric conditions. As a standard atmosphere rarely exists, certain steps are necessary to establish other useful and reliable operating datums for aircraft altitude and height. Altimeters possess a barometric set knob – usually located at the bottom right of the instrument – which allows various static datums to be set; the reason for this datum set capability is described below. Figure 8.4 defines the basic criteria for the various altimeter and height settings:

- Altitude or QNH is defined as the barometric height between the aircraft and mean sea-level (MSL). Therefore an altimeter with a QNH setting indicates the height above mean sea-level.

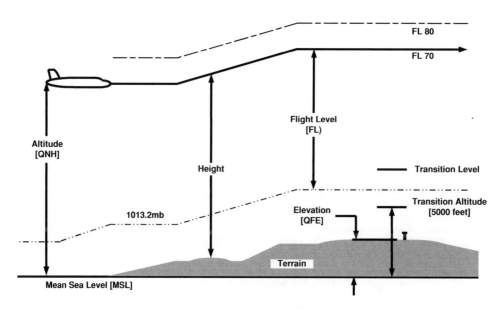

Figure 8.4   Altimeter settings

- Elevation relates to the height of a particular feature, in particular in this case the height of an airfield above sea-level. This is geographically fixed such that the height of the airfield will be fixed in relation to sea-level.
- Height or QFE relates to the height of the aircraft above a particular feature (airport). An altimeter with the QFE correctly set will read zero at the appropriate airfield.
- Above a transitional altitude, which depends upon the country and the height of the local terrain, all aircraft altimeters are set to a nominal standard altimeter setting (SAS) value of 29.92 in Hg/1013.2 mb; this ensures that all the altimeters of aircraft in a locality are set to a common datum and therefore altitude conflicts due to dissimilar datum settings may be avoided. Typically in the UK the transition altitude is around 5000 feet, whereas in the US the figure is 18,000 feet. Initiatives are underway to establish a common figure worldwide in order to standardise operational procedures. Above the transitional altitude, all aircraft altitudes are referred to in terms of flight levels (FL).

The importance of these definitions will become more apparent in later sections when reduced vertical separation minima (RVSM) and auto-land decision height (DH) and decision altitude (DA) criteria are described.

As aircraft systems became more complex and more sophisticated propulsion and flight control laws were adopted, the number of systems that required air data increased. Therefore the provision of air data in various aircraft navigation, flight control and other subsystems required a more integrated approach. The computation tasks involved the following:

- conversion of the sensed parameters into a more useful form; e.g. static pressure in terms of millibars or in inches of mercury into altitude (in feet or metres) and dynamic pressure in inches of mercury into airspeed (knots);

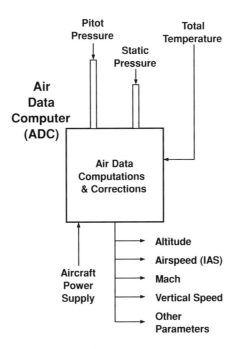

Figure 8.5   Air data computer

- combination of two or more parameters to obtain a third parameter; e.g. airspeed and altitude to obtain Mach, or Mach and temperature to obtain true airspeed;
- to correct for known errors as far as possible.

8.2.6   Air Data Computers (ADCs)

These requirements led to the introduction of one or more air data computers (ADCs), which centrally measured air data and provided corrected data to the recipient subsystems. This had the advantage that while the pilot still had the necessary air data presented to him, more accurate and more relevant forms of the data could be provided to the aircraft systems. Initially this was achieved by analogue signalling means; however, with the evolution of digital data buses in the late 1970s, data were provided to the aircraft subsystems by this means, notably by the use of standard ARINC 429 data buses. The function of an air data computer is to provide the outputs shown in Figure 8.5. The unit contains the capsules necessary to measure the raw air data parameters and the computing means to calculate the necessary corrections. In earlier implementations, analogue computing techniques were utilised. With the advent of cost-effective digital processing, together with low-cost digital data buses, alternative methods of implementation became possible.

Airspeed measurements in the air data computer are derived according to the computations summarised in Figure 8.6 and described below:

- **Indicated airspeed (IAS)**. IAS is the parameter proportional to pitot minus static or dynamic pressure and is directly related to the aerodynamic forces acting on the wings (lift) and

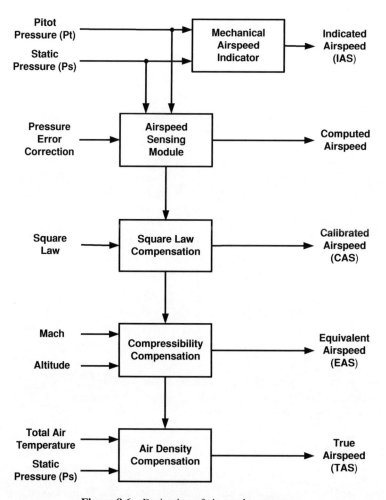

Figure 8.6 Derivation of airspeed parameters

control surfaces. IAS is therefore essential for defining aircraft aerodynamic performance and structural limitations. Mach number is derived from a combination of IAS and altitude and is expressed as a ratio of the aircraft speed to the speed of sound at that particular altitude. Typical reference speeds based upon IAS or Mach are listed below:

V_{stall}:   aircraft stall speed

$V_{rotation}$:   aircraft rotation speed

$V_{gear\ extend}$:   aircraft gear extension maximum

$V_{flaps\ extend}$:   aircraft flaps extend maximum

$V_{maximum\ operating}$:   aircraft maximum operating speed

$M_{maximum\ operating}$:   aircraft maximum operating Mach number

These reference speeds determine the flight envelope defined in the aircraft operating manual. They are used by the aircrew to operate the aircraft within its structural and performance limitations for continued safe operation. For crucial limiting speeds, placards will be placed at prominent positions on the flight deck.

- **Computed airspeed**. Computed airspeed is the IAS corrected for static pressure errors. It is not used as a parameter in its own right but is used as a basis for further corrections and calculations.
- **Calibrated airspeed (CAS)**. CAS is the computed airspeed with further corrections applied for non-linear/square law effects of the airspeed sensing module.
- **Equivalent airspeed (EAS)**. EAS is achieved by modifying CAS to allow for the effects of compressibility at the pitot probe, thereby obtaining corrected airspeed for varying speeds and altitudes.

Typical parameters provided by an ADC are:

- barometric correction;
- barometric corrected altitude;
- altitude rate;
- pressure altitude;
- computed airspeed;
- Mach number;
- true airspeed;
- static air temperature;
- total air temperature;
- impact pressure;
- total pressure;
- static pressure;
- indicated angle of attack;
- overspeed warnings;
- maximum operating speeds;
- maintenance information.

For an extensive reference regarding air data calculations, refer to the Goodrich Corporation Reference [3].

8.2.7 Airstream Direction Detectors

Other air data parameters of interest included the airstream direction detector (ADD). This measures the direction of the airflow relative to the aircraft and permits the angle of incidence – also referred to as angle of attack – to be detected. Typically this information is of use to the flight control system to modify the pitch control laws, or in a stall warning system to warn the pilot of an impending stall. A typical airstream detection sensor is shown in Figure 8.7 (see also [4, 5]).

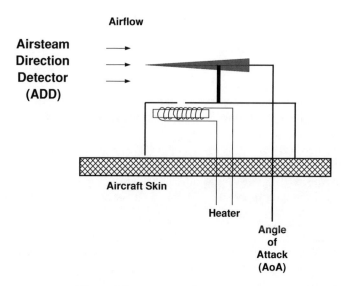

Figure 8.7 Airstream direction detector

8.2.8   Total Aircraft Pitot-Static System

The small bore pneumatic sensing tubes that route the sensed pitot or static pressure throughout the aircraft pose significant engineering and maintenance penalties. As the air data system is critical to the safe operation of the aircraft, it is typical for three or four or more alternate systems to be provided. The narrow bore of the sensing tubes necessitates the positioning of water drain traps at low points in the system where condensation may be drained off – avoiding blockage of the lines due to moisture accumulation. Finally, following the replacement of an instrument or disturbance of any section of the tubing, pitot-static leak and independent checks are mandated to ensure that the sensing lines are intact and leak-free and that no corresponding instrumentation sensing errors are likely to occur.

Flight instrumentation and correct interpretation relating to air data is paramount for safe flight, although a number of accidents have occurred where the flight crew misinterpreted their instruments when erroneous data was presented. Reference [5] gives a good overview of how to understand and counter these effects. Angle of attack measurements and the associated display data have been used in the military fighter community for many years; Reference [6] outlines how to get the best use of this parameter in an air transport setting.

The advent of digital computing and digital data buses such as ARINC 429 meant that computation of the various air data parameters could be accomplished in air data modules (ADMs) closer to the pitot-static sensing points. Widespread use of the ARINC 429 data buses enabled the data to be disseminated rapidly throughout all the necessary aircraft systems. Now, virtually all civil transport aircraft designed since the early 1990s have adopted the air data module implementation. Figure 8.8 shows a modern air data system for an Airbus aircraft.

The Airbus system comprises three pitot and six static sensors. Pitot sensors 1 to 3 and static sensors 1 and 2 on each side of the aircraft are connected to their own air data module. Each of these seven air data modules provides pitot- or static-derived air data to the display, flight

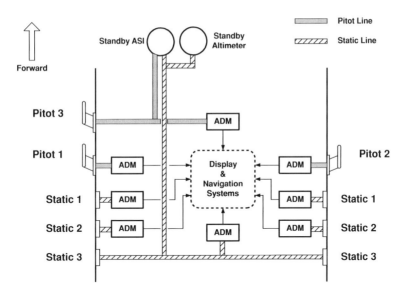

Figure 8.8   Aircraft-level pitot-static architecture

control and navigation systems, among others. Static sensors 3 on each side are connected to a common line and a further air data module. Pitot 3 and the combination of the static 3 sensors are also used to provide pitot and static pressure directly to the aircraft standby airspeed indicator and standby altimeter.

Air data are generally regarded as accurate but in the longer term rather than the short term. The fact that air data sensing involves the use of relatively narrow bore tubing and pneumatic capsules means that there are inherent delays in the measurement of air data as opposed to some other forms of air data. The use of ADMs situated close to, or integrated into, the probes will reduce such errors, as well as eliminating condensation and icing with consequent maintenance benefits.

Air data may be represented as misleading if failures or blockages occur in the pitot-static plumbing system; however, an alert flight crew should be fully capable of resolving any erroneous data that could be presented [5, 6].

8.3   Magnetic Sensors

8.3.1   Introduction

The use of the Earth's magnetic field to sense direction and the use of north-seeking devices to establish the direction of magnetic north for the purposes of navigation is one of the oldest forms of sensor. A north reference is traditionally used to determine an aircraft's bearing, be it true or magnetic. A magnetic sensing device – called a flux valve – is usually located in the outer section of one of the aircraft wings, well clear of any aircraft-induced sources of spurious magnetism. As may be seen in Figure 8.9, the axis of the Earth's magnetic field may be considered to be analogous to a simple bar magnet. This magnetic dipole has its field

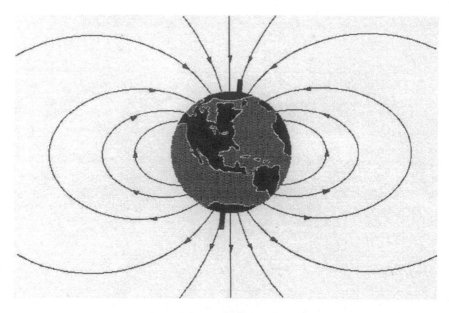

Figure 8.9   Earth's magnetic field

lines originating at a point near the south pole and terminating at a point near the north pole; however, the field is skewed from true geodetic north by ~11.5° in the UK.

8.3.2   Magnetic Field Components

The Earth's field lines enter the Earth at a considerable angle to the local horizontal plane; this angle is called the magnetic angle of inclination. In the US and Europe this angle is around 70°. The Earth's magnetic field (H) is the vector sum of components H_x, H_y, H_z measured in the orthogonal axis set shown in Figure 8.10, in which the angle of declination is shown.

The H_x and H_y components, which are in the local horizontal plane, are used to determine the compass heading with reference to the magnetic north pole. The H_z component is irrelevant for the purposes of simple navigation. However, allowance has to be made for the fact that the magnetic and geodetic poles do not coincide, and also for the fact that there are considerable variations in the Earth's magnetic characteristics across the globe. These factors are measured and mapped across the globe such that the necessary corrections may be applied. The correction term is called the angle of declination and is a corrective angle to be added or subtracted from the magnetic heading to give a true (geodetic) compass heading. Positive angular declinations represent easterly corrections, while negative angles represent westerly corrections. These corrections are measured, charted and periodically updated. Figure 8.11 shows the variation in declination from the world magnetic model for 2000 published by the British Geological Survey in conjunction with the US Geological Survey [7]. Reference [8] updates this model for 2005–2010.

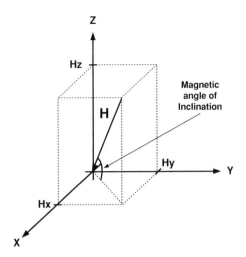

Figure 8.10 Earth's magnetic field components

8.3.3 Magnetic Variation

Perhaps easier to understand is a simplified North American map of declination derived a few years previously, as shown in Figure 8.12. In this figure it may be readily noted that declination for the northeastern US seaboard in the vicinity of Boston would be in the region of 14° west (minus 14°), the corresponding correction for Seattle on the northwestern US seaboard would

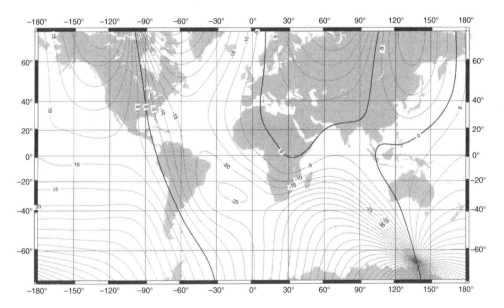

Figure 8.11 World magnetic model 2000

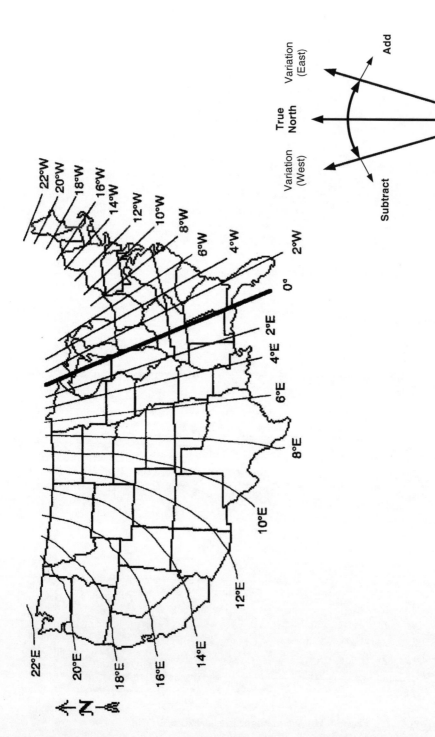

Figure 8.12 North American variation map

be 20° east (plus 20°). When the flight time of ∼4.5 to 5 hours for a typical modern airliner between these points is considered, the magnitude of the variation may be appreciated. This figure also shows how the difference between true and magnetic headings should be calculated.

Early generation flux valves were commonly gimballed in two axes to allow a degree of freedom in pitch and roll, but were fixed in azimuth. More recently, three-dimensional solid-state magnetic sensors have become available which offer a strap-down sensing capability that can fit within the footprint of the existing gimballed design. These strap-down sensors are form-and-fit compatible with the older sensor and can resolve the magnetic field into X, Y and Z components.

8.3.4 Magnetic Heading Reference System

Magnetic sensing therefore provides a very simple heading reference system that may used by virtually all aircraft today, although in transcontinental and transoceanic capable aircraft, inertial and GNSS navigation methods are likely to be preferred. In general aviation aircraft of limited range and performance, it is likely to be the only heading reference available. This type of system is commonly used for navigation even after the late 1960s/early 1970s when inertial platforms became commonly available.

The magnetic sensor or flux-valve described earlier provides magnetic heading, and when combined with a directional gyro (DG), which provides an inertial heading reference, can provide a 'magnetic heading and reference system' (MHRS) as shown in Figure 8.13, where the interrelationship between sensor and display elements may be seen.

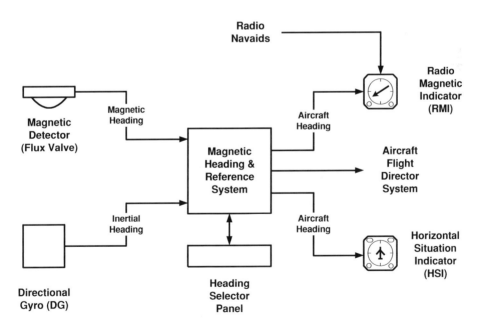

Figure 8.13   Magnetic heading and reference system

8.4   Inertial Sensors

8.4.1   Introduction

Inertial sensors are associated with the detection of motion in a universal (non-Earth) referenced set. Inertial sensors comprise:

- position gyroscopes;
- rate gyroscopes;
- accelerometers.

A useful introduction to inertial navigation may be found in Reference [9].

8.4.2   Position Gyroscopes

Gyroscopes are most commonly implemented as a spinning mass or wheel which tends to hold its position in a space-referenced attitude set due to the principle of conservation of angular momentum. Position gyroscopes or gyros use this property to provide a positional or attitude reference: typically aircraft pitch position, roll position or yaw position (heading). Position gyros are used in heading and reference systems to provide the aircraft with vital information regarding the aircraft attitude for a range of aircraft subsystems.

A simple gyroscope may be represented by the simple spinning wheel in Figure 8.14a. The wheel is rotating in the direction shown around the X-axis, and once it has been spun up to its operating speed will preserve that orientation in space. The degree of 'stiffness' of the gyro will depend upon its angular momentum, in turn depending upon mass and speed of rotation.

In order to preserve the spatial position of a directional gyro, a gimballing system needs to be used as shown in Figure 8.14b. The simple system shown comprises an inner, centre and outer gimbal mechanism. The vertical pivots between the inner and centre gimbals allow freedom of movement in rotational direction A. The horizontal pivots between centre and outer gimbal allow freedom of movement in rotational direction B. Therefore if the outer gimbal is fixed to the aircraft frame of axes, the gyroscope will remain fixed in space as the aircraft moves. Hence pitch, roll and yaw (heading) attitude may be measured.

8.4.3   Rate Gyroscopes

The gyro also has the property of precession when an external force is applied. Referring back to Figure 10.14a: if a force is applied to the forward edge in the Y-axis, shown by the white Arrow A, the actual force will not act as one might imagine by rotating about the Z-axis. Instead, due to the properties of the gyro, the force will actually act at the bottom of the wheel – shown by the black Arrow B. This force will cause the gyro X-axis to tilt counter-clockwise in the direction shown. This property of precession allows body rates to be sensed.

Gyroscopes can therefore be used to provide information relating to the aircraft body rates: pitch rate, roll rate and yaw rate. These rate gyros use the property of gyro precession described. When a gyro is rotating and the frame of the gyro is moved, the gyro moves or precesses due to the effects of the angular momentum of the gyro. By balancing and measuring this precession

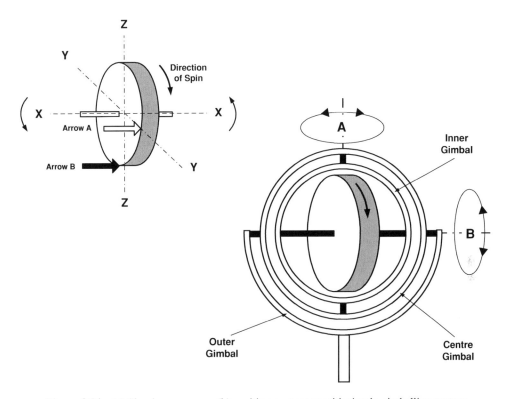

Figure 8.14   (a) Simple gyroscope; (b) position gyroscope with simple gimballing system

force, the applied angular rate of movement applied to the gyroscope frame may be measured. This information, together with the attitude data provided by the position gyros, is crucial in the performance of modern flight control or fly-by-wire (FBW) systems as described in Chapter 10.

In early systems, gyroscopes were air-driven; later electrically-driven gyroscopes became the norm. As these were both rotating devices, bearing friction and wear were major factors in mitigating against high accuracies. In modern systems the gyroscopes used are likely to be laser devices. The principle of operation of a laser ring gyro (LRG) is depicted in Figure 8.15. The LRG uses laser light in the visible or near infra-red wavelength to sense angular motion. Using a coherent, highly stable laser source and a series of mirrors to create a continuous light path, two travelling-wave laser beams are formed independently, one moving in a clockwise and the other in a counter-clockwise direction. When the sensor is stationary in inertial space, both beams have the same optical frequency. When the sensor is rotated around the axis perpendicular to the plane containing the beams and mirrors – shown in the figure – differences occur in the frequency of the two beams. The frequency of each beam alters to maintain the resonance necessary for laser action. The path difference between the beams is very small, ∼1 nanometre (1 nanometre = 1×10^{-9} metres), and therefore a source of high spectral purity and stability is required; typically, helium neon gas lasers are used.

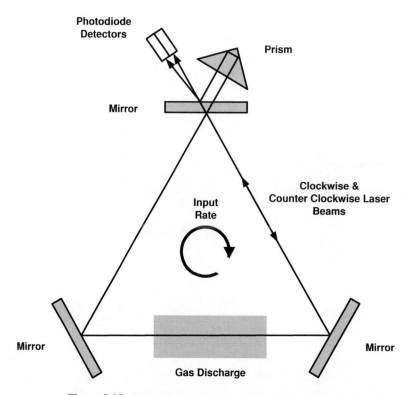

Figure 8.15   Laser-ring gyroscope – principle of operation

The rotational motion is sensed by allowing a small portion of light from each beam to 'leak' through one of the mirrors, and the two beams are combined using a prism to form an interference pattern on a set of photodiodes. The frequency difference between the two beams causes interference fringes to move across the detectors at a frequency proportional to the frequency difference between the beams, and hence proportional to the input angular rate. At very low angular rates the beams can effectively 'lock-in' to the same frequency due to optical back-scattering within the device, and this can cause a dead-band effect. This may be overcome by mechanically dithering or physically vibrating the entire cavity to obviate the 'lock-in' condition. Alternatively, four beams of differing frequencies may be used within the cavity, therefore enabling the 'lock-in' effect to be overcome. LRGs can offer a performance in the region of 0.001°/hour of drift; less complex and cheaper fibre-optic gyroscopes (FOGs) can produce drift rates of the order of 10°/hour or more.

8.4.4   Accelerometers

Accelerometers are devices that measure acceleration along a particular axis. The measurement of acceleration can be integrated using computers used to derive aircraft velocity and position. All accelerometers use the principle of sensing the force on a loosely suspended mass, from which the acceleration may be calculated.

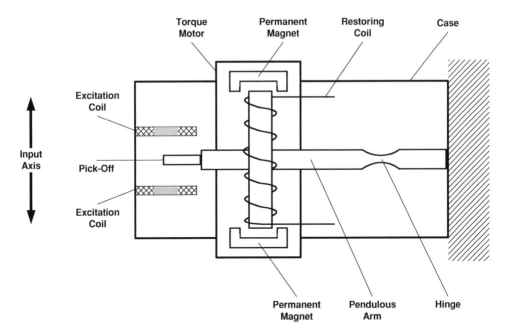

Figure 8.16 Pendulous force feedback accelerometer – principle of operation

A common accelerometer used today is the pendulous force feedback accelerometer shown in Figure 8.16. The device is fixed to the body structure whose acceleration is to be taken – shown as the structure on the right. As the structure and the pendulous arm move, the pick-off at the end of the pendulous arm moves with respect to two excitation coils. By sensing this movement a corrective current is applied to the restoring coil that balances the pick-off to the null position. As the restoring coil is balanced between two permanent magnets above and below, the resulting current in the restoring coil is proportional to the applied acceleration – in this example in the vertical direction. Typical accelerometer accuracies may vary from 50 mg (50×10^{-3} g) down to a few µg (1 µg = 1×10^{-6} g).

8.4.5 Inertial Reference Set

As for the magnetic sensors, inertial sensors may be arranged such that inertial rates are sensed within an orthogonal axis set as shown in Figure 8.17, which is referenced in space.

The inertial platform uses a combination of gyros and accelerometers to provide a platform with a fixed reference in space. By using the combined attributes of position and rate gyros and accelerometers, a stabilised platform provides a fixed attitude reference in space and when fitted in an aircraft can provide information about aircraft body rates and acceleration in all three axes. Suitable computation can also provide useful information relating to velocity and distance travelled in all three axes.

This is a significant achievement in establishing the movement of the aircraft; however, it suffers from a major disadvantage. The aircraft body data are derived relative to a

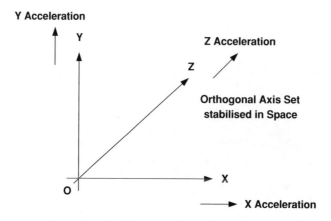

Features of a Mechanical Inertial Platform:

Platform is Gyro Stabilised in Space

Sensitive Accelerometers detect acceleration in the direction of the orthogonal axes: Ox, Oy, Oz

Accelerations are integrated to give first velocity and then position in the Ox, Oy, Oz axes

Platform readings can be transformed to relate to earth rather than spatial axes and coordinates

Figure 8.17   Inertial navigation axes – inertial platform

reference set in space, whereas to be useful in navigating on Earth the system needs to be referenced to a global reference set. The development of the inertial navigation system (INS) provided the additional computation to provide this essential capability. See Figure 8.18 in order to understand how the space-referenced axes set is transformed to an Earth-reference set.

The process by which this axis transformation is accomplished is depicted in Figure 8.19. There are a set of calculations using matrix transformations that are performed in the flow from left to right. These calculations are the same for gimballed and strap-down platforms, except that the latter has a final stage since the sensor set and vehicle axes will not agree. The process is as follows, and is well described in Reference [10].

- The space reference set is transformed to align with the Greenwich Meridian.
- The Greenwich meridian reference set is aligned to true north.
- The true north reference set is aligned to a great Circle.

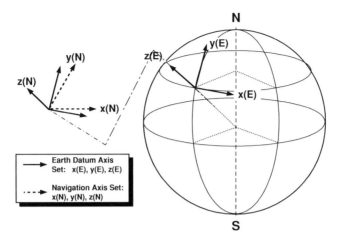

Figure 8.18 Transformation from space to Earth axis set

- The strap-down platform performs calculations to allow for vehicle reference set misalignment. The strap-down platform is sometimes referred to as an 'analytical platform' as all its mechanisation is conducted mathematically.

Once the Earth reference axis set is established, then the platform may sense attitude, attitude rates and accelerations in each axis in a meaningful fashion to fly and navigate the aircraft (see Figure 8.20). Either a gimballed or a strap-down system may provide these data.

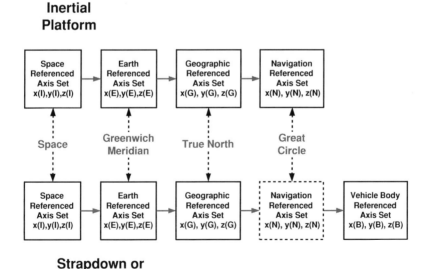

Figure 8.19 Transformation process

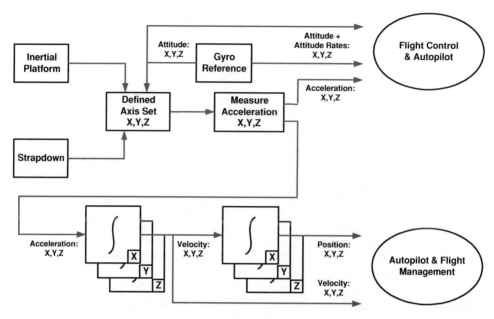

Figure 8.20   Inertial platform sensed and derived data

8.4.6   Platform Alignment

Once aligned, mathematically in the case of the strap-down platform, the platform provides the following data:

- Attitude and attitude rates for all three axes. This provides primary information for the systems associated with flight control.
- Acceleration data for all three axes from which velocity and present position may be derived (calculated). Body accelerations are very useful for flight control and velocity, and present position data are useful for navigation.

The alignment of the INS platform usually takes several minutes and is performed during power-up. The flight crew enter the latitude and longitude of the aircraft's present position and commence the alignment procedure; this information is available to them at the aircraft gate. The alignment procedure involves both levelling and alignment processes. The first task to be accomplished is coarse levelling, followed by fine levelling. At a later stage the gyrocompass process is begun (see Figure 8.21).

The levelling process drives the X and Y axes so that they are parallel with the Earth's surface. Once levelled, the Z axis points vertically towards the centre of the Earth. In a gimballed platform the device uses its own gimbal servos to null the outputs of the horizontal X and Y accelerometers. Fine levelling uses the gyro torque commands to seek the level position (see Figure 8.22).

The gyro-compass process moves the X–Y plane such that the Y axis points north and the X axis points east. This is achieved by rotating the platform to null the output of the Y axis

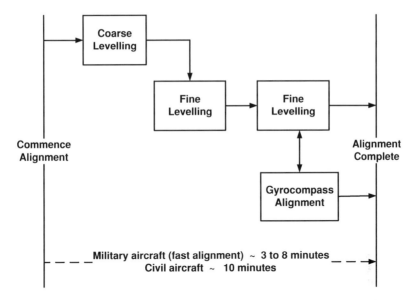

Figure 8.21 Coarse and fine alignment

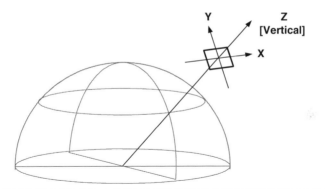

Levelling:

Levelling is the initial process associated with aligning the platform. It involves levelling the X and Y axes such that the platform is perpendicular to the local vertical. At this stage the platform is not necessarily aligned to navigation axes - east and north

The coarse levelling process is performed by using the gimbal servos to rotate the platform such that the outputs of the horizontal accelerometers are nulled. High slew rates ~ several hundred degrees/sec may be achieved

Fine levelling is accomplished by using the gyro torque commands. The rate of levelling is limited by the gyro torquer precession speed and hence this is a slower process

In a strapdown application the 'torquing' effectively performs an axis transformation such that the platform axes are aligned to that of the vehicle

Figure 8.22 Levelling process

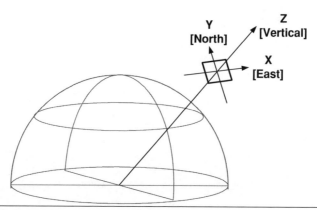

Figure 8.23 Gyro-compassing process

accelerometer. At this point the platform is aligned with a reference set aligned with the Earth and a starting present position defined by the latitude and longitude entered by the flight crew (see Figure 8.23).

The accuracy of the alignment dictates the accuracy that the inertial navigation (IN) platform will provide to the flight crew during flight. This in turn depends upon the accuracy achieved in terms of levelling and the gyro-compassing processes. As part of the process is dependent upon the aircraft latitude, it is difficult to get an accurate alignment above 60°N or below 60°S. In use during flight, the platform calculates movement across the surface of the Earth. It also has to allow for the Earth's rotation variation with latitude. To do this it needs the Earth starting point in terms of latitude and longitude which it was given during the alignment, and it calculates the movement of the platform across the rotating globe that is the Earth. These calculations are complex and their mechanisation is described in full in Reference [10]. Typical information provided for the aircraft by reference to the flight plan residing in the flight management system (FMS) is as follows:

- three-axis accelerations;
- three-axis velocities;
- present position – latitude and longitude;
- distance along and across track;
- angle of drift;

Table 8.1   Typical inertial navigation platform errors

Error type	Sensed error	Navigation error
Tilt/acceleration error	100 micro-radians (∼0.0057 degree) resulting in 100 micro-g acceleration error	Oscillatory Schuler pendulum: Peak error 0.7 nm/hour
Gyro drift error	0.01 degree/hour	Ramp with oscillatory Schuler pendulum superimposed: ∼0.7 nm/hour
Azimuth gyro drift error	0.2 degree/hour	More complex 'square law' function: ∼1 nm/hour (less sensitive in short term)

- time to next waypoint and subsequent waypoints;
- calculated wind-speed and direction, etc.

Because of the limiting factors that apply during platform alignment and some of the complex error functions that apply during the mechanisation process, a typical IN platform will have error characteristics as shown in Table 8.1. The errors are defined in terms of nautical miles (nm) per hour, as the longer the aircraft flies the greater the errors will become. As a rule of thumb a typical IN platform would be expected to accrue errors of ∼1 nm/hour. After a 10-hour flight the errors could have accrued to 10 nm which is not acceptable for the navigation standards demanded today. To obviate these errors the platform has to be updated using other sources. The most accurate way to do this is to use GNSS data, which is discussed elsewhere.

8.4.7   Gimballed Platform

Early inertial navigation systems used a gimballed and gyro-stabilised arrangement to provide a stable platform on which the inertial sensors were located. The platform was stabilised using rate information from the gyros to drive torque motors which stabilised the platform in its original frame of axes, independent of movement of the aircraft. Resolvers provided angular information on the aircraft with relation to the platform, and hence aircraft attitude could be determined. A diagram representing a gimballed, stabilised platform is shown in Figure 8.24.

The inertial sensors located on the stable platform preserve their orientation in space and are:

- X accelerometer X gyro
- Y accelerometer Y gyro
- Z accelerometer Z gyro

Inertial systems have the advantage that they are very accurate in the short term – no settling time is required as for air data. Conversely, inertial sensors have a tendency to drift with time and so become progressively less accurate as the flight continues. As will be seen in other

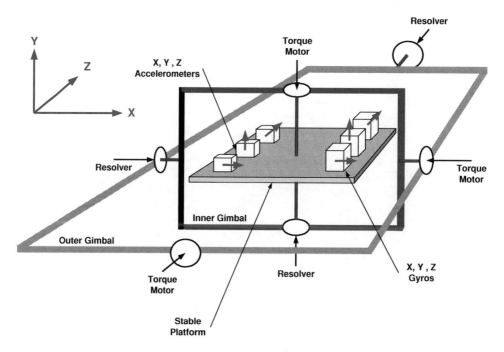

Figure 8.24   Representation of a gimballed platform

chapters, GPS or other satellite-based systems can be used to provide an automatic periodic correction of high accuracy.

The advent of digital computers greatly facilitated the ability to undertake this computation. On an aircraft such as a Boeing 747 Classic, three INSs would be provided and additional calculations performed to establish the best estimate of aircraft position according to information provided by all three systems.

Operational errors accrue due to the fact that the vertical reference (local vertical) has to be updated as the aircraft/platform moves at velocity V, in time t, over the surface of the globe. The product of V and t is known as the 'transport rate', in other words, how far the platform moves in a given time. The correction that needs to be applied is proportional to the transport rate divided by the radius of the earth, R. Corrections are made by using the accelerometer data, once integrated to give the aircraft velocity, V. If the accelerometers are not perfectly aligned with the Earth axes, then components of the gravity force will be detected and integrated to introduce velocity errors

The overall effect is to introduce a sinusoidal error with a period of ∼88 minutes. This is called the Schuler loop error, and whilst careful platform alignment will minimise the error it will always exist, and on a typical civil aircraft the error will be ∼1 nautical mile each side of track with a period of 88 minutes. Advanced processing of the sensors, especially using Kalman filters, enables this error to be minimised on a modern INS. However, it will always be present and can be reduced by allowing the longest possible platform alignment time, thereby minimising the initial vertical alignment errors

8.4.8 Strap-Down System

Many modern inertial systems are strap-down, in other words, the inertial sensors are mounted directly on the aircraft structure and no gimballed platform is required. The inertial signals are resolved mathematically using a computer prior to performing the necessary navigation calculations. The removal of the stabilised platform with its many high-precision moving parts and the use of modern LRG sensor technology considerably increases the system reliability. Strap-down systems are claimed to be around five times more reliable than their stabilised platform predecessors

With the advent of air data modules (ADMs), a combined unit called Air Data and Inertial Reference System (ADIRS) performs all the necessary calculations on air data and inertial data to provide the navigational information that the aircraft requires. Examples of modern ADIRS systems are addressed in the next section.

Notwithstanding the significant problems of providing accurate long-term data for navigation purposes, it should not be forgotten that the INS is the primary source of attitude information for the aircraft, and as such it is a vital piece of equipment. Careful monitoring of the platform-derived aircraft attitude is needed at all times to prevent loss of attitude information, and the provision of an independent accurate secondary attitude source is also important.

8.5 Combined Air Data and Inertial

8.5.1 Introduction

The characteristics of air data and inertial sensing systems have already been described. Because air data sensing involves detecting minor pressure differences of a compressible fluid via a fine pitot-static tube network, there are associated measurement lags. Consequently air data tend to be slow to settle but are accurate in the longer term. Conversely, inertial sensors measurements are instantaneously accurate but accumulate errors over time due to gyro drift or by the accruing of computational errors. Inertial sensors are accurate in the short term but less so in the long term. These contrasting error characteristics are shown in Figure 8.25.

By fusing the two types of sensor information there comes an optimum trade-off point that maximises the benefits of both sensor types. This process is termed sensor fusion, and in civil systems the function of air data computation using ADCs and the computation of inertial data have fused into a combined system called Air Data and Inertial Reference System (ADIRS)

8.5.2 Evolution of Combined Systems

The separate functions are shown before and after the ADIRS concept which was jointly agreed by Boeing and Airbus around the late 1980s. The differences between the two architectures are shown in broad terms in Figure 8.26.

A comparison of functionality may be summarised as follows:

- **Air Data and IRS:** Pitot (Pt) and static (Ps) tubing was routed directly from the appropriate probes to the ADCs. At least two ADCs would be fitted, possibly three. Two total air temperature (TAT) and airstream direction detectors (ADDs) would also be fitted, each

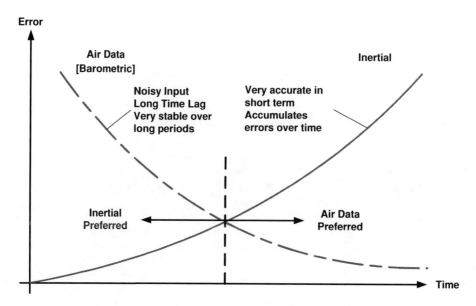

Figure 8.25   Comparison of air data and inertial error characteristics

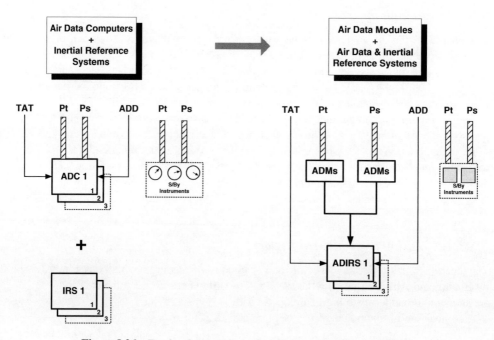

Figure 8.26   Top-level comparison of architectures leading to ADIRS

feeding information in electrical form into the appropriate ADC. Inertial data would be supplied from two (possibly three) inertial reference systems (IRS) depending upon the aircraft configuration. The standby instruments, which in this era would be conventional round-dial instruments including airspeed indicator (ASI) and altimeter, would have their own direct pitot static tubes feeding the instruments. The artificial horizon (AH) would be fed pitch and roll attitude information from an independent attitude source.

- **ADM/ADIRS:** The advent of air data modules (ADMs) enabled the pitot static network to be simplified by placing ADMs closer to the pitot static probes, thereby reducing pitot static tubing; it also permitted the signalling of air data across the recipient aircraft systems including ADIRS using ARINC 429 data buses, as already depicted in Figure 8.8. Figure 8.8 shows a typical Airbus architecture using a total of eight ADMs. Dual TAT and ADD data were also fed into the ADIRS using ARINC 429 buses or other suitable means. Standby instruments received their own dedicated pitot-static tube feed as before. The integration of multiple sensor package data into a common computing source enabled the sensor fusion techniques to be used to the full, using multiple redundant and fault-tolerant computing techniques. Therefore, as increasing availability and accuracy of data was made possible, so the integrity of that data was assured.

8.5.3   Boeing 777 Example

The architecture adopted for the Boeing 777 is portrayed in Figure 8.27.

The B777 has adopted a triplex concept with three pitot (Pt) and three static (Ps) probes which form left (L), centre (C) and right (R) channels. Each of the three pitot (Pt) probes convert their data to digital (ARINC 629) format and feed them to the L, C and R aircraft flight control A629 data buses respectively. Each of the three static (Ps) ADMs accept feeds from both left- and right-side flush static probes and feed the corresponding data on to the L,C and R aircraft flight control A629 data buses. Left and right angle of attack (AoA) data are fed into the left and right Aircraft Information Management System (AIMS) cabinets and thence to the aircraft flight control A629 buses. The TAT probe interfaces into both AIMS cabinets. This arrangement makes all the air data information data available to the aircraft-wide systems.

The users of the data are an Air Data and Inertial Reference Unit (ADIRU) which is the primary unit, and a Secondary Attitude and Air Data Reference Unit (SAARU), which is the secondary. These units contain multi-axes sensor units. The architecture of the B777 ADIRU is shown in Figure 8.28.

The ADIRU has a skewed sensor set of six rate gyros and accelerometers, arranged such that each sensor makes a contribution to X, Y and Z components. Therefore in each axis there are six independent contributors to the associated sensor information, which can be scaled by their respective geometry to the orthogonal axis set. Also, as the contribution of each sensor is determined by the configuration of the sensor set, then the loss of an individual sensor may be readily compensated for. This scheme allows for the loss of one or two sensors in a set to be lost without an associated loss of accuracy.

Furthermore, the ADIRU has four independent channels of computation with three independent ARINC 629 data bus interfaces to the flight systems data buses. This unit is capable of 'graceful degradation' following a series of sensor, computing or data bus interface failures.

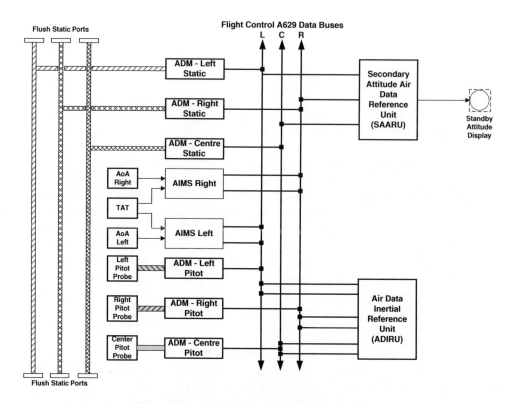

Figure 8.27   B777 ADM and ADIRS configuration

The secondary unit – the SAARU – has two sets of four skewed sensors (rate gyros and accelerometers) with two independent computing channels; it provides further redundancy options, but not to the same level as the ADIRU.

8.5.4   ADIRS Data-Set

In earlier sections the typical data-sets available from the air data system and the inertial reference system were outlined. Although each sensor set was comprehensive in its own right in terms of the data produced, the ADIRS produces yet more data but with the advantage of data fusion combined with powerful computing techniques. A typical ADIRS system's outputs would include the data shown in Table 8.2. This list represents a total of almost 50 data parameters. Reference [11] is an interesting reference to A340-500 polar flights during certification.

8.5.5   Further System Integration

Further integration has been introduced in the last 15 to 20 years using the additional data provided by global navigation satellite systems (GNSSs) – GPS and the like. While GPS and

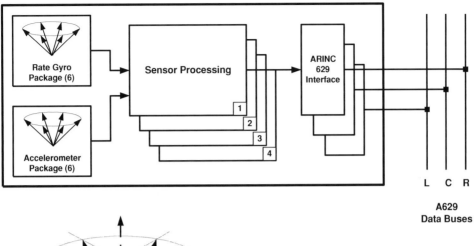

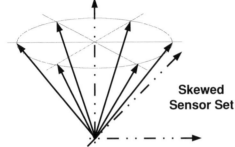

**Skewed
Sensor Set**

Figure 8.28   Boeing ADIRU internal architecture

Table 8.2   Typical outputs from an ADIRS

North velocity	East velocity
Groundspeed	Wind speed
Wind direction	Latitude
Longitude	True heading
True track angle	Magnetic heading
Magnetic track angle	Drift angle
Flight path angle	Inertial altitude
Computed airspeed	Mach no.
Roll attitude	Pitch attitude
Track angle rate	Corrected angle of attack (AoA)
Altitude rate	Altitude
Total air temperature (TAT)	Static air temperature (SAT)
True airspeed	Static pressure (corrected)
Impact pressure	Corrected computed airspeed
Corrected Mach no.	Corrected total pressure
Corrected static pressure	C of G longitudinal acceleration
C of G lateral acceleration	C of G normal acceleration
Flight path acceleration	Vertical speed
Body roll, pitch and yaw attitude	Body pitch, roll and yaw attitude rates
Body longitudinal, lateral and normal accelerations	

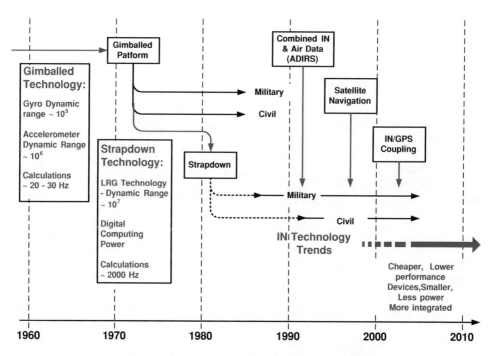

Figure 8.29 Evolution of sensing technologies

other similar systems (GLONASS in Russia; Galileo in the EU; and COMPASS in China) are described in Chapter 9, it is worthwhile at this point to set the history of sensor fusion in context, along with some of the technology drivers of the time (see Figure 8.29).

The story begins in the early 1960s when gimballed technology by way of twin-gyro platforms and later derivatives were in development. Eventually gimballed inertial platforms evolved. Some technology limitations at the time were the following for gimballed platforms:

- gyro dynamic range $\sim 1 \times 10^5$;
- accelerometer dynamic range $\sim 1 \times 10^6$;
- calculation iteration rate ~ 20 to 30 Hz.

These technology limitations of the time imposed significant constraints upon performance. Nevertheless, it was not uncommon for long-range aircraft to be fitted with triple IN systems.

As technology developments evolved, the key technology metrics improved to the following for strap-down platforms:

- LRG technology improving gyro dynamic range to $\sim 1 \times 10^7$;
- increased digital computing power;
- calculations possible at ~ 2000 Hz iteration rate.

These applications were seen in military developments initially, but in the early 1990s ADIRS implementations for civil applications became available.

Further possibilities became available with the availability of GNSS – initially the US Global Positioning System. Other similar navigation system constellations followed. From the late 1990s through to the early 2000s the integration of IN and GNSS (GPS) improved navigational accuracy greatly. INS/GPS coupling was adopted in three basic forms, in descending order of accuracy:

- ultra-coupled INS/GPS – military weapons release;
- highly coupled INS/GPS – military integrated platforms;
- loosely coupled INS/GPS – civil applications.

The essential benefits of this prolonged systems integration evolution is that IN technology has become cheaper, smaller, more reliable, using less power, while becoming more highly integrated with other navigation systems. Similarly, GPS technology has experienced a similar evolution in terms of capability and affordability. The resultant effects in terms of the navigation system are very powerful, offering a very high performance at affordable cost.

8.6   Radar Sensors

Civil aircraft carry a number of radar sensors that permit the aircraft to derive data concerning the flight of the aircraft. The principle radar sensors in use on civil aircraft are:

- radar altimeter;
- weather radar.

8.6.1   Radar Altimeter

The radar altimeter (Rad Alt) uses radar transmissions to reflect off the surface of the sea or the ground immediately below the aircraft. The radar altimeter therefore provides an absolute reading of altitude with regard to the terrain directly beneath the aircraft. This contrasts with the barometric or air data altimeter where the altitude may be referenced to sea-level or some other datum such as the local terrain, where it is referred to as height. The radar altimeter is therefore of particular value in warning the pilot that he is close to the terrain and to warn him if necessary to take corrective action. Alternatively the radar altimeter may provide the flight crew with accurate altitude with respect to terrain during the final stages of a precision approach. Comparison of barometric and radar altitude is shown in Figure 8.30.

Radar altimeters may operate using CW/FM (continuous wave/frequency modulated) or pulsed signal techniques. CW/FM techniques lose accuracy above a range of 5000 feet, so many radio altimeters are limited to an operational range of 5000 feet. Above this, pulsed techniques have to be used. The radar altimeter principle of operation is shown in Figure 8.31. The oscillator and modulator provide the necessary signals to the transmitter and transmit antennae, which direct radar energy towards the terrain beneath aircraft. Reflected energy is received by the receive antenna and passed to a frequency counter. The frequency counter demodulates the received signal and provides a radar altimeter reading to a dedicated display. Alternatively and more usually, the information may be presented on an EFIS. In modern systems, radar altitude will be provided to a range of systems such as FMS, EGPWS, autopilot,

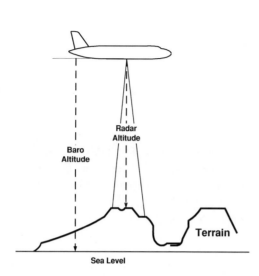

- The radar altimeter measures actual altitude of the aircraft relative to the terrain underneath, as opposed to the air data altimeter which measures the height relative to sea-level or some other datum

- The radar altitude is determined by the time taken for a radar pulse transmitted from the aircraft to return reflected from the terrain

- Most common type used today is the Frequency Modulated Continuous Wave (FMCW)

- Radar altimeters transmit in the range 4250 to 4350 MHz

- Radar altimeters may have dedicated displays; the latest systems display radar altitude on the EFIS

- Military/UAVs may also use laser altimeters

Figure 8.30 Radar altimeter compared with barometric altimeter

and so on, as well as being displayed directly to the flight crew. Radar altimeters usually operate over a maximum range of zero to 5000 feet.

Most radar altimeters use a triangular modulated frequency technique on the transmitted energy, as shown in Figure 8.32, although as mentioned earlier this typically limits the operational range to 5000 feet. The transmitter/receiver generates a continuous wave (CW) signal varying from 4250 to 4350 MHz, modulated at 100 Hz (period = 0.01 sec). Comparison of the frequency of the reflected energy with the transmitted energy – F1 versus F2 in the figure – yields a frequency difference that is proportional to the time taken for the radiated energy to return, hence radar altitude may be calculated.

Radar altimeter installations are calibrated to allow for the aircraft installation delay, which varies from aircraft to aircraft. This allows compensation for the height of the antenna above the landing gear and any lengthy runs of coaxial cable in the aircraft electrical installation. The zero reading of the radar altimeter is set such that it coincides with the point at which the aircraft landing gear is just making contact with the runway. A useful reference regarding radar altimeter installations is Reference [12].

8.6.2   Weather Radar

The weather radar has been in use for over 40 years to alert the flight crew to the presence of adverse weather or terrain in the aircraft's flight path. The weather radar radiates energy in a narrow beam with a beamwidth of ~3° which may be reflected from clouds or terrain ahead of the aircraft. The radar beam is scanned either side of the aircraft centre line to give a radar picture of objects ahead of the aircraft. The antenna may also be tilted in elevation by around ±15° from the horizontal to scan areas above and below the aircraft.

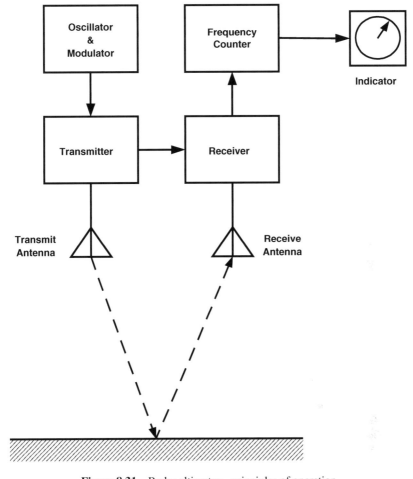

Figure 8.31 Radar altimeter – principles of operation

The principle of operation of a weather radar is shown in Figure 8.33. This shows a storm cloud directly ahead of the aircraft, with some precipitation below and also steadily rising terrain. Precipitation can be indicative of severe vertical wind-shear which can cause a hazard to the aircraft.

The radar beam is directly ahead of the aircraft with the antenna in its mid or datum position and will detect the storm cloud through which the aircraft is about to fly. By referring to his weather radar display, the pilot will be able to see whether the storm cells can be avoided by altering course left or right. The use of the antenna tilt function is crucial. In the example given, if the antenna is fully raised, the crew will not gain complete information relating to storm cloud precipitation as they will only be looking at the top of the cloud formation. If the antenna is fully depressed, the radar will detect the rising terrain in preference to the storm cloud or precipitation ahead. For this reason, many weather radars incorporate an automatic tilt feature so that the radar returns are optimised for the flight crew depending upon the returns that are received.

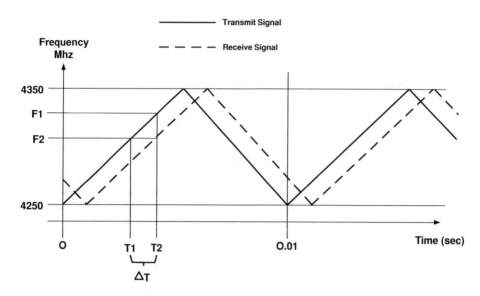

Figure 8.32   Radar altimeter frequency modulation technique

Most modern weather radars can use Doppler processing to detect turbulence ahead of the aircraft. This is a very useful feature, as maximum wind-shear does not necessarily occur coincidentally with the heaviest precipitation. In fact, some of the most dangerous wind-shear can occur in clear air with the aircraft flying nowhere near any clouds or precipitation.

The radar picture is usually displayed on a dedicated radar display or overlaid on the pilot or first officer's navigation display. Displays are typically in colour, which helps the flight crew to interpret the radar data. Displays have various selectable range markers and

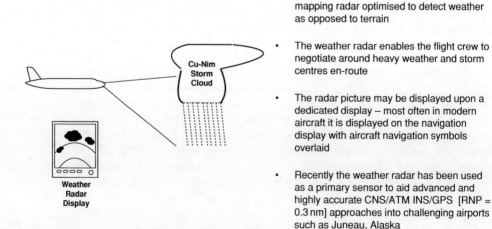

- Weather radar is a limited version of the mapping radar optimised to detect weather as opposed to terrain

- The weather radar enables the flight crew to negotiate around heavy weather and storm centres en-route

- The radar picture may be displayed upon a dedicated display – most often in modern aircraft it is displayed on the navigation display with aircraft navigation symbols overlaid

- Recently the weather radar has been used as a primary sensor to aid advanced and highly accurate CNS/ATM INS/GPS [RNP = 0.3 nm] approaches into challenging airports such as Juneau, Alaska

Figure 8.33   Operation of weather radar

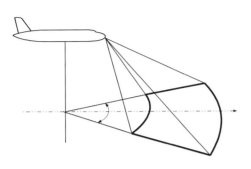

- Mapping radar uses a specially shaped beam to 'paint' the terrain ahead of the aircraft

- A detailed radar map of the terrain and other man-made features ahead of the aircraft may be compiled

- By positioning the mapping radar antenna at a fixed depression angle, some degree of terrain avoidance guidance may be provided

- Mapping radars are most commonly used in military aircraft as part of the offensive weapons suite

- Mapping radars are used on civil aircraft to illustrate terrain in a terrain-avoidance mode

Figure 8.34   Use of weather radar in mapping/terrain avoidance mode

are usually referenced to the aircraft heading. Separate displays may be provided for weather or turbulence modes. Useful information regarding operational use of weather radar may be found in Reference [13], which provides pilot guidance for the operation of the Honeywell RDR-4B weather radar. The transmitter operates at 9.345 GHz and the system has three basic modes of operation:

- weather and map with a maximum range of 320 nm;
- turbulence (TURB) mode out to 40 nm;
- wind-shear detection out to 5 nm.

Weather radars may also be used in a terrain avoidance mode in certain circumstances, as shown in Figure 8.34. When flying precision approaches into airfields surrounded by high terrain, the weather radar can provide an independent cross-check on the information provided by the Terrain Avoidance Warning System (TAWS). The captain can view TAWS information on his navigation display (ND) while being monitored by the first officer with a weather display overlaid on his ND. In Reference [14] the text explains why the weather radar was added to the aircraft minimum equipment list (MEL) for such flights.

References

[1] Pitot Static Sensors 0851 (2012) 4245A LIT 10/12, UTC Aerosystems, Rosemount Aerospace Inc.
[2] Total Temperature Sensors Technical Report 5755 (2002) LIT 08/02 Goodrich Corporation, Rosemount Aerospace Inc.
[3] Air Data Handbook (2002) 4081 LIT 08/02, Sensor Systems, Rosemount Aerospace Inc.
[4] Angle of Attack Sensor 0861 (2012) 4000C LIT 1/12, Goodrich Corporation, Rosemount Aerospace Inc.
[5] Boeing Aero No 12, October 2000, *Operational use of Angle of Attack.*
[6] Boeing Aero No 8, October 1999, *Erroneous Flight Instruments.*

[7] British Geological Survey (2000) *The Derivation of the World Magnetic Model 2000.* British Geological Survey Technical Report WM/00/17R.

[8] The US/UK World Magnetic Model 2005–2010, NOAA Technical Report NESDIS/NGDC-1.

[9] IEE Electronics & Communication Engineering Journal (2000) *Modern Inertial Navigation Technology and its Application.*

[10] Kayton, M. and Fried, W.R. (1998) *Avionics Navigation Systems*, 2nd edn, John Wiley & Sons Inc., Chichester.

[11] Airbus FAST 31 Magazine (2002) *A340-500 Arctic Flights*, December 2002. Available at Airbus.com/support/publications [accessed April 2013].

[12] Airbus FAST 49 Magazine (2012) *Radar Altimeter Systems, Correct Maintenance Practices,* January 2012. Available at Airbus.com/support/publications [accessed April 2013].

[13] Pilot's Handbook – Honeywell Weather Radar RDR-4B.

[14] Cramer, M. (1998) RPN RNAV Terminal Areas – Procedures Design and Approval for Juneau, Alaska. FITEC '98.

9

Communications and Navigation Aids

9.1 Introduction

9.1.1 Introduction and RF Spectrum

The onboard aircraft sensors are described in Chapter 8. Sensors are those that are onboard or are autonomous to the aircraft and which do not require the assistance of a third party. However, the aircraft also uses a number of other systems, either for communications or for navigational assistance, that depend upon external agencies in terms of communications, navigation beacons, transmitters, satellite constellations and other support. Communications systems comprise the following equipment and capabilities:

- high frequency (HF) radio transmit/receive;
- very high frequency (VHF) radio transmit/receive and aircraft communications and reporting system (ACARS);
- ultra high frequency (UHF) radio transmit/receive, mainly used in military communications;
- satellite communications (SATCOM) including passenger telephone communications. UHF SATCOM may be used on some civil derivative platforms used in a military context;
- aircraft transponder and air traffic control (ATC) mode A/C and S (also known in the military environment as identification friend or foe/secondary surveillance radar (IFF/SSR));
- traffic collision and avoidance system (TCAS);
- communications control system (CCS).

Common navigation aids and approach aids are:

- very high frequency omni-range (VOR);
- distance measuring equipment (DME);
- automatic direction finding (ADF);
- tactical air navigation system (TACAN), primarily used by the military;
- VOR/TACAN (VOR/TAC);

Civil Avionics Systems, Second Edition. Ian Moir, Allan Seabridge and Malcolm Jukes.
© 2013 John Wiley & Sons, Ltd. Published 2013 by John Wiley & Sons, Ltd.

- instrument landing system (ILS);
- microwave landing system (MLS);
- global navigation satellite systems (GNSSs), of which the Global Positioning System (GPS) is the most notable, although others will be described.

All these navigation systems aid and supplement the onboard sensors described in Chapter 8. The integrated total of all these systems provide the impressive navigational capabilities enjoyed by the civil air transport aircraft of today.

The radio frequency (RF) spectrum extends from 10 kHz (1×10^4 Hertz) up to 10 GHz (1×10^{10} Hertz), and is shown in a simplified form in Figure 9.1. This spectrum, stretching over five spectral decades, covers the range in which most of the civil aircraft communications and navigation equipment operate. For military aircraft the spectrum will be wider, as attack radars, electronic warfare (EW) and infra-red sensors also need to be included. The wide frequency coverage of this spectrum, and the nature of radio wave propagation, means that the performance of different equipment varies according to the conditions of operation. The figure distinguishes between communications and navigation aids. Therefore care has to be

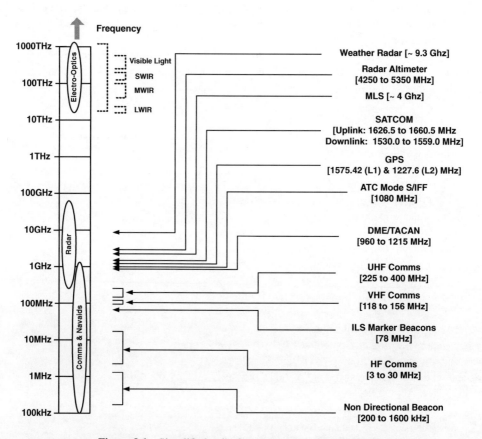

Figure 9.1 Simplified radio frequency spectrum – civil use

Table 9.1   Broad categorisation of radio frequency bands

Name	Abbreviation	Frequency
Very low frequency	VLF	3 to 30 kHz
Low frequency	LF	30 to 300 kHz
Medium frequency	MF	300 kHz to 3000 kHz (3 MHz)
High frequency	HF	3 to 30 MHz
Very high frequency	VHF	30 to 300 MHz
Ultra high frequency	UHF	300 to 3000 MHz (3 GHz)
Super high frequency	SHF	3 to 30 GHz
Extremely high frequency	EHF	30 to 300 GHz

taken when designing and operating the aircraft that any mutual interference effects are kept to a minimum.

The broad categorisation of radio frequency bands is shown in Table 9.1.

Some of the higher frequencies are also categorised by a letter designation as shown in Table 9.2. Note that this designation method is not contiguous as for notation applied in Table 9.1. Several bands overlap and this designation system, as well as being historical, tends to categorise bands with similar properties.

9.1.2   Equipment

It is worth summarising the equipments listed in Figure 9.1 from the viewpoint of increasing frequency. In broad terms this also represents a historical overview, as equipments have evolved to utilise higher frequencies over time – a natural result of technology development:

- **Non-directional beacon (NDB)**. This navigation aid provides bearing information for the aircraft automatic direction finding (ADF) systems, ADF 1 and ADF 2.
- **HF communications**. As aircraft ranges have increased, so has the need for HF communications. Most aircraft have two HF systems fitted, some with high-frequency data link (HFDL).

Table 9.2   Letter designation of higher frequency bands

Letter designation	Frequency range (GHz)
L	0.39 to 1.55
Ls	0.90 to 0.95
S	1.55 to 5.20
C	3.90 to 6.20
X	5.20 to 10.90
Xb	6.25 to 6.90
K[1]	10.90 to 17.25
Ku	15.35 to 17.25
Ka	33.00 to 36.00
Q	36.00 to 46.00

- **ILS marker beacons**. ILS marker beacons are used to provide information regarding the aircraft's progress down the ILS glideslope.
- **VHF communications**. VHF communications are used extensively by the civil aviation community. A standard fit would include two or more, probably three, VHF transmitter/receivers (TRs). In certain military applications more VHF sets would be fitted. This frequency band also incorporates the ILS equipment.
- **UHF communications**. UHF communications are used exclusively by military operators. Many military communications systems will incorporate combined V/UHF sets.
- **DME/TACAN**. DME and TACAN tend to be favoured by the civil and military communities respectively. A typical civil fit would include two DME equipments, usually paired with VHF omni-range (VOR) equipment.
- **Air traffic control (ATC) mode S/identification friend or foe (IFF)**. Usually two ATC equipments are fitted.
- **GPS**. Two GPS equipments are fitted – more in specialist military applications.
- **SATCOM.** Civil aircraft, particularly long-range aircraft, usually have SATCOM fitted. Military aircraft will additionally have UHF SATCOM fitted, operating at the top end of the UHF range.
- **Microwave landing system (MLS)**.
- **Radar altimeter**. Normally a dual fit; possibly triplex for military applications.
- **Weather radar**.

9.1.3 Antennae

A typical civil aircraft antenna complement is shown in Figure 9.2, in which the Boeing 777 is depicted.

The number of antennae required onboard an aircraft to handle all the sensors, communications and navigation aids is considerable. This is compounded by the fact that many of the key equipments may be replicated in duplicate or triplicate. This is especially true of VHF, HF, VOR and DME equipments. Figure 9.2 shows typical antenna locations on a Boeing 777 aircraft; this is indicative of the installation on most civil aircraft operating today, particularly those operating transoceanic routes. Due to their operating characteristics and transmission properties, many of these antennae have their own installation criteria. SATCOM antennae that communicate with satellites will have the antennae mounted on the top of the aircraft so as to have the best coverage of the sky. ILS antennae associated with the approach and landing phase will be located on the forward, lower side of the fuselage. Others may require continuous coverage while the aircraft is manoeuvring and may have antennae located on both upper and lower parts of the aircraft; multiple installations are commonplace.

9.2 Communications

9.2.1 Simple Modulation Techniques

In aviation, communications between the aircraft and the ground (air traffic/local approach/ground handling) have historically been by means of voice communication. More recently, data link communications have been introduced due to their higher data rates and in some cases superior operating characteristics. As will be seen, data links are becoming widely

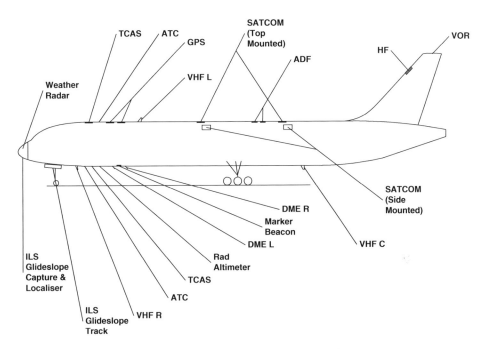

Figure 9.2 Boeing 777 typical antenna complement

used in the HF and VHF bands for basic communications, but also to provide some of the advanced reporting features required by FANS.

After selecting the appropriate communications channel on the channel selector, the pilot transmits a message by pressing the transmit button which connects the pilot's microphone into the appropriate radio. The voice message is used to modulate the carrier frequency, and it is this composite signal that is transmitted.

A typical voice signal is shown in the lower part of Figure 9.3, whilst the amplitude modulated (AM) signal that is transmitted is shown in the upper portion. The recipient receiver demodulates the incoming signal to recover the original voice component. The advantage of this very simple method of transmission is that it is extremely easy to use – all the pilot has to do is speak. A disadvantage is that it occupies a wide bandwidth, typically ~5 kHz, and that speech is not a particularly efficient method of using time and bandwidth compared with data link applications.

The frequency components associated with amplitude modulation are summarised in Figure 9.4. The simple AM case is shown on the left where it can be seen that the carrier is accompanied by upper and lower sidebands (SB). The example shows the spectrum that would be produced when a carrier of 2100 kHz (2.1 MHz) is amplitude modulated by a 1 kHz tone. The three constituent elements are:

- lower SB (LSB) at (carrier – tone) frequency = 2100 – 1 = 2099 kHz
- carrier component at 2100 kHz
- upper SB (USB) at (carrier + tone) frequency = 2100 + 1 = 2101 kHz

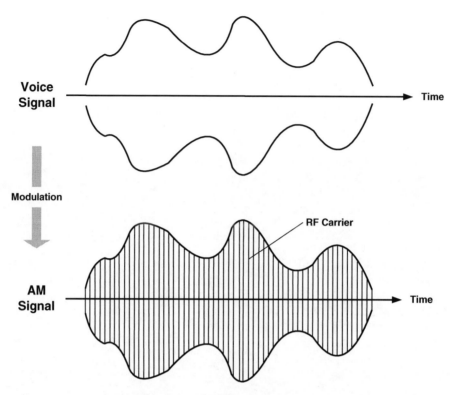

Figure 9.3 Amplitude modulation (AM)

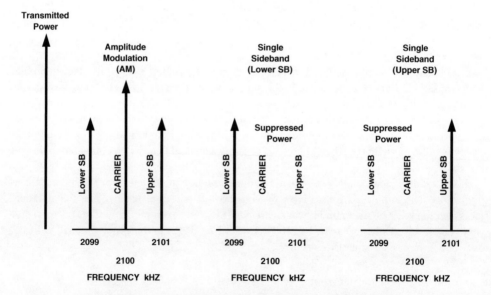

Figure 9.4 Amplitude modulation and single sideband operation

It can be seen that energy is wasted in that power is being transmitted on both SBs and carrier, while the effective signal could be decoded from either LSB or USB. Therefore the technique of single sideband (SSB) has been developed which transmits either the upper or lower SB while suppressing the carrier. This SSB operation can yield effectively eight times more signal power than AM without any power increase at the transmitter. The SSB techniques are used especially in high frequency (HF) communications; the USB is used extensively for aviation, while the LSB is used for other services such as amateur radio.

The principles of single sideband LSB and USB operation are shown in the centre and right-hand diagrams of Figure 9.4 respectively. In complex systems much more sophisticated forms of modulation are used; these are outside the scope of this book.

9.2.2   HF Communications

High frequency (HF) covers the communications band between 3 and 30 MHz and is a very common communications means for land, sea and air. The utilised band is HF SSB/AM over the frequency range 2.000 to 29.999 MHz using a 1 kHz (0.001 MHz) channel spacing. The primary advantage of HF communications is that this system offers communication beyond the line-of-sight. This method does, however, suffer from idiosyncrasies with regard to the means of signal propagation.

Figure 9.5 shows that there are two main means of propagation, known as the sky wave and the ground wave. The sky wave method of propagation relies upon single or multiple path bounces between the Earth and the ionosphere until the signal reaches its intended location. The behaviour of the ionosphere is itself greatly affected by radiation falling upon the Earth, notably solar radiation. Times of high sunspot activity are known to adversely affect the ability of the ionosphere as a reflector. It may also be affected by the time of day and other atmospheric

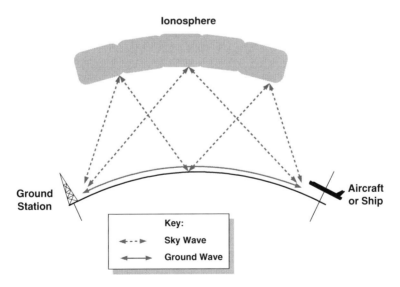

Figure 9.5   HF communications signal propagation

conditions. The sky wave as a means of propagation may therefore be severely degraded by a variety of conditions, occasionally to the point of being unusable.

The ground wave method of propagation relies upon the ability of the wave to follow the curvature of the Earth until it reaches its intended destination. As for the sky wave, the ground wave may on occasions be adversely affected by atmospheric conditions. Therefore on occasions HF voice communications may be corrupted and prove unreliable, although HF data links are more resistant to these propagation upsets as described below.

HF communications are one of the main methods of communicating over long ranges between air and ground during oceanic and wilderness crossings when there is no line-of-sight between the aircraft and ground communications stations. For reasons of availability, most long-range civil aircraft are equipped with two HF sets with an increasing tendency also to use HF data link if polar operations are contemplated.

HF data link (HFDL) offers an improvement over HF voice communications due to the bit encoding inherent in a data link message format, which permits the use of error-correcting codes. Furthermore, the use of more advanced modulation and frequency management techniques allow the data link to perform in propagation conditions where HF voice would be unusable or incomprehensible. A HFDL service is provided by ARINC using a number of ground stations. These ground stations provide coverage out to ~2700 nm and on occasion provide coverage beyond that. Presently, HFDL ground stations are operating at the following locations (see also Figure 9.6):

1. Santa Cruz, Bolivia.
2. Reykjavik, Iceland.
3. Shannon, Ireland.
4. Auckland, New Zealand.

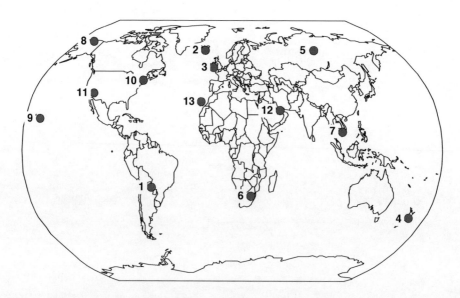

Figure 9.6   HFDL ground stations

  5. Krasnoyarsk, Russia.
  6. Johannesburg, South Africa.
  7. Hat Yai, Thailand.
  8. Barrow, Alaska, US.
  9. Molokai, Hawaii, US.
 10. Riverhead, New York, US.
 11. San Francisco, California, US.
 12. Bahrain.
 13. Gran Canaria, Canary Islands.

9.2.3   VHF Communications

Very high frequency (VHF) voice communication is probably the most heavily used method of communication used by civil aircraft for radio-telephony operations. The VHF band for aeronautical applications operates in the frequency range 118.000 to 135.975 MHz with channel spacing in recent years of 25 kHz (0.025 MHz). More recently, in order to overcome frequency congestion and take advantage of digital radio technology, channel spacing has been reduced to 8.33 kHz (0.00833 MHz) permitting three times more radio channels in the available spectrum.

   The VHF band also experiences limitations in the method of propagation. Except in exceptional circumstances, VHF signals will only propagate over line-of-sight; that is, the signal will only be detected by the receiver when it has line-of-sight or can 'see' the transmitter. VHF transmissions possess neither of the qualities of HF transmission, and accordingly neither sky wave nor ground wave properties apply. This line-of-sight property is affected by the relative heights of the radio tower and aircraft, as illustrated in Figure 9.7.

   The formula that determines the line-of sight range for VHF transmissions is as follows:

$$R = 1.2\sqrt{H_t} + 1.2\sqrt{H_a}$$

where R is the range in nautical miles, H_t is the height of the transmission tower in feet, and H_a is the height of the aircraft in feet. Therefore for an aircraft flying at 35,000 feet, transmissions will generally be received by a 100-ft-high radio tower if the aircraft is within a range of around 235 nautical miles.

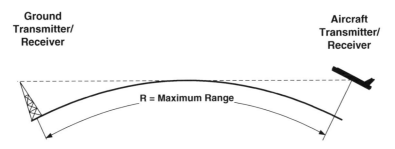

Figure 9.7   VHF signal propagation

Additionally, VHF transmissions may be masked by terrain, for example by a range of mountains. These line-of-sight limitations also apply to equipments operating in higher frequency bands. This means that VHF communications and other equipment operating in the VHF band or above, such as the navigation aids VOR and DME, may not be used except over large land masses, and then only when there is adequate transmitter coverage. Most long-range aircraft have three VHF equipments, with one usually being assigned to ACARS (ARINC Communications and Reporting System) transmissions, although not necessarily dedicated to that purpose. The requirements for certifying the function of airborne VHF equipment are given in Reference [1], while Reference [2] specifies the necessary minimum operational performance standards (MOPS).

Both HF and VHF communications incorporate a feature termed SELective CALling (SEL-CAL). It enables a ground controller to place selective controls to an individual aircraft. If the ground controller wishes to establish communication with an aircraft on a selected frequency, s/he selects a code that relates specifically to the aircraft and initiates the transceiver on a frequency known to be monitored by the crew. When the encoded SELCAL message is received by the aircraft, the message is decoded and if the correct coding sequence is detected, the crew are alerted by a visual or aural annunciator – almost like a front door bell. The flight crew can then communicate normally with the ground station.

A number of VHF data links (VHFDL) may be used and these are discussed in more depth later. ACARS is a specific variant of VHF communications operating on 131.55 MHz that utilises a data link rather than voice transmission. As will be seen during the discussion on future air navigation systems, data link rather than voice transmission will increasingly be used for air to ground, ground to air, and air to air communications as higher data rates may be used while at the same reducing flight crew workload. ACARS is dedicated to downlinking operational data to the airline operational control centre. The initial leg is by using VHF communications to an appropriate ground receiver; thereafter the data may be routed via landlines or microwave links to the airline operations centre. At this point it will be allowed access to the internal airline storage and management systems: operational, flight crew, maintenance, and so on.

Originally only four basic event parameters were transmitted: OUT-OFF-ON-IN, abbreviated to OOOI:

- OUT: Aircraft is clear of the gate and ready to taxi.
- OFF: Aircraft has lifted off the runway.
- ON: Aircraft has landed.
- IN: Aircraft has taxied to the ramp area.

Now comprehensive data such as fuel state, aircraft serviceability, arrival and departure times, weather, crew status, and so on, are also included in the data messages. ACARS was introduced to assist the operational effectiveness of an airline; future data link applications will allow the transfer of more complex data relating to air traffic control routing and flight planning. Onboard the aircraft, ACARS introduces a dedicated management unit, control panel and printer to provide the interface with the flight crew for formatting, dispatching, receiving and printing messages. This, together with existing VHF equipment and an interface with the flight management system (FMS), forms a typical system as shown in Figure 9.8.

It is also now possible to format the data and transmit it via SATCOM, which provides data via a satellite link over long oceanic crossings where the VHF line of sight is insufficient. This

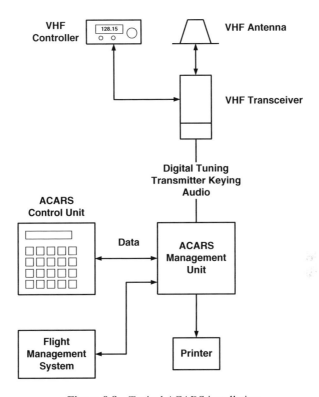

Figure 9.8   Typical ACARS installation

facility enabled the investigators to recover a modicum of data during the investigation into the Air France 447 accident before the crash recorders were recovered.

All aircraft and air traffic control centres maintain a listening watch on the international distress frequency 121.5 MHz. In addition, military controllers maintain a listening watch on 243.0 MHz in the UHF band. This is because the UHF receiver could detect harmonics of a civil VHF distress transmission and relay the appropriate details in an emergency (second harmonic of 121.5 MHz ($\times 2$) = 243.0 MHz; these are the international distress frequencies for VHF and UHF bands respectively).

9.2.4   SATCOM

Satellite communications provide a more reliable method of communications, originally using the INternational MARitime SATellite (INMARSAT) organisation satellite constellation which was developed for maritime use. Now satellite communications, abbreviated to SATCOM, form a useful component of aerospace communications over a range of different frequency bands and provided by a number of service providers.

The principles of operation of SATCOM are shown in Figure 9.9. The aircraft communicates via the INMARSAT constellation and remote ground station by means of C-Band uplinks and

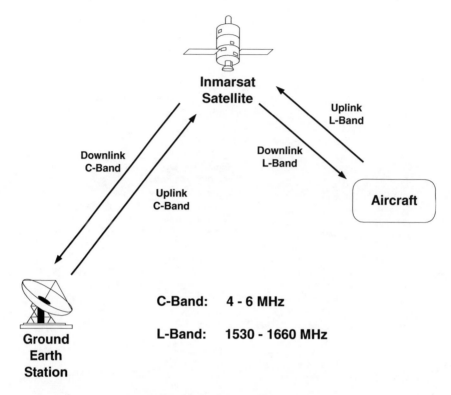

Figure 9.9   SATCOM principles of operation

downlinks to/from the ground stations and L-Band links to/from the aircraft. In this way communications are routed from the aircraft via the satellite to the ground station and on to the destination. Conversely communications to the aircraft are routed in the reverse fashion. Therefore, provided the aircraft is within the area of coverage or footprint of a satellite, then communication may be established.

The airborne SATCOM terminal transmits on frequencies in the range 1626.5 to 1660.5 MHz and receives messages on frequencies in the range 1530.0 to 1559.0 MHz. Upon power-up, the radio frequency unit (RFU) scans a stored set of frequencies and locates the transmission of the appropriate satellite. The aircraft logs on to the ground station network so that any ground stations are able to locate the aircraft. Once logged on to the system, communications between the aircraft and any user may begin. The satellite to ground C-band uplink/downlink is invisible to the aircraft, as is the remainder of the Earth support network.

The coverage offered by the INMARSAT constellation was a total of four satellites in 2001. Further satellites from different competitors have been launched more recently. The INMARSAT satellites are placed in geostationary orbit above the Equator in the locations shown in Figure 9.10:

- Two satellites are positioned over the Atlantic: AOR-W at 54° West and AOR-E at 15.5° West.

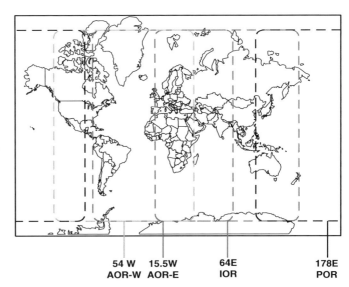

Limitations of SATCOM:

• **The geostationary nature of the INMARSAT satellite constellation means that SATCOM is ineffective at latitudes greater than 80° North or 80° South, i.e. in the polar regions**

• **The aircraft installation is also a significant limiting factor**

54 W	15.5W	64E	178E
AOR-W	AOR-E	IOR	POR

Figure 9.10   INMARSAT satellite coverage around 2001

- One satellite is positioned over the Indian Ocean: IOR at 64° East.
- One satellite is positioned over the Pacific Ocean: POR at 178° East.

This represents the coverage provided by the four I-3 satellites launched in the 1996–97 timeframe. In 2005 to 2008, three I-4 satellites were commissioned, which offer a higher bandwidth Broadcast Global Area Network (BGAN) capability.

Blanket coverage is offered over the entire footprint of each of these satellites. In addition there is a spot beam mode that provides cover over most of the land mass residing under each satellite. This spot beam coverage is available to provide cover to lower capability systems that do not require blanket oceanic coverage.

The geostationary nature of the satellites does impose some limitations. Due to low grazing angles, coverage begins to degrade beyond 80° North and 80° South, and fades completely beyond about 82°. Therefore no coverage exists in the extreme polar regions, a fact assuming more prominence as airlines seek to expand northern polar routes. A second limitation may be posed by the performance of the onboard aircraft system in terms of antenna installation, and this is discussed shortly. Nevertheless, SATCOM is proving to be a very useful addition to the airborne communications suite, and promises to be an important component as procedures compatible with future air navigation system (FANS) are developed.

A typical SATCOM system typically comprises the following units:

- satellite data unit (SDU);
- radio frequency unit (RFU);
- amplifiers, duplexers/splitters;
- low gain antenna;
- high gain antenna.

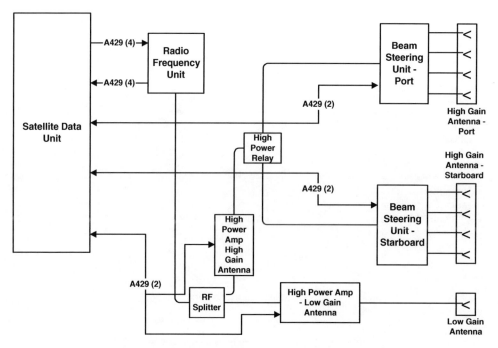

Figure 9.11   SATCOM system using conformal antennae

A typical SATCOM system as installed on the B777 is shown in Figure 9.11. This example uses extensive ARINC 429 data buses for control and communication between the major system elements. This configuration demonstrates the use of two high gain conformal antennae which are mounted on the upper fuselage at positions approximately ±20° respectively from the vertical. Conformal antennae lie flush with the aircraft skin, offering negligible additional drag. Alternatively, the system may be configured such that a single top-mounted antenna may be mounted on the aircraft spine. Both systems have their protagonists and opponents. Claims and counter-claims are made for which antenna configuration offers the best coverage. Conformal configurations reportedly suffer from fuselage obscuration dead-ahead and dead-stern, while the top-mounted rival supposedly suffers from poor coverage at low grazing angle near the horizon. Whatever the relative merits, both configurations are widely used by airlines today.

9.2.5   Air Traffic Control (ATC) Transponder

As a means to aid the identification of individual aircraft and to facilitate the safe passage of aircraft through controlled airspace, the ATC transponder allows ground surveillance radars to interrogate aircraft and decode data which enable correlation of a radar track with a specific aircraft. The principle of transponder operation is shown in Figure 9.12. A ground-based primary surveillance radar (PSR) will transmit radar energy and will be able to detect an aircraft by means of the reflected radar energy – termed the aircraft return. This will enable the

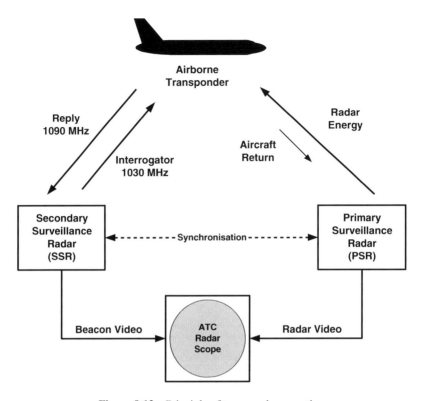

Figure 9.12   Principle of transponder operation

aircraft return to be displayed on an ATC console at a range and bearing commensurate with the aircraft position. Coincident with the primary radar operation, a secondary surveillance radar (SSR) will transmit a series of interrogation pulses that are received by the onboard aircraft transponder. The transponder aircraft replies with a different series of pulses which give information relating to the aircraft, normally aircraft identifier and altitude. If the PSR and SSR are synchronised – usually by being co-boresighted – then both the presented radar returns and the aircraft transponder information may be presented together on the ATC console. Therefore the controller will have aircraft identification (e.g. BA123) and altitude presented alongside the aircraft radar return, thereby greatly improving the controller's situational awareness.

The system is also known as identification friend or foe (IFF)/secondary surveillance radar (SSR), and this nomenclature is in common use in the military field.

Onboard the aircraft, the equipment fit is as shown in Figure 9.13. The main elements are:

- ATC transponder controller unit for setting modes and response codes;
- dedicated ATC transponder unit;
- an ATC antenna unit with an optional second antenna. It is usual to use both upper and lower mounted antennas to prevent blanking effects as the aircraft manoeuvres.

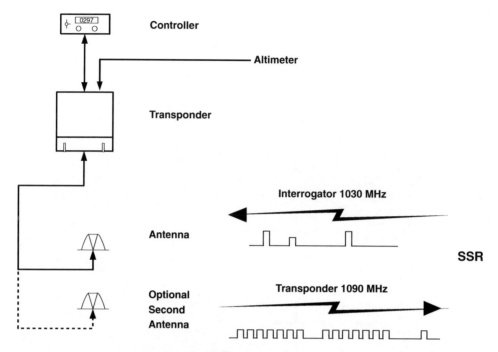

Figure 9.13   Airborne transponder equipment

The SSR interrogates the aircraft by means of a transmission on the dedicated frequency of 1030 MHz that contains the interrogation pulse sequence. The aircraft transponder replies on a dedicated frequency of 1090 MHz with a response that contains the reply pulse sequence with additional information suitably encoded in the pulse stream.

In its present form the ATC transponder allows aircraft identification – usually the airline call-sign – to be transmitted when using Mode A. When Mode C is selected the aircraft will respond with its identifier together with altitude information.

More recently, an additional mode – Mode S or Mode Select – has been introduced with the intention of expanding this capability. In ATC Mode S the SSR uses more sophisticated mono-pulse techniques that enable the aircraft azimuth bearing to be determined more quickly. Upon determining the address and location of the aircraft, it is entered into a roll-call file. This, together with details of all the other aircraft detected within the interrogator's sphere of operation, forms a complete tally of all of the aircraft in the vicinity. Each Mode S reply contains a discrete 24-bit address identifier. This unique address, together with the fact that the interrogator knows where to expect the aircraft from its roll-call file, enables a large number of aircraft to operate in a busy air traffic control environment – see Section 9.2.6. ATC Mode S has other features that enable it to provide the following:

- air-to-air as well as air-to-ground interrogation;
- the ability for aircraft to autonomously determine the precise whereabouts of other aircraft in their vicinity.

Mode S is an improved conventional secondary radar operating at the same frequencies (1030/1090 MHz). Its 'selectivity' is based on unambiguous identification of each aircraft by the unique 24-bit addresses. This acts as its technical telecommunications address, but does not replace the Mode A code. There are also plans for recovery of the A and C codes via Mode S.

Apart from this precise characterisation of the aircraft, Mode S protects the data it transmits, thanks to the inclusion of several parity bits that mean that up to 12 erroneous bits may be tolerated by the application of error detection and correction algorithms. For transmission, these parity bits are superimposed on those of the Mode S address.

When used together with TCAS, ATC Mode S provides an important feature for FANS: that of automatic dependent surveillance – address (ADS-A). This capability will assist the safe passage of aircraft when operating in a direct routing mode [3].

9.2.6   Traffic Collision Avoidance System (TCAS)

The Traffic Collision Avoidance System (TCAS) was developed in prototype form during the 1960s and 1970s to help aircraft avoid collisions. It was certified by the FAA in the 1980s and has been in widespread use in the US in its initial form. TCAS is based upon beacon-interrogator and operates in a similar fashion to the ground-based SSR already described. The system comprises two elements: a surveillance system and collision avoidance system. TCAS detects the range, bearing and altitude of aircraft in the near proximity for display to the pilots.

TCAS transmits a Mode C interrogation search pattern for Mode A and C transponder equipped aircraft, and receives replies from all such equipped aircraft. In addition, TCAS transmits one Mode S interrogation for each Mode S transponder equipped aircraft, receiving individual responses from each one. It will be recalled that Mode A relates to range and bearing, while Mode C relates to range, bearing and altitude, and Mode S to range, bearing and altitude with a unique Mode S reply. The aircraft TCAS equipment comprises a radio transmitter and receiver, directional antennae, computer and flight deck display. Whenever another aircraft receives an interrogation it transmits a reply, and the TCAS computer is able to determine the range depending upon the time taken to receive the reply. The directional antennae enable the bearing of the responding aircraft to be measured. TCAS can track up to 30 aircraft but only display 25, the highest priority targets being the ones that are displayed.

TCAS is unable to detect aircraft that are not carrying an appropriately operating transponder or which have unserviceable equipment. A transponder is mandated if an aircraft flies above 10,000 feet or within 30 miles of major airports; consequently all commercial aircraft and the great majority of corporate and general aviation aircraft are fitted with the equipment.

TCAS exists in two forms: TCAS I and TCAS II. TCAS I indicates the range and bearing of aircraft within a selected range, usually 15–40 nm forward, 5–15 nm aft and 10–20 nm on each side. The system also warns of aircraft within ± 8700 feet of the aircraft's own altitude.

The collision avoidance system element predicts the time to, and separation at, the intruder's closest point of approach. These calculations are undertaken using range, closure rate, altitude and vertical speed. Should the TCAS ascertain that certain safety boundaries will be violated, it will issue a Traffic Advisory (TA) to alert the crew that closing traffic is in the vicinity via the display of certain coloured symbols. Upon receiving a TA the flight crew must visually identify the intruding aircraft and may alter their altitude by up to 300 feet. A TA will normally be advised between 20 and 48 seconds before the point of closest approach with a simple audio warning in the flight crew's headsets: 'TRAFFIC, TRAFFIC'. TCAS I does not offer any de-confliction solutions, but does provide the crew with vital data in order that they may determine the best course of action.

TCAS II offers a more comprehensive capability with the provision of Resolution Advisories (RAs). TCAS II determines the relative motion of the two aircraft and determines an appropriate course of action. The system issues a RA via Mode S advising the pilots to execute the necessary manoeuvre to avoid the other aircraft. A RA will usually be issued when the point of closest approach is within 15 and 35 seconds and the de-confliction symbology is displayed coincident with the appropriate warning. A total of ten audio warnings may be issued. Examples are:

- 'CLIMB, CLIMB, CLIMB'
- 'DESCEND, DESCEND, DESCEND'
- 'REDUCE CLIMB, REDUCE CLIMB'

Finally, when the situation is resolved:

- 'CLEAR OF CONFLICT'.

TCAS II clearly requires a high level of integration between the active equipment. Figure 9.14 shows the interrelationship between:

- TCAS transmitter/receiver;
- ATC Mode S transponders;
- VSI display showing vertical guidance for TAs and RAs;
- optional horizontal situational indicator for RAs that could be the navigation display;

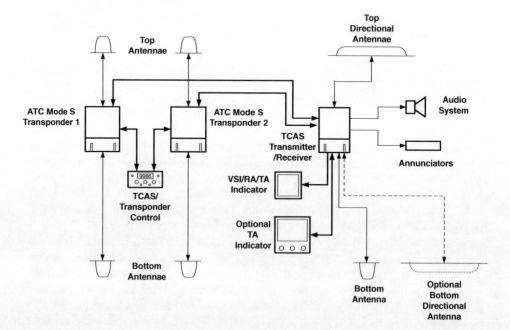

Figure 9.14   Typical TCAS architecture

- audio system and annunciators;
- antennae for ATC Mode S and TCAS.

This is indicative of the level of integration required between ATC Mode S transponders, TCAS, displays and annunciators. It should be noted that there are a variety of display options and the system shown does not represent the only TCAS option.

Further changes were introduced to TCAS II, known as TCAS II Change 7. This introduced software changes and updated algorithms that alter some of the TCAS operating parameters. Specifically, Change 7 includes the following features:

- elimination of nuisance warnings;
- improved RA performance in a multi-aircraft environment;
- modification of vertical thresholds to align with reduced vertical separation minima (RVSM);
- modification of RA display symbology and aural annunciations.

The Change 7 modifications became mandatory in Europe for aircraft with 30 seats or more from 31 March 2001, and for aircraft with more than 19 seats from 1 January 2005. The rest of the world will be following a different but broadly similar timescale for implementation. Change 7 is not mandated in the US but it is expected that most aircraft will be equipped to that standard in any case. References [3] to [6] relate to certification and performance requirements for TCAS II and Mode S.

9.3   Ground-Based Navigation Aids

9.3.1   Introduction

As aviation began to expand in the 1930s, the first radio navigation systems were developed. Initially these were installed at the new growing US airports, and it is interesting to note that the last of these early systems was decommissioned as recently as 1979.

The most prominent was the 'radio range' system developed in Italy by Bellini and Tosi, which was conceived as early as 1907. The operation of the Bellini-Tosi system relied upon the transmission of Morse characters A (dot-dash) and N (dash-dot) in four evenly-spaced orthogonal directions. When flying the correct course the A and N characters combined to produce a humming noise which the pilot could detect in his earphones. Deviation from the desired course would result in either the A or N characters becoming most dominant, signifying the need for corrective action by turning left or right as appropriate.

Following WWII the International Civil Aviation Organisation (ICAO) produced international standards that led to the definition of the very high frequency omni-range (VOR) system which is widespread use today and is described below. The key ground-based navigation beacons in use today are:

- non-directional beacon (NDB);
- very high frequency omni-range (VOR);
- distance measuring equipment (DME);
- TACtical Air Navigation (TACAN);
- VOR/TACAN (VOR/TAC).

Reference [7] is a useful reference for ground-based navigation aids.

9.3.2 Non-Directional Beacon

Automatic direction finding (ADF) involves the use of a loop direction-finding technique to establish the bearing to a radiating source. This might be to a VHF beacon or a non-directional beacon (NDB) operating in the 200 to 1600 kHz band. Non-directional beacons in particular are the most prolific and widespread beacons in use today; they are so-called as there is no directional information provided to the flight crew, unlike VOR and TACAN. The aircraft ADF system comprises integral sense and loop antennae that establish the bearing of the NDB station to which the ADF receiver is tuned. The bearing is shown on the radio-magnetic indicator (RMI) or electronic flight instrument system (EFIS) as appropriate.

9.3.3 VHF Omni-Range

The VOR system was accepted as standard by the US in 1946 and later adopted by the International Civil Aviation Organisation (ICAO) as an international standard. The system provides a widely used set of radio beacons operating in the VHF frequency band over the range 108 to 117.95 MHz with 100 kHz spacing. Each beacon emits a Morse code modulated tone which may be provided to the flight crew for the purposes of beacon identification.

The ground station radiates a cardioid pattern that rotates at 30 revolutions per minute, generating a 30 Hz modulation at the aircraft receiver. The ground station also radiates an omni-directional signal which is frequency-modulated with a 30 Hz reference tone. The phase difference between the two tones varies directly with the bearing of the aircraft. At the high frequencies at which VHF operates, there are no sky wave effects and the system performance is relatively consistent. VOR has the disadvantage that it can be severely disrupted by adverse weather – particularly by electrical storms – and as such may prove unreliable on occasions.

Overland in the North American continent and Europe, VOR beacons are widely situated to provide an overall coverage of beacons. Usually these are arranged to coincide with major airway waypoints and intersections in conjunction with DME stations – see below – such that the aircraft may navigate for the entire flight using the extensive route/beacon structure. By virtue of the transmissions within the VHF band, these beacons are subject to the line-of-sight and terrain-masking limitations of VHF communications. Reference [8] lays out a method of complying with the airworthiness rules for VOR/DME/TACAN.

Total error experienced by the VOR system is ascribed to various sources and the root sum squared figure taken to establish the overall error (see Table 9.3). Typical values for 95% error probability are described in Reference [7]. This gives a total system error of $\pm 4.5°$. Later Doppler VOR (DVOR) installations which are now widely used will offer accuracies at least ten times better than the example quoted above. The implications of VOR accuracy will be further examined later.

9.3.4 Distance Measuring Equipment

Distance measuring equipment (DME) is a method of pulse ranging used in the 960–1215 MHz band to determine the distance of the aircraft from a designated ground station. The aircraft equipment interrogates a ground-based beacon and upon the receipt of retransmitted pulses – unique to the onboard equipment – is able to determine the range to the DME beacon (see

Table 9.3 Typical VOR error budget

Error component	Ascribed value
Radial signal error (E_g): (based in practice upon measured beacon data)	$\pm 1.4°$
Airborne component error (E_a): (based upon the accuracy achieved by typical avionics systems – modern equipment will achieve higher accuracy)	$\pm 3.0°$
Instrument error (E_i): (based upon an analogue system – digital systems would achieve better)	$\pm 2.0°$
Flight technical error (E_f): (assumed to be independent of the other variables – probably pessimistic)	$\pm 2.3°$
Total error: $= \sqrt{[(E_g)^2 + (E_a)^2 + (E_i)^2 + (E_f)^2]}$	$= \sqrt{[(1.4)^2 + (3.0)^2 + (2.0)^2 + (2.3)^2]} = \pm 4.5°$

Figure 9.15). DME beacons are able to service requests from a large number of aircraft simultaneously, but are generally understood to have the capacity of handling $\sim$200 aircraft at once. Specified DME accuracy is reportedly better than $\pm 3\%$ or ± 0.5 nm, whichever is the greater [7]. Precision DME beacons installed at airports will usually be far more accurate.

DME and TACAN beacons are paired with ILS/VOR beacons throughout the airway route structure, in accordance with the table set out in Appendix C of Reference [7]. This is arranged

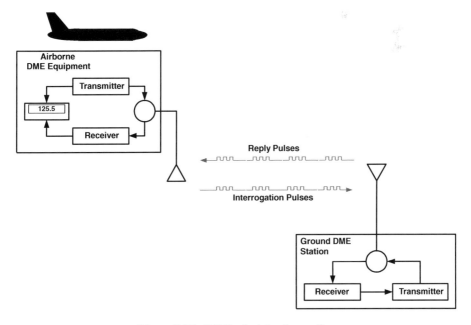

Figure 9.15 DME principle of operation

such that aircraft can navigate the airways by having a combination of VOR bearings to, and DME distance-to-run to the next beacon in the airway route structure. A more recent development – scanning DME – allows the airborne equipment to scan a number of DME beacons rapidly, thereby achieving greater accuracy by taking the best estimate of a number of distance readings. This combination of VOR/DME navigation aids has served the aviation community well in the US and Europe for many years; however, it does depend upon establishing and maintaining a beacon structure across the land mass or continent being covered. The system does provide a backup navigational system of adequate accuracy in the event of the loss of GNSS signals. New developments in developing countries are likely to skip this approach in favour of a GNSS system.

9.3.5 TACAN

TACAN (TACtical Air Navigation) is a military omni-bearing and distance measuring equipment that uses similar techniques for distance measurement as DME. The bearing information is accomplished by amplitude modulation achieved within the beacon which imposes 15 Hz and 135 Hz modulated patterns and transmits these data together with 15 Hz and 135 Hz reference pulses. The airborne equipment is therefore able to measure distance using DME interrogation techniques while using the modulated data to establish bearing.

TACAN beacons operate in the frequency band 960–1215 MHz as opposed to the 108–118 MHz used by DME. This means that the beacons are smaller, making them suitable for shipborne and mobile tactical use. Some airborne equipment has the ability to offset to a point remote from the beacon, which facilitates recovery to an airfield when the TACAN beacon is not colocated. TACAN is reportedly accurate to within ±1% in azimuth and ±0.1 nm in range, so it offers accuracy improvements over VOR/DME.

9.3.6 VOR/TAC

As most military aircraft are equipped with TACAN, some countries provide VOR/TAC beacons with combine VOR and TACAN beacons. This allows interoperability of military and civil air traffic. Military users use the TACAN beacon while civil users use the VOR bearing and TACAN (DME) distance measuring facilities. This is especially helpful for large military aircraft, such as transport or surveillance aircraft, since they are able to use civil air lanes and operational procedures during training or on transit between theatres of operations.

9.4 Instrument Landing Systems

9.4.1 Overview

The availability of landing systems to assist aircraft to perform precision approaches in bad weather or low visibility has been around for many years. Typical systems are:

- instrument landing system (ILS);
- microwave landing system (MLS);
- global navigation satellite system (GNSS) aided systems or GPS overlays.

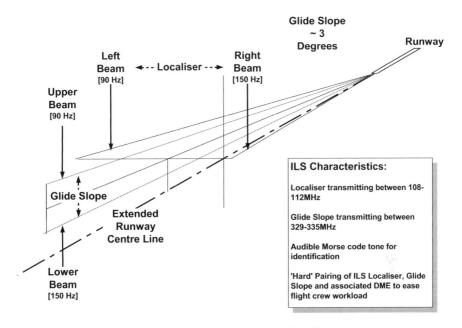

Figure 9.16   ILS glide slope and localiser

9.4.2   Instrument Landing System

The instrument landing system (ILS) is an approach and landing aid that has been in widespread use since the 1950s (see Figure 9.16). The main elements of ILS include:

- **Localiser**: A localiser antenna usually located close to the runway centre line to provide lateral guidance. 40 operating channels are available within the band 108 to 112 MHz. The localiser provides left and right lobe signals that are modulated by different frequencies (90 Hz and 150 Hz) such that one signal or the other will dominate when the aircraft is off the runway centre line. The beams are arranged such that the 90 Hz modulated signal will predominate when the aircraft is to the left, while the 150 Hz signal will be strongest to the right. The difference in the depth of signal modulation is used to drive a cross-pointer deviation needle such that the pilot is instructed to 'fly right' when the 90 Hz signal is strongest and 'fly left' when the 150 Hz signal dominates. When the aircraft is on the centre line the cross-pointer deviation needle is positioned in the central position. This deviation signal is proportional to azimuth out to $\pm 10°$ of the centre line.
- **Glideslope**: A glide slope antenna located beside the runway threshold to provide lateral guidance. Forty operating channels are available within the frequency band 329–335 MHz. As for the localiser, two beams are located such that the null position is aligned with the desired glide slope, usually set at a nominal 3° (usually $\sim$3°). In the case of the glide slope, the 150 Hz modulated signal predominates below the glide slope and the 90 Hz signal is stronger above. When the signals are balanced the aircraft is correctly positioned on the glide slope and the glide slope deviation needle is positioned in a central position. As for the localiser needle, the pilot is provided with 'fly up' or fly down' guidance to help him

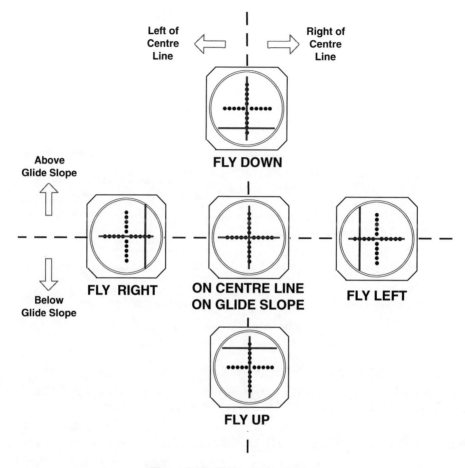

Figure 9.17   ILS guidance display

to acquire and maintain the glide slope. Figure 9.17 illustrates how guidance information is portrayed for the pilot according to the aircraft position relative to the desired approach path. On older aircraft this would be shown on a dedicated deflection display; on modern aircraft with digital cockpits this information is displayed on the primary flight display (PFD). More aggressive ILS approaches have recently been developed with steeper glide slopes, such as at London City airport.

The ILS localiser, glide slope and DME channels are connected such that only the localiser channel needs to be tuned for all three channels to be correctly aligned.

- Marker beacons are located at various points down the approach path to give the pilot information as to what stage of the approach has been reached. These are the outer, middle and inner markers. Location of the marker beacons are:
  - Outer marker approximately 4 to 7 nm from the runway threshold

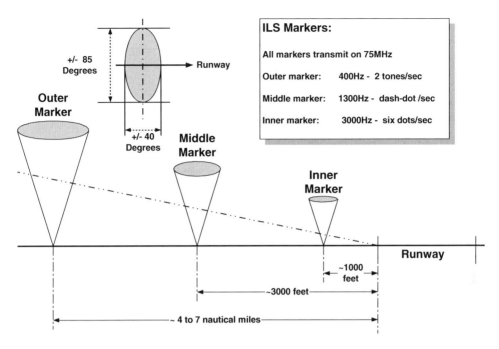

Figure 9.18 ILS approach markers

– Middle marker ~3000 feet from touchdown.
– Inner marker ~1000 feet from touchdown.
  The high approach speeds of most modern aircraft render the inner marker almost super-
  fluous and it is seldom installed.
• The marker beacons are all fan beams radiating on 75 MHz and provide different Morse
  code modulation tones which can be heard through the pilot's headset and may also cause
  visual cues on the aircraft direction indicator (ADI). The layout of the marker beacons with
  respect to the runway is as shown in Figure 9.18. The beam pattern is ±40° along track and
  ±85° across track. The overall audio effect of the marker beacons is to convey an increasing
  sense of urgency to the pilot as the aircraft nears the runway threshold.

A significant disadvantage of the ILS system is its susceptibility to beam distortion and multipath effects. This distortion can be caused by local terrain effects, large manmade structures,
or even taxiing aircraft can cause unacceptable beam distortion, with the glide slope being the
most sensitive. At times on busy airfields and during periods of limited visibility, this may
preclude the movement of aircraft in sensitive areas, which in turn can lead to a reduction
in airfield capacity. More recently, interference by high-power local FM radio stations has
presented an additional problem, although this has been overcome by including improved
discrimination circuits in the aircraft ILS receiver.

Standard ILS glide slope approaches have been based upon a nominal 3.0° angle of approach.
Recent developments have lead to the approval of steeper and more integrated approaches
[9, 10].

9.4.3   Microwave Landing System

The microwave landing system (MLS) is an approach aid that was conceived to redress some of the shortcomings of ILS. The specification of a time-reference scanning beam MLS was developed through the late 1970s/early 1980s, and a transition to MLS was envisaged to begin in 1998. However, with the emergence of satellite systems such as GPS, there was also a realisation that both ILS and MLS could be rendered obsolete when such systems reach maturity. In the event, the US civil community is embarking upon higher accuracy developments of the basic GPS system: the Wide Area Augmentation System (WAAS) and Local Area Augmentation System (LAAS), that will be described later. In Europe, the UK, the Netherlands and Denmark have embarked upon a modest programme of MLS installations at major airports.

MLS operates in the frequency band 5031.0 to 5190.7 MHz and offers some 200 channels of operation. It has a wider field of view than ILS, covering ±40° in azimuth and up to 20° in elevation, with 15° useful range coverage. Coverage is out to 20 nm for a normal approach and up to 7 nm for back azimuth/go-around. The collocation of a DME beacon permits 3-D positioning with regard to the runway, and the combination of higher data rates mean that curved arc approaches may be made, as opposed to the straightforward linear approach offered by ILS. This offers advantages when operating into airfields with confined approach geometry and tactical approaches favoured by the military. For safe operation during go-around, precision DME (P-DME) is usually used for a more precise back azimuth signal.

A ground-based MLS installation comprises azimuth and elevation ground stations, each of which transmit angle and data functions that are frequency shift key (FSK) modulated and which are scanned within the volume of coverage already described. The MLS scanning function is characterised by narrow beam widths of around 1–2° scanning at high slew rates. Scanning rates are extremely high at 20,000°/sec, which provides data rates that are around ten times greater than is necessary to control the aircraft. These high data rates are very useful in being able to reject spurious and unwanted effects due to multiple reflections, for example. Typical coverage in azimuth and elevation for a MLS installation is shown in Figure 9.19.

9.4.4   GNSS Based Systems

Global navigation satellite systems offer the prospect of precision approaches. However GPS along with other GNSS systems are not permitted to be used as sole source sensors; they have to be augmented by more conventional ground-based techniques. In such a fashion, GPS overlay procedures may be flown that overlay the precision that GPS offers, with integrity being provided by other sensors. For example, a GPS overlay procedure may be executed together with a dual DME procedure. These procedures may incorporate differential GPS techniques using either a space-based augmentation system (SBAS) using space-based assets, or a ground-based augmentation system (GBAS) using local assets (see Section 9.5).

9.5   Space-Based Navigation Systems

9.5.1   Introduction

The ground-based navigation beacons techniques were prevalent from the 1960s through to the 1990s when satellites became commonly available. The use of global navigation satellite

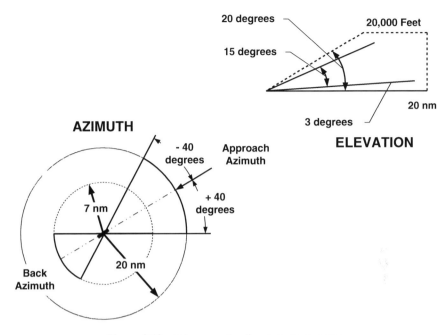

Figure 9.19 Microwave landing system coverage

systems (GNSS), to use the generic name, offers a cheap and accurate navigational means to anyone who possesses a suitable receiver. Although the former Soviet Union developed a system called GLONASS, it is the US Global Positioning System (GPS) that is the most widely used.

The major systems that are in the commissioning or maintenance process today are as follows:

- Global Positioning System (GPS) – US.
- GLONASS – Russia.
- Galileo – European Union (EU).
- COMPASS – China.

9.5.2 Global Positioning System

For the purposes of outlining the principles of GNSS operation, GPS will be used. However, these principles equally apply to the other systems, although there may be subtle differences between them.

GPS is a US satellite-based radio navigational, positioning, and time transfer system operated by the Department of Defense (DoD). The system provides highly accurate position and velocity information and precise time on a continuous global basis to an unlimited number of properly equipped users. The system is unaffected by weather and provides a worldwide common grid reference system based on the Earth-fixed coordinate system. For its Earth model, GPS uses the World Geodetic System of 1984 (WGS-84) datum.

The GPS constellation has evolved over time, including the following phases:

- **Block I** – a series of concept validation satellites, 11 in total – launched from 1978 to 1985.
- **Blocks II/IIA/IIR/IIR-M/IIF** – launched following the concept validation phase and progressively including technical enhancements and more radiated frequencies.

In general, the life of the GPS satellites has exceeded the predicted life in a mildly embarrassing manner, recently leading to some backwards-compatibility system issues. At the time of writing there were a total of 32 operational satellites available. For detailed information, readers should refer to United States Naval Observatory (USNO) resources.

The Department of Defense declared initial operational capability (IOC) of the US GPS on 8 December 1993. The FAA has granted approval for US civil operators to use properly certified GPS equipment as a primary means of navigation in oceanic and certain remote areas. GPS equipment may also be used as a supplementary means of international flight rules (IFR) navigation for domestic en-route, terminal operations and certain instrument approaches.

GPS comprises three major components or segments, as characterised in Figure 9.20:

- **The control segment** which embraces the infrastructure of ground control stations, monitor stations and ground-based satellite dishes that exercise control over the system.
- **The space segment** that includes the satellite constellation, presently around 25 satellites, which form the basis of the network.

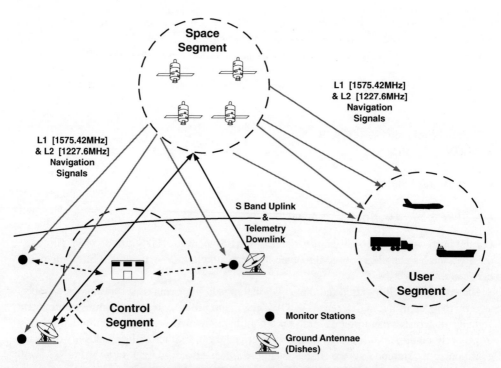

Figure 9.20   Principles of GPS satellite navigation

- **The user segment** that includes all the users: ships, trucks, automobiles, aircraft and hand-held sets. In fact, anyone in possession of a GPS receiver is part of the user segment.

The baseline satellite constellation downlinks data in two bands: L1 on 1575.42 MHz and L2 on 1227.60 MHz. Additional signals have been added as the GPS constellation satellites have been renewed. Potential interference between different satellite constellations needs to be carefully addressed [11].

GPS operation is based on the concept of ranging and triangulation from a group or constellation of satellites in space, which act as precise reference points. A GPS receiver measures distance from a satellite using the travel time of a radio signal. Each satellite transmits a specific code, called Course/Acquisition (CA), which contains information on the satellite's position, the GPS system time, and the health and accuracy of the transmitted data. Knowing the speed at which the signal travelled (approximately 186,000 miles per second) and the exact broadcast time, the distance travelled by the signal can be computed from the time difference of arrival.

The GPS constellation of a minimum of 24 satellites is designed so that a minimum of five are always observable by a user anywhere on Earth. The receiver uses data from a minimum of four satellites above the mask angle (the lowest angle above the horizon at which it can use a satellite).

GPS receivers match each satellite's CA code with an identical copy of the code contained in the receiver's database. By shifting its copy of each satellite's code in a matching process, and by comparing this shift with its internal clock, the receiver can calculate how long it took the signal to travel from a particular satellite to arrive. By then correlating the inputs from several satellites, more than four and usually five, a position on the surface of the Earth may be established. The value derived from this method of computing distance is called a pseudo-range because it is not a direct measurement of distance, but a measurement derived from time. Pseudo-range is subject to several error sources, for example, ionospheric and tropospheric delays and multipath. In addition to knowing the distance to a satellite, a receiver needs to know the satellite's exact position in space; this is known as its ephemeris. Each satellite transmits information about its exact orbital location. The GPS receiver uses this information to establish the precise position of the satellite. Using the calculated pseudo-range and position information supplied by the satellite, the GPS receiver mathematically determines its position by triangulation. The GPS receiver needs at least four satellites to yield a three-dimensional position (latitude, longitude and altitude) and time solution. The GPS receiver computes navigational values such as distance and bearing to a waypoint, ground speed, and so on, by using the aircraft's known latitude/longitude and referencing these to a database built into the receiver.

The GPS receiver verifies the integrity (availability) of the signals received from the GPS constellation through a process called receiver autonomous integrity monitoring (RAIM) to determine whether a satellite is providing corrupted information. At least one other satellite, in addition to those required for navigation, must be in view for the receiver to perform the RAIM function. Therefore performance of the RAIM function needs a minimum of five satellites in view, or four satellites and a barometric altimeter (baro-aiding) to detect an integrity anomaly. For receivers capable of doing so, RAIM needs six satellites in view (or five satellites with baro-aiding) to isolate the corrupt satellite signal and remove it from the navigation solution.

RAIM messages vary somewhat between receivers; however, generally there are two types. One type indicates that there insufficient satellites available to provide RAIM integrity monitoring. Another type indicates that the RAIM integrity monitor has detected a potential error that exceeds the limit for the current phase of flight. Without the RAIM capability, the pilot has no assurance of the accuracy of the GPS position. Areas exist where RAIM warnings apply and which can be predicted, especially at higher latitudes. This represents one of the major shortcomings of GPS and the reason it cannot be used a sole means of navigation.

The geometry of the GPS satellites favours accurate lateral fixes. However, because a number of the visible satellites may be low in the sky, determination of vertical position is less accurate. Baro-aiding is a method of augmenting the GPS integrity solution by using a non-satellite input source to refine the vertical (height) position estimate. GPS-derived altitude should not be relied upon to determine aircraft altitude since the vertical error is sufficiently large to be unacceptable. To ensure that baro-aiding is available, the current altimeter setting must be entered into the receiver as described in the operating manual.

GPS offers two levels of service: Standard Positioning Service (SPS) and Precise Positioning Service (PPS). SPS provides, to all users, horizontal positioning accuracy of 100 metres or less with a probability of 95%, and 300 metres with a probability of 99.99%. PPS is more accurate than SPS. However, this is intended to have a selective availability function, limiting access to authorised US and allied military, federal government, and civil users who can satisfy specific US requirements. At the moment the selective availability feature is disabled, making the PPS capability available to all users pending the availability of Differential GPS (DGPS) solutions to improve the SPS accuracy. This step was taken until the development of differential or augmented GPS systems that provide high accuracy to civil users while preserving the accuracy and security that military users demand. The availability of multiple GNSS systems will provide even further improved accuracy as the size of the useable constellation is increased, and will also improve the vertical resolution problem described earlier. Typical accuracies are shown in Table 9.4.

There are considerable concerns regarding the susceptibility of GPS to spoofing, jamming or other interference [12 to 16].

9.5.3   GLONASS

The former Soviet Union GLONASS system (GLObal'naya NAvigatsionnaya Sputnikovaya Sistema) commenced with the first launch in 1982 and eventually a full constellation of 24

Table 9.4   Comparison of GPS accuracies

Mode	Typical accuracy (metres)	Remarks
Standard GPS with selective availability	±100 m	Now disabled
Standard GPS without selective availability	±7.62 m	Standard usage – aerospace applications
Differential GPS	± 1–2 m	Standard usage – aerospace applications
OmniStar Differential GPS	±0.1–0.2 m	Specialised usage – Global Hawk

satellites was declared operational in 1995. The disintegration of the Soviet Union in 1991 meant that low priority was afforded to the GLONASS programme. By 2001 the GLONASS constellation had degraded to only six operational satellites with a consequent severe reduction in capability, entirely due to the lack of financial support during the immediate post-Soviet era. The launch of later generation GLONASS-M and GLONASS-K satellites increased the constellation to a declared constellation of 24 operational plus 4 spare satellites by May 2012. References [17, 18] relate to the future prospects of GLONASS modernisation.

There has been a significant emphasis by both the US and Russian authorities to ensure an orderly progression of the development and interoperability between both GPS and GLONASS systems. Dual GPS/GLONASS receivers are becoming available on the market for terrestrial use, although the authors are not aware at the time of writing whether any dual applications have been certified for civil air transport applications. Nevertheless, the increase of the available GNSS satellites to a combined constellation of over 50 satellites should surely address some of the availability and integrity issues, as well as improving accuracy. System diversity in terms of satellites and differing modulation techniques may also address some of the integrity concerns outlined above.

9.5.4   Galileo

The European Union (EU) also decided to embark upon a European-based GNSS. This venture has suffered significant setbacks due to budgetary and political impediments. The Galileo programme is funded by the EU and managed by the European Space Agency (ESA). The number of satellites and the composition of the constellation have been subject to continual debate and politics over recent years. Two experimental satellites, GIOVE-A and GIOVE-B, were launched in 2005 and 2008 respectively. Eventually it was agreed that the constellation would comprise a total of 27 satellites plus 3 spares. A primary motivation for Galileo was that the US could potentially block or limit access to GPS and therefore Europe needed its own capability. The rationale behind this was that the US could potentially re-enable Selective Availability (SA), which was disabled upon the direction of US President Clinton in May 2000. In fact, later GPS satellites from Block IIF onwards are unable to support the re-instatement of SA. Additionally, more recent standardisation features ensure that GPS and Galileo will continue to be inter-operable as a multiple GNSS solution. Present plans envisage that the Galileo constellation will be commissioned by 2019.

9.5.5   COMPASS

The Chinese COMPASS or BeiDou navigation satellite constellation is in the process of installation. The early development models, a total of four BeiDou-1 satellites, have been available since 2003. The full constellation, BeiDou-2, will comprise a total of 35 satellites organised as follows:

- five geostationary satellites (GEOS)
- 27 satellites in medium Earth orbits (MEOS)
- three satellites in inclined geostationary orbits (IGSO) [19].

It is anticipated that a full worldwide BeiDou capability will be available by 2020.

9.5.6   Differential GPS

One way of overcoming the problems of selective availability is to employ a technique called Differential GPS (DGPS). Differential techniques involve the transmission of a corrected message which is derived from users located on the ground. The correction information is sent to the user who can apply the corrections and reduce the satellite ranging error. The two main techniques are:

- **Wide area DGPS**. The wide area correction technique involves networks of data collection ground stations. Information is collected at several ground stations which are usually located more than 500 miles apart. The correction information derived by each station is transmitted to a central location where the satellite corrections are determined. Corrections are sent to the user by geostationary satellites or other appropriate means. The wide area augmentation system being developed in the US is called the Wide Area Augmentation System (WAAS) which is outlined below. This technique is also known as the space-based augmentation system (SBAS)
- **Local area DGPS**. The corrections are derived locally at a ground reference site. As the position of the site is accurately known, the satellite inaccuracies can be determined and transmitted locally to the user, in this case by line-of-sight VHF data link. The local area DGPS system under development in the US is called the Local Area Augmentation System (LAAS) and is described below. This technique is also termed locally based augmentation system (LBAS).

9.5.7   Wide Area Augmentation System (WAAS/SBAS)

The operation of WAAS/SBAS, shown in Figure 9.21 is described as follows:

- WAAS is a safety-critical system that augments basic GPS and has been deployed in the contiguous US, Hawaii, Alaska and parts of Canada.
- WAAS has multiple wide-area reference stations that are precisely surveyed and monitor the outputs from the GPS constellation.
- These reference stations are linked to wide-area master stations where corrections are calculated and the system integrity assessed. Correction messages are uplinked to geostationary Earth orbit (GEO) satellites that transmit the corrected data on the communications L1 band to aircraft flying within the WAAS area of coverage. Effectively the GEO satellites act as surrogate GPS satellites.
- Originally expected to produce accuracies of the order ±7.5 m, the system improves the GPS accuracy to around ±1 m most of the time, a considerable improvement over the 'raw' signal. This level of accuracy is sufficient for Cat I approach guidance.

9.5.8   Local Area Augmentation System (LAAS/LBAS)

The operation of LAAS as shown in Figure 9.22 is described below:

- LAAS is intended to complement WAAS but at a local level.

SBAS – Space Based Augmentation System

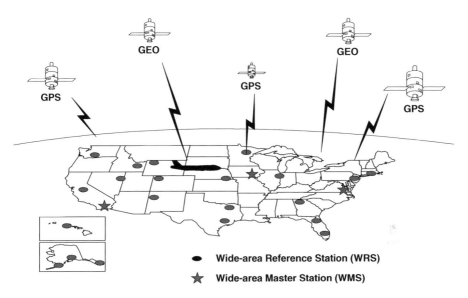

Figure 9.21 Wide Area Augmentation System

GBAS – Ground Based Augmentation System

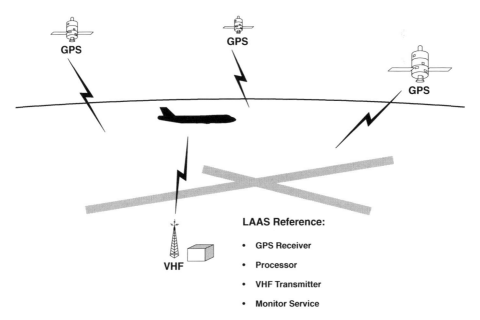

Figure 9.22 Local Area Augmentation System

- LAAS works on similar principles except that local reference stations transmit correction data direct to user aircraft on VHF. As such the LAAS coverage is limited by VHF line-of-sight and terrain masking limitations.
- LAAS is expected to improve the GPS accuracy to less than 1 metre. This level of accuracy is sufficient to permit Cat II and Cat III approaches. A particular benefit of the LAAS is that is not rigidly structured like an ILS approach, enabling curved approaches to be used where the terrain demands it.

Implementation is expected to begin as part of the FAA NextGen plan to improve air traffic control services. According to present plans it is expected that LAAS will be deployed at up to 143 airfields throughout the US. It is the anticipation of LAAS implementation that has caused the US to modify its stance upon the implementation of MLS as an approach aid successor to ILS, the space-based GPS system being seen as more flexible than ground-based MLS.

9.6   Communications Control Systems

The control of the aircraft suite of communications systems, including internal communications, has become an increasingly complex task. This task has expanded as aircraft speeds and traffic density have increased and the breadth of communications types expanded. The communications control function is increasingly being absorbed into the flight management function as the management of communications type, frequency selection and intended aircraft flight path have become more interwoven. Now the flight management system can automatically select and tune the communications and navigation aids required for a particular flight leg, reducing crew workload and allowing the crew to concentrate more on managing the onboard systems.

A communications control system (CCS) architecture typical of an A320 type aircraft is shown in Figure 9.23.

The basic equipment comprises a dual fit of communications (COMMS) and navigation (NAV) elements: VHF 1 and 2 and HF 1 and 2 for COMMS and VOR/DME 1 and 2; ILS 1 and 2; ADF 1 and 2 for NAV functions. Additional optional equipment may include:

- VHF 3 for ARINC Communication and Reporting System (ACARS);
- Air Traffic Service Unit (ATSU) for FANS A operations;
- a multi-mode receiver (MMR) which integrates ILS/GNSS(GPS)/MLS and GLS capabilities into one unit.

Control of the COMMS and NAV suite is excercised using a radio management panel (RMP); two RMPs are provided on the centre pedestal, one each for the captain and first officer. The RMP has radio selection buttons and frequency windows so that the flight crew can select and monitor the radio frequencies being used and also the next frequency to be selected in the flight phase. It also contains a standby navigation capability should both multifunction control and display units (MCDUs) and both flight management guidance computers (FMGCs) fail. Normally the FMGCs will automatically control the radio and navigation equipment that the flight crew have selected.

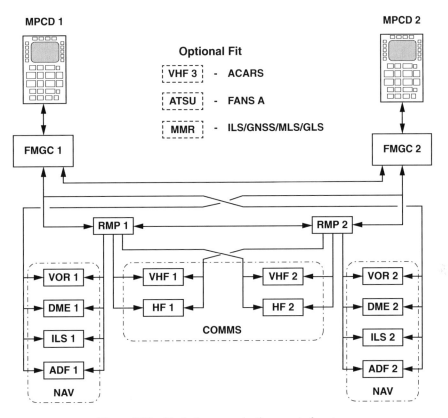

Figure 9.23 Typical communications control system

In addition to the RMP each pilot will have an audio control panel (ACP) located next to the RMP on the centre pedestal. The ACP allows the flight crew to mix and control the volume of all the equipments being used as follows:

- VHF 1 and 2.
- HF 1 and 2.
- VOR 1 and 2 identification tones.
- ADF 1 and 2 identification tones
- ILS identification and ILS marker tones; MLS identification tones (if fitted).
- SELective CALling (SELCAL) selections and annunciations.
- Internal communications such as the interphone and the PA system.

References

[1] Advisory Circular AC 20-67B (1986) *Airborne VHF Communications Installations.*
[2] RTCA DO-186 (1984) *Minimum Operational Performance Standards (MOPS) for Radio Communications Equipment operating with the Radio Frequency Range 117.975 to 137.000 MHz.*

[3] RTCA DO-181, *Minimum Operational Performance Standards for Air Traffic Control Radar beacon System/Mode Select (ATCRBS/Mode S) Airborne Equipment*.

[4] Advisory Circular 129-55A (1993) *Air Carrier Operational Approval and Use of TCAS II*.

[5] Advisory Circular AC 20-131A (1993) *Air Worthiness Approval of Traffic Alert and Collision Avoidance Systems (TCAS II) and Mode S Transponders*.

[6] RTCA DO-185, *Minimum Operational Performance Standards for Traffic Alert and Collision Avoidance Systems (TCAS) Airborne Equipment*.

[7] Kayton, M. and Fried, W.R. (1997) *Avionics Navigation Systems*, 2nd edn. John Wiley & Sons, Ltd., Chichester.

[8] Advisory Circular AC 00-31A (1982) *National Aviation Standard for the Very High Frequency Omni-directional Radio Range (VOR)/Distance Measuring Equipment (DME)/Tactical Air Navigation (TACAN) Systems*.

[9] *Airbus FAST 40 Magazine (2007) A318 Steep Approach Capability*, July 2007. Available at Airbus.com/support/publications [accessed April 2013].

[10] *Boeing AERO Magazine (2003) New Approach Options B730-600/-700/-800/-900*. April 2003.

[11] *GPS World (2010) GNSS RF Compatibility Assessment*. December 2010.

[12] *GPS World (2010) GNNS Vulnerability and Alternative PNT*. July 2010.

[13] *GPS World (2010) Spoofing Detection and Mitigation*. September 2010.

[14] *GPS World (2011) GPS Integrity*. November 2011.

[15] *GPS World (2012) Detecting False Signals*. April 2012.

[16] *GPS World (2012) Personal Privacy Jammers*. April 2012.

[17] *GPS World (2011) GLONASS – Developing Strategies for the Future*. April 2011.

[18] *GPS World (2011) GLONASS Modernisation*. November 2011.

[19] *GPS World (2010) The Strategic Significance of Compass*. December 2010.

10

Flight Control Systems

10.1 Principles of Flight Control

The task of flying and navigating the modern commercial aircraft has become more difficult and stressful with crowded skies and busy airline schedules. To ease the pilot's task the functional complexity of flight control and guidance has increased. Whereas Concorde was the first civil aircraft to have a fly-by-wire system with mechanical backup, Airbus introduced a fly-by-wire system on to the A320 family and a similar system has been carried forward to the A330/340. Boeing's first fly-by-wire system on the Boeing 777 was widely believed to be a response to the Airbus technology development. This chapter on flight control systems examines some of the key principles of flight control and indicates differences between the Airbus and Boeing philosophies and implementations.

The impact of advanced autopilot and flight management functions is also considered, as is the interaction between them.

10.1.1 Frame of Reference

The frames of reference for the motion are referenced to the aircraft and comprise the axis set shown in Figure 10.1.

- **The X-axis** represents the direction of motion of the aircraft. Axial forces due to aircraft thrust or drag operate in this direction. Differences in axial force will result in axial acceleration or deceleration, causing an increase or decrease in aircraft forward (axial) velocity. Rotation around the X-axis relates to the aircraft roll or bank angle, and rates of change such as aircraft body roll rate which will be important in flight control.
- **The Y-axis** portrays the direction of lateral movement and forces. Normally on a civil aircraft lateral forces and accelerations are not large, but they need to be taken into account. Rotation around the Y-axis results in changes in pitch angle which results in the aircraft climbing or descending. Body pitch rate may also be an important consideration in some modes of flight control operation.

Civil Avionics Systems, Second Edition. Ian Moir, Allan Seabridge and Malcolm Jukes.
© 2013 John Wiley & Sons, Ltd. Published 2013 by John Wiley & Sons, Ltd.

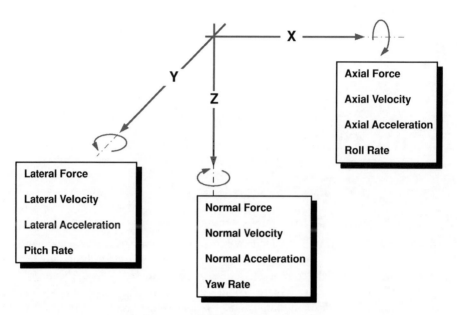

Figure 10.1   Flight control – frame of reference

- **The Z-axis** portrays the direction of normal force and acceleration. Lift and aircraft weight are major forces that act in this vertical direction. Changes in normal force lead to normal accelerations and velocities; vertical velocity is an important parameter to be considered during climb and descent. Rotation around the Z-axis results in changes in yaw, the direction in which the aircraft is pointing, usually portrayed in navigation terms as changes of angle in azimuth or heading. Body yaw rates will be taken into account in flight control laws.

The sign convention is usually that force, velocity and acceleration act in the direction of the axes arrows. Positive body rates are represented by a right rotation around the axis; positive roll would therefore be represented by right wing down.

10.1.2   Typical Flight Control Surfaces

A typical primary flight control system for a civil transport aircraft is shown in Figure 10.2. Although the precise number of control surfaces and means of flight control computer implementation may vary from aircraft to aircraft, all modern FBW systems accord to this generic form. The key control surfaces are as follows:

- Pitch control is usually effected by four powered flight control actuators powering four elevator sections.
- Pitch trim is undertaken by means of two tailplane horizontal stabiliser (THS) actuators – operating as normal and standby systems – which move the entire horizontal tailplane surface or stabiliser (or stabilator in US parlance).

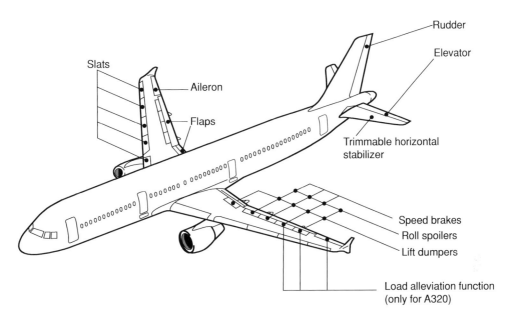

Figure 10.2 Typical flight control system – civil aircraft

- Roll control is invoked by using the left and right ailerons, augmented as required by the extension of a number of spoilers on the inboard wing sections.
- Yaw control is provided by means of two or three rudder sections.
- Both wing sets of spoilers may be extended together to perform the following functions:
  - Inboard spoiler sections to provide a speed-brake function in flight, allowing the aircraft to be rapidly slowed to the desired airspeed, usually during descent.
  - Use of all spoilers in a ground spoiler or lift dump function during the landing roll, enabling the aircraft to rapidly reduce lift during the early portions of the landing run.

Direct inputs from the pilot's controls or inputs from the autopilot feed the necessary guidance signals into a number of flight control computers depending upon the system architecture. These computers modify the flight control demands according to a number of aerodynamic and other parameters such that effective and harmonised handling characteristics are achieved.

In addition, secondary flight control or high lift augmentation is provided by leading edge slats and trailing edge flaps which are extended for take-off and landing as appropriate. On the Boeing 777, the aileron and flap functions are combined by the use of two inboard flaperons, whereas conventional ailerons are used outboard. Operation of the speed-brakes, flaps and slats are initiated by dedicated control levers located on the flight deck central console (see Figure 10.3).

Many aircraft use mechanical push rods and bell cranks or cables and pulleys to convey the pilot's control demands from the flight deck to the flight control architectures. The manner in which this is achieved is described in Reference [1]. The advent of FBW technology allows much of this complicated mechanical linkage to be replaced by computing. There is a trade-off between the linkage and weight saved versus the increased cost and complexity of certifying

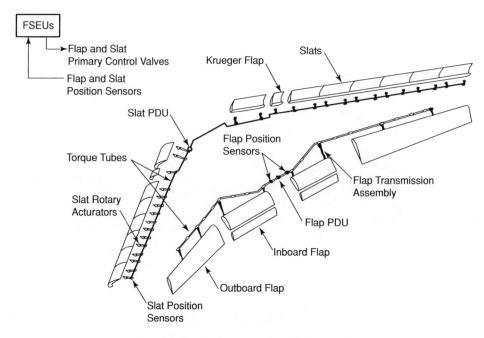

Figure 10.3   Typical secondary flight controls

a high-integrity flight control system, with all that entails in terms of development cost. To date, the Gulfsteam G650 is the smallest aircraft to have adopted full FBW technology in all axes.

10.2   Flight Control Elements

10.2.1   *Interrelationship of Flight Control Functions*

There is sometimes some confusion regarding the interrelationship of primary flight control, autopilot/flight director system and flight management system functions, for which Figure 10.4 hopefully provides adequate clarification. This confusion is understandable as these systems are highly integrated and it not always easy to ascertain what the various functions are doing.

These separate but intertwined functions may be described as three nested control loops, each with their own distinct tasks: a portrayal that a control engineer will find easy to comprehend. These are loops are described as follows:

- Primary flight control or fly-by-wire (FBW) – ATA Chapter 27, Flight Control.
- Autopilot flight director system (AFDS) – ATA Chapter 22, Autoflight Control.
- Flight management system (FMS) – ATA Chapter 34, Navigation.

The FBW system comprising the inner loop is concerned with controlling the attitude of the aircraft. Inputs from the pilot's controls (control column, or in Airbus systems, side-stick), rudders and throttles determine, via the aircraft dynamics, how the aircraft will respond

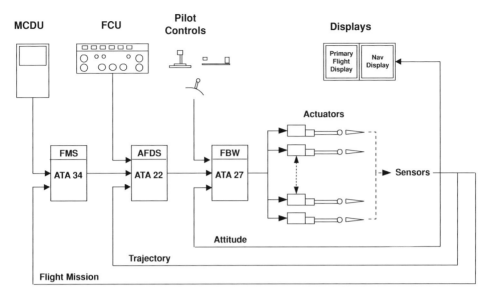

Figure 10.4   Interrelationship of flight functions

at various speeds and altitudes throughout the flight envelope. Inertial and air data sensors determine the aircraft response and close the pitch, roll and yaw control loops to ensure that the aircraft possess well-harmonised control characteristics throughout the flight regime. In some aircraft relaxed stability modes of operation may be invoked by using the fuel system to modify the aircraft centre of gravity, reducing trim drag and reducing aerodynamic loads on the tailplane or stabiliser. The aircraft pitch, roll and yaw (azimuth) attitude or heading are presented on the primary flight display and navigation display.

The autopilot flight director system (AFDS) performs additional control loop closure to control the aircraft trajectory. The AFDS controls the speed, height and heading at which the aircraft flies. Navigation functions associated with specific operations such as heading hold and heading acquire are also included. Approach and landing guidance is provided by coupling the autopilot to the ILS or MLS approach systems. The control and indication associated with these multiple autopilot modes is provided by a flight mode selector panel (FMSP) or flight control panel (FCP), which enables the selection of the principal modes and also provides information confirming that the modes are correctly engaged and functioning properly.

The final outer loop closure is that undertaken by the flight management system (FMS) that performs the navigation or mission function, ensuring that the FBW and AFDS systems position the aircraft at the correct point in the sky to coincide with the multiple waypoints that characterise the aircraft route from departure to destination airfield. The pilot interface with the FMS to initiate and monitor the aircraft progress is via a multifunction control and display unit (MCDU), also known more loosely as the control and display unit (CDU).

As the functions migrate from inner to outer loops, the functionality increases as the integrity decreases. Therefore FBW may be flight-critical, implemented in a triplex architecture; autopilot functions major, implemented in dual/dual architecture; and FMS functions mission-critical, implemented in a dual architecture. This is a subtle point, but one the flight control systems integrator needs to understand.

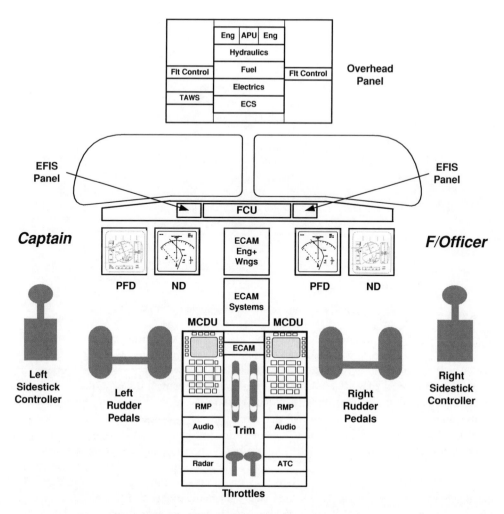

Figure 10.5 Typical flight deck portrayal

10.2.2 Flight Crew Interface

To understand how the flight control system is controlled by the flight crew, it is necessary to understand how they drive the flight control system. A typical flight deck portrayal is shown in Figure 10.5.

Figure 10.5 is based upon an early Airbus FBW six-screen implementation comprising two primary flight displays (PFDs), two navigation displays (NDs) and two Electronic Crew Alerting and Monitoring (ECAM) displays. That fact is not critical to develop the various dependencies and interactions of the flight control system. A Boeing 777 may look similar, with the exception that Boeing have retained a conventional control column and yoke in place of the Airbus side-stick controllers. Also, as aircraft models have evolved through time, the number, size and layout of the display surfaces has also evolved. The key features of the generic flight deck philosophy shown in Figure 10.5 are as follows:

- **Overhead panel**: An overhead panel containing a number of aircraft system control panels by which the crew exercise control over that particular system. Each panel will comprise push buttons, annunciator lights and other information that enable the crew to select the operating mode of the appropriate system. Particular overhead control panels of interest for flight control are: electrics, hydraulics, and flight control system selections. Additional aircraft system data including system synoptic, fault and maintenance data may be displayed on the ECAM displays on an 'as-requested' basis. This would include the provision of system fault data following failure. The general philosophy of the overhead panel is 'light out while operative', which allows the flight crew to tend to routine tasks but attracts their attention when a failure has occurred.
- **PFD and ND**: Flight direction information enabling the flight crew to ensure that the aircraft is flying an attitude profile that is safe is given in terms of speed, altitude and pitch roll and yaw attitude displayed on the captain and first officer's PFD and ND displays. The data on these displays and the data sources are selected using the EFIS panels left and right.
- **Manual flight control**: Manual flight crew inputs are fed into the system by the side-stick controller (Airbus) or control yoke (Boeing) and by the rudder pedals.
- **Autopilot control**: Flight crew control of the aircraft when the autopilot is engaged is exercised via the flight control unit (FCU) located on the glare-shield between the two EFIS control panels. This panel allows the flight crew to select various autopilot modes of operation, various autopilot datums such as airspeed, altitude and speed/Mach, and also capture certain autopilot datums such as ILS glide-slope and localiser. This enables the aircraft trajectory to be flown as required by air traffic control.
- **Flight management system (FMS)**: Following FMS selection the mission profile will be flown according to the waypoints and timing allocated in the flight plan. This controls the air-craft mission and the primary man–machine interface is provided by the FMS multipurpose control and display units (MCDUs).

Derivatives of Figures 10.4 and 10.5 will be introduced later in this chapter and Chapter 11 to underline the differences and the interrelationship between FBW, the autopilot flight director system (AFDS) and the flight management system (FMS).

10.3   Flight Control Actuation

A key element in the flight control system, increasingly so with the advent of fly-by-wire and active control units, is the power actuation. Actuation has always been important to the ability of the flight control system to attain its specified performance. The development of analogue and digital multiple control lane technology has put the actuation central to performance and integrity issues. The actuator is also fundamentally important in terms of the 'muscle' or gain it provides to the flight control loop as whole – see Reference [1]. Finally, as will be seen, advances in technology such as digital computing/FBW as well as other electrical power components have enabled new and more sophisticated solutions.

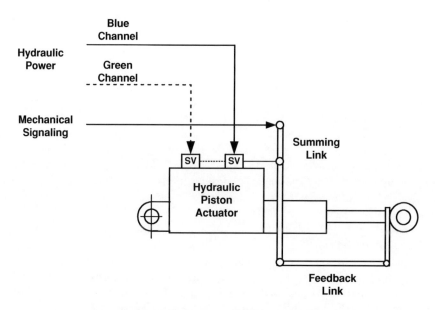

Figure 10.6   Conventional linear actuator

10.3.1   Conventional Linear Actuation

A common conventional linear actuator used in powered flight controls would be of the type shown in Figure 10.6. This type of actuator would usually be powered by one of the aircraft hydraulic systems – in this case the blue channel is shown. In certain crucial applications a dual hydraulic supply from another aircraft hydraulic system may be used. The hydraulic supply uses a mechanically operated servo-valve (SV) that allows the input pressure to be applied to the appropriate side of the piston ram. As the pilot feeds a mechanical demand to the flight control surface actuator, the summing link moves depending upon the direction of the input. As the bottom pivot is fixed, then the summing link also applies a mechanical input to the SV allowing pressure to the appropriate side of the piston and causing the actuator ram to extend or retract accordingly. As the left-hand portion of the ram is fixed to the aircraft structure, the ram extension moves the control surface in the desired manner, causing the aircraft to manoeuvre.

As the ram moves, so does the collar secured to the ram; this in turn moves the lower pivot point in the same direction as the ram. This has the effect of backing off the original input such that the SV reaches a null position; at this point the actuator ram has reached the position the pilot was demanding.

10.3.2   Linear Actuation with Manual and Autopilot Inputs

As aircraft acquired autopilots to reduce pilot workload, then it became necessary to couple electrical as well as mechanical inputs to the actuator as shown in Figure 10.7. The manual (pilot) input to the actuator acts as before when the pilot is exercising manual control. When the

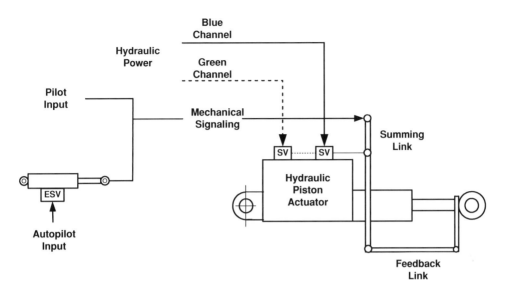

Figure 10.7   Manual and autopilot inputs

autopilot is engaged, electrical demands from the autopilot computer drive an electrical input which takes precedence over the pilot's demand. The actuator itself operates in an identical fashion as before, with the mechanical inputs to the summing link causing the SV to move. When the pilot retrieves control by disengaging the autopilot, the normal mechanical link to the pilot through the aircraft control run is restored.

Simple electrical demand signals are inputs from the pilots that are signalled by electrical means. For certain non-critical flight control surfaces it may be easier, cheaper and lighter to utilise an electrical link. An example of this is the airbrake actuator used on the BAe 146; simplex electrical signalling is used and in the case of failure the reversion mode is aerodynamic closure.

In most cases where electrical signalling is used, this will at least be duplex in implementation, and for fly-by-wire systems signalling is likely to be triplex or quadruplex. In general, those systems that extensively use simplex electrical signalling do so for autostabilisation. In these systems the electrical demand is a stabilisation signal derived within a computer unit. The simplest form of autostabilisation used on older civil aircraft was the yaw damper, which damped out the cyclic cross-coupled oscillations that occur in roll and yaw known as 'Dutch Roll'.

10.3.3   Screwjack Actuation

The linear actuator already described is generally acceptable for the high bandwidth/fast-acting requirements of most flight control surfaces. However, there are others, such as the tailplane horizontal stabiliser (THS), that are required to support huge loads but whose dynamic performance is not critical. In these situations a screwjack actuator is used, illustrated in Figure 10.8.

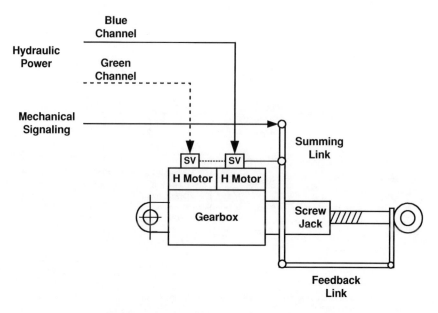

Figure 10.8   Typical screwjack actuator (THS)

The principle of operation is similar to the linear actuator except the SV controls the hydraulic fluid to drive a hydraulic motor(s), which in turn drives a gearbox, thereby extending the actuator to attain a new demanded position. As before, a mechanical feedback link helps the system to decide when the demanded position has been achieved. This arrangement is widely used on civil transport aircraft today.

10.3.4   Integrated Actuation Package

In the UK, the introduction of powerful new AC electrical systems paved the way for the introduction of electrically-powered power flying controls. Four-channel AC electrical systems were used on the military Avro Vulcan B2 and Handley Page Victor V-Bombers, and subsequently the civil Vickers VC10 transport aircraft utilised flight control actuators powered by the aircraft AC electrical system rather than centralised aircraft hydraulic systems.

Figure 10.9 shows the concept of operation of this form of actuator known as an integrated actuator package (IAP). The operation of demand, summing and feedback linkage is similar to the conventional linear actuator already described. The actuator power or 'muscle' is provided by a three-phase constant speed electrical motor driving a variable displacement hydraulic pump. The hydraulic pump and associated system provide hydraulic pressure to power the actuator ram. When little demand is required of the actuator, the variable displacement pump backs off the hydraulic delivery to pressurise the ram in its quiescent state. When a control demand is applied, the delivery of the pump is increased to provide the necessary pressure to enable the ram to extend to the demanded position. Once reached, the pump reverts to idling until the next demand is received.

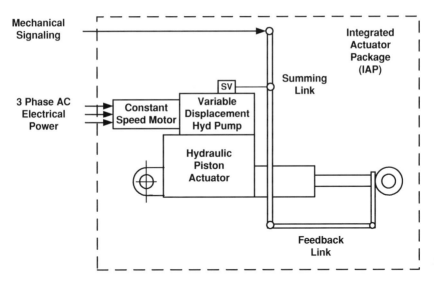

Figure 10.9   Integrated actuator package

Figure 10.10 depicts an overview of a typical IAP used on the Vickers VC-10 flight control system. A total of 11 such units were used in the VC-10 system to power each of the following flight control surfaces:

- ailerons – 4 sections;
- elevators – 4 sections;
- rudder – 3 sections.

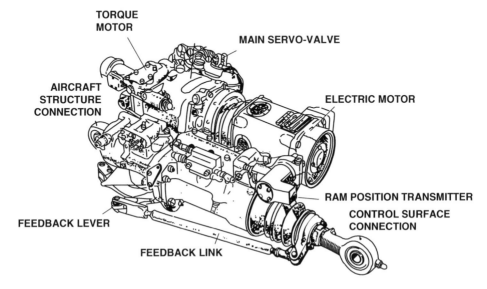

Figure 10.10   VC-10 integrated actuator package (IAP)

The power consumption of each of the IAPs is in the region of 2.75 kVA and at the time of writing they are still flying today in the Royal Air Force's VC-10 Tanker fleet. The units are powered by a constant frequency, split-parallel, 115 VAC 3-phase electrical system.

10.3.5 FBW and Direct Electrical Link

The advent of fly-by-wire (FBW) flight control systems in civil aircraft commencing with the Airbus A320 introduced the need for a more sophisticated interface between the flight control systems and actuation. Most first-generation FBW aircraft may operate in three distinct modes that can be summarised in general terms as follows:

- **Full FBW mode**. This mode encompasses the full FBW algorithms and protection and is the normal mode of operation.
- **Direct electrical link mode**. This mode will usually provide rudimentary algorithms or possibly only a direct electrical signalling capability in the event that the primary FBW mode is not available.
- **Mechanical reversion mode**. This provides a crude means of flying the aircraft – probably using a limited number of flight control surfaces following the failure of FBW and direct electrical link modes. In later implementations such as the Airbus A380, no mechanical reversion is provided.

The interface with the actuator is frequently achieved by means of an actuator control electronics (ACE) unit that closes the control loop electrically around the actuator, rather than mechanical loop closure as hitherto described (see Figure 10.11). The digital FBW or direct link demands from the flight control system are processed by the ACE which supplies an analogue command to the actuator SV. This allows aircraft systems hydraulic power to be supplied to the appropriate side of the ram piston, moving the ram to the desired position.

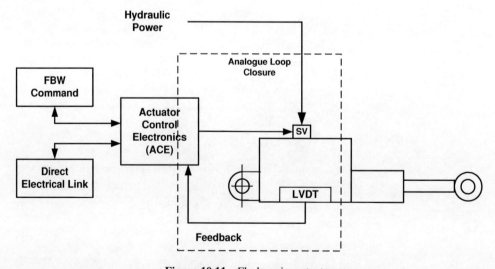

Figure 10.11   Fly-by-wire actuator

In this implementation the ram position is detected by means of a linear variable differential transformer (LVDT), which feeds the signal back to the ACE where the loop around the actuator is closed. Therefore ACE performs two functions: conversion of digital flight control demands into analogue signals, and analogue loop closure around the actuator.

10.3.6   Electrohydrostatic Actuation (EHA)

The move towards more-electric aircraft has coincided with another form of electrical actuation: the electrohydrostatic actuator (EHA), which uses state-of-the-art power electronics and control techniques to provide more efficient flight control actuation. The conventional actuation techniques described so far continually pressurise the actuator, whether or not there is any demand. In reality, for much of the flight, actuator demands are minimal, and this represents a wasteful approach as lost energy ultimately results in higher energy off-take from the engine and hence higher fuel consumption.

The EHA seeks to provide a more efficient form of actuation where the actuator only draws significant power when a control demand is sought; for the remainder of the flight the actuator is quiescent, as shown in Figure 10.12. The EHA accomplishes this by using the 3-phase AC power to feed power drive electronics, which in turn drive a variable speed pump together with a constant displacement hydraulic pump. This constitutes a local hydraulic system for the actuator in a similar fashion to the IAP, the difference being that when there is no demand, the only power drawn is that to maintain the control electronics. When a demand is received from the ACE the power drive electronics is able to react sufficiently rapidly to drive the variable speed motor and hence pressurise the actuator so that the associated control surface may be moved to satisfy the demand. Once the demand has been satisfied, then the power electronics resumes its normal dormant state. Consequently power is only drawn for the aircraft buses bars while the actuator is moving, representing a great saving in energy. The ACE closes the control loop around the actuator electrically as previously described.

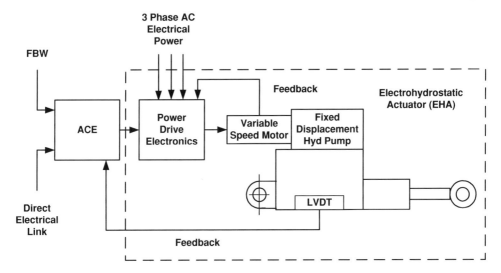

Figure 10.12   Electrohydrostatic actuator (EHA)

EHAs are being applied across a range of aircraft and unmanned air vehicle (UAV) developments. The Airbus A380 and Lockheed Martin F-35 Joint Strike Fighter (JSF) both use EHAs in the flight control system. For aircraft such as the A380 with a conventional 3-phase, 115 VAC electrical system, the actuator uses an inbuilt matrix converter to convert the aircraft 3-phase AC power to 270 VDC to drive a brushless DC motor, which in turn drives the fixed displacement pump. The Royal Aeronautical Society Conference, *More-Electric Aircraft*, 27–28 April 2005, is an excellent reference for more-electric aircraft and more-electric engine developments where some of these solutions are described [2].

10.3.7 *Electromechanical Actuation (EMA)*

The electromechanical actuator or EMA replaces the electrical signalling and power actuation of the electro-hydraulic actuator with an electrical motor and gearbox assembly applying the motive force to move the ram. EMAs have been used on aircraft for many years for such functions as trim and door actuation; however, the power, motive force and response times have been less than that required for flight control actuation. The three main technology advancements that have improved the EMA to the point where it may be viable for flight control applications are: the use of rare-earth magnetic materials in 270 VDC motors; high-power solid-state switching devices; and microprocessors for lightweight control of the actuator motor.

As the EHA is the more-electric replacement for linear actuators, so the electromechanical actuator (EMA) is the more-electric version of the screwjack actuator as shown in Figure 10.13. The concept of the EMA is identical, with the exception that the power drive electronics drive a brushless DC motor operating a reduction gear that applies rotary motion, allowing the jack ram to extend or retract to satisfy input demands. EMAs are therefore used to power the tailplane horizontal stabiliser (THS) on civil aircraft and flap and slat drives, and also find a use

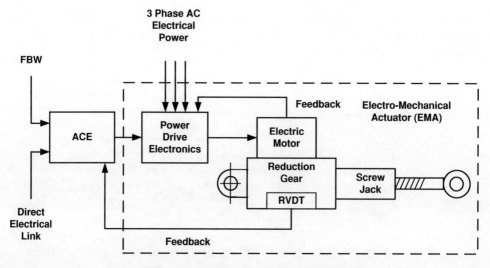

Figure 10.13   Electromechanical actuator

Table 10.1 Typical applications of flight control actuators

Actuator type	Power source	Primary flight control	Spoilers	Tailplane horizontal stabiliser	Flaps and slats
Conventional linear actuator	Aircraft hydraulic systems: B/Y/G or L/C/R[1]	X	X		
Conventional screwjack actuator	Aircraft hydraulic or electrical systems[2]			X	X
Integrated actuator package (IAP)	Aircraft electrical system (115 VAC)	X	X		
Electrically signalled hydraulic actuator	Aircraft hydraulic systems	X	X		
Electrohydrostatic actuator (EHA)	Aircraft electrical system[3,4]	X	X		
Electromechanical actuator (EMA)	Aircraft electrical system[3]			X	X

[1] B/Y/G = Blue/Green/Yellow of L/C/R = Left/Center/Right (Boeing).
[2] For THS and Flaps and Slats, both hydraulic and electrical supplies are often used for redundancy.
[3] 3-phase VAC to 270 VDC matrix converter used in civil aircraft.
[4] 270 VDC aircraft electrical system used on F-35/JSF.

in helicopter flight control systems. A major concern regarding the EMA is the consideration of the actuator jamming case, and this has negated their use in primary flight controls on conventional aircraft.

10.3.8 Actuator Applications

Most of these actuation types are used in civil aircraft today. Table 10.1 lists how the various actuator types may be used for different actuation tasks on a typical civil airliner.

10.4 Principles of Fly-By-Wire

10.4.1 Fly-By-Wire Overview

The flight control and guidance of civil transport aircraft has steadily been getting more sophisticated in recent years, as described in the introduction. Whereas Concorde was the first civil aircraft to have a fly-by-wire system, Airbus introduced a digital fly-by-wire system onto the A320 family, and a similar system has been carried forward to the A330/340 family and beyond to A380 and A350. Boeing's first fly-by-wire system on the Boeing 777 was widely believed to be a response to the Airbus technology development; Boeing also carried the concept forward onto the B787. The key differences between the Airbus and Boeing philosophies and implementations are described below.

In Figure 10.4 the interrelationship between FBW, AFDS and FMS was shown, while Figure 10.5 depicted in generic form the controls and displays associated with the flight control loops. Figure 10.14 specifically highlights the control and displays associated with manual flight. The motif in the top left-hand corner shows the loop associated with FBW. The main diagram shows those controls and displays in use.

The pilot manually controls the aircraft in pitch and roll using the side-stick controller (Airbus) or control yoke (Boeing). The rudder pedals are used to input yaw demands, while the throttles are operated manually to control airspeed.

The pitch and roll attitude are shown on the PFD while yaw attitude or heading is displayed on the ND. Aircraft altitude and airspeed are shown on the PFD. The nature of the displays on the PFD and ND is selected using the EFIS control panels on either side of the glare-shield. Engine parameters are displayed on the upper of the ECAM displays, as will any system limitations, cautions or warnings.

The various systems needed to operate the flight controls (selection of the FBW mode and selection of the electrical and hydraulic systems) are accessed using the appropriate panels on the overhead panels, and system synoptic displays may be called up as required on the lower ECAM display. On Boeing aircraft the engine and system displays are shown on the Engine Indication and Crew Alerting System (EICAS), which performs a virtually identical function to the Airbus ECAM.

10.4.2   Typical Operating Modes

Whilst it is impossible to generalise, the approach to the application of control laws in a FBW system and the various reversionary modes does have a degree of similarity. The application of normal, alternate and direct control laws, and in the final analysis, mechanical reversion, often follow the typical format outlined in Figure 10.15. The authority of each of these levels may be summarised as follows:

- **Normal laws**: Provision of basic control laws with the addition of coordination algorithms to enhance the quality of handling and protection to avoid the exceedance of certain attitudes and attitude rates. Double failures in computing, sensors or actuation power channels will cause reversion to the alternate mode.
- **Alternate laws**: Provision of the basic control laws but without many of the additional handling enhancement features and protection offered by the normal mode. Further failures cause reversion to the mechanical mode.
- **Direct laws**: Direct relationship from control stick to control surface, manual trimming, with certain limitations depending upon aircraft and flight control system configuration. In certain specific cases, crew intervention may enable re-engagement of the alternate mode. Further failures result in reversion to mechanical mode.
- **Mechanical reversion**: Rudimentary manual control of the aircraft using pitch trim and rudder pedals to facilitate recovery of the aircraft electrical system or land the aircraft as soon as is practicable.

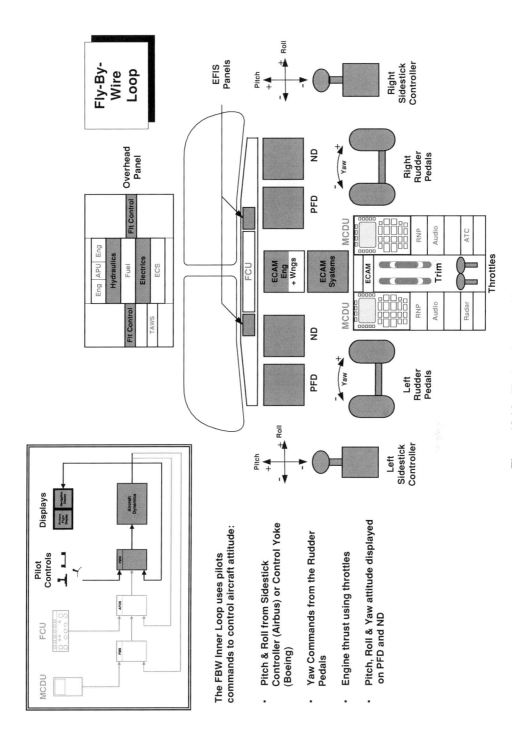

Figure 10.14 Fly-by-wire control loop

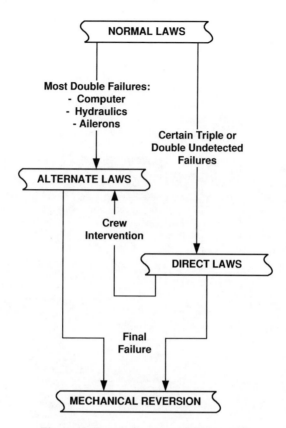

Figure 10.15   Fly-by-wire operating modes

10.4.3   Boeing and Airbus Philosophies

The importance and integrity of aspects of flight control lead to some form of monitoring function to ensure the safe operation of the control loop. Also, for integrity and availability reasons, some form of redundancy is usually required. Figure 10.16 shows a top-level comparison between the Boeing and Airbus FBW implementations.

In the Boeing philosophy, shown in simplified form on the left of Figure 10.16, the system comprises three primary flight computers (PFCs), each of which has three similar lanes with dissimilar hardware but the same software. Each lane has a separate role during an operating period, and the roles are cycled after power-up. Voting techniques are used to detect discrepancies or disagreements between lanes, and the comparison techniques used vary for different types of data. Communication with the four actuator control electronics (ACE) units is by multiple A629 flight control data buses. The ACE units directly drive the flight control actuators. A separate flight control DC system is provided to power the flight control system. The schemes used on the Boeing 777 will be described in more detail later in this chapter.

The Airbus approach is shown on the right of Figure 10.16, which depicts the A330/340 configuration. Five main computers are used: three flight control primary computers (FCPCs) and two flight control secondary computers (FCSCs). Each computer consists of command

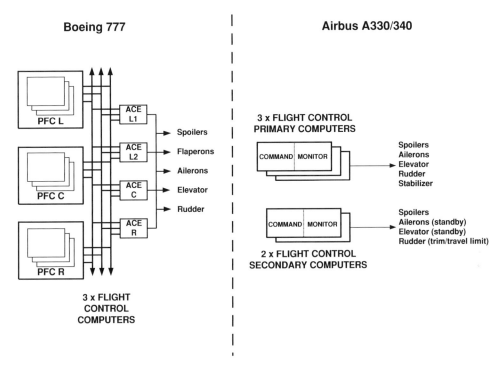

Figure 10.16   Comparison of Boeing and Airbus FBW architectures

and monitor elements with different software. The primary and secondary computers have different architectures and different hardware. Command outputs from the FCSCs to ailerons, elevators and the rudder are for standby use only. Power sources and signalling lanes are segregated.

10.5   Boeing 777 Flight Control System

10.5.1   Top Level Primary Flight Control System

Boeing ventured into the FBW field with the Boeing 777, partly, it has been said, to counter the technology lead established by Airbus with the A320. Whatever the reason, Boeing approached the task in a different manner and have developed a solution quite distinct to the Airbus philosophy. References [3, 4] give a detailed description of the B777 FBW system.

The B777 PFC is outlined at a system level in Figure 10.17. The drawing shows the three primary flight control computers (PFCs), four actuator control electronics (ACEs) and three autopilot flight director computers (AFDCs) interfacing with the triple-redundant A629 flight control buses. The AFDCs have terminals on both the flight control and A629 data buses. Attitude and information is provided by the ADIRU, and SAHRU and air data by the air data modules (ADMs). The three control and display units (CDUs) and the left and right Aircraft Information Management System (AIMS) cabinets provide the flight deck interface. In total there are 76 ARINC 629 couplers on the flight control buses.

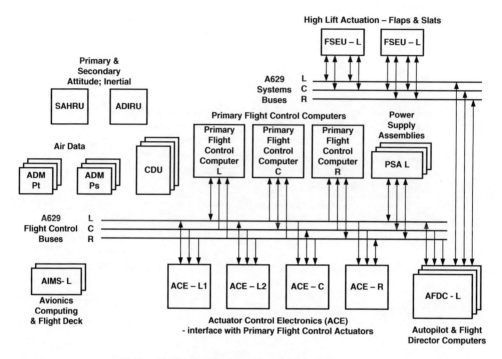

Figure 10.17   Boeing 777 primary flight control system

10.5.2   *Actuator Control Unit Interface*

The PFC system comprises the following control surface actuators and feel actuators:

- Four elevators: left and right inboard and outboard.
- Elevator feel: left and right.
- Two rudder: upper and lower.
- Four ailerons: left and right inboard and outboard.
- Four flaperons: left and right inboard and outboard.
- 14 spoilers: seven left and seven right.

The flight control actuators are interfaced to the three A629 flight control data buses by means of four actuator control electronics (ACE) units. These are:

- ACE Left 1;
- ACE Left 2;
- ACE Centre;
- ACE Right.

These units interface in turn with the flight control and feel actuators in accordance with the scheme shown in Table 10.2.

Table 10.2 ACE to PFC interface

ACE L1	ACE L2	ACE C	ACE R
ROB aileron	LOB aileron	LIB aileron	RIB aileron
LOB aileron	RIB aileron	ROB aileron	LIB aileron
		Upper rudder	Lower rudder
LIB elevator	LOB elevator	ROB elevator	RIB elevator
	L elevator feel	R elevator feel	
Spoiler 2	Spoiler 5	Spoiler 1	Spoiler 3
Spoiler 13	Spoiler 4	Spoiler 7	Spoiler 6
	Spoiler 11	Spoiler 8	Spoiler 9
	Spoiler 10	Spoiler 14	Spoiler 12

The ACE units contain the digital-to-analogue and analogue-to-digital elements of the system. A simplified schematic for an ACE unit is shown in Figure 10.18. Each ACE unit has a single interface with each of the A629 flight control data buses and the unit contains the signal conversion to interface the 'digital' and 'analogue' worlds.

The actuator control loop is shown in the centre-right of the diagram. The actuator demand is signalled to the power control unit (PCU), which moves the actuator ram in accordance with the control demand and feeds back a ram position signal to the ACE, thereby closing

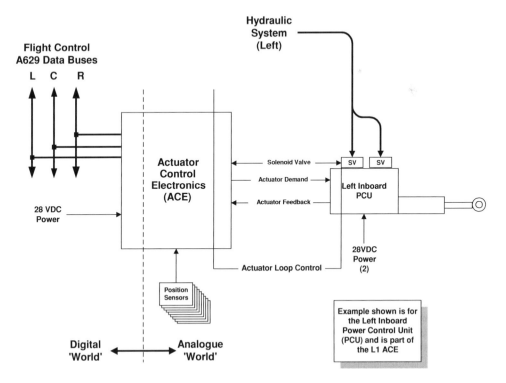

Figure 10.18 Actuator control electronics interface

the actuator control loop. The ACE also interfaces to the solenoid valve with a command to energise the solenoid valves to allow – in this example – the left hydraulic system to supply the actuator with motive power, and at this point the control surface becomes 'live'.

10.5.3   Pitch and Yaw Channel Overview

The pitch elevator control laws exercise control over the four PCUs which drive the left and right elevators to provide the aircraft with primary pitch control. The pitch portion of the control column is mechanically connected to position transducers (2), elevator feel units/actuators (2), and force transducers (2) as well as the back-drive actuator units (2) (see Figure 10.19).

Pitch demands are therefore fed into the ACE units either from the pilot transducers or as FCC or AFDC commands from the A629 data buses as appropriate and converted into analogue demands for the four elevator PCUs. Position feedback from each of the PCUs closes the control loop for each actuator.

The AFDCs interface with two back-drive actuators to align the mechanical transducers with the autopilot demands when the autopilot is engaged, to ensure that no disagreement between pilot and autopilot inputs may persist. In this way, no out-of-trim conditions exist when the autopilot is disengaged or becomes disconnected, for example, after a detected fault.

The yaw command from the pilot's rudder pedals is mechanically connected to the pedal position transmitters (4), and via a feel and centring unit to the rudder feel unit. Pilot yaw

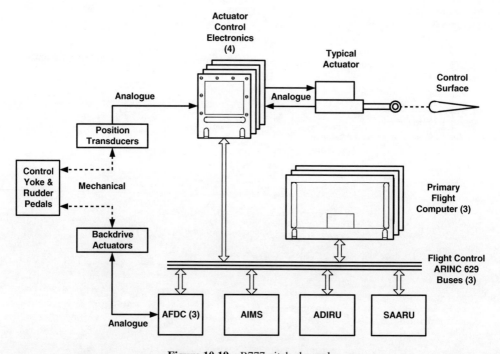

Figure 10.19   B777 pitch channel

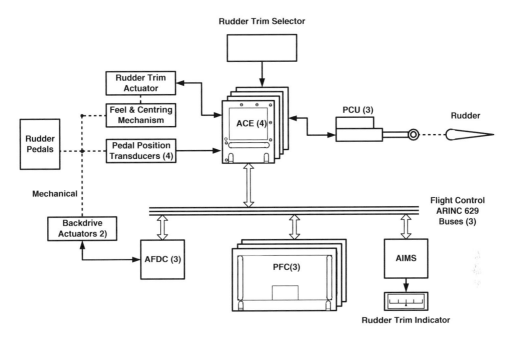

Figure 10.20 B777 yaw channel

demands are therefore fed into the ACEs and thence to the three rudder PCUs. As for the other channels, two back-drive units permit the mechanical assemblies to be backed off by the yaw channel AFDC commands (see Figure 10.20).

A rudder trim selector allows the flight crew to apply rudder trim via the ACE units. Rudder trim indication is fed via the AIMS to the rudder trim indicator on the flight deck.

10.5.4 Channel Control Logic

The flight control computations are carried out in the primary flight computers (PFCs) shown in Figure 10.21. The operation of the PFCs has been briefly described earlier in the chapter, but will be recounted and amplified in this section.

Each PFC has three A629 interfaces with each of the A629 flight control buses, giving a total of nine data bus connections in all. These data bus interfaces and how they are connected and used form part of the overall Boeing 777 PFCS philosophy. The three active lanes within each PFC are embodied in dissimilar hardware. Each of the three lanes is allocated a different function as follows:

- **PFC command lane**: The command lane is effectively the channel in control. This lane will output the flight control commands on the appropriate A629 bus (e.g. PFC left will output commands on the left A629 bus).
- **PFC standby lane**: The standby lane performs the same calculations as the command lane but does not output the commands onto the A629 bus. In effect the standby lane is a 'hot

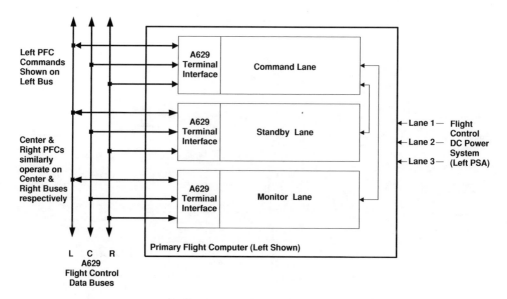

Figure 10.21   B777 channel control logic

standby', ready to take command in the event that the command lane fails. The standby lane only transmits cross-lane and cross-channel data on the A629 data bus.

- **PFC monitor lane**: The monitor lane also performs the same calculations as the command lane. The monitor lane operates in this way for both the command lane and the standby lane. Like the standby lane, it only transmits cross-lane and cross-channel data on the A629 data bus.

Figure 10.21 indicates that on the data bus, each PFC will only transmit aircraft control data on the appropriate left, centre or right A629 data bus. Within each PFC the command, standby and monitor lanes will be in operation as previously described, and only the command channel – shown as the upper channel in Figure 10.21 (left command channel for left PFC) – will actually transmit command data. Within this PFC and A629 architecture, cross-lane comparisons are conducted via the like bus (in this case the left bus). Cross-channel comparisons are conducted via the unlike buses (in this case the centre and right buses).

This use of standard A629 data buses to implement the flight control integration and to host the cross-lane and cross-channel monitoring is believed to be unique in flight control. There are effectively nine lanes available to conduct the flight control function. In the event that a single lane fails, then only that lane will be shut down. Subsequent loss of a second lane within that channel will cause that channel to shut down, as simplex control is not permitted.

Minimum equipment list (MEL) criteria allow the aircraft to be operated indefinitely with one lane out of nine failed. The aircraft may be dispatched with two out of nine lanes failed for ten days. The aircraft may be operated for a day with one PFC channel wholly inoperative.

10.5.5   Overall System Integration

A top-level portrayal of the entire B777 PFCS is shown in Figure 10.22. This shows the entire system linked via the triplex A629 flight control data buses: Left, Centre and Right. The key elements are as follows:

- air data;
- displays;
- attitude (including air data);
- AFDS;
- FBW;
- actuators;
- FMS.

This enables the reader to relate all of the PFCS elements to previous discussions.

10.6   Airbus Flight Control Systems

10.6.1   Airbus FBW Evolution

The first airbus FBW aircraft was the A320 which was first certified in 1988. Since then the A320 family has expanded to include the A318, A319 and A321; the A330 and A340 aircraft have entered service and the A380 is about to follow at the time of writing. In that time the number of flight control actuators has increased with the size of the aircraft as may be seen in Table 10.3.

The Airbus family FBW has evolved historically from the A320 family through the A330/340 family to the latest A380 aircraft. The A320 and A330 are twin-engine, while the A340 is a four-engine aircraft. Figure 10.23 clearly illustrates this progression in a simplified form. In this diagram the shaded portion represents the FBW or primary flight control system, while the units shown below represent the associated autopilot and flight management system (FMS) functions.

On the A320 family the autopilot and FMS functions are provided by standalone units. On the A330/340, flight guidance is provided by the flight management and guidance computers (FMGCs) that embody both autopilot and guidance functions. On the A380 integration has progressed, with the autopilot function being subsumed into the FCS with the flight management computer (FMC) as standalone. Although the name of the computers has changed from application to application, a clear lineage may be seen, with the A380 complement being:

- 3 × flight control primary computers (FCPC);
- 3 × flight control secondary computers (FCSC);
- 2 × flight control data concentrators (FCDC);
- 2 × flap/slat control computers (FSCC).

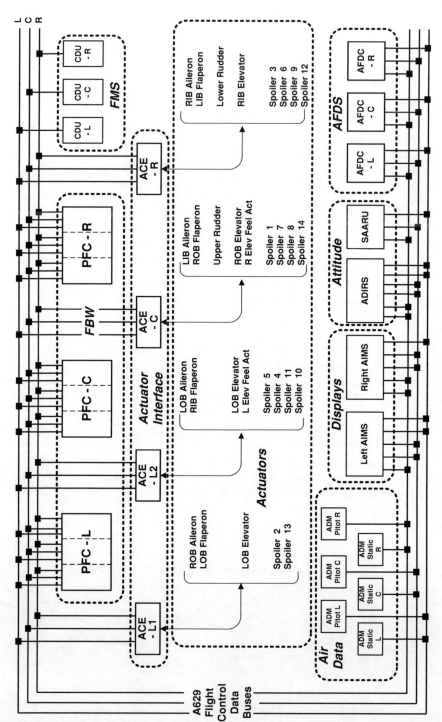

Figure 10.22 Total B777 PFCS

Table 10.3 Airbus family – roll effectors

Airbus model	Spoilers per wing	Ailerons/actuators per wing
A320 family	5	1/2
A330/340 family	6	2/4
A380	8	3/6

10.6.2 A320 FBW System

A schematic of the A320 flight control system is shown in Figure 10.24. The flight control surfaces are all hydraulically powered and are tabulated as follows:

- Electrical control:
  - elevators   2
  - ailerons   2
  - roll spoilers   8
  - tailplane trim   1
  - slats   10
  - flaps   4
  - speedbrakes   6
  - lift dumpers   10
  - trims

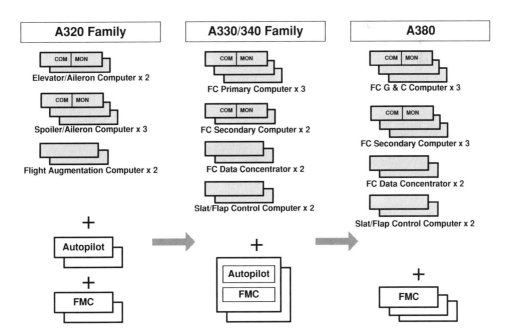

Figure 10.23 Evolution of Airbus fly-by-wire systems

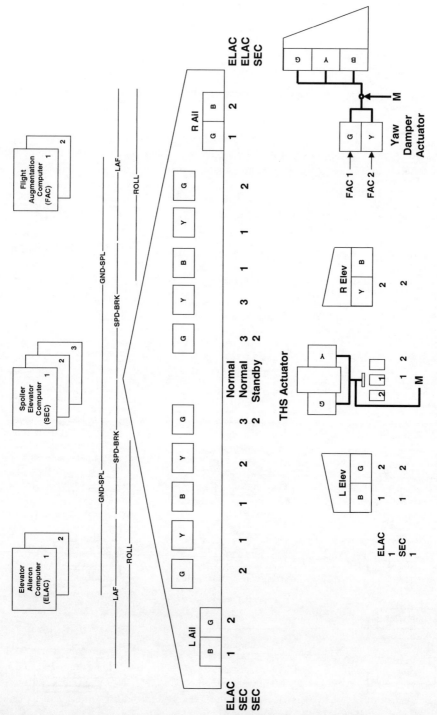

Figure 10.24 A320 flight control system

- Mechanical control:
  - rudder
  - tailplane trim (reversionary mode).

The aircraft has three independent hydraulic power systems: blue (B), green (G) and yellow (Y). Figure 10.24 shows how these systems respectively power the hydraulic flight control actuators.

A total of seven computers undertake the flight control computation task as follows:

- **Two elevator/aileron computers (ELACs)**. The ELACs control the aileron and elevator actuators according to the notation in Figure 10.24.
- **Three spoiler/elevator computers (SECs)**. The SECs control all of the spoilers and in addition provide secondary control to the elevator actuators. The various spoiler sections have different functions as shown, namely:
  - Ground spoiler mode: all spoilers.
  - Speed-brake mode: inboard three spoiler sections.
  - Load alleviation mode: outboard two spoiler sections (plus ailerons). This function has recently been disabled and is no longer embodied in recent models.
  - Roll augmentation: outboard four spoiler sections.
- **Two Flight Augmentation Computers (FACs)**. These provide a conventional yaw damper function, interfacing only with the yaw damper actuators.

In the very unlikely event of the failure of all computers, it is still possible to fly and land the aircraft – this has been demonstrated during certification. In this case the tailplane horizontal stabiliser (THS) and rudder sections are controlled directly by mechanical trim inputs and the rudder pedals – shown as M in the diagram – which allow pitch and lateral control of the aircraft to be maintained.

Another noteworthy feature of the Airbus FBW systems is that they do not use the conventional pitch and roll yoke. The pilot's pitch and roll inputs to the system are by means of a side-stick controller, and this has been widely accepted by the international airline community.

In common with contemporary civil aircraft, the A320 is not an unstable aircraft like some fighter aircraft. Instead the aircraft operates with a longitudinal stability margin of around 5% of aerodynamic mean chord, or around half what would normally be expected for an aircraft of this type. This is sometimes termed relaxed stability. The A320 family can claim to be the widest application of civil FBW, with over 5000 examples delivered.

10.6.3   A330/340 FBW System

The A330/340 FBW system bears many similarities to the A320 heritage, as might be expected.

The pilot's inputs to the flight control primary computers (FCPCs) and flight control secondary computers (FCSCs) are by means of the side-stick controller. The Flight Management Guidance and Envelope Computers (FMGECs) provide autopilot pitch commands to the FCPC. The normal method of commanding the elevator actuators is via the FCPC, although they can be controlled by the FCSC in a standby mode. Three autotrim motors may be engaged via a clutch to drive the mechanical input to the THS.

For the pitch channel, the FCPCs provide primary control and the FCSCs the backup. Pilot's inputs are via the rudder pedals directly, or in the case of rudder trim, via the FCSC to the rudder trim motors.

The yaw damper function resides within the FCPCs rather than the separate Flight Augmentation Computers (FACs) used on the A320 family. Autopilot yaw demands are fed from the FMGECs to the FCPCs.

There is a variable travel limitation unit to limit the travel of the rudder input at various stages of flight. As before, the three hydraulic systems feed the rudder actuators and two yaw damper actuators.

Therefore although the implementation and notation of the flight control computers differs between the A320 and A330/340, a common philosophy can be identified between the two families. The overall flight control system elements for the A330/340 are:

- **Three flight control primary computers (FCPCs)** – the function of the FCPCs has been described.
- **Two flight control secondary computers (FCSCs)** – similarly, the function of the secondary computers has been explained.
- **Two flight control data concentrators (FCDCs)** – the FCDCs provide data from the primary and secondary flight computers for indication, recording and maintenance purposes.
- **Two slat/flap control computers (SFCCs)** – the SFCCs are each able to control the full-span leading-edge slats and trailing-edge flaps via the hydraulically driven slat and flap motors.

Spoiler usage on the A330/340 differs from that on the A320. There is no load alleviation function and there are six pairs of spoilers versus the five pairs on the A320. Also the functions of the various spoiler pairs differ slightly from the A320 implementation. However, overall, the philosophy is the same.

10.6.4   A380 FBW System

The electrical and hydraulic power derived for the A380 flight control actuators is summarised in Figure 10.25.

The A380 flight control actuator configuration is shown in detail in References [1, 2]. Many of the actuators are powered only by the aircraft green (LH side powered by engines 1 and 2) and yellow (RH side powered by engines 3 and 4) hydraulic systems. However, many are powered by a combination of conventional hydraulic and electrohydrostatic actuators (see Figure 10.26).

The use of the various actuation types may be summarised as follows:

- The two outboard aileron surfaces and six spoiler surfaces on each wing are powered by conventional hydraulic actuators – yellow or green system.
- The mid and inboard aileron surfaces and the inboard and outboard elevator surfaces are powered by both hydraulic and EHAs, each of which can drive the surface in the event of a failure of the other.

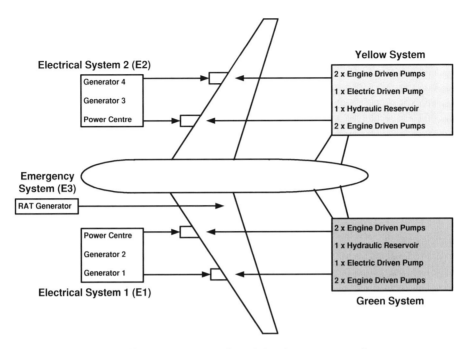

Figure 10.25 A380 hydraulic and electric power generation

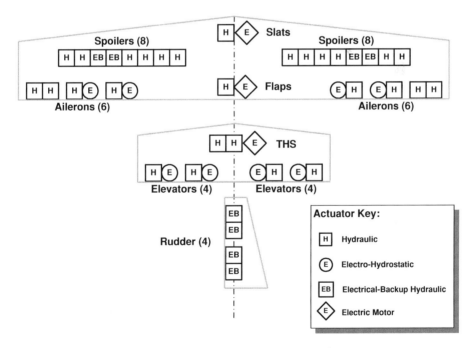

Figure 10.26 A380 flight control actuation

- Two spoiler surfaces (5 and 6 on each wing) and both rudder sections are powered by electrical backup hydraulic actuators (EBHAs) which combine the features of hydraulic actuators and EHAs.
- The tailplane horizontal stabiliser (THS) actuator is powered independently from green and yellow channels and from E2.

For completeness, the diagram also shows the flap and slat drives. Slats may be powered by green or E1; flaps may be powered from green or yellow channels.

EBHAs receive a hydraulic input from the appropriate channel (green or yellow) and electrical channel (E1 or E2, or exceptionally E3 AC essential ram air turbine (RAT)). In the case of the rudder, the upper surface is powered by green and yellow, E1 and E2 AC 2, and the lower surface is powered by green and yellow, E1 and E3. EBHAs are capable of two modes of operation:

- **Normal – hydraulic mode**: In the normal mode the actuator receives hydraulic power from the appropriate green or yellow hydraulic system and the SV moderates the supply to the actuator according to the FBW computer demand.
- **Backup – EHA mode**: In the backup mode the actuator operates like an EHA. Electrical power is received from the aircraft AC electrical system and the FBW computer feeds demands to the EHA control package. The rotational direction and speed of the electrical motor determine the direction and rate of travel of the actuator ram.

A top-level schematic of an EBHA is shown in Figure 10.27. The combination of multiple redundant FBW computing resources (three primary and three secondary flight control computers) and the actuator hydraulic and electrical power architectures mean that the aircraft is not fitted with a mechanical reversion.

10.7 Autopilot Flight Director System

10.7.1 Autopilot Principles

The workload in modern flight decks, together with the precision with which the aircraft needs to be flown, makes an autopilot an essential part of the avionics system. More correctly, the autopilot function is part of the autopilot flight director system (AFDS) which combines flight director and autopilot functions. In flight director mode, trajectory calculations are undertaken and cues fed to the displays for the pilot to follow the necessary flight path while flying the aircraft manually. When the autopilot is engaged the control loop is closed and the autopilot flies the aircraft.

The generic portrayal of an autopilot channel is shown in Figure 10.28. This comprises a forward loop where autopilot inputs are processed according to a defined set of control laws. The autopilot demands derived from the control laws drive servomotors or actuators which impart these demands to the flight control surfaces. Inner loop feedback is provided to ensure that the servomotors achieve and maintain the demanded position with the desired response rate and stability.

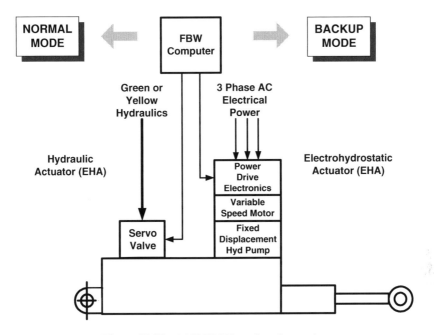

Figure 10.27   A380 EBHA modes of operation

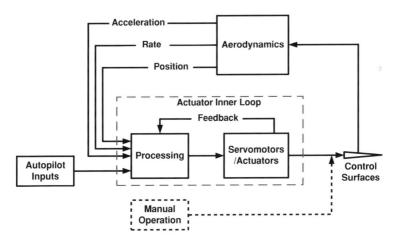

Figure 10.28   Typical autopilot control loop

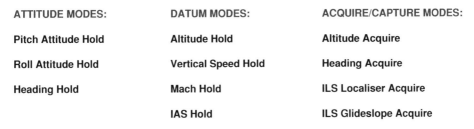

A typical modern aircraft may have 20 or 30 autopilot modes to embrace a wide range of flight conditions. For simplicity it may be considered that there are three basic types of autopilot mode as outlined below:

- **Attitude modes**. The autopilot attitude modes allow the pilot to maintain the present aircraft attitude in pitch, roll and yaw, either singly or in combination.
- **Datum modes**. Datum modes permit the pilot to set a given flight datum that he wants the aircraft to fly. Datum modes include altitude hold, vertical speed hold, Mach hold and indicated airspeed (IAS) hold. The pilot sets the appropriate datum be means of the flight control unit (FCU) located on the glare-shield.
- **Acquire/capture modes**. The acquire and capture modes give more authority to the autopilot, with the pilot selecting a particular trajectory that he wants the aircraft to adopt when he reaches a certain stage of the flight. Typical modes are altitude acquire, heading acquire and ILS localiser/glideslope acquisition and capture.

10.7.2   Interrelationship with the Flight Deck

In the same way that the flight deck controls were described for the FBW system in Figure 10.14, so Figure 10.29 shows how the various controls are used when the autopilot is engaged.

The motif in the top left-hand corner of Figure 10.29 shows that the autopilot loop has closed the inner FBW loop and is now flying the aircraft. The pilot now controls the aircraft by selecting the appropriate autopilot modes on the FCU and altering the necessary datums that determine the trajectory that the aircraft will fly. Inputs using the manual controls will not affect the aircraft.

The display of information is as before. The pitch and roll attitude are shown on the PFD, while yaw attitude or heading is displayed on the ND. Aircraft altitude and airspeed are shown on the PFD. Depending upon the autopilot mode, additional symbols may be shown on the PFDs and NDs. The nature of the displays on the PFD and ND are selected using the EFIS control panels on either side of the glare-shield. Engine parameters are displayed on the upper of the two ECAM panels.

A typical FCP layout is shown in Figure 10.30. The FCP provides the following functions:

- provides a means by which autopilot or flight director may be selected;
- enables engagement of the major autopilot modes by providing mode selection buttons for key modes of operation;
- provides display windows for important autopilot parameters and datums;
- enables reference speed or datum adjustment.

The example given has the following features: speed window (knots or Mach); heading window (degrees); vertical speed window (feet per second); altitude window (feet). A number of mode selection buttons are shown that allow the pilot to select these and other modes in conjunction, to provide automatic control as required throughout various phases of flight.

An instinctive cut-out (ICO) button will also be provided amongst the pilot's controls. The purpose of this device is to allow the pilot to disengage the autopilot rapidly in an emergency such as an uncommanded pitch movement or trim runaway.

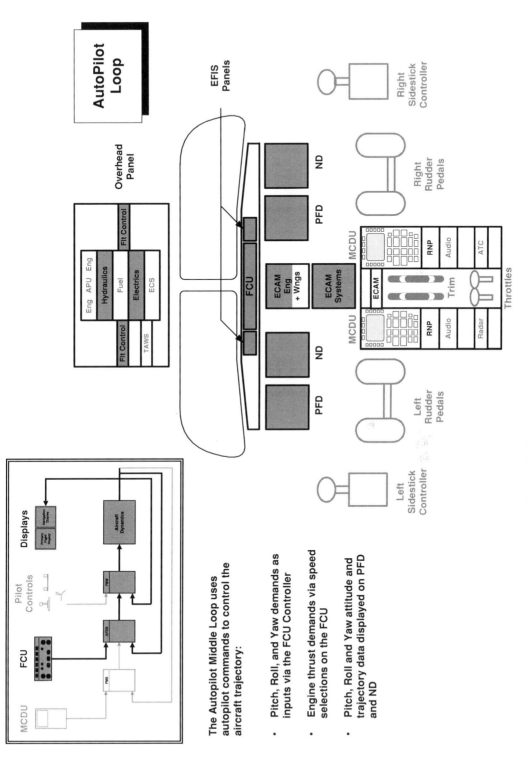

Figure 10.29 Flight deck showing autopilot loop

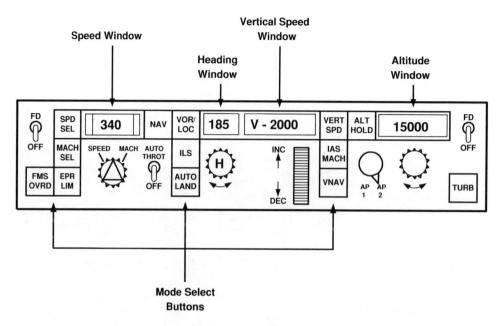

Figure 10.30 Typical FCP layout

As will be described in the next chapter, the FMS may be coupled into the autopilot to provide 3-D LNAV plus VNAV guidance to enable the aircraft to fly the necessary procedural profiles as required en-route and in busy terminal areas.

10.7.3 Automatic Landing

As the importance of maintaining airline schedules in adverse weather conditions has increased, so too has the importance of automatic landing. At its most refined, automatic landing allows the aircraft to land in virtually zero visibility conditions. The various stages of auto-land are defined by criteria presented in Figure 10.31. These criteria may be summarised as follows:

- **Category I (Cat I)**. This relates to a decision height (DH) of not less than 200 ft and visibility not less than 2600 ft (or not less than 1800 ft (545 metres) runway visual range (RVR) where the airfield/runway is equipped with dedicated measuring devices).
- **Category II (Cat II)**. DH not less than 100 ft and RVR not less than 1200 ft (364 m).
- **Category IIIA (Cat IIIA)**. DH less than 100 ft and RVR not less than 700 ft (212 m). Also described as 'see to land'.
- Category IIIB (Cat IIIB). **DH less than 50 ft, RVR not less than 150 ft (45.4 m). Also described as 'see to taxi'.**

The introduction of a Cat III auto-land capability poses further constraints for the aircraft systems – see References [5, 6]. Simply put these are:

- Three independent AC and DC electrical power channels.
- Three independent flight control channels.

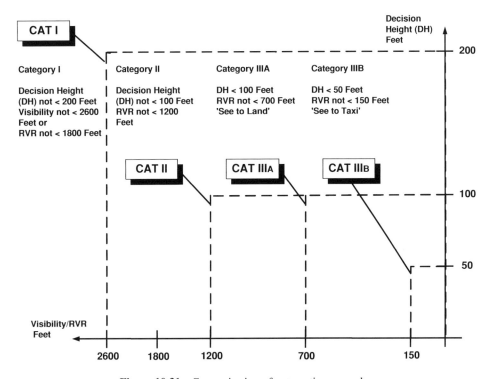

Figure 10.31   Categorisation of automatic approaches

- Precision approach guidance – ILS or MLS. The categorisation of the ground-based equipment is equally as important as the aircraft mounted equipment.
- Autotrim. It is vital in all autopilot applications that the autopilot has to be 'in trim'.

It should also be noted that the GPS augmentation systems will also be capable of providing auto-land quality flight guidance once the systems are fully operational. It is intended that the Local Area Augmentation System (LAAS) and Wide Area Augmentation System (WAAS) GPS augmentation systems will be capable of providing Cat II/III and Cat I guidance respectively (see Chapter 11).

10.8   Flight Data Recorders

10.8.1   Principles of Flight Data Recording

All aircraft involved in passenger operations and which are above 12,500 lb (5670 kg) are mandated to carry a flight data recording system, and the requirements for these systems are becoming steadily more rigorous. Older flight data recorders (FDRs) recorded various aircraft functions such as vertical acceleration, heading, airspeed, altitude, and so on, in analogue form using a stylus and moving oscillographic foil medium composed of steel or steel alloy. In modern systems the recording medium is ruggedised to withstand shock, fire and long-term immersion in seawater using digital solid-state electronics memory. In the US, present regulations now dictate that a digital flight data recorder (DFDR) must be used [7]. A cockpit voice recorder (CVR) is used to record crew and ATC conversations using a 'hot mike' located on the flight deck.

The overall system is known as a digital flight data recording system (DFDRS); this comprises the equipment, sensors, wiring and other installed equipment necessary to perform the function. Where a dedicated sensor has to be installed to provide the DFDR function then it forms part of the DFDRS. Where a sensor is already installed for another purpose, such as a lateral accelerometer for automatic flight control, it does not comprise part of the system.

The main elements of a DFDRS are as follows:

- Digital flight data recorders (DFDRs).
- Flight data acquisition unit (FDAU) or a digital flight data acquisition unit (DFDAU). The FDAU has the ability to collect, sample, condition and digitise analogue signals and provide the data to the DFDR in a digital data stream according to the requirements of ARINC 573. The DFDAU has the ability to receive both analogue parameters and digital data streams and convert them to the DFDR digital data format in accordance with ARINC 717.
- Underwater locating device (ULD) in the form of a sonar locator beacon.

The number of parameters that the DFDRS is required to record has progressively increased in recent years [8]. In essence, all aircraft with provision for carrying 10 passengers or more have to have a DFDRS fitted in accordance with Reference [6]. The key provisions of Reference [7] relate to the standard of DFDRS to be fitted to a compliance schedule depending upon the date of aircraft manufacture:

- All aircraft manufactured before 11 October 1991 had to be retrofitted with a system capable of recording 34 parameters by 20 August 2000. All aircraft manufactured between 11 October 1991 and 20 August 2000 had to be similarly equipped.
- All aircraft manufactured between 20 August 2000 and 19 August 2002 were mandated to be fitted with a system capable of recording 57 parameters.
- All aircraft manufactured after 19 August 2002 must have a system capable of recording 88 parameters.

DFDRs, known colloquially in the media as the 'black box recorder', are actually painted bright orange to aid recovery at a crash scene. The reason for increasing the number of recorded parameters was because the causes of many accidents were not being identified due to the paucity of data, and the schedule outlined above is an attempt to redress this shortcoming. There is still a problem that relates to powering the DFDRS; where electrical power is lost to part or all of the system, so too is the recorded data, and consideration is being given to alternative electrical power sources to prevent this happening. Digital flight data recording – if recoverable – is also very useful in terms of capturing data 'snap shots' in the event of any abnormal operation or system upsets. It enables technical investigators to establish rapidly the root cause of an aircraft incident. In a number of recent incidents it has helped to establish and confirm the utter professionalism of the flight crew to deliver the passengers safely in the event of a dire flying emergency. References [9, 10] give examples of data recording and data acquisition. Reference [11] includes an example of the data recording outputs from the Quantas (Queensland and Northern Territories Air Service) QF32 A380 incident.

10.8.2   Data Recording Environments

Data recording environments have rapidly evolved in line with technology over the past few decades. Survivability levels have increased as indicated below over recent years:

- Impact: 100 g → 1000 g → 1700 g → 3400 g.
- Penetration: none → 500 lbs; 10 foot drop; $^1/_4$ inch pin.
- Static crush: None → 5000 lbs for 5 minutes in each axis and diagonal.
- 1100° fire test increased from 30 to 60 minutes duration; flame coverage increased from 50% to 100%.
- Added low-temperature fire of 10 hours at 260°C.
- Seawater immersion: Increased from 36 hours to 30 days.
- Added hydrostatic pressure – equivalent to depth of 20,000 feet.

10.8.3   Future Requirements

There are considerable concerns that present data capture methods – powerful though they may be – are not going to capture all the requirements of future incidents. Present systems rely upon a single recording entity, and if electrical power is lost the entire recording capability is also lost. Future systems may also be required to provide a video history of the flight deck environment and dual power supplies and wiring to provide additional sources of data (see Figure 10.32).

Recent views in Europe have suggested that the following should be considered:

- ED-112 (Europe) giving new performance requirements for flight recorders;
- automatically deployable recorders;

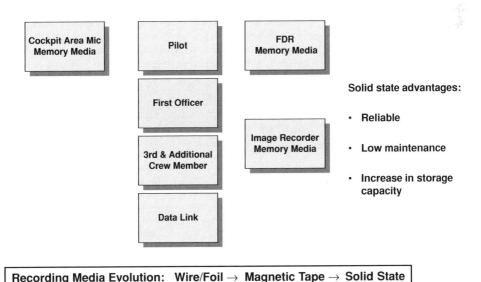

Figure 10.32   Possible new data gathering areas

- combined recorders;
- recorder-independent power supplies;
- image recording – ambient conditions in the cockpit; crew activity; instruments and control panels;
- CNS/ATM – the flight path of the aircraft is authorised, directed or controlled, and relayed over data link rather than voice communication.

In the US the following ideas are being developed. Two redundant multifunction recorders installed in forward and tail sections provide operational as well as accident investigation benefits:

- increases the recovery probability of all types of information – audio; aircraft system data; image; CNS/ATM;
- aft location: maximizes impact survivability;
- forward location: maintains wiring integrity; colocated with crew.

References

[1] Moir, I. and Seabridge, A. (2008) *Aircraft Systems: Mechanical, Electrical and Avionics Subsystems Integration*, 3rd edn. John Wiley & Sons, Ltd., Chichester.

[2] *More-Electric Control Surface Actuation* (2004) Royal Aeronautical Society Conference, 'More-Electric Aircraft', London, 27–28 April.

[3] Tucker, B.G.S. (1993) *Boeing 777 Primary Flight Control Computer System – Philosophy and Implementation*. RAeS Conference, Advanced Avionics on the A330/A340 and the Boeing 777, November 1993.

[4] McWha, J. (1995) *777 – Ready for Service*. RAeS Conference, 'The Design & Maintenance of Complex Systems on Modern Aircraft', April 1995.

[5] AC 120-67 (1997) *Criteria for Operational Approval of Auto Flight Guidance Systems*. Federal Aviation Authority.

[6] AC 120-28D (1999) *Criteria for Approval of Category III Weather Minima for Takeoff, Landing and Rollout*. Federal Aviation Authority.

[7] AC 20-141 (1999) *Airworthiness and Operational Approval of Digital Flight Data Recording Systems*. Federal Aviation Authority.

[8] EUROCAE ED-55 (1990) *Minimum Performance Specification for Flight Data Recorder System*.

[9] *Avionics Magazine* (2011) *Data Acquisition*. July 2011.

[10] *FAST Magazine* (2011) *Flight Data Recorder*. August 2011.

[11] Australian Transport Safety Bureau (2010) Aviation Occurrence Investigation – AO-2010-089. *In-flight Uncontained Engine Failure*. Australian Transport Safety Bureau.

11

Navigation Systems

11.1 Principles of Navigation

11.1.1 Basic Navigation

The description of basic navigation parameters may be summarised by the following, as outlined in Figure 11.1:

- An aircraft will be flying at a certain altitude relative to a barometric datum (barometric altitude) or terrain (radar altitude).
- The aircraft may be moving with velocity components in the X (V_x), Y (V_y) and Z (V_z) axes. The aircraft speed through the air mass may be characterised as either indicated airspeed (IAS) or Mach number (M). Its speed relative to the ground is determined by true airspeed (TAS) in still air conditions.
- The aircraft will be flying on a certain heading; however, the prevailing wind-speed and direction will modify this to the aircraft 'track'. The aircraft track represents the aircraft path across the terrain and will lead to the aircraft's destination or next waypoint. Wind-speed and direction will modify the aircraft speed over the ground to 'groundspeed'.
- The aircraft heading will be defined by a bearing to magnetic (compass) north or to true north relating to Earth-related geographical coordinates.
- The aircraft will be flying from its present position, defined by latitude and longitude, to a waypoint also determined by latitude and longitude.
- A series of flight legs – defined by way points – will determine the aircraft designated flight path from the departure airfield to the destination airfield. As has already briefly been described, there are a variety of sensors and navigation techniques which may be used solely or in combination to navigate the aircraft throughout the mission.

A historical set of references can be found at the end of this chapter. These references embrace many of the historic rules and procedures that were invoked going back as far as 1969, and some procedures or parts thereof may be equally applicable today. They are not necessarily key to any discussion related to the avionics systems of 2012 and beyond. The

Civil Avionics Systems, Second Edition. Ian Moir, Allan Seabridge and Malcolm Jukes.
© 2013 John Wiley & Sons, Ltd. Published 2013 by John Wiley & Sons, Ltd.

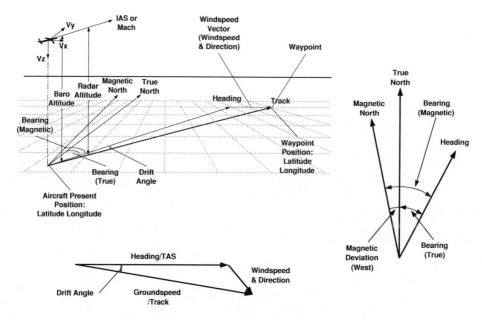

Figure 11.1   Basic navigation parameters

authors have in general limited themselves to references from the start of the new millennium (i.e. 2000 onwards).

The relationship of the different axis sets is shown in Figure 11.2. These may be characterised as:

- **Earth datum set:** As shown in Figure 11.2, the Earth axis reference set comprises the orthogonal set E_x, E_y, E_z where:
  - E_x represents true north.
  - E_y represents east.
  - E_z represents the local gravity vector or vertical.
- **The orthogonal aircraft axis set where:**
  - A_x is the aircraft longitudinal axis (corresponding to the aircraft heading).
  - A_y is the aircraft lateral axis.
  - A_z is the aircraft vertical axis (corresponding to E_z).

For navigation purposes, the accuracy with which the aircraft attitude may be determined is a key variable for navigation systems within which the velocity components need to be resolved into aircraft axes. Similarly, attitude is used for inertial navigation (IN) axis transformations. The aircraft axes in respect of flight control have been discussed in Chapter 10 (Flight Control), and IN alignment requirements described in Chapter 8 (Sensors).

The navigation function therefore performs the task of manoeuvring the aircraft from a known starting point to the intended destination using a variety of sensors and navigation aids and via a number of specified waypoints.

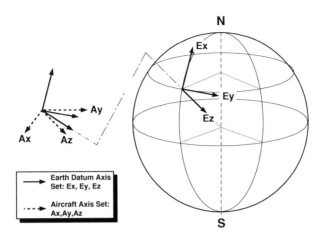

Figure 11.2   Earth-related coordinates

The classic method of navigation for many years has been to use a combination of magnetic and inertial directional gyros, used together with airspeed information derived from the air data sensors, to navigate in accordance with the parameters shown in Figure 11.1. This dead-reckoning system is subject to errors in both the heading system and the effects of en-route winds, which can cause along-track and across-track errors. In the 1930s it was recognised that the use of radio beacons and navigation aids could significantly reduce these errors by providing the flight crew with navigation assistance related to precise points on the ground.

Aircraft navigation systems in use today utilise a combination of different techniques according to which is most appropriate at the time. These techniques may be summarised as follows:

- navigation aids – Section 11.1.2;
- air data and inertial navigation – Section 11.1.3;
- global navigation satellite systems – Section 11.1.4.

11.1.2   Navigation using Ground-Based Navigation Aids

For many years the primary means of navigation over land, at least in continental Europe and the North American continent, was by means of radio navigation routes defined by VHF omni-ranging/distance measuring equipment (VOR/DME) beacons as shown in Figure 11.3. By arranging the location of these beacons at major navigation or crossing points, and in some cases airfields, it was possible to construct an entire airway network which could be used by the flight crew to define the aircraft flight from take-off to touch-down. Other radio frequency aids include distance measuring equipment (DME) and non-directional beacons (NDBs). Figure 11.3 portrays:

- **Three VOR/DME beacon pairs:** VOR 1/DME 1, VOR 2/DME 2 and VOR 3/DME 3 which define the waypoints 1 to 3. These beacons represent the aircraft intended waypoints

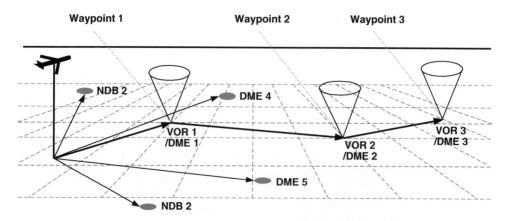

Figure 11.3   Navigation using navigation aids

1, 2 and 3 as it proceeds down the intended flight plan route – most likely an identified airway. When correctly tuned, the VOR/DME pairs successively present the flight crew with bearing to and distance from the next waypoint.

- **Off-route DME beacons:** DME 4 and DME 5, which may be used as additional means to locate the aircraft position using the DME fix obtained where the two DME 4 and DME 5 range circles intersect. DME/DME fixes are a key attribute in the modern navigation system.
- **Off-route NDB beacons** may be used as an additional means to determine the aircraft position by obtaining a cross-fix from the intersection of the bearings from NBD 1 and NDB 2. These bearings are derived using the aircraft ADF system.

Thus in addition to using navigation information from the 'paired' VOR/DME beacons that define the main navigation route, position fix, cross-fix, range or bearing information may also be derived from DME or NDB beacons in the vicinity of the planned route by using automatic direction-finding techniques. As has already been described in Chapter 9, a major limitation of the radio beacon navigation technique results from line-of-sight propagation limitations at the frequencies at which both VOR and DME operate. As well as the line-of-sight and terrain masking deficiencies, the reliability and accuracy of the radio beacons can also be severely affected by electrical storms.

Due to the line-of-sight limitations of these radio beacons, these navigation techniques were only useable overland where the beacon coverage was sufficiently comprehensive or close off-shore routes where the beacons could be relied upon.

11.1.3   Navigation using Air Data and Inertial Navigation

The availability of inertial navigation systems (INS) to the civil aviation community during the late 1960s added another dimension to the navigation equation. The flight crew were now able to navigate by autonomous means using an onboard INS with inertial sensors. By aligning the platform to Earth-referenced coordinates and present position during initialisation, it was now possible to fly for long distances without relying upon VOR/DME beacons overland.

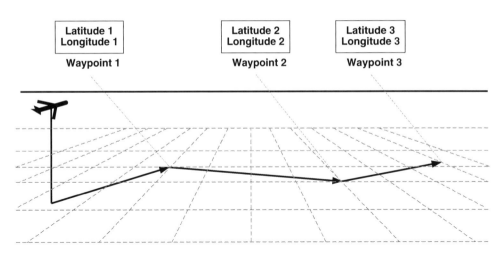

Figure 11.4   Navigation using air data and inertial navigation

Waypoints could be specified in terms of latitude and longitude as fixed reference points on the globe, more suited to the aircraft's intended flight path rather than a specific geographical feature or point in a radio beacon network (see Figure 11.4).

The specifications in force at this time also offered an INS solution to North Atlantic crossings. The inertial solution required serviceable dual INS and associated computers to be able to undertake the crossing. There were also limitations on the latitudes at which the ground alignment could be performed – 76° north or south – as attaining satisfactory alignment becomes progressively more difficult the nearer to the poles the INS becomes.

Historically, the requirements for operating an INS as sole means of navigation for a significant portion of the flight were carefully documented. These requirements may be summarised as follows:

- The ability to provide the following functions:
  - Valid ground alignment at all latitudes appropriate for the intended use of the INS.
  - The display of alignment status to the crew.
  - Provision of the present position of the aircraft, in suitable coordinates: usually latitude from +90° (north) to −90° (south) and longitude from +180° (east) to −180° (west).
  - Provision of information on destinations or waypoints.
  - Provision of data to acquire and maintain the desired track and the ability to determine deviation from the desired track (across-track error).
  - Provision of information needed to determine the estimated time of arrival (ETA).
- The ability to comply with the following requirements:
  - ±20 nm across track and ±25 nm along track.
  - The ability to maintain this accuracy on a 95% probability basis at representative speeds and altitudes and over the desired latitude range.
  - The ability to compare the INS position with visual fixes or by using TACAN, VOR, DME or ground radar (air traffic control).

- The provision of a memory or in-flight alignment means. Alternatively the provision of a separate electrical power source – usually a dedicated stand-alone battery – with the ability to support the INS with full capability for at least 5 minutes in the event of an interruption to the normal power supply.

For reasons of availability and accuracy, systems were developed with dual and triple INS installations. A typical triple INS would be representative of an INS installation before the availability of satellite sensors in the 1980s. By this time the gimballed IN platform would have been replaced by a more reliable strap-down system similar to the Litton LTN-92 system available at the time.

As has already been described in Chapter 8, air data and IN sensor systems began to be combined into the Air Data and Inertial Reference System (ADIRS) to take advantage of air data and IN sensor fusion.

11.1.4 Navigation using Global Navigation Satellite Systems

The foregoing techniques were prevalent from the 1960s through until 1990s when satellite navigation became commonly available. The use of global navigation satellite systems (GNSS), to use the generic name, offers a cheap and accurate navigational means to anyone who possesses a suitable receiver (see Figure 11.5). Although the former Soviet Union developed a system called GLONASS, it is the US Global Positioning System (GPS) that is the most widely used. The principles of satellite navigation using GPS have already been described in Chapter 9. GPS receivers may be provided for the airborne equipment in a number of ways:

- Stand-alone GPS receivers – most likely to be used for GPS upgrades to an existing system.
- Multichannel (typically 12-channel) global navigation satellite system (GNSS) receivers – many aircraft utilise this approach.

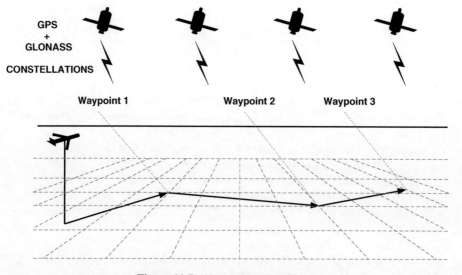

Figure 11.5 Navigation using GNSS

- GPS receivers integrated into a multifunction receiver unit called a multi-mode receiver (MMR) where the GPS receiver function is integrated into one LRU along with VOR and ILS receivers.

The use of GNSS systems such as GPS and GLONASS, including accuracy enhancing techniques such as differential GPS, have already been outlined in Chapter 9.

11.1.5   Flight Technical Error – Lateral Navigation

Lateral NAVigation (LNAV) systems allow the aircraft to operate within any desired course within the coverage of station-referenced signals (VOR, DME), or within the limits of a self-contained system capability (IRS, GPS), or a combination of these. RNAV systems have a horizontal 2-D capability using one or more of the onboard navigational sensors to determine a flight path by navigation aids or waypoints referenced to latitude and longitude. In addition the RNAV system provides guidance cues or tracking of the flight path. Many modern RNAV systems include a 3-D capability to define a vertical flight path based upon altimetry, and some include a full aircraft and engine performance model.

The performance of pre-RNAV systems has historically been defined according the criteria in Figure 11.6:

- along-track error;
- across-track error;
- flight technical error (FTE).

The total navigation error is the root sum squares (RSS) of these elements for a given navigation means or phase of flight.

The availability of the navigation capability is defined at 99.999% and the integrity requirement for misleading navigation information is set at 99.9999%. Navigation accuracies and FTE are typically demonstrated within the values given in Table 11.1 for differing phases of flight.

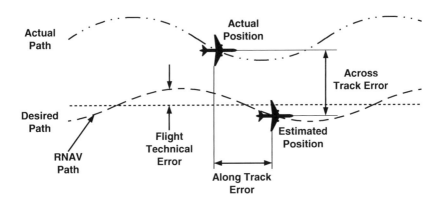

Figure 11.6   Flight technical error – lateral

Table 11.1   Demonstrated accuracy and FTE per phase of flight

Airspace/operation	Accuracy (95%)	FTE
Oceanic/en-route remote	±3.8 nm	±1.0 nm
Domestic en-route	±2.8 nm	±1.0 nm
Terminal	±1.7 nm	±0.5 nm
Approach – VOR/DME	±0.5 nm	±0.125 nm
Approach – multisensor	±0.3 nm	±0.125 nm

11.1.6   Flight Technical Error – Vertical Navigation

Vertical NAVigation (VNAV) capability further enhances flight operations by specifying a vertical flight path to be flown coincident with the lateral navigation guidance, providing a 3-D path for the aircraft to follow. As for the lateral guidance case, path deviation and guidance for tracking the path are provided. Key elements in VNAV are:

- computed vertical flight paths as the basis for aircraft guidance;
- assured repeatability of performance;
- tailoring the flight path to aircraft/engine performance – full performance VNAV;
- providing situational awareness cues to the flight crew.

VNAV capabilities have been defined by the error components defined in Figure 11.7 and listed below:

- flight path definition error;
- altimetry error (99.7%);
- FTE.

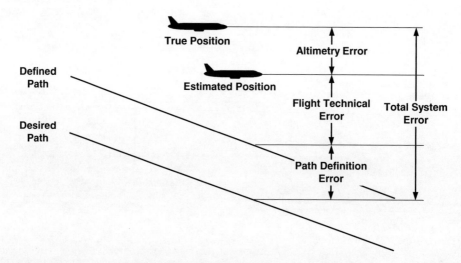

Figure 11.7   Flight technical error – vertical

The total demonstrated VNAV system error is the RSS of these values, depending upon altitude and flight trajectory conditions given below in Table 11.2 below.

Table 11.2   Demonstrated VNAV performance

Airspace operation	Level 99.7%	Level intercept FTE	Climb descent 99.7%	Climb descent FTE
At or below 5000 ft	±50 ft	±150 ft	±100 ft	±200 ft
5000 to 10,000 ft	±50 ft	±240 ft	±150 ft	±300 ft
Above 10,000 ft	±50 ft	±240 ft	±220 ft	±300 ft

Note: This performance is consistent with the 1000 feet RVSM criteria.

Reference [1] describes a scheme for the display of LNAV and VNAV steering commands.

11.2   Flight Management System

11.2.1   Principles of Flight Management Systems (FMS)

It is clear from the foregoing description of the aircraft navigation functions that navigation is a complex task, and with modern integrated systems, accuracy expectations are high, as is the need to reduce flight crew workload. FMS functionality has increased rapidly over the last two decades as requirements have increased, and many more enhancements are in prospect as the future features required by air traffic management (ATM) are added.

In essence the FMS performs the task of coalescing the inputs from the various navigation sensors in an optimum manner, navigating precisely to pre-programmed flight plans and procedures in space and time and providing an effective man–machine interface for the flight crew.

A typical FMS will embrace dual computers and dual multifunction control and display units (MCDUs) as shown in Figure 11.8.

This diagram is key to depicting the integration of the navigation functions described above. Inputs, usually duplicated for reasons of availability and integrity, are shown on the left. These are:

- dual INS/IRS;
- dual navigation aid sensors: VOR/DME; DME/DME;
- dual GNSS sensors – usually GPS;
- dual or triplex air data sensors – latterly combined with INS/IRS in an ADIRS;
- dual inputs from onboard sensors relating to fuel and time.

These inputs are used by the FMS to perform the necessary navigation calculations and provide information to the flight crew via a range of display units:

- electronic flight instrument system (EFIS);
- communications control system (CCS);
- interface with the autopilot/flight director system to provide the flight crew with flight direction or automatic flight control in a number of predefined modes.

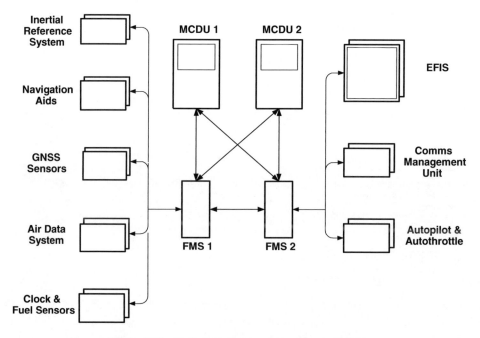

Figure 11.8 Typical flight management system (FMS)

11.2.2   FMS Crew Interface – Navigation Display

The FMS to crew interface is shown in Figure 11.9. The key interface with the flight crew is via the following displays:

- **Captain's and first officer's navigation displays** (ND), part of the electronic flight instrument system (EFIS). The navigation displays may display information in a variety of different ways.
- **Control and display units 1 and 2,** part of the FMS. The CDUs both display information and act as a means for the flight crew to enter data manually.

Reference [2] describes the FMS lineage of the Airbus A320/A330/A340 families.

The FMS computers perform all the necessary computations and display the appropriate navigation parameters on the appropriate display. The navigation displays contain the navigation and steering information necessary to fly the intended route. These are colour displays and can operate in a number of different formats, depending upon the phase of flight. These are:

- **Expanded approach mode**. This display shows the selected runway heading and the lateral deviation from the runway centre line. The expanded approach mode displays 80° ($\pm$40°) of the compass rose with the aircraft symbol and localiser deviation bar at the bottom. The display is orientated with heading at the 12 o'clock position. The glide-slope deviation is shown at the right of the display (see Figure 11.10a).

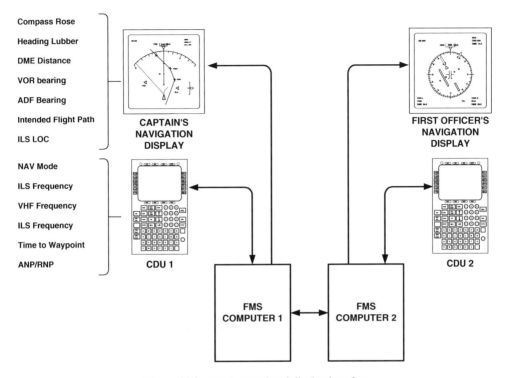

Figure 11.9 FMS control and display interface

- **Centre approach mode**. The centre approach mode shows 360° of the compass rose with the aircraft symbol in the centre. Otherwise the display is heading-orientated as before, and the ILS deviation symbology is the same as for the expanded mode (see Figure 11.10b).
- **Expanded map mode**. This display shows the portion of the flight plan within the selected range (up to 640 nautical miles). The expanded portion of the map displays 80° of the compass rose with the aircraft symbol at the bottom. The display is orientated with track at the 12 o'clock position. The display shows elements of the flight plan showing intended track between waypoints and also displays the relevant radio beacons – in this example VOR and TACAN (see Figure 11.10c).
- **Centre map mode**. This shows 360° of the compass rose with the aircraft symbol in the centre. The display is track-orientated as for the expanded map mode and also shows elements of the flight plan track and relevant beacons (see Figure 11.10d).
- **Plan mode**. This display shown in Figure 11.10e is used for three purposes:
  - to view a flight plan stored in the FMS computers.
  - to amend a flight plan stored in the FMS computers.
  - to create a new flight plan.

The plan mode displays 80° of the compass rose. The plan mode is orientated with true north at the 12 o'clock position. This mode effectively represents a 'bird's eye' view of the intended flight plan, and its representation of information is more similar to that displayed in

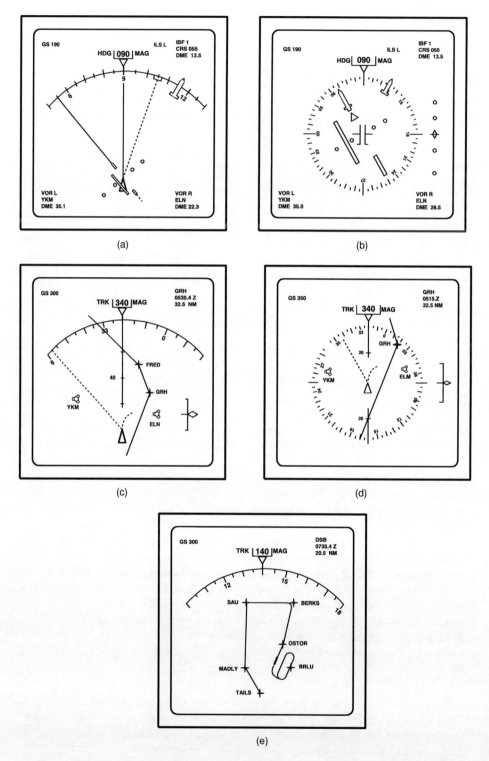

(a)

(b)

(c)

(d)

(e)

Figure 11.10   Approach displays – expanded, centre and plan modes (see text for details)

the published terminal procedures (SIDS, STARS, etc.), examples of which are given later. Colour is used extensively to accentuate the visual impact of the displays.

11.2.3 FMS Crew Interface – Control and Display Unit

The FMS CDU is the key flight crew interface, with the navigation system allowing the flight crew to enter data as well as having vital navigation information displayed. A typical FMS CDU is shown in Figure 11.11.

The CDU has a small screen on which alpha-numeric information is displayed, in contrast to the pictorial information displayed on the EFIS navigation displays. This screen was a cathode ray tube (CRT) monochrome display in early systems; later systems use colour active matrix liquid crystal displays (AMLCDs) (see Chapter 12, Flight Deck Displays). The tactile keyboard has alpha-numeric keys in order to allow manual entry of navigation data – perhaps inserting final alterations to the flight plan – as well as various function keys by which specific navigation modes may be selected. The line keys at the sides of the display are soft keys that allow the flight crew to enter a menu-driven system of subdisplays to access more detailed information. On many aircraft the CDU is used to portray maintenance status and to execute test procedures using the soft keys and the menu-driven feature. Finally, there are various annunciator lights and a lighting control system. Examples of the data displayed on the CDU are shown in Figure 11.12.

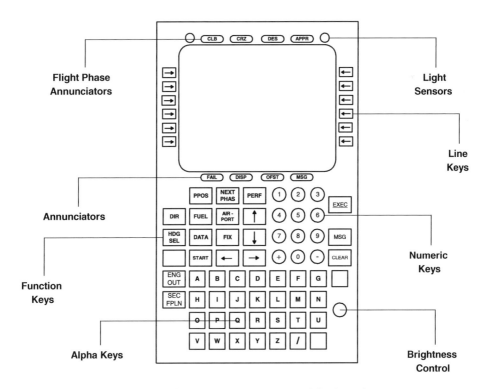

Figure 11.11 Typical FMS control and display unit

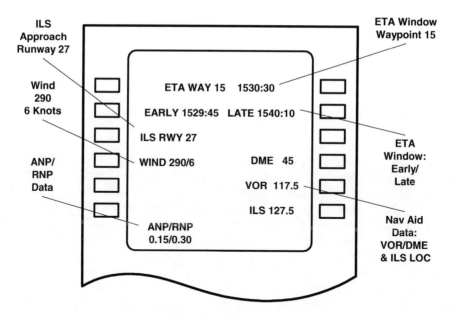

Figure 11.12 Typical FMS CDU display data

The CDU example shown displays the following:

- An ETA waypoint window that shows the estimated time of arrival (ETA) at the waypoint, in this case waypoint 15.
- Early/late timing information which represents the earliest and latest times the aircraft can reach the waypoint, given its performance characteristics.
- Information on the runway – an ILS approach to runway 27.
- Wind information for the approach – wind bearing 290°.
- Information on the navigation aids being used: VOR, DME and ILS/LOC.
- ANP/RNP window. This compares the actual navigation performance (ANP) of the system against the required navigation performance (RNP) for the flight phase and navigation guidance being flown. In this case the ANP is 0.15 nm against a RNP of 0.3 nm and the system is operating well within limits.

The flight deck level integration is portrayed in Figure 11.13. This corresponds to earlier diagrams but in this case, as the motif in the top left-hand corner shows, it is the FMS loop that is in control. This is the outer loop that closes around the FBW and AFDS loops to provide mission control. As previously, the active control elements are shown with emphasis.

The FMS outer loop uses FMS inputs and autopilot commands to control the aircraft according to the desired mission flight path:

- Pitch, roll and yaw demands as inputs via the FCP controller.
- Engine thrust demands via speed selections on the FCP or as modified by the FMS flight plan.

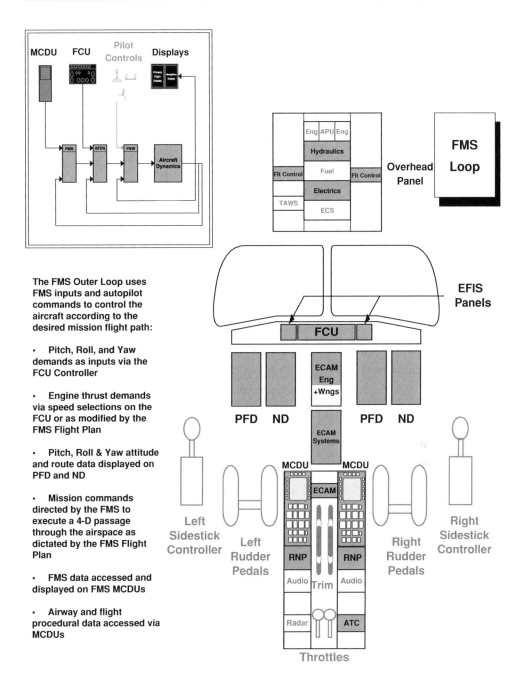

The FMS Outer Loop uses FMS inputs and autopilot commands to control the aircraft according to the desired mission flight path:

• Pitch, Roll, and Yaw demands as inputs via the FCU Controller

• Engine thrust demands via speed selections on the FCU or as modified by the FMS Flight Plan

• Pitch, Roll & Yaw attitude and route data displayed on PFD and ND

• Mission commands directed by the FMS to execute a 4-D passage through the airspace as dictated by the FMS Flight Plan

• FMS data accessed and displayed on FMS MCDUs

• Airway and flight procedural data accessed via MCDUs

Figure 11.13   FMS loop closure

- Pitch, roll and yaw attitude and route data displayed on PFD and ND.
- Mission commands directed by the FMS to execute an LNAV, VNAV or 4-D passage through the airspace as dictated by the FMS flight plan.
- FMS data accessed and displayed on FMS MCDUs.
- Airway and flight procedural data accessed via MCDUs.

11.2.4   FMS Functions

The functions of the FMS at a top level are shown in Figure 11.14. This diagram displays an overview of the functions performed by the FMS computers. These may be functions summarised as follows:

- **Navigation computations and display data**. All the necessary navigation computations are undertaken to derive the navigation or guidance information according to the phase of flight and the sensors utilised. This information is displayed on the EFIS navigation display or the FMS CDU. Flight director and steering commands are sent to the autopilot for the flight director with the pilot in the loop or for the engagement automatic flight control modes.
- **Navigation sensors**. INS, GPS, VOR, ILS, ADF, TACAN and other navigation aids provide dual sensor information to be used for various navigation modes.
- **Air data**. The ADCs or ADIRS provides the FMS with high-grade corrected air data parameters and attitude information for use in the navigation computations.
- **Fuel state**. The fuel quantity measurement system and the engine-mounted fuel flow-meters provide information on the aircraft fuel quantity and engine fuel flow. The calculation of fuel

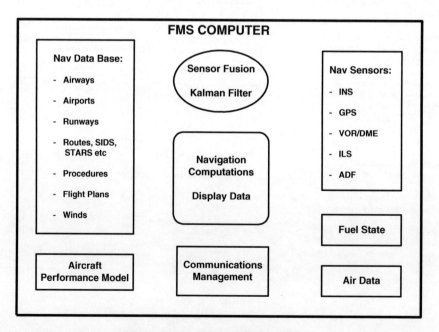

Figure 11.14   Top-level FMS functions

use and total fuel consumption are used to derive aircraft and engine performance during the flight. When used together with a full aircraft performance model, optimum flight guidance may be derived which minimises fuel consumed.

- **Sensor fusion and Kalman filter**. The sensor information is fused and validated against other sources to determine the validity and degree of fidelity of the data. By using a sophisticated Kalman filter the computer is able to determine the accuracy and integrity of the navigation sensor and navigation computations and determine the actual navigation performance (ANP) of the system in real time.
- **Communications management**. The system passes information to the communication control system regarding communication and navigation aid channel selections which have been initiated by the FMS in accordance with the requirements of the flight plan.
- **Navigation database**. The navigation database contains a wide range of data that are relevant to the flight legs and routes the aircraft may expect to use. This database will include the normal flight plan information for standard routes that the aircraft will fly, together with normal diversions. Navigation databases are organised in accordance with Reference [3]. They will be regularly updated and maintained. A comprehensive list of these items includes:
  - Airways.
  - Airports including approach and departure information, airport and runway lighting, obstructions, limitations, airport layout, gates, and so on.
  - Runways including approach data, approach aids, category of approach – Cat I or Cat II/III, decision altitudes.
  - Routes, clearance altitudes, SIDS, STARS and other defined navigation data.
  - Procedures including notification of short-term airspace restrictions or special requirements.
  - Flight plans with standard diversions.
  - Wind data – forecast winds and actual winds derived throughout flight.
  - Aircraft performance model. The inclusion of a full performance model adds to the system's ability to compute four-dimensional (x, y, z, time) flight profiles and at the same time make optimum use of the aircraft energy to optimise and minimise fuel burn.

The FMS provides the essential integration of all of these functions to ensure that the overall function of controlling the navigation of the aircraft is attained. As may be imagined, this does not merely include steering information to direct the aircraft from waypoint to waypoint. The FMS also controls the tuning of all of the appropriate aircraft receivers to navigation beacons and communications frequencies via the communications control units, and many other functions besides. The flight plan that resides within the FMS memory will be programmed for the entire route profile, including all eventualities, including emergencies. More advanced capabilities include 3-D navigation and the ability to adjust the aircraft speed to reach a waypoint within a very small time window (typically ± 6 seconds). The various levels of performance and sophistication are summarised in Table 11.3 and described in detail later in the chapter.

11.2.5   FMS Procedures

Although the foregoing explanations have concentrated on performance enhancements, the assistance that the FMS provides the flight crew in terms of procedural displays cannot be

Table 11.3 Summary of FMS capabilities

Function	Capability
LNAV	The ability to navigate laterally in two dimensions (2-D).
VNAV	The ability to navigate laterally in two dimensions plus the ability to navigate in the vertical plane. When combined with LNAV this provides 3-D navigation.
4-D navigation	The ability to navigate in 3-D plus the addition of time constraints for the satisfaction of time of arrival requirements at a waypoint.
Full performance-based navigation	The capability of 4-D navigation together with the addition of an aircraft-specific performance model. By using cost indexing techniques, full account may be made of the aircraft performance in real time during flight allowing optimum use of fuel and aircraft energy to achieve the necessary flight path.
Future air navigation system (FANS)	The combination of the full performance model together with all the advantages that FANS will confer, eventually enabling the concept of 'free flight'.

forgotten. Several typical scenarios are included in this section to illustrate the importance of the procedural information. Chronologically, as the aircraft conducts the flight, these examples include, with reference to Figure 11.15:

- standard instrument departure (SID);
- en-route procedures;
- standard terminal arrival requirements (STAR);
- ILS approach.

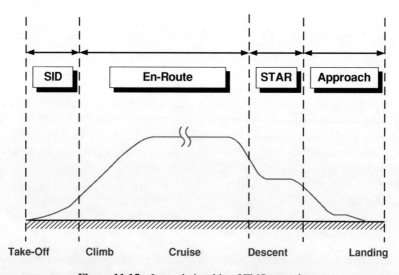

Figure 11.15 Interrelationship of FMS procedures

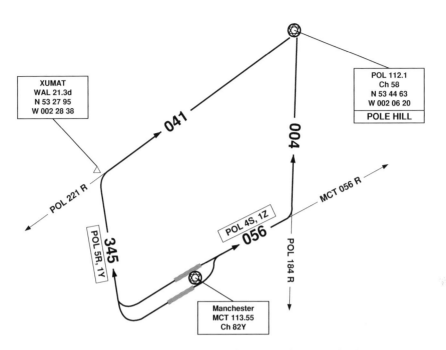

Figure 11.16   Typical standard instrument departure (SID)

11.2.6   Standard Instrument Departure

The SID shown in Figure 11.16 is a simplified version of the Manchester airport Pole Hill departure to the northwest. There are VOR/DME beacons at the airport runway 06/24 (MCT) and Pole Hill (POL) that, together with the waypoint at XUMAT, form the basis of the various instrument departures, of which there are four in all. POL 1Y (runway 24L) and POL 5R (runway 24R) depart to the west and POL 4S (runway 06L) and POL 1Z (runway 06R) depart to the east. The waypoints at XUMAT and Pole Hill also have latitude and longitude specified. The SID also specifies – not shown on this diagram for reasons of clarity – height gates at various points on the departure path to ensure that departing aircraft climb in a structured and orderly manner. Departure speeds are also regulated; on this chart all aircraft are limited to a maximum of 250 kts below FL100 unless otherwise advised.

It may be seen that there are a large number of options available to the flight crew within even this relatively simple example. For instance, an aircraft climbing out on POL5R/1Y from runways 24L or 24R could fly to intercept the POL 221 radial before following the Pole Hill VOR/DME, or could elect to fly to inertial latitude/longitude waypoints at XUMAT and then Pole Hill. An FMS will therefore contain all the SIDs for all the airports that the aircraft may expect to use, both for normal and alternate departures and also including procedural information for possible diversion airfields.

11.2.7   En-Route Procedures

There are a number of FMS en-route procedures – these are specified in ARINC 424-15 – in all more than 20 different procedures are specified. These vary from the simplest procedure

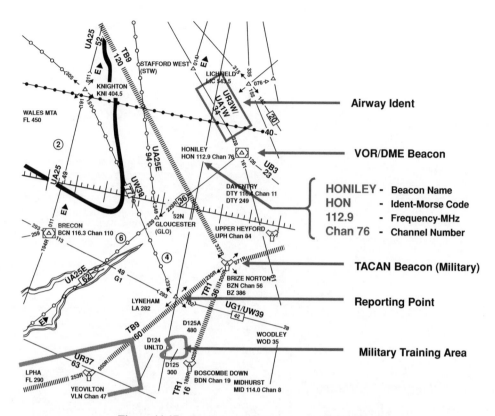

Figure 11.17   Typical route structure over SW England

involving following a course to a radio beacon, to some of the more complex associated with holding patterns in terminal areas. These procedures apply to the standing ATC route structures. Figure 11.17 gives an example of a typical route structure over the southwest of England.

Several examples of these procedures, which are self-explanatory, are depicted in Figure 11.18. These en-route FMS procedures are:

- course to radial;
- DME arc turn;
- procedure turn;
- holding pattern.

11.2.8   Standard Terminal Arrival Routes

Standard terminal arrival routes (STAR) are defined for all airfields for approaches from various directions to accommodate aircraft arriving from different departure airfields and arriving at different runways, to allow for wind changes. Alternate procedures are also published to allow for airfield and terminal area equipment unserviceability.

The example shown in Figure 11.19 is a simplified version of the STAR for Manchester airport for occasions when the airport VOR is not operational. All arrivals are routed via

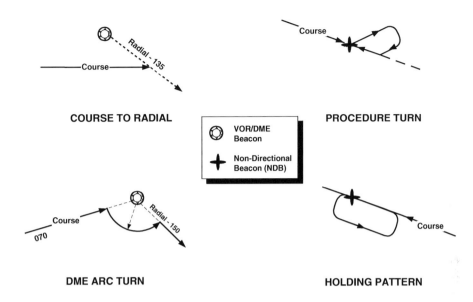

Figure 11.18 Several examples of en-route FMS procedures

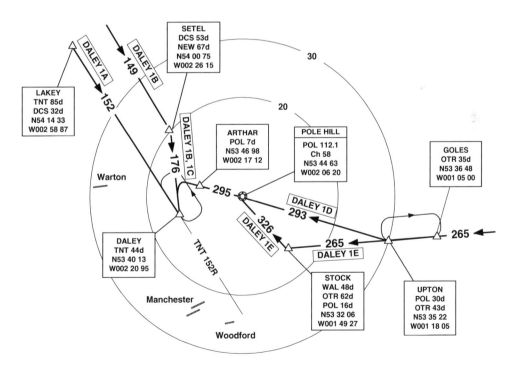

Figure 11.19 Typical standard terminal arrival route (STAR)

various waypoints to the Pole Hill VOR/DME beacon described in the SID in Figure 11.16. However, whereas all the departures via Pole Hill were at 5000 feet, the arrivals are all at FL80 (8000 feet) or above. There are four main routes to Pole Hill:

- DALEY 1A from the northwest for flights at FL150 or above. High-altitude arrivals are told to expect clearance to FL200 by 10 nm before LAKEY.
- DALEY 1B, also from the northwest for flights at FL140 or below.
- DALEY 1D from the east for flights at FL90 and above. High-altitude arrivals are told to expect clearance to FL200 by 10 nm before GOLES.
- DALEY 1E from the east for aircraft at FL80. There is an additional holding area between GOLES and UPTON for aircraft approaching from the east. Aircraft arriving from these main arrival routes are spatially separated both laterally and vertically.

If traffic is busy, aircraft arriving at Pole Hill may be directed into a holding pattern at the DALEY waypoint which is located about 20 nm north-northwest of the airfield. At this point aircraft will be flying at an altitude of between FL70 and FL140 with a speed not exceeding 230 kts. If clearance is available, traffic will be advised to fly direct from Pole Hill to the airport holding area waypoint AMLET, before being cleared for approach (see Figure 11.20).

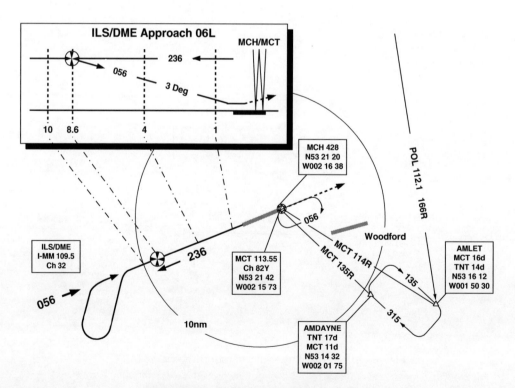

Figure 11.20   Typical ILS approach procedure

11.2.9 ILS Procedures

Having negotiated a passage through one of the STARs, the aircraft will be ready to perform an approach to the active runway. In this case a simplified example of an ILS/DME approach to Runway 06L is shown in Figure 11.20.

Aircraft from Pole Hill will arrive at AMLET on the 166° radial and take up a position at 6000 feet in the terminal holding pattern defined by the AMLET and DAYNE waypoints. When cleared, the aircraft will be instructed to descend to 3000 feet on the 236° outbound course from the airfield (MCH/MCT). At around 11 miles from the airfield it will perform a procedural manoeuvre to position itself inbound on the 056° heading, by which time it should have acquired the airport DME/ILS localiser. At a point ~8.6 nm from the runway threshold, the aircraft should capture the glide-slope and descend on a 3° glide-slope to the touchdown point. Should the aircraft need to execute a missed approach it will be vectored back to the terminal area holding zone via the MCT 114° radial to AMLET where it will hold at 5000 feet to avoid conflict with incoming traffic while awaiting clearance for another approach.

11.2.10 Typical FMS Architecture

So far we have described in general terms the function and operation of the FMS. A realistic architecture showing the integrated navigation sensor suite that feeds the FMS is shown in Figure 11.21.

The key attributes of this architecture are as follows:

- Twin FMS computers interfacing with the NDs and MCDUs as already described.
- FMS system interfacing with the autopilot/AFDS, as described in Chapter 10.
- Dual sources of air and inertial reference data (ADM/ADIRS or ADC/IRS).
- Dual fit of VOR/DME; ILS; ADF, Rad Alt and GPS receivers. In some installations some of these receivers may be integrated into a single multi-mode receiver (MMR) unit.
- Interface with TCAS and TAWS.
- An interface will also be provided with the flight data recorder and crash survivable memory.

This architecture is typical of a medium-range twin-engine transport aircraft. Longer-range variants may have a third set of ADM/ADIRS or ADC/IRS fitted for reasons of availability.

11.3 Electronic Flight Bag

11.3.1 EFB Functions

The global switch (or perhaps revolution) towards portable IT data solutions clearly offers immense potential for the flight crew to dispense with their heavy flight bags and also tap into a vast information database with immense display potential. Aeronautical publications are often stored and promulgated in electronic form that permits the data to be accessed, edited and updated almost anywhere in the world and at any time.

Typical functions provided by an electronic flight bag (EFB) may include the following:

- **Airport moving map**: An airport moving map provides high-resolution airport diagrams with the aircraft position superimposed – akin to the car SATNAV that we are all familiar

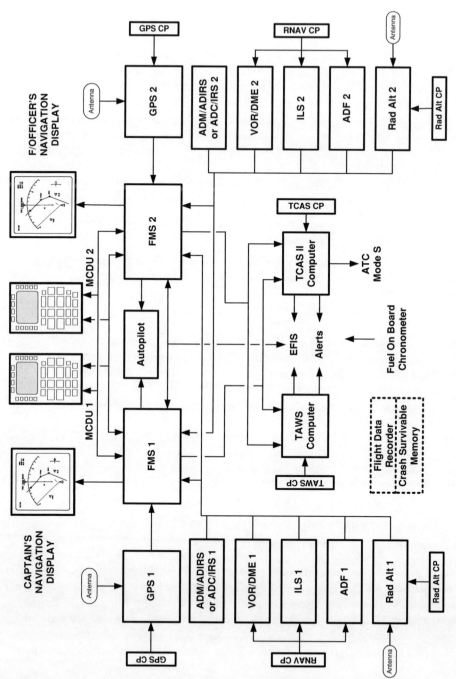

Figure 11.21 Typical navigation system architecture

with. Airport databases with GPS system-level tracking accuracy depict the airport ground environment and layout with a high degree of fidelity and visual detail.

- **Electronic charts**: Electronic charts offer clear, concise airport charts and procedures such as SIDs, en-route procedures, STARs and so on, as have already been outlined in the FMS description. Worldwide coverage may be customised to form a dedicated subset for individual airlines/operators or aircraft type.
- **Electronic documents**: Electronic documents offer the potential to call up a wide range of aircraft documentation, whether it be operational, procedural or technical in nature.
- **Onboard performance**: Databases with lookup tables and performance variations or limitations may be provided for a given aircraft status or configuration, which enable any vital performance issues to be noted and progressed.
- **Electronic logbook**: The electronic logbook works in association with onboard technical fault logs, often down to submodule level; system performance and operational logs record aircraft performance.
- **Electronic flight folder**: The concept of an electronic flight folder enables the digitising of all of the specific briefing information relative to a specific flight: NOTAMs, weather or other key information.
- **Video surveillance**: The ability of the flight crew to call up and display video sensor data from cameras located around the aircraft, to show either external or internal video data.

This is a dynamic area of avionics systems development, and highly competitive because of the content – that is, the embedded or implicit software functions contained therein – not to mention the need to future-proof the applications by providing a durable and robust forward development path. It is also evident that many of the functions outlined above could have flight integrity implications in terms of what decision the flight crew make, in which case the hardware and software implementations must be subject to the prevailing development disciplines.

11.3.2   EFB Implementation

The definitions of an EFB in terms of hardware implementation encompass the following general attributes:

- **Class 1.** A portable Class 1 EFB, such as a laptop, generally only has an interface to airplane power and must be stowed below 10,000 feet. This represents the lowest level of capability.
- **Class 2.** The hardware does not allow direct connection to airplane systems. It must have the ability to be removed by the crew without tools and without leaving the flight deck – typically modern solutions offer a docking solution. As before, this is an electronic portable device and not a fixed installation. A separate airplane interface module may allow connection to aircraft systems only with suitable partitioning. Certain airlines are adopting later versions of the Apple iPad for this class of application. Clearly, the robustness and security of these applications require significant attention and may not offer the same level of integrity as Class 3 solutions without significant certification effort.
- **Class 3**. Class 3 features installed avionics covered by type design approval, with a type designed software partition. It is a fixed part of the aircraft installation. It may have a Class 2

partition of functions, but depending upon the implementation may not necessarily require high-integrity software.

References [4, 5, 6] provide an overview of the EFB.

11.4   Air Traffic Management

11.4.1   Aims of Air Traffic Management

The rapidly increasing air traffic density is leading to a pressing need to improve the air traffic management (ATM) system by all available means, and move on from the techniques and technologies that have served the industry for the last 40 years. This evolution will embrace the use of new technologies mixed with existing capabilities to offer improved air traffic management. The aims of ATM may be summarised as follows:

- to maintain or increase levels of safety;
- to allow dynamic accommodation of user-preferred 3-D and 4-D flight trajectories;
- to improve the provision of information to users in terms of weather, traffic situation and services;
- to increase user involvement in ATM decision-making, including air–ground computer dialogues;
- to organise airspace in accordance with ATM procedures;
- to increase system capacity to meet traffic demand;
- to accommodate a full range of aircraft types and capabilities;
- to improve navigation and landing capabilities to support advanced approach and departure procedures;
- to create, to the maximum extent possible, a seamless continuum of airspace where boundaries are transparent to the user.

To this end the air traffic control authorities, airline industry, regulatory authorities and airframe and equipment manufacturers are working to create products that will uphold the performance requirements arising from the ICAO Future Air Navigation Systems (FANS) committee reports of the early 1990s.

11.4.2   Communications, Navigation, Surveillance

The areas where improvements may be made relate to communications, navigation and surveillance, commonly referred to as CNS. The key attributes of these improvements may be briefly summarised as:

- **Communication**: The use of data links to increase data flow and permit the delivery of complex air traffic control clearances.
- **Navigation**: The use of GNSS in conjunction with other navigational means to improve accuracy and allow closer spacing of aircraft.
- **Surveillance**: The use of data links to signal aircraft position and intent to the ground and other users.

These headings form a useful framework to examine the CNS/ATM improvements already made and those planned for the future.

Communications

The main elements of improvement in communications are:

- air–ground VHF data link for domestic communications;
- air–ground SATCOM communications for oceanic communications;
- high-frequency data link (HFDL);
- 8.33 kHz VHF voice communications to improve utilisation of VHF radio frequency spectrum.

Navigation

A number of navigational improvements are envisaged and in many cases already implemented:

- Introduction of required navigation performance (RNP) and actual navigation performance (ANP) criteria. These define absolute navigational performance requirements for various flight conditions and compare them with the actual performance the aircraft system is capable of providing.
- Reduced vertical separation minima (RVSM).
- Differential GPS (DGPS) enhancements:
  - WAAS
  - LAAS
- Protected ILS.
- Introduction of microwave landing systems (MLS) as an interim step before the introduction of precision GNSS approaches.
- Polar routes.

Surveillance

Surveillance enhancements include the following:

- TCAS II;
- ATC Mode S;
- automatic dependent surveillance – address (ADS-A);
- automatic dependent surveillance – broadcast (ADS-B).

11.4.3 NextGen

In the US the FAA have initiated a programme to infuse some of the above technologies and derivatives thereof to provide a development path to further implement ATM aims. This programme is the Next Generation air transport system (NextGen). It is developing and improving a range of technologies to address present shortcomings. A task force identified the following problem areas which required focus, and the technologies most likely to meet the aims. The aims of these improvements were as follows:

- **Airport surface operations**: Improve the management of airport arrivals, departures, taxi and ramp operations by expanding surveillance coverage and implementing real-time sharing of these data.
- **Runway access**: Improve the use of closely-spaced parallel, converging and intersecting runways during reduced visibility conditions.
- **Metroplex access**: Improve the capacity and efficiency that affects multiple airports near large metropolitan areas including Chicago, New York/New Jersey and southern California.
- **Cruise**: Improve efficiency and reduce flight delays via increased availability of special activity airspace (SAA) information, improved flow management around choke-points and more flexible routings.
- **Access to the National Air Transport System (NATS)**: Improve efficiency and safe access to low-altitude airspace and smaller airports by publishing precision approaches and adding surveillance and weather/traffic services at these locations, not often served by radar.
- **Data comm**: Improve traffic efficiency through air-to-ground digital data communication of revised pre-departure clearances, controller–pilot data link, and multiple aircraft weather re-routes.
- **Integrated ATM**: Integrate ATM solutions across pre-flight and in-flight to minimise delay and maximise the ability of operators to achieve business objectives in the context of overall system efficiency.

11.4.4   Single European Sky ATM Research (SESAR)

The European Union has a similar programme called Single European Sky ATM Research (SESAR) with similar objectives. However, the European solution is more challenging in some respects as the air traffic density is probably the highest of anywhere in the world. This, combined with the number of independent states and air traffic control entities, compounds a difficult problem. The management approach to address these issues has been configured in the following phases:

- **Definition phase (2004 to 2008)**: The delivery of a master plan defining the content, development and deployment of the next generation of ATM systems.
- **Development phase (2008 to 2013)**: The production of the required new generation of technology systems, components and operational procedures as defined in the master plan.
- **Deployment phase (2014 to 2020)**: The large-scale production and implementation of the new ATM infrastructure.

Key contributors include a variety of stakeholders that are organised as a SESAR Joint Understanding (SESAR JU) with the SESAR stakeholders, as briefly listed below:

- founding members: Eurocontrol and the EC.
- airport operators – 3;
- air navigation service providers – 6;

- ground and aerospace manufacturing – 5;
- aircraft manufacturers – 2;
- aircraft equipment suppliers – 2.

11.5   Performance-Based Navigation

11.5.1   Performance-Based Navigation Definition

In the stages of FANS around the late 1990s, the concept of required navigation performance (RNP) was introduced into the aerospace community to define the navigational accuracy necessary to execute specific flight routes and procedures. These have already been outlined in Section 11.2. As time has progressed and more options have evolved, the aerospace community decided in 2006 to more closely, and perhaps more logically, define these routes and procedures by invoking the term performance-based navigation (PBN). This resulted in a more precise definition of area navigation (RNAV) and RNP procedures (see Figure 11.22 for a breakdown of these functions).

Figure 11.22 shows the major subdivision of PBN into two distinct categories:

- Area navigation (RNAV). RNAV instigates procedures where the procedures may be invoked without onboard monitoring and alerting.
- Required navigation performance (RNP). RNP incorporates procedures that require onboard monitoring and alerting.

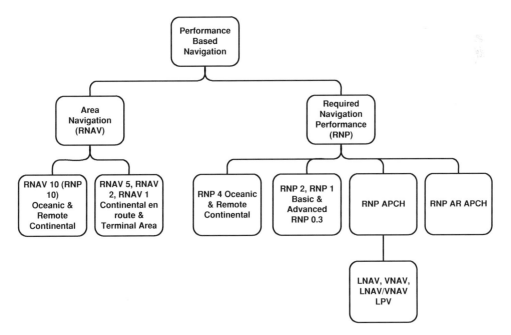

Figure 11.22   PBN definitions

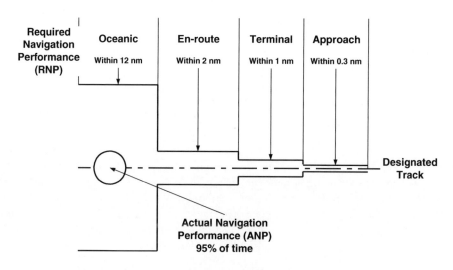

Figure 11.23 Top-level RNP definitions

References [7] to [10] and [11] to [16] are articles published by Airbus and Boeing respectively that provide additional insight as to their approach to implementing PBN. Readers may find it useful to read the appropriate articles together with the chapter text.

11.5.2 Area Navigation (RNAV)

RNAV may be subdivided into two categories with reference to Figure 11.23 for an overview:

- RNAV 10 (equivalent to RNP 10) for oceanic and remote continental regions;
- RNAV 5, RNAV 2, RPNAV 1 (formerly PRNAV) for continental en-route and terminal areas.

RNAV standards within Europe

Two RNAV standards are being developed in Europe. These are:

- **Basic RNAV (BRNAV)**: BRNAV was introduced in 1988 and is equivalent to RNP-5 for RNAV operations. Navigation may be accomplished by using the following means:
  - DME/DME.
  - VOR/DME with a 62 nm VOR range limit.
  - INS with radio updating or limited to 2 hours since last on-ground position update.
  - GPS with limitations.
  - From 2005, primary sources of navigation will be DME/DME, VOR/DME, and GPS. Advisory Circular AC 90-96, *Approval of US Operators and Aircraft to Operate under Instrument Flight Rules (IFR) in European Airspace designated for Basic Area Navigation (BRNAV)*, 20 March 1998, approved the operation of US aircraft in European air space under the application of existing Advisory Circulars.

- **Precision RNAV (PRNAV)**: PRNAV is intended to be introduced at some time in Europe. PRNAV will invoke the use of navigation under RNP-1 accuracy requirements or better.

The effect of using RNP techniques allied to the navigational capabilities of a modern aircraft needs to be examined in the context of the key areas of error defined above, namely:

- **Path definition error**: the difference between the desired flight path and actual flight path.
- **Path steering error**: the ability of the pilot and/or autopilot system to conform to the defined flight path (flight technical error or FTE).
- **Position estimation error**: the ability of the navigation system to estimate position, and the level of integrity which may be assured in that process.

Path definition error/RNP fixes

The process of introducing higher-precision navigation criteria has led to a critical review of the nature of fixes presently allowed and their viability within the RNP framework. Certain types of fix are rigidly defined and relate to the position of a beacon or as defined by inertial coordinates – today usually GPS-derived. Examples are given of these fixes in Figure 11.24; these show that they are absolute in their characterisation:

- initial fix (IF);
- track to a fix (TF);
- constant radius to a fix (RF);
- holding to fix (HX).

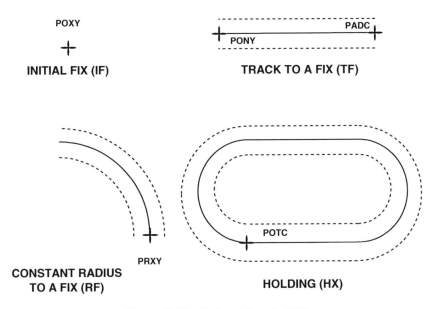

POXY

INITIAL FIX (IF)

PADC

PONY

TRACK TO A FIX (TF)

PRXY

**CONSTANT RADIUS
TO A FIX (RF)**

POTC

HOLDING (HX)

Figure 11.24   Preferred fixes for RNP

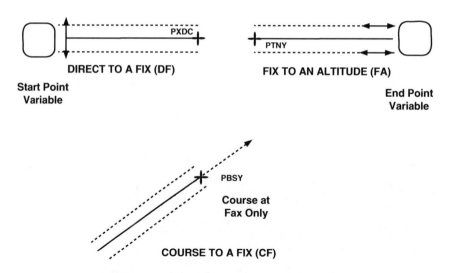

Figure 11.25 Fixes that are discouraged for RNP

Each of these fixes is unambiguously defined to a particular point in the route structure; there are no conditional constraints that apply. Fixes with this type of characterisation are ideal for specifying a robust route structure.

By contrast, fixes that are conditional or depend to some degree upon a time conditional or dynamic constraint are less favoured in the proposed route structure. The examples shown in Figure 11.25 all have conditional constraints. The direct to a fix (DF) example has a variable starting point that will affect the aircraft dynamics when arriving at the fix (DF) point. Similarly, fix to an altitude (FA) will depend upon an individual aircraft performance or ability to climb to an assigned altitude, and other key navigation parameters may be suppressed. In essence, although the start point is defined, the end point is indefinite. Finally, the example given for course to a fix (CF) only relates to the course that the aircraft is flying when the designated fix point is reached. The future course of the aircraft beyond this point will depend upon errors accumulated in the previous leg due to accuracy of the previous fix, wind variation and other sources. The course to be followed following this fix will not necessarily agree with the aircraft's defined route/flight plan.

It follows that to define and maintain a robust and repeatable route structure, the emphasis needs to be place upon those navigation legs that are most rugged in their ability to minimise flight path error. The combination of such flight legs may lead to a structure similar to that shown in Figure 11.26. In this example, a series of robust legs is combined to define a highly repeatable approach comprising arrival, downwind, crosswind and approach legs. The repeatability of this philosophy has been amply demonstrated in a number of terminal approach examples, including Frankfurt, Germany; Schiphol, the Netherlands; and San Francisco and Boston in the US, to name a few. Not only can these approaches be flown with greater navigational accuracy, but the precise alignment of arrival and departure routes can be aligned to greatly improve noise abatement and the overall environmental impact upon communities in the airport locality.

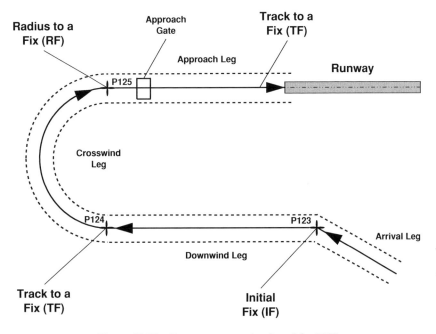

Figure 11.26 Route structures developed for RNP

Path steering error

The assumed values for FTE are given in Table 11.4:

Table 11.4 Assumed values for path steering FTE (5% probability) [2]

Flight phase	Manual (nm)	Flight director (nm)	Autopilot (nm)
Oceanic	2.0	0.5	0.25
En-route	1.0	0.5	0.25
Terminal	1.0	0.5	0.25
Approach	0.50	0.25	0.125

Studies on recent aircraft suggest that actual values achieved in modern aircraft using digital autopilots are in fact much better than the assumed values (see Table 11.5).

Table 11.5 Actual values for path steering error [2]

	Manual flight with map display (nm)	LNAV with flight director coupled (nm)	LNAV with autopilot coupled (nm)
En-route	0.502–0.918	0.111–0.232	0.055–0.109
Terminal	0.208–0.402	0.073–0.206	0.068–0.088

Position estimation error

As well as the data given in the tables above, Reference [2] gives a comprehensive overview of the considerations to be addressed when considering RNP/RNAV and of future likely developments. The specified increases in traffic density described so far only affect the lateral spacing of aircraft; vertical spacing is another consideration. The desire to reduce the vertical separation between aircraft is addressed in another aspect of FANS, reduced vertical separation minima (RVSM), which has been introduced worldwide.

11.5.3   Required Navigation Performance (RNP)

Within the RNP definition of PBN, there are the following subdivisions which are discussed in Section 11.5.4.

- RNP-4 for oceanic and remote continental routes.
- RNP-2 and RNP-1, and basic and advanced RNP-0.3.
- Two further subdivisions determine precision approaches: RNP APCH and RNP AR APCH.

Actual navigation performance

The actual navigation performance (ANP) of the aircraft navigation system is represented by a circle that defines the accuracy of the aircraft navigation system for 95% of the time. The value of ANP is derived by taking the value of all of the navigation sensors and statistically weighing them against the other sensors. After a period of time a degree of confidence is established in which are the most accurate sensors and therefore the ANP value is established. The 95% probability circle is that which is compared with RNP to decide whether the navigation system performance is good enough for the route segment being flown. The ANP and RNP values are displayed on the FMS CDU such that the flight crew can readily check on the navigation system status. Should the ANP exceed the RNP value for a given route sector for any reason – for example, due to a critical navigation sensor failing – the crew are alerted to the fact that the system is not maintaining the accuracy necessary. This will result in the aircraft reverting to some lower capability navigational means; in an approach guidance mode it may necessitate the crew executing a go-around and re-initiating the approach using a less accurate guidance means.

Required navigation performance

The RNP defines the lateral track limits within which the ANP circle should be constrained for various phases of flight. The general requirements are:

- For oceanic crossings the RNP is ±12 nm, also be referred to as RNP-12.
- For en-route navigation the RNP is ±2 nm (RNP-2).
- For terminal operations the RNP is ±1 nm (RNP-1).
- For approach operations the RNP is ±0.3 nm (RNP-0.3).

Other specific RNP requirements may apply in certain geographical areas, such as RNP-4 and RNP-10 (see Figure 11.23).

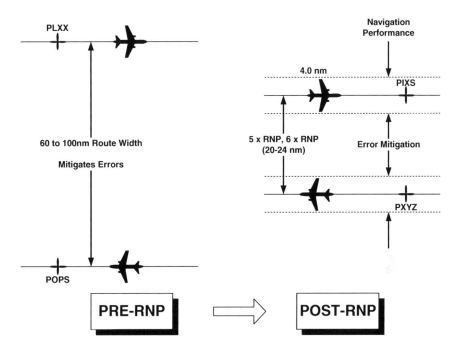

Figure 11.27   Comparison of historical and RNP definitions of route structure

It is clear that this represents a more definitive way of specifying aircraft navigational performance, versus the type of leg being flown, than has previously been the case. Other more specific criteria exist: RNP-5 (also known as BRNAV or area navigation) has already been introduced in parts of the European airspace with the prospect that RNP-1 (also known as PRNAV or precision navigation) will be introduced in a few years. There are new precision approaches in existence – noticeably those in Juneau, Alaska – where RNP-0.15 is required for mountainous terrain. The characteristics of this approach will be described later.

The effect of using RNP to specify the route structure versus the historical approach is shown graphically in Figure 11.27.

The pre-RNP approach is shown on the left of the diagram. The route width ranges from 60 nm up to 100 nm. This enormous buffer is required to mitigate against the combined effect of navigation errors, navigation performance, route, traffic density, surveillance, communications and ATC. This approach considers the worst combination of factors that may apply. The RNP approach, together with other FANS improvements in surveillance and communications, leads to the situation shown on the right.

The post-RNP philosophy defines the route separation and error mitigation buffer in terms of multiples of the RNP performance. For the 4.0 nm example route (RNP-4) shown, the route width is ~20 to 24 nm depending upon the RNP multiple used. This compares most favourably with the pre-RNP route structure.

Similar spacing considerations apply along the route. As the navigational performance also affects along-track accuracy, similar track-wise improvements in separation may be achieved.

11.5.4   Precision Approaches

In 2007, ICAO documented two additional types of approaches:

- RNP approaches (RNP APCH) charted RNAV (GNSS) or RNAV (GPS).
- RNP with Authorization Required Approaches (RNP AR APCH) charted RNAV (RNP). In this context AR refers to 'Authorisation Required.' It is equivalent to the US Special Aircraft & Aircrew Authorisation Required (SAAAR) operations.

Key characteristics of RNP AR procedures are as follows:

- RNP values less than or equal to 0.3 nm (down to 0.1 nm).
- Curved flight path before and after the final approach fix (FAF).
- Protection areas limited to $2 \times$ RNP laterally without any additional buffer. These approaches offer improved flexibility to the airline, particularly to airfields with challenging terrain and in poor weather.

Such approaches may be public or private. It is the responsibility of the operator to apply for authorisation to fly such procedures for a particular aircraft/runway combination [17].

The judicious application of the FANS techniques described in this chapter has allowed major advances to be made and approaches designed which would not previously have proved possible. One early example was the new approach designed for RW26 at Juneau, Alaska. The airport at Juneau is surrounded by high mountains and the approach to RW26 requires flying northwest up a glacial valley that embraces the Gastineau Channel.

Figure 11.28 depicts the horizontal profile of the approach to RW26. Before the design and implementation of the RNP approach this access to Juneau was impossible in all but the best weather conditions. The procedures have been designed to accommodate RNP-0.15, RNP-0.2 and RNP-0.3 approaches, which allow decision heights (DHs) of 337, 437 and 1810 feet respectively. The approaches were designed by Alaska Airlines in conjunction with Boeing and the FMS manufacturer, Smiths Industries. A full description of the considerations involved in developing the approach is given in Reference [6].

Using the new approach allowed shorter arrival and departure procedures when flying to Juneau from Seattle – about 5 minutes inbound and 6 minutes outbound. In terms of annual savings it has been estimated that about 245 flight hours or 740,000 pounds of fuel could be saved, assuming present flight schedules.

The use of RNP procedures also allows a more stable approach path – see Figure 11.28. Usually, a conventional approach would involve a series of short descents followed by level flight as the aircraft 'stepped down' the approach path. Using improved guidance the aircraft is able to follow a gradual approach path starting at the outer waypoint (MARMN), some 15 miles out, and continue this descent down to the DH. This reduces crew workload and allows the flight crew to spend more time monitoring the status of the approach rather than continually changing power and trim settings. This is a further benefit of an RNP-based approach to add to the other benefits.

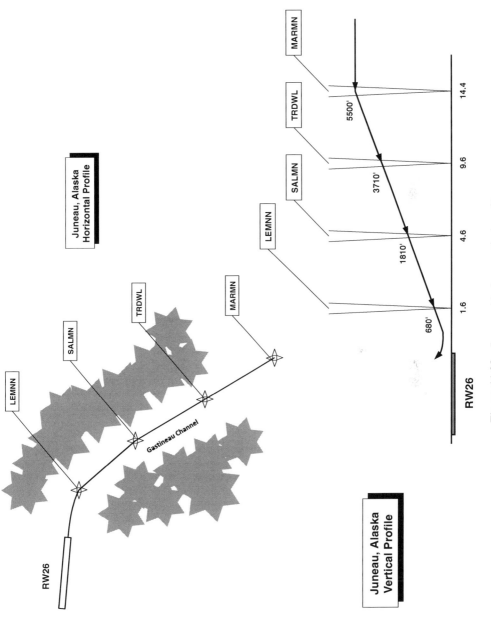

Figure 11.28 Juneau, Alaska – vertical profile

The aircraft used to fly this approach is a Boeing 737. The minimum navigation equipment fit required to execute this approach is listed below:

- dual FMS;
- dual IRS;
- dual GPS;
- dual EFIS.

Experience showed that the weather radar provided terrain information which was most useful to monitor the status of the approach. As a result, serviceable weather radar was added to the minimum equipment fit.

Reference [17] explains this particular approach in more detail – one of the first to be developed. References [18, 19, 20] provide other useful information.

11.6 Automatic Dependent Surveillance – Broadcast

ADS-B may be used to transmit 4-D position and flight plan intent based upon GNSS position, using line-of-sight VHF communications. Either Mode S or digital VHF radio is used to transmit the data. ADS-B requires a cockpit display of traffic information.

An extended demonstration project in the Louisville area conducted by the FAA in conjunction with United Parcels Services (UPS) and others have helped to demonstrate the ADS-B concept and paved the way to certification. UPS Aviation Technologies have secured certification for new displays that can provide improved runway, aircraft and weather information, which facilitates the movement of aircraft in the terminal area. UPS have predicted that this technology will yield a 20% capacity increase at Louisville, since reducing the arrival spacing of aircraft by 20 to 30 seconds can reduce the nightly sorting operation by around 30 minutes. Time savings of this proportion are critical in maintaining timely delivery of parcels on time.

ADS-B is being widely introduced as a means of improving traffic flow and awareness. Key areas where major development is underway are the US, Europe, Canada and Australia. Figure 11.29 shows the timescales for these four major developments. The triangular milestones represent deployment of the necessary ground infrastructure; the pentagonal milestones represent aircraft equipage with the necessary avionics. Although these four programmes represent the largest underway, some estimates suggest that as many as 25 smaller projects may be underway worldwide [21, 22].

11.7 Boeing and Airbus Implementations

Although the FANS requirements are similar in nature, Boeing and Airbus have adopted different approaches for the functional implementation as outlined below.

11.7.1 Boeing Implementation

The Boeing approach to providing a systems capability has been to provide software upgrades for older models that impart additional navigation and communications functionality to existing

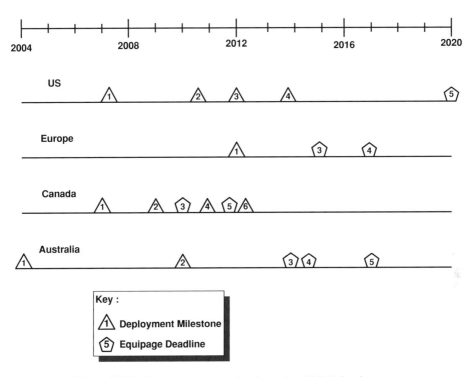

Figure 11.29 Programme timescales for major ADS-B developments

avionics hardware. This was extended to aircraft such as the Boeing 777 and Boeing 767-400 and the newer models including the B787 and B747-8. This approach may be summarised by reference to Figure 11.30, which shows a typical Boeing display/FMS interface. The major features are:

- Conventional navigation information displayed on the captain and first officer's navigation displays (NDs).
- On the Boeing 747-400, when the FANS 1 package is loaded, the captain's CDU acts as a host to display additional ATC communications functions as well as the existing navigation functions.
- On the Boeing 737 aircraft with the appropriate software upgrade loaded, additional guidance information is shown on the primary flight display (PFD) as described in Reference [5]. The PFD is enhanced by the display of lateral and vertical deviation scales along the bottom and right-hand edge of the PFD. By flying the aircraft to maintain zero lateral and vertical deviation (the same as for an ILS approach) deviations from the desired flight path are minimised. This enhancement enables the same flight technical error deviation to be achieved in flight director (FD) as in autopilot mode.

The stated performance of the Boeing aircraft in RNP navigation modes is quoted in References [4] to [7].

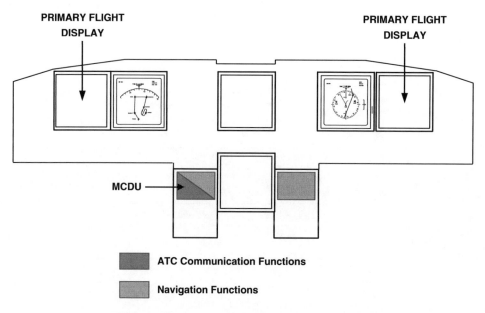

Figure 11.30   Boeing FANS 1 – display of information

11.7.2   Airbus Implementation

Airbus have taken a slightly different approach to implementing their FANS upgrade by adding additional equipment, rather than by adding software functionality to existing hardware. The units added include:

- An Air Traffic Services Unit (ATSU) incorporating ARINC Communications and Reporting System (ACARS) and airline operation communication (AOC) functions.
- Replacement of existing radios with VHF digital radios (VDR).
- Introduction of dual Data-link Control & Display Units (DCDUs) located on the centre console (see Figure 11.31).
- An upgraded FMS module hosted in the Flight Management Guidance and Envelope Computer (FMEGC). Initially this was fitted to the A330/340 family with a view to moving towards a FANS-B version implementing aeronautical telecommunications network (ATN) functionality.

11.8   Terrain Avoidance Warning System (TAWS)

While TCAS is designed to prevent air-to-air collisions, the ground proximity warning system (GPWS) is intended to prevent unintentional flight into the ground. Controlled flight into terrain (CFIT) is the cause of many accidents. The term describes conditions where the crew are in control of the aircraft, but due a lack of situational awareness they are unaware that they are about to crash into the terrain. The GPWS originally took data from the radio altimeter and the barometric vertical speed indication generated a series of audio warnings when a hazardous

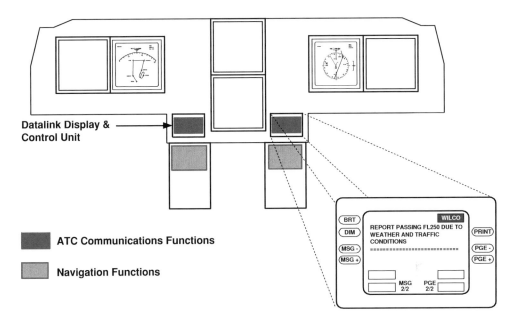

Figure 11.31   Airbus FANS A – display of information

situation was developing. In subsequent developments this has now been superseded by a more generic title: Terrain Avoidance Warning System (TAWS), where these data are combined with those provided by other sensors [23].

TAWS uses radar altimeter information together with other information relating to the aircraft flight path. Warnings are generated when the following scenarios are unfolding:

- flight below the specified descent angle during an instrument approach;
- excessive bank angle at low altitude;
- excessive descent rate;
- insufficient terrain clearance;
- inadvertent descent after take-off;
- excessive closure rate to terrain – the aircraft is descending too quickly or approaching higher terrain.

Inputs are taken from a variety of aircraft sensors and compared with a number of performance-based algorithms that define the safe envelope within which the aircraft is flying. When key aircraft dynamic parameters deviate from the values defined by the appropriate guidance algorithms, then appropriate warnings are generated.

The installation of the original GPWS equipment for all airliners flying in US airspace was mandated by the FAA in 1974, since when the number of CFIT accidents has dramatically decreased. Within about four years the system was adopted worldwide on the majority of passenger-carrying airliners. In time, enhanced versions have become available. Enhanced GPWS (EGPWS) offered a much greater situational awareness to the flight crew as more

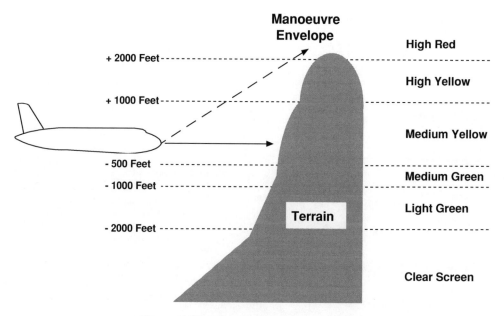

Figure 11.32 Principle of operation of TAWS

quantitative information is provided to the flight crew together with earlier warning of the situation arising. It used a worldwide terrain database that is compared with the aircraft's present position and altitude. Within the terrain database the Earth's surface is divided into a grid matrix with a specific altitude assigned to each square within the grid representing the terrain at that point. EGPWS has now been superseded by TAWS.

The aircraft intended flight path and manoeuvre envelope for the prevailing flight conditions are compared with the terrain matrix and the result graded according to the proximity of the terrain as shown in Figure 11.32. Terrain responses are graded as follows:

- No display for terrain more than 2000 feet below the aircraft.
- Light green dot pattern for terrain between 1000 and 2000 feet below the aircraft.
- Medium green dot pattern for terrain between 500 and 1000 feet below the aircraft.
- Medium yellow dot pattern for terrain between 1000 feet above and 500 feet below the aircraft.
- Heavy yellow display for terrain between 1000 and 2000 feet above the aircraft.
- Heavy red display for terrain more than 2000 feet above the aircraft.

This type of portrayal using colour imagery is very similar to that for the weather radar and is usually shown on the navigation display. It is far more informative than the audio warnings given by earlier versions of GPWS. The TAWS also gives audio warnings but much earlier than those given by the earlier system. These earlier warnings, together with the quantitative colour display, gives the flight crew a much better overall situational awareness in respect of terrain and more time to react positively to their predicament than did previous systems.

References

[1] *Boeing AERO Magazine* (2001) Lateral & Vertical Navigation Deviation Displays. October 2001.

[2] *Airbus FAST Magazine* (2009) FMS Past, Present & Future. September 2009.

[3] ARINC 424-15 (2005) *Navigation System Data Base*, November 2005.

[4] *Boeing AERO Magazine* (2003) Electronic Flight Bag. May 2003.

[5] *Boeing AERO Magazine* (2010) Class 2 EFB. January 2010.

[6] *Boeing AERO Magazine* (2012) Operational Efficiency of Dynamic Navigational Charting. May 2012.

[7] *Airbus FAST Magazine* (2007) The Future Air Navigation System FANS B. July 2007.

[8] *Airbus FAST Magazine* (2008) New Flight Operations Documentation, A380. July 2008.

[9] *Airbus FAST Magazine* (2011) FANS for A320, A330, A340 Families. August 2011.

[10] Airbus Training Symposium, December 2008, *Training for RNP*.

[11] Boeing FMS RNAV Workshop (2000) *General Information on the Functional and Technical Aspects of Required Navigation Performance (RNP) Area Navigation (RNAV) and Applications*. D. Nakamura, 9 February 2000.

[12] *Boeing AERO Magazine* (2003) B737-700 Technical Demonstration Flights in Bhutan. July 2003.

[13] *Boeing AERO Magazine* (2008) Operational Benefits of Performance Based Navigation.

[14] *Boeing AERO Magazine* (2009) Contribution of Flight Systems to PBN.

[15] *Boeing AERO Magazine* (2011) Improving Runway Safety with Flight Deck Enhancements. January 2011.

[16] *Boeing AERO Magazine* (2011) Equipping a Fleet for RNP. January 2011.

[17] Kramer, M. (1998) *RNP RNAV Terminal Areas – Design and Approval Procedures for Juneau, Alaska*. Smiths Industries (now GE Aviation), FITEC 98.

[18] *Boeing AERO Magazine* (2003) GNLSS Landing Systems. January 2003.

[19] *Airbus FAST Magazine* (2009) Implementing RNP AR. July 2009.

[20] Airbus Flight Operations & Services (2009) *Getting to Grips with RNP AR*. February 2009.

[21] *Aviation Week & Space Technology*, 5 March 2012.

[22] *Boeing AERO Magazine* (2010) New Air Traffic Surveillance Technology.

[23] Advisory Circular AC 25-23 (2000) *Airworthiness Criteria for the Installation Approval of a Terrain Awareness and Warning System (TAWS) for Part 25 Airplanes*, 22 May 2000.

Historical References (in Chronological Order)

Advisory Circular AC 25-4, *Inertial Navigation Systems (INS)*, 18 February 1966.

Advisory Circular AC 121-13, *Self-Contained Navigation Systems (Long Range)*, 14 October 1969.

Advisory Circular AC 90-45A, *Approval of Area Navigation Systems for Use in the US National Airspace System*, 21 February 1975.

Advisory Circular AC 120-33, *Operational Approval for Airborne Long-Range Navigation Systems for Flight within the North Atlantic Minimum Navigation Performance Specifications Airspace*, 24 June 1977.

Advisory Circular AC 00-31A, *National Aviation Standard for the Very High Frequency Omni-directional Radio Range(VOR)/Distance Measuring Equipment (DME)/Tactical Air Navigation Systems*, 20 September 1982.

Advisory Circular AC 20-129, *Airworthiness Approval of Vertical Navigation (VNAV) Systems for use in the National Airspace System (NAS) and Alaska*, 12 September 1988.

Advisory Circular AC 90-94, *Guidelines for using Global Positioning System Equipment for IFR En-Route and Terminal Operations and for Non-Precision Approaches in the US National Airspace System*, 14 December 1994.

Advisory Circular AC 20-130A, *Airworthiness Approval of Navigation or Flight Management Systems Integrating Multiple Sensors*, 14 June 1995.

Technical Standing Order (TSO) C-129a, *Airborne Supplementary Navigation Equipment using Global Positioning System (GPS)*, 20 February 1996.

Advisory Circular AC 90-96, *Approval of US Operators and Aircraft to Operate under Instrument Flight Rules (IFR) in European Airspace designated for Basic Area Navigation (BRNAV)*, 20 March 1998.

Advisory Circular AC 90-97, *Use of Barometric Vertical Navigation (VNAV) for Instrument Approach Operations using Decision Altitude*, 19 October 2000.

12

Flight Deck Displays

12.1 Introduction

This chapter discusses flight deck displays and associated technologies. The 1970s saw a revolutionary change on the flight deck with the reduction in crew members from three to two, eliminating the flight engineer and subsuming his role into the Electronic Centralised Aircraft Monitor (ECAM) – Airbus, or Engine Indicating and Crew Alerting System (EICAS) – Boeing. This was facilitated by the advent of the 'glass' flight deck, then implemented using cathode ray tube (CRT) technology, now using active matrix liquid crystal displays (AMLCDs). The system moding controls, switches and indicators were moved to the overhead panels, allowing ready access by both remaining crew members. Initially the display formats followed the traditional structures of their previous electromechanical counterparts, but with the underlying principle of 'only display what you need to know when you need to know it'. The current generation of flight deck has continued to evolve. Today the role of the flight crew has changed from being pilots to being managers of systems. Basic piloting skills are rarely used; the majority of the flight is spent in managing the aircraft trajectory in airspace in which other aircraft are traversing on potentially conflicting flight paths, into crowded terminal areas.

The air transportation system is complex and new technologies continue to be deployed to accommodate the movement of large numbers of people and goods in ever-denser skies in a safe and efficient manner. This chapter discusses the head-up display (HUD), enhanced vision systems (EVS), synthetic vision systems (SVS), and combinations thereof, as a means to increase crew situational awareness and provide additional margins of safety for low visibility approaches in support of Next Generation Air Transport System (NextGen) operations.

The modern flight deck is extremely smart in operation and has brought great advances in safety. However, when things go wrong they can become quite challenging very quickly. Arguably the time is approaching for another revolution. The next generation flight deck needs to be crew-centric, providing a simple and intuitive human–machine interface which helps to keep the decision-making safe. However, more automation must not be at the loss of transparency.

Civil Avionics Systems, Second Edition. Ian Moir, Allan Seabridge and Malcolm Jukes.
© 2013 John Wiley & Sons, Ltd. Published 2013 by John Wiley & Sons, Ltd.

The final paragraphs of this chapter explore display system architectures, current and potential future display technologies, and usability aspects under the different and challenging flight deck ambient lighting environments.

For a more detailed discussion of the subject material, see Reference [1].

12.2   First Generation Flight Deck: the Electromagnetic Era

This section identifies a few of the more significant highlights and pioneers in the evolution of manned flight, to illustrate some of the concepts that have strongly influenced the present-day civil flight deck. For further information see Reference [2].

12.2.1   Embryonic Primary Flight Instruments

The 1918 advertisement in Figure 12.1 for Smith's Aviation Instrument Board illustrates what arguably can be said to be the first primary flight display (PFD). It has almost all the required features of the modern PFD, namely altimeter, airspeed indicator, attitude (by way of vertical rate-of-climb indicator and lateral inclinometer), engine rev counter and time, but lacks a compass. The style of the instruments bears witness to the trade of S. Smith and Sons (later to become Smith Industries, now part of GE Aviation), namely clock and watch makers. The diligent reader should take a close look at the human factors aspects, which are sadly lacking.

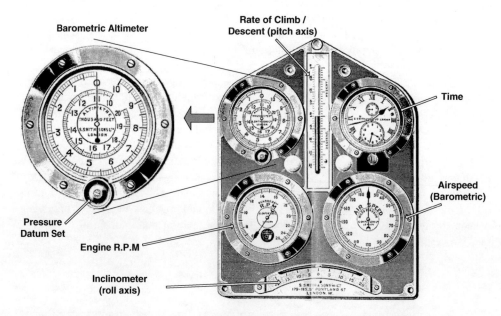

Figure 12.1   S. Smith and Sons Aviation Instrument Board. Copyright S. Smith and Sons

12.2.2   The Early Pioneers

The Wright Brothers

The first flight of a heavier-than-air machine took place in Kitty Hawk, US, on 17 December 1903. The only flight instrumentation was an anemometer (to measure wind speed), a weather vane (to measure angle-of-incidence) and a scarf between the landing skids (to indicate sideslip) [3].

Bleriot

As aviators gained more flying experience, they recognised the need for additional information. Bleriot added an engine rev counter and a compass for his historic crossing of the English Channel on 25 July 1909.

Lindbergh

On 22 May 1927 Charles Lindbergh landed his Ryan monoplane, the *Spirit of St Louis*, at Le Bourget, Paris, after a 28-hour flight across the Atlantic. Lindbergh equipped his aircraft with the most advanced instruments available at the time, but the human-machine interface was crude and his forward view extremely limited (see Figure 12.2). Dominating the instrument panel was a compass, which together with the altimeter and airspeed indicator were his primary

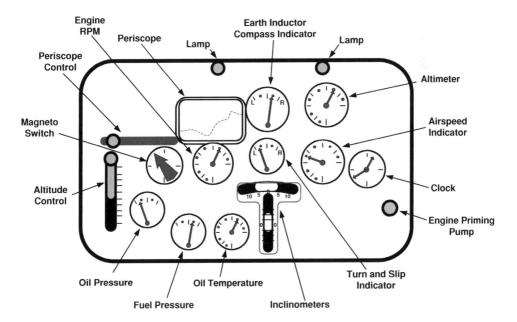

Figure 12.2   Flight deck of the *Spirit of St Louis*

means of navigation. To the top left of this group was a periscope through which Lindbergh could see forward along the fuselage past the engine – arguably the first enhanced vision system. The lower centre of the instrument panel carried lateral and fore–aft inclinometers. There was only one gyroscopic instrument, a turn and slip indicator. The other instruments and controls were associated with the engine [4].

Lieutenant James Doolittle

Lieutenant James Doolittle was convinced that the recently developed gyroscopic instruments together with radio navigation aids were the key to instrument flying. He pioneered the techniques for 'blind' flying, and in September 1929 after more than 100 proving flights, flew his Consolidated NY-2 single-engined biplane, from take-off, round the airfield circuit and back to landing with the cockpit hooded. Doolittle followed the indications of a radio guidance system, a precursor to today's instrument landing system (ILS), and used the recently developed Kollsman precision altimeter and Sperry gyroscopic artificial horizon to maintain the correct attitude and to descend steadily and at a precise rate to land his aircraft successfully at Mitchell Field.

His Sperry artificial horizon, shown in Figure 12.3, illustrates a very important principle. The aircraft representation (symbol) is 'fixed' in the centre of the instrument with its wings horizontal on each side. The artificial horizon moves relative to the fixed aircraft symbol to indicate pitch and roll attitude and therefore will be aligned with the real horizon as seen by the pilot if he looks forward through the windshield (i.e. 'inside looking out'). However, for many

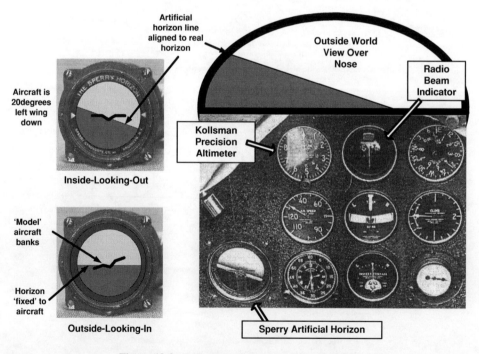

Figure 12.3   Instrument flying 'inside looking out'

years there were a number of aircraft with alternative representations in which the horizon was fixed and the aircraft indicated pitch and roll attitude relative to it: 'outside looking in'. The Russian aircraft industry continued to adopt outside-looking-in until the fall of the Soviet Union.

12.2.3   The 'Classic' Electromechanical Flight Deck

The 1970s heralded the mass market for civil transport aircraft, typified by the introduction of the Boeing 747 Jumbo Jet. At its introduction the flight deck of the 'classic' 747, shown in Figure 12.4, had a set of conventional electromechanical instruments and was operated by a three-man crew, the captain, the first officer and the flight engineer. The plethora of instruments illustrates the fundamental limitations of the technology available at the time. There was very little automation. If the crew needed to be aware of an aircraft system parameter, then a dedicated electromechanical instrument was required to display it all of the time [5, 6].

Civil air transport operational procedures had evolved considerably since the mid-1950s, and after a number of mid-air collisions aircraft were generally constrained by air traffic control to fly straight-line segments (known as airways) between VOR/DME radio navigation beacons. The primary flight instruments were structured and optimised into a 'Basic T' set of four instruments to provide the crew with good situational awareness to support these operational procedures (see Figure 12.5).

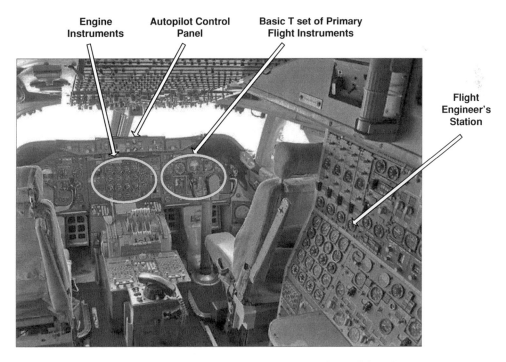

Figure 12.4   Boeing 747 classic flight deck. Copyright Olivier Cleynen

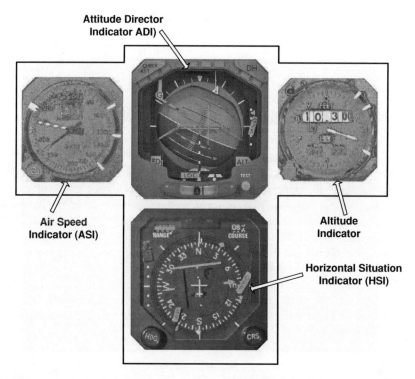

Figure 12.5 'Basic T' set of primary flight instruments. Reproduced with permission from GE Aviation (formerly Smiths Aerospace)

The Attitude Director Indicator (ADI)

The ADI combines the functions of three earlier generation instruments into one:

- artificial horizon;
- ILS direction indicator;
- turn and slip indicator.

The Horizontal Situation Indicator (HSI)

The HSI combines the functions of three earlier generation instruments into one:

- compass (gyro and magnetic);
- radiomagnetic indicator;
- distance measuring indicator.

The Altimeter

The altimeter contains an electromechanical servo-mechanism driven by signals from the air data computer to indicate barometric altitude by means of a pointer and a counter (numerical read-out). A knob with a numerical read-out sets the barometric pressure datum.

The Airspeed Indicator (ASI)

As with the altimeter, the airspeed indicator is driven by an electromechanical servo-mechanism slaved to the air data computer.

The ADI and HSI are commonly found in 5ATI format, that is, having a physical form factor in accordance with the air transport indicator (ATI) standard with a 5-inch square faceplate. The other instruments are commonly found in 3ATI format (3-inch square faceplate). The instruments were available from a number of manufacturers, have standardised functionality and are certified to comply with Technical Standards Orders (TSOs) published by the FAA [7] to [10].

A third set of primary flight instruments, known as standby instruments, usually in 3ATI format, provide fully independent information to aid the flight crew to resolve discrepancies. The standby instruments must be totally segregated from the primary instruments and operate on separate electrical power, usually the DC essential bus.

12.3   Second Generation Flight Deck: the Electro-Optic Era

This section traces some significant major milestones in the evolution of the glass flight deck. To include all the various flight deck iterations would be overwhelming; the examples chosen address the salient points of the evolutionary process.

12.3.1   The Advanced Civil Flight Deck

Research into the Advanced Civil Flight Deck (ACFD) began in the UK in the mid-1970s at BAe Weybridge, sponsored by the Department of Trade and Industry (DTI) with contributions from GEC and Smiths Industries. Complementary research programmes were being undertaken in the US at NASA Langley and Boeing as part of the US Supersonic Transport (SST) programme. When the SST programme was cancelled, the work continued into an evaluation phase in a Boeing 737 for the Terminal Configured Vehicle (TCV) programme.

The ACFD ground-based simulator, shown in Figure 12.6, had seven displays arranged in a side-by-side configuration across the flight deck, plus two outboard electronic library displays. The objective was to develop a two-man flight deck, subsuming the role of the flight engineer into the Electronic Centralised Aircraft Monitor (ECAM) system. The underlying principle was 'only display what you need to know when you need to know it'.

The displays were generated in 625 line 25:50 Hz interlaced video (TV standard) on monochrome (white on black) 6 inch × 4.5 inch landscape (4:3 aspect ratio) cathode ray tubes (CRTs). The research activities included extensive human factors evaluation by pilots, and demonstrated the viability of an all-glass two-man crew flight deck, and the side-by-side configuration of primary flight displays (PFD), navigation displays, the ECAM displays (in this case three) and even an elementary electronic library.

In 1980 the ACFD research programme moved to a flight demonstration phase. Two displays (the primary flight and navigation displays) were installed into the captain's side of a BAC 1-11 civil transport aircraft operated by the Royal Aircraft Establishment (RAE) at Bedford. The programme was launched with monochrome CRTs (green on black), but quickly transitioned to 6.25 × 6.25 inch useable screen area square colour displays (see Figure 12.7), once it was established that colour shadow mask CRTs could be ruggedised to withstand the civil air transport environment.

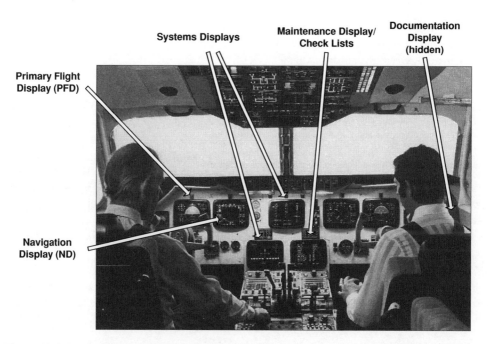

Figure 12.6 The Advanced Civil Flight Deck ground-based simulator. Reproduced with permission of BAE Systems Electronic Systems in Rochester, Kent

The display suite was fully integrated with the aircraft systems, with the right-seat crew-member acting as the safety pilot. The aircraft first flew in the spring of 1981 and made an extensive series of test and demonstration flights in Europe and the US, inviting guests from aircraft manufacturers, airlines and research organisations to fly the aircraft. Most pilots found the displays intuitive and easy to use; they adapted quickly to the side-by-side PFD/ND configuration.

The display formats mirrored closely the style of the electromechanical instruments they were replacing, to facilitate transition to the new media. The primary flight display (PFD) preserved the 'Basic T' configuration of attitude, airspeed, attitude and heading. Counter-pointer presentations of speed and height were retained, but in addition digital read-outs of speed and height were presented on the horizontal centre line and adjacent to the attitude ball so the pilot could rapidly assess his primary flight data without having to scan the whole display. The navigation display (ND) provided two format styles, the compass rose which preserved the original electromechanical HSI style, and the map format which presented the planned route, updated with present position, in real time. This latter demonstrated the real freedom of the media to present an image that hitherto the pilot had had to form in his head. The map could be oriented heading-up, track-up or north-up (for planning purposes). Weather radar data could be added to facilitate re-routing to avoid storm centres. Navigational information relative to the planned route was presented on each side of the map.

12.3.2   The Boeing 757 and 767

The Boeing 757 and 767 aircraft were the first to enter service with a 'glass' flight deck (see Figure 12.8). The electronic flight instrument system (EFIS) comprised an electronic ADI (EADI) with a 5 inch × 4 inch colour CRT display in portrait mode, and an electronic

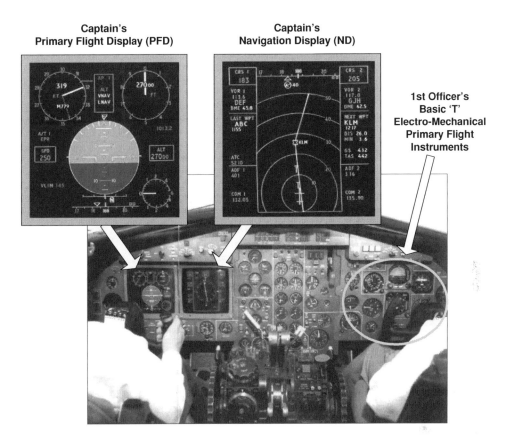

Figure 12.7   BAC 1-11 Advanced Civil Flight Deck demonstrator (April 1981). Reproduced with permission from GE Aviation (formerly Smiths Aerospace). For a colour version of this figure, please refer to the colour plates

HSI (EHSI) with a 5 inch × 6 inch CRT display in landscape mode. The displays were placed one above the other with a conventional electromechanical airspeed indicator and altimeter to the left and right respectively, preserving the 'Basic T' configuration. Conventional electromechanical instruments were used for the altimeter, airspeed indicator and vertical speed indicators. Two centrally mounted CRT displays, known as the Engine Indicating and Crew Alerting System (EICAS), positioned one above the other, provide engine data and cautions [11].

12.3.3   The Airbus A320, A330 and A340

The Airbus A320, which entered service in March 1988, was the first civil transport aircraft to adopt the side-by-side PFD/ND configuration with six 6.25 × 6.25 inch colour CRT displays installed on the flight deck (see Figure 12.9). The A330 has a similar configuration. The two ECAM displays are installed one above the other in the centre of the flight deck.

The primary flight display (PFD), shown in Figure 12.10, presents airspeed and altitude as two tape scales positioned either side of the attitude 'ball'. A numerical read-out is positioned

Airspeed | Altitude | Standby | Electronic | Electronic
Indicator (ASI) | Indicator | Instruments | Engines | Systems
(conventional) | (conventional) | (conventional) | Display | Display

Electronic
ADI
(EADI)

Electronic
HSI
(EHSI)

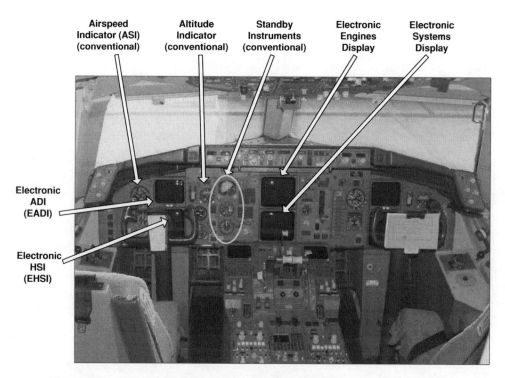

Figure 12.8   Boeing 757 and 767 flight deck (September 1982). Copyright Bill Abbott

as a window in the tape scales on the horizontal centre axis of the attitude ball. This allows
the pilot to acquire the key flight parameters quickly without having to scan the whole display
surface. The tape scales facilitate the acquisition of selected speed and height cues, which are
presented on and attached to the moving scales. The moving scales also offer a representation
of rate of change, although this is recognised to be inferior to a circular scale and pointer
representation. A vertical speed scale therefore augments the height tape. At the bottom of the
PFD format the upper segment of a compass rose provides heading and lateral guidance cues.
Autopilot mode annunciations are incorporated along the top of the display.

12.3.4   The Boeing 747-400 and 777

In June 1990 Boeing adopted the side-by-side, six-display configuration for its -400 variant of
the 747 aircraft. The two systems displays are positioned one above the other in the vertically
extended centre portion of the instrument panel [12].

The displays are 6.7 × 6.7 inch and employ active matrix liquid crystal display (AMLCD)
technology. The display formats are similar to those of the Airbus A320/330, with strip speed
and height scales positioned either side of the attitude ball. The Boeing 747-400 operates with
a two-man crew, the flight engineer's panel functions having been absorbed into the aircraft
systems and presented and controlled through the two Engine Indicating and Crew Alerting
System (EICAS) displays in conjunction with the overhead panel.

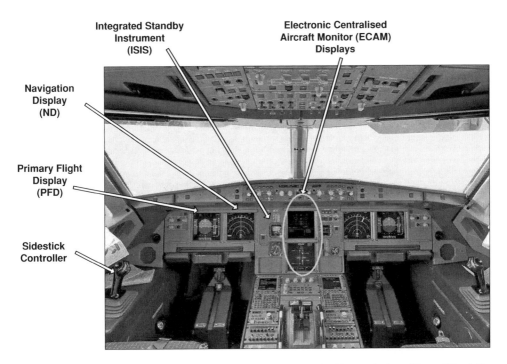

Figure 12.9 Airbus A320 flight deck (March 1988). Reproduced with permission of Jason Wood. For a colour version of this figure, please refer to the colour plates

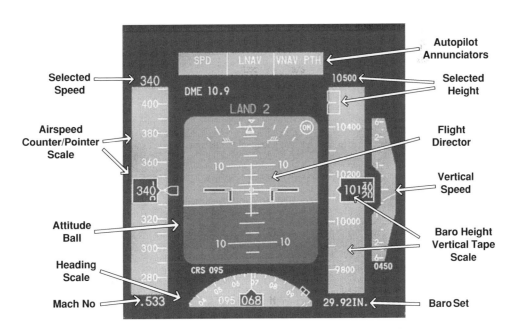

Figure 12.10 Airbus A320 primary flight display. For a colour version of this figure, please refer to the colour plates

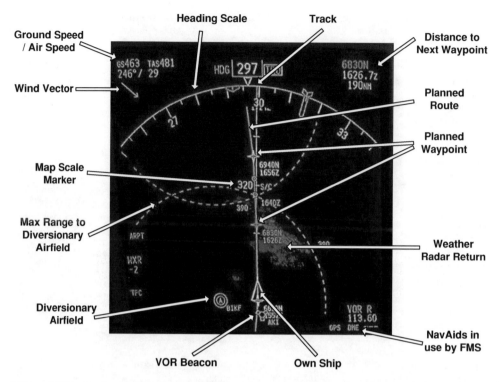

Figure 12.11 Boeing 777 navigation display with weather radar. Copyright TriplET. For a colour version of this figure, please refer to the colour plates

The Boeing 777 (June 1995) adopts a similar flight deck configuration and display size, but has a significantly different avionics system architecture. The displays themselves were the first in a wide-body civil transport aircraft to utilise AMLCD technology. The Boeing 777 navigation display (ND) shown in Figure 12.11 provides planned and actual route together with lateral guidance command cues optionally overlaid with weather radar data.

12.3.5 The Airbus A380

The flight deck of the Airbus A380, shown in Figure 12.12, maintains a strong family likeness to the early Airbus flight decks; Airbus desired to demonstrate that flying the A380 was just like flying any other aircraft in the Airbus family. The flight control system would make the aircraft handle in the same way through the same side-stick controller, implementing the same control laws. Air crew would be able to transition relatively simply from previous Airbus aircraft to the new type.

The EFIS/ECAM comprises six displays like previous generation Airbus aircraft; two side-by-side PFDs and NDs for each crewmember and two vertically disposed ECAM displays in the centre console. However, the display format changed from 6.25 inch square to 6 × 8 inch rectangular portrait displays using commercial off-the-shelf (COTS) AMLCD technology. Simply, it had been found to be no longer necessary on performance grounds, nor cost-effective,

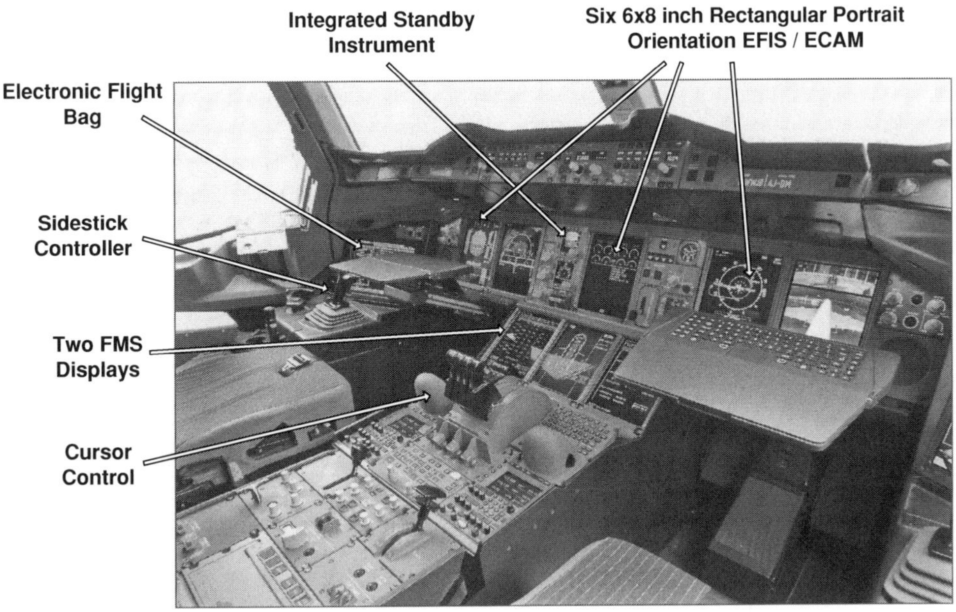

Figure 12.12   Airbus A380 flight deck (2007). Reproduced with permission from Airbus. For a colour version of this figure, please refer to the colour plates

to fabricate uniquely sized, aerospace quality AMLCDs. The increased area facilitates a full compass on the PFD and additional route information the ND, but in all other respects the formats and human–machine interface are familiar to those trained to fly earlier Airbus aircraft types. In the centre console the flight management system now offers the same size displays as the EFIS/ECAM to facilitate easier route planning. The most novel feature of the flight deck is the provision of the electronic flight bag or electronic library accessed by the crew through two large area, rectangular, landscape-format displays outboard of each crewmember station.

12.3.6   The Boeing 787

The flight deck of the Boeing 787, shown in Figure 12.13, is an evolutionary progression from the Boeing 777. However, it takes advantage of improved high-resolution COTS AMLCD technology to implement the flight deck displays in five 15.1 inch rectangular landscape-format displays, replacing the six square-format screens on the Boeing 777. The new outboard captain and first officer displays consolidate the once stand-alone primary flight display, clocks, and flight information, into a single primary flight display with a mini-map and auxiliary display functions. The total number of flight deck line-replaceable units has reduced from 22 on the Boeing 777 to just 12 on the Boeing 787.

A dual installation of the Rockwell Collins HGS-6000 head-up display provides enhanced situational awareness for 'eyes-out' precision flying in poor visibility landing and take-off (see Section 12.6). The dual HUD is certificated as a standard fit, not an option: a first for a Boeing commercial aircraft.

Head-Up
Display

Five Large Area 15.1 inch Diagonal
Multifunction Split-Screen Displays

Electronic Flight
Bag

Conventional Yoke
Control Inceptor

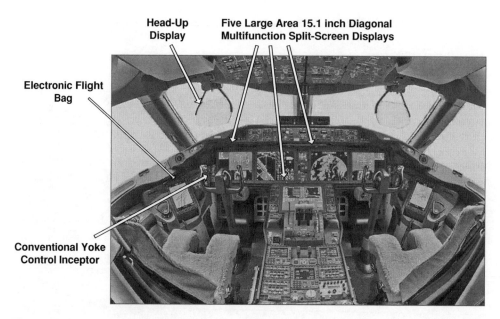

Figure 12.13   Boeing 787 flight deck (2011). Copyright Alex Beltyukov. For a colour version of this figure, please refer to the colour plates

The flight management system and the aircraft maintenance systems, synoptic pages and electronic checklists have been consolidated across three central multifunction displays (MFDs) that allow for up to five half-page display areas plus a permanent display of engine performance data and caution/warnings (previously the EICAS on the Boeing 777). The three MFDs are interfaced through a multifunction keyboard and the cursor control device first introduced on the Boeing 777. Each navigation display can be presented half- or full-width on each MFD, with images as wide angle as 1280 nautical miles or zoomed in to 0.5 nautical miles, presenting gates, taxiways and runways on the integrated airport moving map. The navigation display can be augmented with a vertical situation display and terrain display for RNP-0.1 approach capability (see Section 12.8.8).

A dual class 3 electronic flight bag is standard-fit, allowing a paperless interface presenting video surveillance, aircraft performance data computation, navigational charts, electronic documents and logbooks.

Most significantly, the Boeing 787 continues the Boeing tradition of a conventional control yoke. Despite being a fly-by-wire aircraft, Boeing remain firmly resolute in their position (as indeed they did with the Boeing 777) that the handling characteristics should be similar to those of a conventionally rigged aircraft, including feedback into the yoke of all aerodynamic and control forces by either crew member. Further discussion is to be found in Section 12.12.

12.3.7   The Airbus A350

When Airbus launched the A350 they announced it would have the same flight deck as the A380, maintaining the strong family likeness across the entire Airbus fleet. The customer community reacted adversely and Airbus was forced to change and implement a flight deck with six large area rectangular screens and a head-up display.

12.4 Third Generation: the Next Generation Flight Deck

The current generation flight deck has been enormously enriched and improved by digital technology, large-area high-resolution colour displays and a high degree of automation. The role of the flight crew has changed from being pilots to being managers of systems. Basic piloting skills are used rarely, mostly in just the few minutes of take-off and landing. The majority of the flight is spent in managing the aircraft trajectory in four dimensions in airspace in which other aircraft are traversing on potentially conflicting flight paths, into crowded terminal areas, avoiding bad weather, with ever more complex air traffic management procedures.

12.4.1 Loss of Situational Awareness in Adverse Operational Conditions

Although the latest generation flight decks have fewer, but larger displays, the increased area has been used to present even more information on split screens. Display formats have evolved considerably, but the genesis of these formats still bears a significant resemblance to their electromechanical forebears, none more so than the primary flight display.

The modern flight deck is extremely smart in operation and has brought great advances in safety. Integrated flight decks make it far easier for the operator, provided everything is working well, but when things go wrong they can become quite challenging very quickly. A benign situation can very rapidly turn into a serious or even catastrophic outcome if the crew fail to recognise the situation in which they suddenly and unexpectedly find themselves, and frequently without knowledge of what caused them to get there. Many complex functions are available with sophisticated degraded modes, but they are seldom used even in the simulator, and when key information cannot be simply accessed in an emergency then there is a greater risk of human error. Information overload under stress can render making timely decisions almost impossible.

The loss of Air France flight AF447 is a recent high-profile object lesson. The primary cause of the accident was the obstruction of the pitot probes by ice crystals which caused airspeed to be denied to the flight deck, the autopilot to disconnect, and the flight control system to degrade to alternate law without full envelope protection. Bennett [13] argued that the crew responded inappropriately and commanded a trajectory that caused the aircraft to exceed its normal flight envelope and enter a high-speed, high-altitude stall. The crew failed to appreciate fully the situation in which they found themselves and therefore were unable to recover from it. The flight accident report reveals that contributory factors were flight crew training, their lack of familiarity with manual flying, and their belief that automation technology would normalise the situation.

There have been at least two occasions where Russian pilots flying modern Western electronic ADIs (Boeing 737) have become disorientated in clear sky, night conditions when suddenly confronted with an unexpected situation and banked their aircraft in the wrong direction in an attempt to recover to wings-level, with fatal consequences. A contributory factor could have been that Russia continued to adopt the 'outside-looking-in' convention for the ADI until recent times (see Section 12.2.2).

12.4.2 Research Areas

The flight crew capability limit has possibly been reached in the current generation flight deck and the time is approaching for a more radical approach. The next generation flight deck needs to be crew-centric, recognising that pilots are not necessarily computer engineers. The human–machine interface needs to be simple and intuitive, giving the pilot what he needs to know and helping him to keep the decision-making safe. More automation can be harnessed to ease the crew workload and increase situational awareness to manage flight missions more efficiently, but this must not be at the expense of loss of transparency. Research areas currently under evaluation include:

- pilot task analysis;
- intention detection;
- workload scheduling;
- cognitive resource management;
- user error anticipation;
- adaptive interfaces;
- drowsiness monitoring;
- stress detection;
- perseverance and funnelisation.

Aviation flight safety is dependent upon collaborative decision-making and accurate situational awareness. Studies into authority sharing, data fusion and organisation psychology will play an important part in defining what future flight-deck users will need to help them cope with mission demands while remaining alert and capable of reacting quickly, correctly and safely. Further research is required into what is needed to ensure consistency between the crew's mental representation of system behaviour and real system behaviour in order to support decision-making models and decision aids.

12.4.3   Concepts

The next generation flight deck is likely to take advantage of emerging display technologies, touch screens, haptic feedback, three-dimensional views and sound, active noise reduction, gesture recognition and integrated biosensors. A view from Thales, ODICIS (One DIsplay for a Cockpit Interactive Solution), is conceptualised in Figure 12.14 [14]. It envisages a single, contiguous large-area 3-D display screen with touch-screen interface and a wide-angle colour head-up display.

The head-down display area envisages opening windows superimposed on a view of the horizontal flight profile. These windows pop up as required to suit the current phase of flight and the status of the aircraft, on the basis of displaying what the crew need to know when they need to know it. The windows may be sized and positioned using familiar drag and drop techniques. The flight profile extends into the head-up display area in a totally intuitive manner. The images make full use of the emerging enhanced vision and synthetic vision technologies discussed in Section 12.8.

Rear micro-projectors, flexible direct-view screens, touch-screen and direct voice command, wide eye-box colour HUDs, head-worn displays and ever higher graphic quality and image processing will transform what is possible on the flight deck. These technologies are discussed in Section 12.11.

By 2030 the next generation flight deck will be safer, simpler, easier to train for and possibly single-crew capable during periods of low workload (e.g. during the cruise), augmented by a second crew member during take-off and landing.

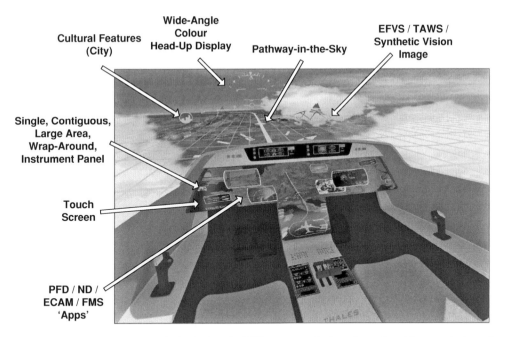

Cultural Features (City)

Wide-Angle Colour Head-Up Display

Pathway-in-the-Sky

EFVS / TAWS / Synthetic Vision Image

Single, Contiguous, Large Area, Wrap-Around, Instrument Panel

Touch Screen

PFD / ND / ECAM / FMS 'Apps'

Figure 12.14 Generation 3.0 flight deck – ODICIS conceptual view. Reproduced by permission of Thales Avionics. Copyright Felix and Associés. For a colour version of this figure, please refer to the colour plates

12.5 Electronic Centralised Aircraft Monitor (ECAM) System

The Airbus ECAM system continuously monitors aircraft systems behaviour and reports back to the central computer, checking that all systems are operating within normal limits. The 'need to know' logic ensures that the displayed data are matched to the need for information.

12.5.1 ECAM Scheduling

In normal operation one display continuously shows engine data, total fuel state, outside air temperature and 'MEMO' items. The other display shows synoptic diagrams appropriate to the current flight phase, such as pre-flight, take-off, climb, cruise, descent, approach and after landing. At any time, the crew can manually select any desired aircraft system for routine checking.

12.5.2 ECAM Moding

The ECAM system has four display modes: automatic flight phase, advisory and failure-related modes, and manual.

- **The failure-related mode** has precedence. As soon as any warning is triggered, the appropriate system synoptic diagram is displayed. The malfunction is clearly described in plain language/checklist form and corrective action is prescribed. Any resultant operating limitations are displayed as a reminder.

- **The automatic advisory mode** ensures that the relevant data are brought to the crew's notice, should any parameter drift out of the normal operating range, well before a warning level is reached. In this case, the parameter value is flashing. Systems trends are thus automatically monitored, and the crew is warned of possible trouble well in advance, without having to refer to the overhead panel. The ECAM provides the foresight that ensures efficient command.

The ECAM provides the crew with an immediate analysis of systems state in the case of failure as well as during corrective action, and eliminates the need to diagnose failures and their implications from the overhead panel indications. When an alert occurs, the ECAM shows the affected system in red or amber, with the corrective action required in plain language in blue. The ECAM logic identifies the primary, or 'generating', failure which appears first in a special 'boxed' format; the crew therefore deal with it first.

The appropriate caution/warning light flashes on the warning panel. Lights for services or functions affected by secondary or 'consequential' failures are identified by an asterisk. The appropriate push-button selectors illuminate on the overhead panel. The crew press each button to set the component it controls to the required configuration. As each action is performed, the blue instruction on the ECAM display turns white, and the synoptic display is reconfigured accordingly. Actions performed automatically, needing no crew action, are displayed in white.

12.5.3   ECAM Pages

Figure 12.15 shows example ECAM pages.

12.5.4   Qantas Flight QF32

The Airbus A380 Qantas flight QF32 incident is worthy of note. The primary failure was an uncontained rotor burst of the IP turbine disc in No. 2 (inboard port) engine four minutes after taking off from Singapore Changi. Shrapnel from the exploding engine punctured part of the wing. The ECAM system reported the following secondary failures:

- Engines Nos 1 and 4 operating in a degraded mode.
- GREEN hydraulic system – low system pressure and low fluid level.
- YELLOW hydraulic system – engine No. 4 pump errors.
- Failure of the alternating current (AC) electrical Nos 1 and 2 bus systems.
- Flight controls operating in alternate law.
- Wing slats inoperative.
- Flight controls – ailerons partial control only, reduced spoiler control.
- Landing gear control and indicator warnings.
- Multiple brake system messages.
- Engine anti-ice and air data sensor messages.
- Multiple fuel system messages, including a fuel jettison fault and C of G messages.
- Auto-thrust and auto-land inoperative.

It took the crew more than 50 minutes to assess the condition of the aircraft and execute ECAM recommendations before returning to Changi where they made a successful landing 30 knots faster than normal after lowering the landing gear under gravity.

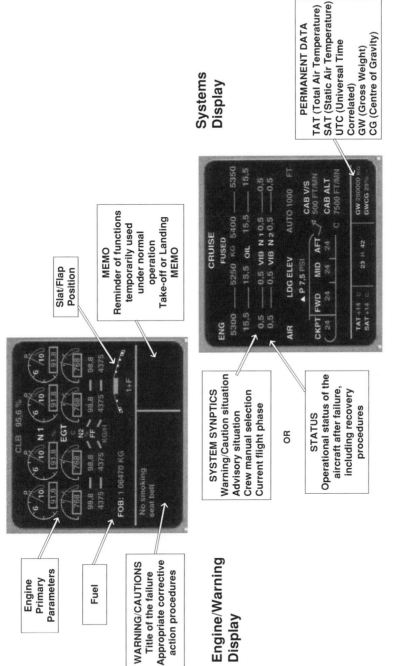

Figure 12.15 ECAM example pages

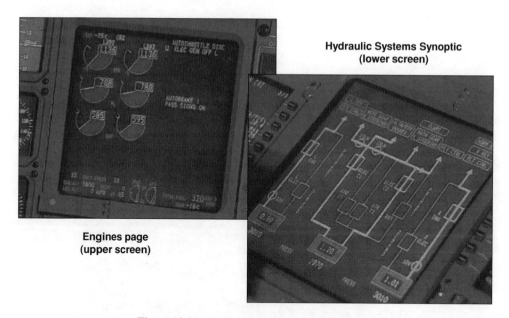

Figure 12.16   Boeing 777 EICAS example pages

12.5.5   *The Boeing Engine Indicating and Crew Alerting System (EICAS)*

The Boeing (EICAS) operates in a similar manner. Figure 12.16 shows some actual screen shots of the upper and lower screens on the Boeing 777.

12.6   Standby Instruments

The availability and integrity of flight-critical information is always augmented with independent, segregated standby instruments.

In early glass flight decks, the standby instruments were the traditional set of three miniature (usually 3ATI) dedicated attitude, airspeed and altimeter electromechanical instruments. More recently these have been superseded by electronic instruments using AMLCD technology with integral solid-state sensors. Piezo-resistive pressure sensors have replaced the aneroid capsules. Micro-machined rate sensors and accelerometers have replaced the rotating gyroscopes.

The integrated standby instrument system (ISIS) shown in Figure 12.17 is in 3ATI form factor and has a high-resolution full-colour AMLCD with a useable screen area of 2.4 × 2.4 inches. It is self-contained with integral sensors, operates from the aircraft 28 VDC essential bus, and consumes 15 watts at full brightness. The ISIS provides a display of:

- attitude;
- indicated airspeed and Mach number;
- baro-corrected altitude and baro-set;
- slip/skid;
- heading.

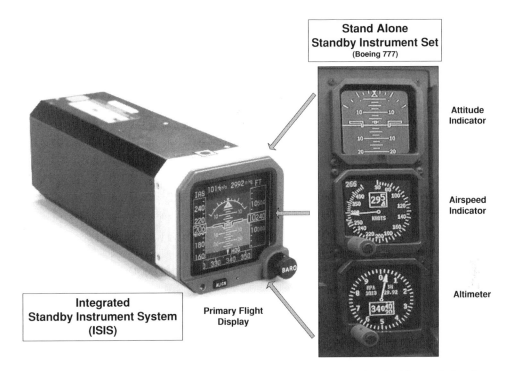

Figure 12.17   3ATI integrated standby instrument system (ISIS). Reproduced with permission from GE Aviation (formerly Smiths Aerospace)

The display formats are designed to be compatible with the primary EFIS formats to reduce pilot adaptation time and workload.

The improved reliability of the solid-state instrument over the earlier three dedicated electromechanical instruments means that a single ISIS displaying a composite PFD format can be installed without any system integrity penalties, as seen on recent Airbus and Boeing aircraft.

12.7   Head-Up Display Visual Guidance System (HVGS)

This section will use HVGS as the generic name for a head-up display visual guidance system and HUD for the optical projection system. HUD manufacturers use other terms. BAE Systems and Honeywell use the terminology Visual Guidance System (VGS). Rockwell Collins Flight Dynamics markets Head-up Guidance Systems (HGS). Thales Avionics markets Head-up Flight Display Systems (HFDS).

12.7.1   Introduction to Visual Guidance Systems

The principle attribute of the HVGS is to provide the pilot with conformal flight information that is in 'contact analogue' with his real world view. The HVGS provides guidance and flight data so the pilot can maintain complete awareness of his situation with respect to all the flight critical parameters without having to look inside the flight deck. Contact analogue symbology is by definition aligned to the real world and by its nature is intuitive.

The HVGS image is collimated (focused at infinity) by the HUD optical system and then projected into the pilot's sightline by a partially reflective combiner so that the image appears to the pilot to be superimposed on his outside world view. Being collimated, the projected image remains in perfect registration with the outside world, irrespective of the pilot's head motion (provided his eyes remain within the HUD projection porthole).

12.7.2   HVGS on Civil Transport Aircraft

In 1990, the Flight Safety Foundation (FSF) analysed more than 1000 accidents and concluded that a properly functioning HVGS operated by a correctly trained flight crew might have prevented or positively influenced the outcome of 30% of the accidents. Subsequently the International Federation of Air Line Pilots' Associations (IFALPA) recommended that all jet transport airplanes be equipped with a HVGS; a dual installation was preferred for monitoring purposes and redundancy.

A HVGS has been available as optional equipment on several civil aircraft models, but until recently has been standard equipment on only a few, including some Boeing and several Gulf-stream and Embraer business jets, usually as single (captain) installations. That is changing – more than 30 airlines worldwide have equipped at least portions of their fleets with a HVGS. Boeing and Airbus now offer a dual HVGS installation as standard fit on the Boeing 787 and the Airbus A350, and both companies are certificating a HVGS as an option fit on earlier aircraft types.

In January 2004 the FAA lowered the approach minima for an aircraft equipped with a HUD and an EVS – called an enhanced flight vision system (EFVS). The FAA ruling allows a properly trained pilot to continue to execute an approach at a non-precision (Cat I) airport down to an altitude of 100 ft (Cat II) provided he can see the required cues in his EFVS. In 2010 that credit was extended to approach and landing with a runway visual range (RVR) of 1000 ft (see Section 12.8).

12.7.3   HVGS Installation

A typical civil HVGS installation, shown in Figure 12.18, comprises the following line-replaceable units (LRUs):

- **display guidance computer**, which receives data from aircraft sensors and generates display symbology;
- **overhead projector**, which includes a CRT or AMLCD plus collimation optics that projects the image onto the combiner;
- **combiner**, a holographic optical element (glass plate or plastic plate), mounted behind the windshield, that reflects the projected image toward the pilot's eyes, while allowing ambient light to pass through;
- **control panel**, which is used by the flight crew to select HVGS modes and features, and to enter data that the computer does not receive from aircraft sensors;
- **annunciator panel**, which provides HVGS status information and warning information.

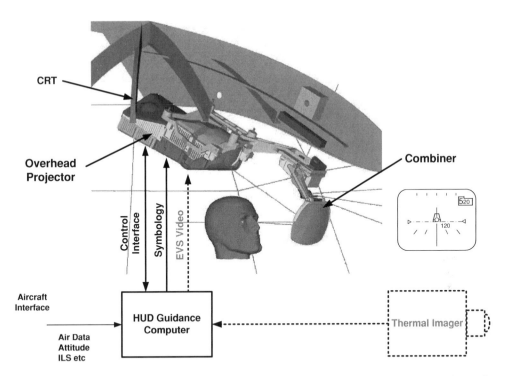

Figure 12.18 Typical HVGS installation. Reproduced with permission of BAE Systems Electronic Systems in Rochester, Kent

The overhead projector and combiner are mounted on the cabin roof. The combiner can be stowed (folded) when not required. See Section 12.11.6 for an overview of HUD optical principles and technologies.

12.7.4   HVGS Symbology

For the civil airline pilot the most important feature of the HVGS is its ability to display the flight path vector in contact-analogue with the outside world. The flight path vector provides an instantaneous and intuitive indication of the direction of flight. During approach and landing the pilot can simply fly the desired glide path angle and place the flight path vector on the intended touchdown zone (TDZ). The flight path vector enables the pilot to 'spot' land the aircraft and provides important cues if drift or sideslip develop prior to touchdown. In low visibility approaches the pilot can fly head-up using visual ILS direction cues. His attention is directed at precisely the right point in space to visually acquire the runway at approach minima. If the minima criteria are not satisfied, he can rapidly and confidently execute a missed approach. Using the flight path vector in conjunction with other cues on the HVGS, the pilot is able to precisely control the aircraft manually, and routinely maintain desired airspeed to within 2 knots, heading to within 2°, altitude to within 50 ft and pitch to within 0.5°. This degree of control is difficult to achieve using head-down instruments [15, 16].

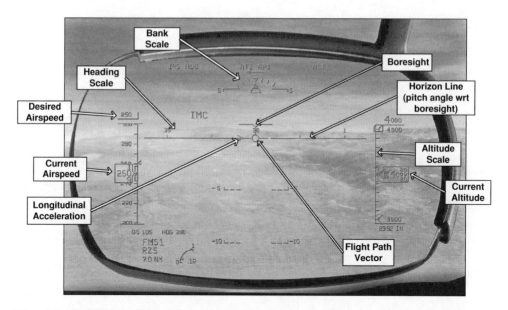

Figure 12.19 Typical HVGS symbology. Reproduced with permission from Rockwell-Collins. For a colour version of this figure, please refer to the colour plates

Typical HVGS symbology, shown in Figure 12.19, provides the pilot with all the flight performance data found head-down: airspeed, altitude, attitude heading, and so on. Flight path, attitude, heading and directional guidance are conformal; other information is presented in a similar format to that found on the PFD to aid assimilation and transition between head-down and head-up.

The symbology comprises:

- The **flight path vector** in the form of a small airplane symbol. It provides a conformal display of the longitudinal and lateral instantaneous and inertial flight path. It indicates where the aircraft is going, not where it is pointed. In level flight the flight path symbol will overlay the horizon line. On the approach, the flight path symbol can be used to control the actual touchdown point of the aircraft. On a correct approach both the glideslope reference line and flight path symbol overlay the TDZ.
- The **boresight symbol**, which has the appearance of a seagull towards the top of the display, is where the aircraft is physically pointed.
- The **longitudinal acceleration caret** (potential flight path angle) is a symbol that aligns with the wing on the flight path symbol when there is no longitudinal acceleration. Inertial acceleration, air mass acceleration and throttle setting drive this symbol to indicate the correct throttle movement in steady air and wind shear. When the throttles are advanced, giving excess thrust, this thrust can be turned into increased speed or a steeper climb angle.
- The **horizon line** represents a level plane through the aircraft's present position. It is always parallel with the outside world horizon, and at moderate heights overlies this horizon.
- **Heading** is marked conformally every 10° on the horizon line.
- The **desired approach speed** is entered by the crew into the system, and is displayed on the top of the airspeed scale.

- The actual **airspeed** appears in the box halfway down the airspeed scale.
- The **altitude scale, bugs and trend arrows** similarly appear on the right.
- **Annunciator** boxes appear along the top of the display and indicate the autopilot modes, both current and armed.

During an ILS approach, additional symbology appears to aid precision manual piloting of the aircraft or monitor an ILS coupled approach (see also Figure 12.24).

- The **instrument landing system (ILS) scales** appear on the bottom in the case of localiser, and on the right in the case of glideslope.
- The **guidance cue** is generated by the system from the ILS beams and aircraft sensors through the aerodynamic model of the aircraft. If the pilot manoeuvres the aircraft such that the flight path marker overlays the guidance cue, the system will accurately take the aircraft down to TDZ.
- The **glideslope reference line** angle is selected by the pilot. If an ILS is installed, the glideslope reference line will be set to the same value. If the glideslope reference line is set at 3° and this line overlays the visual TDZ, then the aircraft must be on a 3° glideslope.
- The difference between the desired approach speed and the actual aircraft speed is displayed as a **'speed error worm'** on the flight path symbol as a tape emanating from the 'left wing' of the flight path symbol. If the speed is below the set speed, the tape extends below the left-hand wing of the flight path symbol, and above it if the speed is too high.

Further material is available in Chapter 7 of Reference [1].

12.8   Enhanced and Synthetic Vision Systems

12.8.1   Overview

As weather and visibility conditions deteriorate it becomes increasingly difficult to conduct flight operations in the same manner and rate as in visual meteorological conditions (VMC). Whilst current technology provides solutions to many of the problems caused by low visibility, the potential now exists to enhance visual references for the flight crew or provide them with artificial graphical depictions of visual references.

Enhanced vision systems (EVS) help mitigate reduced visibility as a limiting factor in flight operations. EVS uses sensors to 'see' the environment along the flight path. In January 2004 the FAA granted operational flight credit for aircraft equipped with a HUD and an EVS. The FAA refers to a HUD+EVS as an enhanced flight vision system (EFVS). EASA uses the term EVS as equivalent to EFVS. This chapter will use the FAA nomenclature: EFVS. The flight credit (defined in US Code of Federal Regulations 14 CFR §91.175) allows a properly trained pilot to continue a non-precision approach beyond the normal Cat I approach minima of 200 ft, provided he can 'see' the required visual cues head-up with the aid of the EFVS. Head-down displays of EVS are used in some light aircraft and helicopters for general hazard awareness, but these are not currently allowed operational credit.

Synthetic vision systems (SVS) provide an aid for runway location and other objects. A SVS is a computer-generated means to display a synthetic image of the external scene topography from the perspective of the flight deck. The image is derived from aircraft attitude, high-precision navigation solution, and a database of terrain, obstacles and relevant cultural features. SVS can provide an enhanced intuitive view of the flight environment along with a depiction of hazardous terrain and obstacles and airport features with precision navigation guidance.

12.8.2 EVS, EFVS and SVS Architecture Diagrams

To highlight the distinctions between them, EVS, EFVS and SVS system architecture diagrams are shown in Figure 12.20.

EVS, SVS and combined vision systems (CVS) may be displayed on any of the existing flight deck displays: PFD, ND, MFD or Class 3 Electronic Flight Bag (EFB). However, EFVS must be displayed on a conformal HUD.

- **Primary flight display (PFD)**: the EVS, SVS or CVS image is typically merged into the sky/ground shading of the attitude direction indicator.
- **Navigation display (ND)** or multifunction display (MFD) as one of many stand-alone type formats available.
- **Electronic flight bag**: While similar in concept to the MFD, an EFB implementation is limited due to the installation constraints and may present unique certification challenges such as alignment or positioning concerns relating to the EFB installation.

FAA Advisory Circular AC 20-167, Airworthiness Approval of Enhanced Vision Systems, Synthetic Vision Systems, Combined Vision Systems and Enhanced Flight Vision Systems provides further guidance for installing and demonstrating acceptable means of compliance [17]. Flight guidance system (FGS) criteria may be found in documents such as FAA Advisory Circular (AC) 25.1329A [18].

12.8.3 Minimum Aviation System Performance Standard (MASPS)

EUROCAE Working Group 79 (WG-79) and RTCA Special Committee (SC) 213 are jointly tasked with developing Minimum Aviation System Performance Standards (MASPS) for enhanced vision systems (EVS), enhanced flight vision systems (EFVS), synthetic vision systems (SVS) and combined vision systems (CVS). Deliverables are shown in Table 12.1 [19].

12.8.4 Enhanced Vision Systems (EVS)

An enhanced vision system (EVS) is an electronic means to provide the flight crew with a sensor-derived or enhanced image of the external scene through the use of passive imaging sensors such as forward-looking infra-red, and/or low light-level image-intensifying devices or active imaging sensors such as millimetre wave radiometry.

Unlike the pilot's external view, the enhanced vision image will be a monochrome presentation. Some, but not all, of the depth cues found in the natural view can still be present in the imagery. The quality of the enhanced vision image and the level of enhanced vision sensor performance depends upon the atmospheric and external visible and non-visible energy source conditions. Gain settings of the sensor, and brightness or contrast settings of the HUD, can significantly affect image quality.

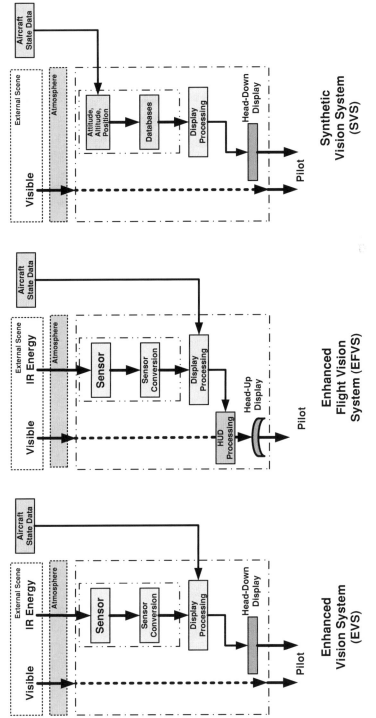

Figure 12.20 Comparison of EVS, EFVS and SVS systems architectures

Table 12.1 WG-79/SC-213 MASPS

Product	Description	Due date
MASPS RTCA-DO-315	Enhanced Vision Systems, Enhanced Flight Vision Systems, Synthetic Vision Systems and Combined Vision Systems.	Completed Dec 2008
MASPS RTCA-DO-315A	EFVS for approach, landing and rollout (1000 ft RVR)	Completed July 2011
MASPS RTCA-DO-315B	SVS for approach	Completed June 2011
MASPS RTCA-DO-315C	SVS/CVS for approach, SA CAT I or CAT II (LPV) or lower standard minima CAT I	Planned for July 2013
MASPS RTCA-DO-XXX	Vision Systems for approach, landing and rollout (300 ft RVR)	Planned for July 2012
MASPS RTCA-DO-XXXA	Vision Systems for taxi (300 ft RVR)	Planned Dec 2013

Passive Thermal Imagers

A passive thermal imager comprises the following elements – see Figure 12.21:

- optics, to collect and focus the thermal energy onto the detector:
- detector, a focal plane array that converts the thermal energy into electrical signals;
- electronics to amplify and process the signals into a video image;
- display, to present the information to the user.

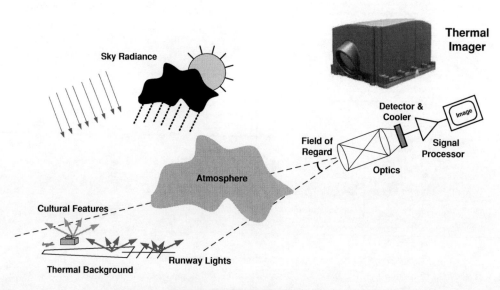

Figure 12.21 Thermal imager

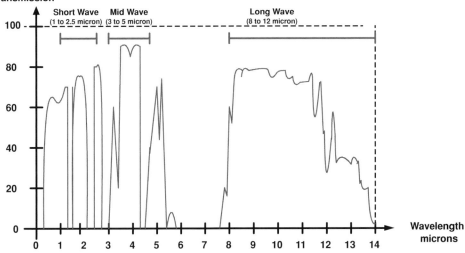

Figure 12.22 Atmospheric absorption

Electromagnetic radiation (EMR) is emitted and/or reflected by all bodies over a range of frequencies as a function of their temperature. Real materials can be good reflectors or good emitters, but not both. The different temperatures and different emissivity of real materials facilitate the generation of an image from thermal differences in the external scene.

EMR energy will radiate from a point source in accordance with the inverse square law. At the ranges of interest the energy incident on a detector will be small. The thermal signature will be impacted by cloud and rain. All EMR is scattered or absorbed as it propagates. Visible light (0.39–0.77 μm) is primarily scattered. The atmospheric absorbers of infra-red (thermal) energy are primarily H_2O, CO_2 and O_3. Infra-red (IR) propagation is severely affected over major portions of the spectrum (see Figure 12.22).

Propagation effects constrain long-range IR detection into three distinct bands:

- long-wave IR (8–12 μm) for imaging the thermal background;
- mid-wave IR (3–5 μm) for imaging cultural features (hot bodies);
- near or short-wave IR (0.9–2.5 μm) for imaging in low light/starlight conditions.

A detector assembly consists of a detector fabricated as a focal plane array situated behind the optics. There are two types of IR radiation detectors:

- **Photon detectors**: When IR radiation falls on detector material, usually a semiconductor, then individual electrons are expelled from the atoms of the material and increase its electrical conductivity in proportion to the number of impinging (IR) photons.
- **Thermal (pyroelectric) detectors**: Absorb radiant energy, increasing the detector temperature which produces changes in the bulk physical properties of the material.

Photon detectors are narrow band. Detector materials can be chosen to match the wavelength of interest; no single photon detector material is optimised for all applications. Typical choices are:

- **Medium wave infra-red (MWIR) 3–5 μm**: Indium antimonide (InSb) is the choice of most designers because it provides the best performance over a wide range of conditions. In order to minimise the number of free electrons moving randomly in the detector material (the dark current), it is necessary to cool the array.
- **LWIR 8–12 μm**: Mercury cadmium telluride (HgCdTe). Long wave infra-red (LWIR) energy is generally better able to penetrate weather, but is more expensive and has to be cooled to within a few degrees of absolute zero.
- **Near IR 0.9–1.7 μm**: Indium gallium arsenide (InGaAs) is used as a detector in night-vision goggles, also known as image intensifiers (IIT) since they amplify the starlight illuminated scene.

A micro-bolometer is a wide band pyroelectric uncooled thermal sensor. IR radiation heats the detector material, changing its electrical resistance. This resistance change is measured to create an image. A micro-bolometer consists of an array of pixels, each being made up of an IR absorbing element and a readout circuit. The absorbing material must be thermally isolated from the readout circuit. Amorphous silicon, vanadium oxide and barium strontium titanate (capacitive) are the commonly used materials. The quality of images created from micro-bolometers has continued to improve. The current technology limit is around 320 × 240 pixels. It is a broad-band detector. Its response is generally limited by the optics 'window', not by the device itself. It has a slower response than a photon detector, of the order of 10 ms, and lower sensitivity.

Examples of current technology thermal imagers can be found from Elbit Kollsman (photon-detector-based products) and Max-Viz (pyro-detector-based products).

Active Millimetric Imaging Radar

The longer range penetration of weather by forward-looking infra-red (FLIR) is limited, however, when certain fog densities are encountered. In such cases millimetric wave imaging radar (MWIR) does often exhibit more effective penetration but at a significant increase in cost and power and at reduced resolution. The MWIR radar scans the forward scene. Echo data from the antenna are in the form of range return and elevation and azimuth angles which must then be processed into a form that is familiar to the pilot and consistent with his normal visual perspective of the world.

12.8.5 Enhanced Flight Vision Systems (EFVS)

The purpose of the EFVS sensor is to provide a visual advantage over the pilot's out-the-window view. In low visibility conditions the EFVS is used to visually acquire the references required to operate below the decision height/decision altitude (DH/DA) or minimum descent height/minimum descent altitude (MDH/MDA) at a longer distance than can be seen with the unaided eye.

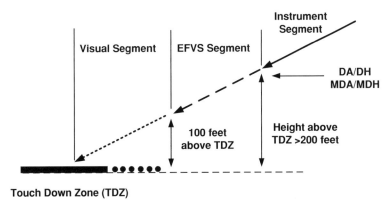

Figure 12.23   EFVS approach minima and visual transition points

MASPS RTCA-DO-135 published in December 2008 extended the visible approach minima for EFVS-equipped aircraft from Cat I down to the Cat II approach minima of 100 ft (see Figure 12.23).

Instrument Approach with EFVS

For EFVS operations, the sensor imagery and required flight information and symbology is displayed on a HUD so that the pilot flies both the instrument and visual segments head-up, eliminating head-down to head-up transition and visual accommodation time. The EFVS display must be conformal, that is, the sensor imagery, aircraft flight symbology and other cues that are referenced to the imagery and external scene are aligned with the external view (see Figure 12.24).

EFVS operations require the pilot to accomplish several visual-based judgment and control tasks in quick succession. These include using the imagery, flight reference information, and eventually the outside view at the same time. The pilot is able to look for the outside visual references in the same location as they appear in the EFVS image and readily see them as soon as visibility conditions permit, without delays or distraction due to multiple head-up and head-down transitions.

At DH/DA (or MDH/MDA), the pilot makes a decision whether to continue the descent using the same visual reference criteria as he would for an unaided approach but using enhanced flight visibility augmenting his natural vision, namely:

- The aircraft shall be continuously in a position from which a descent to landing can be made:
  - on the intended runway
  - at a normal rate of descent
  - using normal manoeuvres.
- The flight visibility may not be less than that visibility prescribed in the instrument approach procedure using natural vision and describes the average forward horizontal distance at which prominent unlighted objects may be seen and identified by day, and prominent lighted objects may be seen and identified by night.

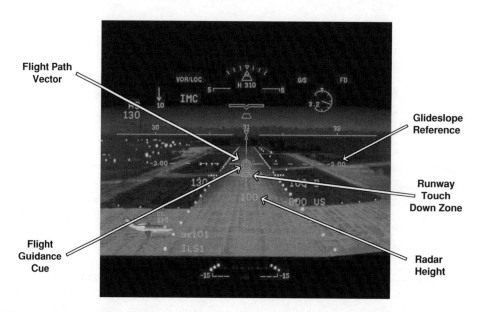

Figure 12.24 EFVS HUD image (sensor plus flight guidance symbology). Reproduced with permission from Rockwell-Collins

At 100 feet above the TDZ, the visual transition point, the pilot makes a second determination about whether the flight visibility is sufficient to continue the approach and distinctly identify the required visual references using his natural vision.

MASPS RTCA-DO-135A adds system design criteria for use of EFVS to landing with reported visibility as low as 1000 ft RVR (or 300 m). Specific features include a touchdown zone requirement and use of a radio altimeter and flare prompt. Vertical flight path guidance, flare guidance, pilot monitoring display requirements, and design assurance requirements are also addressed. MASPS RTCA-DO-XXX planned for July 2012 will extend this further to 300 ft RVR.

FAA Advisory Circular AC 90-106 provides operational requirements for EFVS, conducted below DH/DA or MDA down to 100 ft above TDZ [20]. Current regulations do not require the EFVS to be stowed. The pilot should, however, be able to easily and quickly declutter the EFVS or remove the sensor image at any time he deems it necessary or appropriate, without having to look away from the HUD.

FAA regulations do not require that the sensor image and flight information from the EFVS be presented to the non-flying pilot. EASA regulations require a separate repeater display located in or very near the primary field of view of the non-flying pilot. Operators may elect to equip with dual EFVS displays. Both the Boeing 787 and the Airbus A350 are equipped with dual HUD installations.

MASPS RTCA-DO-315 provides minimum EFVS system performance requirements. The key requirements can be summarised as follows:

- **Field of regard (FoR)**: 20° horizontal and 15° vertical minimum.
- **Harmonisation**: The accuracy of the integrated EFVS and HUD image shall be better than 5 mr.

- **Sensor sensitivity**: The sensor sensitivity shall be at least a noise-equivalent temperature difference (NETD) of 50 milliKelvins (mK) at R_{max} from 200 ft height above TDZ elevation with a typical 3° glideslope.
- **Sensor resolution**: The EFVS shall adequately resolve the following visual references of a 60 ft wide runway environment from a range of 200 ft height above the TDZ with a typical 3° glide slope:
  - Approach light system, if installed, or runway threshold, identified by at least the beginning of the runway landing surface, threshold lights, or runway end identifier lights.
  - Touchdown zone, identified by at least runway touchdown zone landing surface, touchdown zone lights, touchdown zone markings, or runway lights.
- **Latency**: Should be no greater than 100 msec.

12.8.6   Synthetic Vision Systems (SVS)

A synthetic vision system (SVS) is an electronic means to display a computer-generated image of the applicable external topography from the perspective of the flight deck (inside-looking-out) that is derived from aircraft attitude, altitude, position, and a coordinate-referenced database. Currently, the application of synthetic vision systems is through a primary flight display (PFD).

MASPS RTCA-DO-315B provides system design criteria for the use of SVS for lower-than-standard Cat I ILS minima.

FAA Advisory Circular AC 23-26 provides guidance on synthetic vision and pathway depictions on the primary flight display for Part 23 aircraft (small airplanes) [21].

Synthetic vision coordinate databases are usually referenced to Digital Terrain Elevation Data (DTED), developed by the National Imagery and Mapping Agency. There are three DTED levels with the following terrain spacing:

- DTED Level 0 uses 30 arc-second spacing (nominally 1 km).
- DTED Level 1 uses 3 arc-second spacing (approximately 100 m).
- DTED Level 2 uses one arc-second spacing (approximately 30 m).

Terrain Avoidance Warning System (TAWS)

TAWS (previously known as EGPWS) was the precursor to SVS and uses a worldwide terrain database which is compared with the aircraft's present position and altitude. Within the terrain database the Earth's surface is divided into a grid matrix with a specific altitude assigned to each square within the grid, representing the terrain at that point. The aircraft's intended flight path and manoeuvre envelope for the prevailing flight conditions are compared with the terrain matrix and the result graded according to the proximity of the terrain. This type of portrayal using coloured imagery is very similar to that for the weather radar and is usually shown on the navigation display. Terrain responses are graded as follows:

- more than 2000 ft below the aircraft – clear display;
- 2000 to 1000 ft below the aircraft – light-density green dot pattern;
- 1000 to 500 ft below the aircraft – medium-density green dot pattern;
- 500 ft below to 1000 ft above the aircraft – medium-density yellow dot pattern;

- 1000 to 2000 ft above the aircraft – high-density yellow dot pattern;
- more than 2000 ft above the aircraft – high-density red dot pattern.

Vertical Situation Display (VSD)

Boeing augmented the display of TAWS data with a vertical situation display (VSD). The VSD works in conjunction with the TAWS terrain-mapping feature to provide flight crews with an intuitive presentation of the vertical situation relative to the surrounding terrain and the final approach descent path. It graphically represents a view of the vertical profile or side view, of terrain and flight path data that follows the current track of the aircraft. When selected by the flight crew, it appears at the bottom of the navigation display (see Figure 12.25).

The VSD depicts terrain information from the TAWS or other onboard sources from another perspective. The TAWS generates a lateral view of the surrounding terrain and provides terrain proximity alerting. The VSD depicts the vertical dimension of the terrain which will allow crews to recognise possible terrain conflicts more readily before a TAWS alert is generated [22].

SVS Research Initiatives

Early research into intuitive 3-D and 4-D graphical information displays coined the term 'highway-in-the-sky'. 'Highway-in-the-sky' itself first flew experimentally on a Calspan

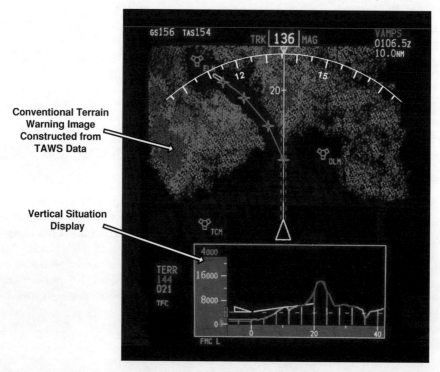

Figure 12.25 TAWS and VSD on a navigation display. For a colour version of this figure, please refer to the colour plates

Laboratory C130 aircraft in 1983. Many research programmes further developed the 'highway-in-the-sky' concept to generate an accurate 3-D representation of the terrain viewed from the flight deck.

In 1995 the University of Munich, Germany, carried out a significant programme of work together with VDO Luftfahrtgerate Werk of Frankfurt. An aircraft was equipped with 4-D displays and evaluated taxiway, take-off, approach, landing and terrain avoidance manoeuvres with much success [23]. The experimental primary flight display replaced the simple sky/ground shading of the traditional PFD format with a 3-D perspective of the outside world terrain. It promoted better situational awareness for the pilot and an early prediction of danger, especially the proximity of terrain. A flight-path predictor symbol indicated the anticipated motion of the aircraft based on its current manoeuvre. The desired pre-planned flight path (the 'highway-in-the-sky') was experimentally flown in a number of styles.

In a similar timeframe NASA conducted flight tests of a SVS and EFVS in a Boeing 757 at Vail, Colorado, US. The test database included terrain and obstructions, landing and approach patterns, and runway surfaces, and evaluated blending images generated by an SVS with images derived from infra-red sensors and millimetre-wave radar. The objective was to mature the technology sufficiently to reduce the technical risk and provide regulatory and design guidance to support the introduction of SVS/EFVS advanced cockpit vision technologies in terminal manoeuvring area NextGen operations [24].

Certificated Synthetic Vision Systems

Executive jets were first to embrace the technology led by Gulfstream, who certificated Honeywell's SmartView™ Integrated Primary Flight Display (IPFD) in their G500 aircraft in 2007. The IPFD provides an easy-to-understand, intuitive, 3-D perspective view that replicates what the pilot would see on a clear day. Figure 12.26 shows some typical images. The terrain is colour-coded in accordance with TAWS conventions and the warning PULL-UP is clearly visible (together with an audio warning). It will be seen that the primary flight symbology more closely resembles that of a head-up display and includes a clear representation of the aircraft flight path vector to aid precision flying.

In the business jet retrofit and forward-fit markets, synthetic vision is one of the applications most in demand. Often larger than the original equipment, new displays can deliver to pilots of legacy platforms greater situational awareness as well as the foundation for technologies needed for NextGen operations. Spurred on by the popularity of the iPad and other tablets, there is considerable interest in using touch-screen and larger displays. Aside from its use on fixed wing aircraft, SVS technology is also being increasingly used on rotor-craft. Many companies are entering the market alongside Honeywell and Rockwell Collins, including CMC Electronics, Elbit Kollsman, Esterline, Garmin, Universal Avionics Systems and others.

However, in the civil transport market, take-up of synthetic vision has been relatively slow. Operators are looking for operational flight credit in a similar vein to that given for EFVS to provide the catalyst for introduction. MASPS RTCA-DO-315C is planned for July 2013.

Initiatives to obtain operational credit for SVS and reduce visibility minima are targeted at validating the synthetic runway scene. Honeywell are evaluating providing additional assurance by comparing the runway position from two independent airport databases; one from

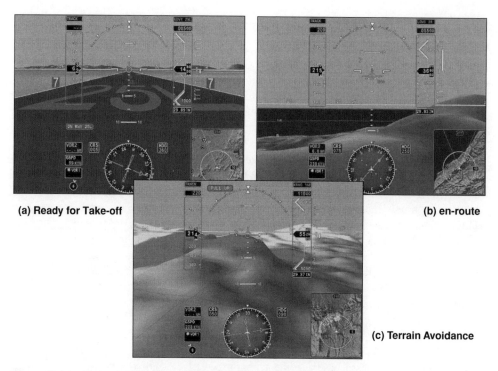

(a) Ready for Take-off

(b) en-route

(c) Terrain Avoidance

Figure 12.26 Honeywell SmartView™ SVS. Reproduced with permission from Honeywell Aerospace. For a colour version of this figure, please refer to the colour plates

Honeywell's runway-awareness-and-advisory system, the other from Jeppessen's navigational database. Honeywell also plan to independently verify the position of the runway indicator by cross-comparing the SBAS-derived aircraft position with the aircraft's inertial reference system (IRS) position and cross-correlate that with ILS guidance cues.

Rockwell Collins plan to use its weather radar to provide addition assurance. The weather radar can recognise the runway signature and compute height-above-threshold and range-to-threshold to provide independent position correlation assurance between the 'real world' and the 'synthetic world'.

Further studies include the possibility of using the weather radar to detect obstacles and runway incursions, in addition to deriving groundspeed using Doppler principles [25].

12.8.7   Combined Vision Systems

A combined vision system (CVS) is a selective blending of EVS and SVS superimposed and correlated on the same display. The current integration concepts typically utilise a database-driven synthetic picture for higher altitudes, and an enhanced real-time sensor image for lower altitudes down to the TDZ. For example, on an approach, most of the arrival and/or the procedure turn would utilise the SVS picture, but somewhere between the final approach fix and the runway, the picture would gradually transition from SVS to EVS.

HUD Symbology

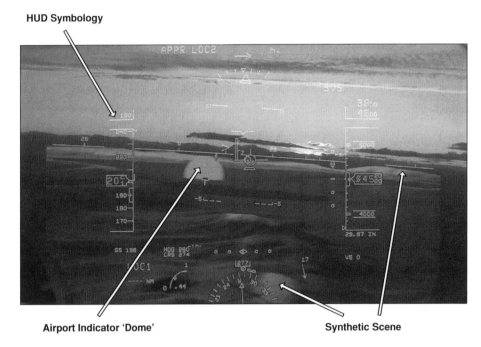

Airport Indicator 'Dome'   **Synthetic Scene**

Figure 12.27 Rockwell Collins HUD with synthetic vision. Reproduced with permission from Rockwell-Collins. For a colour version of this figure, please refer to the colour plates

Honeywell and Rockwell Collins are approaching the subject of CVS from different directions. Both companies are seeking landing credit for lowered instrument approach minima using synthetic vision combined with enhanced vision [26].

Rockwell Collins have demonstrated an increase in flight-path control accuracy when using synthetic vision to fly an instrument approach on their HUD, maintaining that using a HUD avoids the difficult head-down to head-up transition prior to the DH transition to natural vision. SVS on a HUD is limited to monochrome (green) and needs to be conformal to a high degree of precision. It must not obscure the pilot's natural vision of the required reference features. Figure 12.27 shows SVS on a Rockwell Collins HUD indicating the ability to 'see' through the ground cloud layer. The HUD option has always been significantly more expensive than SVS on a HDD. To address this, Rockwell Collins are developing a lower-cost compact HUD using waveguide optics to inject the imagery directly into the holographic combiner (see Section 12.11.6).

Honeywell has been progressively fine-tuning the blending of SVS and EVS images on their PFD, decluttering the pitch reference ladder so it does not obscure runway lights and modifying the EVS 'box' to be translucent over the SVS image and in the same colour as the SVS background (see Figure 12.28). Honeywell claims that pilots who are highly experienced with HUDs perform equally well with head-down technology. SVS on a head-down PFD is not conformal. It is relatively small compared with the actual scene it represents: a compression ratio of more than 2:1 is typical.

PFD Symbology
(HUD style)

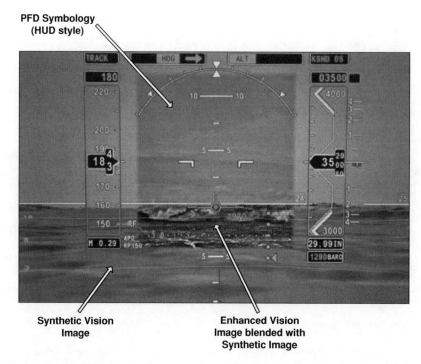

Synthetic Vision
Image

Enhanced Vision
Image blended with
Synthetic Image

Figure 12.28 Honeywell primary flight display with SVS and EVS overlay. Reproduced with permission from Honeywell Aerospace. For a colour version of this figure, please refer to the colour plates

12.9   Display System Architectures

12.9.1   Airworthiness Regulations

Airworthiness regulations in respect of flight deck display systems for civil transport aircraft are to be found in JAR/CS part 25, in the following paragraphs:

- 25.1303 Flight and navigation instruments.
- 25.1309 Equipment, systems and installations.
- 25.1323 Airspeed indicating systems.
- 25.1333 Instrument systems.
- 25.1334 Flight director systems.

FAA Technical Service Order TSOc113 provides minimum standards for Airborne Multipurpose Electronic Displays [29].

12.9.2   Display Availability and Integrity

FAA Advisory Circular AC 25-11 provides specific display availability and data integrity guidelines, summarised in Table 12.2 [30]. See also references [27] & [28].

Table 12.2 Display data integrity requirements

Parameter	Criticality	Probability of:	
		Complete loss (availability)	Misleading data (integrity)
Attitude	Critical	All = extremely improbable Primary = improbable	Extremely improbable
Airspeed	Critical	All = extremely improbable Primary = improbable	Extremely improbable
Barometric altitude	Critical	All = extremely improbable Primary = improbable	Extremely improbable
Vertical speed	Essential	Improbable	
Rate-of-turn	Non-essential		
Slip/skid	Essential	Improbable	Improbable
Heading	Essential	Extremely improbable	Improbable
Navigation	Essential	Improbable	Improbable
Powerplant	Critical	Extremely improbable	Extremely improbable
Crew alerting	Essential	Improbable	Improbable
Check lists			Improbable
Weather radar	Non-essential		Improbable

1. All loss: loss of parameter on primary displays and standby instruments.
2. Primary loss: loss of parameter on primary displays (excluding the standby instruments).
3. Misleading data: potentially hazardously misleading data on any primary display.

12.9.3 Display System Functional Elements

The constituent elements to be found in a display system are shown in Figure 12.29 and comprise:

- **Data collector/concentrator:**
  - Acquires the data to be displayed from the other onboard systems.
  - Selects the most appropriate data sources.
  - Performs data integrity checks.
- **Display management processor:**
  - Determines the display mode, submode and elements to be displayed.
  - Translates the above information into graphics data and commands.
- **Symbol/graphics processor constructs the symbology and graphics comprising:**
  - Alpha-numeric characters in a range of font styles and sizes.
  - Special symbols, pointers and icons.
  - Lines of various widths and styles.
  - Circles, ellipses and arcs of various widths and styles.
  - Area shade infill.

(a) Dumb Display Architecture

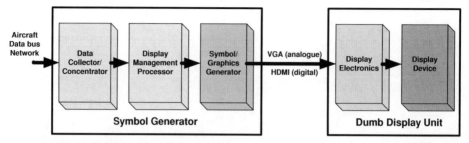

(b) Semi-Smart Display Architecture

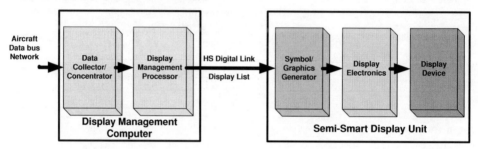

(c) Integrated Display Architecture

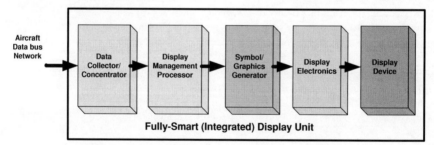

Figure 12.29   Basic display system functional elements

- **Display unit comprises:**
  - The display device itself.
  - Display device electronic support circuits.

These components can be and have been combined in a variety of ways to construct a display suite.

12.9.4   Dumb Display Architecture

The dumb display architecture is shown in Figure 12.29a. The display unit contains only the necessary electronics to support and drive the display device itself. This is the equivalent to

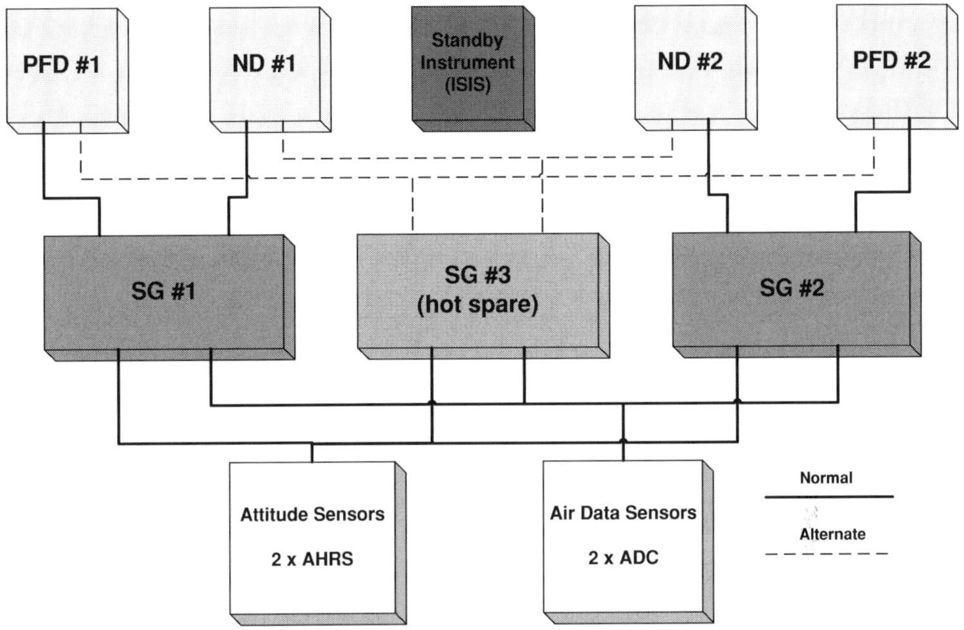

Figure 12.30 Typical early EFIS architecture

the VDU or monitor in a desktop computer installation. In the early CRT-based EFIS systems, the interface between the symbol generator (SG) and the display unit (DU) was a complex set of X, Y and Z analogue signals to describe a hybrid stroke plus raster image on the CRT faceplate. Today the interface to a digital display device such as an AMLCD will be a dedicated high-bandwidth video link encoded with uncompressed red, green and blue (R:G:B) pixel data using such standards as VGA (analogue) or a digital high-definition multimedia interface (HDMI). HDMI can support high-definition images of 1900×1200 pixels at data rates up to 10 Gbps at distances up to 30 m. A fibre-optic version is available.

Early electronic flight instrument systems (EFIS) providing primary flight (PFD) and navigation (ND) display functions of the type installed on the Boeing 757 and 767 were designed as distributed digital systems with dumb displays. Typically they had three symbol generators operating with four display units as shown in Figure 12.30. Each SG is able to simultaneously produce both the PFD and ND formats. Each display unit sources its display format from a normal or an alternate SG source. Each SG sources its inputs from both left-hand and right-hand sensors.

In normal operation the third symbol generator is a 'hot spare'. In the event that one symbol generator fails, then the affected display units select their inputs from the 'hot spare'. The formats presented on PFD1 and ND1 remain independent of those on the other side of the flight deck. If one DU fails, then the driving SG will reconfigure its output so that the image on the adjacent DU will be a composite PFD/ND display with all requisite primary flight data.

Independent and segregated standby attitude, altimeter and airspeed instruments are installed in the flight deck to aid in resolving discrepancies and to provide critical flight information in the event of complete failure of the primary avionic systems.

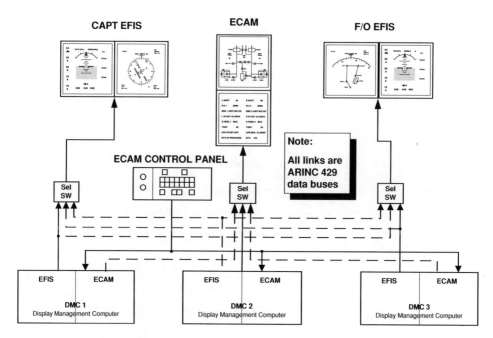

Figure 12.31 Airbus A330/340 EFIS and ECAM architecture

12.9.5 Semi-Smart Display Architecture

The semi-smart display architecture shown in Figure 12.29b overcomes the need for a specialised dedicated high bandwidth analogue/digital interface to the display unit. The graphics processor is placed within the display unit. The display management processor assembles and compiles the image into a 'display list' of instructions in a universal graphics language such as OpenGL to be executed by the graphics processor in the display unit. Bandwidth requirements are of the order of 100–300 Mbps and can be accommodated on an aerospace-type digital data bus.

The Airbus A330/340 EFIS/ECAM architecture shown in Figure 12.31 employs semi-smart display units (DUs). The avionics architecture is a traditional 'federated' system, that is, separate line-replaceable units (LRUs) are provided for each major identifiable function (e.g. air data, inertial navigation, flight control, radios). In the Airbus A330/340 these systems communicate with the displays system via dedicated multi-drop ARINC 429 serial data buses. Three DMCs drive the six DUs. Each DMC is able to source four display formats: the PFD, ND and the two ECAM displays. Each DMC to the DU data link is a customised dedicated 100 kbps ARINC 429 data bus. A three-way switch selects the DMC source to the each DU.

12.9.6 Fully Smart (Integrated) Display Architecture

In the fully integrated smart display architecture shown in Figure 12.29c, all the display elements including the application software are contained within the display unit itself. Data bandwidth requirements are much reduced.

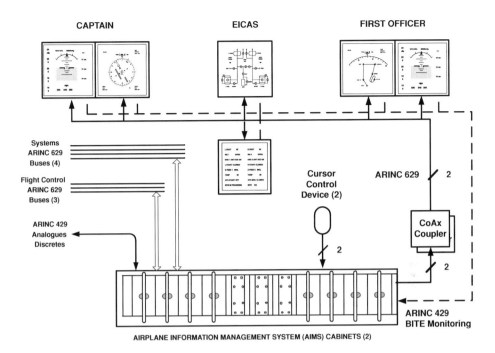

Figure 12.32   Boeing 777 AIMS display architecture

The architecture of the Boeing 777 aircraft information management system (AIMS) as shown in Figure 12.32 illustrates an integrated modular avionics (IMA) architecture with smart display units.

The six smart integrated display units contain all the electronics to manage, process and generate display formats from data transferred to them from the AIMS cabinet via 2 Mbps ARINC 629 data bus networks. The data bus is connected to all DUs; two data buses provide redundancy. The AIMS cabinet provides a centralised and redundant computing resource on which the avionic functions of the classic federated architecture reside as software applications. A comprehensive software operating system assures a safe and secure partitioned operating environment. The AIMS cabinet manages failure of internal functions by reconfiguring applications to run on the remaining functional computing resources. DU failure is managed as described earlier, by presenting combined formats on remaining DUs.

12.10   Display Usability

12.10.1   Regulatory Requirements

Advisory material and guidelines concerning display usability and legibility can be found in:

- FAA AC 25-11   Advisory Circular, transport category airplane electronic display systems [30].
- SAE AS 8034   Minimum performance standards for Electronic Displays [31].

- SAE ARP 1068   Flight Deck Instrumentation and Display Design Objectives for Transport Aircraft, Society of Automotive Engineers Inc. [32].
- SAE ARP 1874   Design objectives for CRT Displays for Part 25 (Transport) Aircraft, Society of Automotive Engineers Inc. [33].
- SAE ARP 4256   Design Objectives for Liquid Crystal Displays for Part 25 (Transport) Aircraft, Society of Automotive Engineers Inc. [34].

12.10.2   Display Format and Symbology Guidelines

Advisory Circular AC 25-11 provides extensive recommendations on display formats and symbology. The key recommendations are summarised here (see also chapter 12 in [1]).

- The display should convey information in a simple uncluttered manner.
- Display elements should be natural, intuitive and not dependent on training or adaptation for correct interpretation.
- Symbols or messages should be logically and consistently positioned and collocated with associated information. Warning flags should be generated when immediate recognition and corrective or compensatory action is required by the crew; the warning colour is red; cautions in amber, etc.
- Careful attention should be given to symbol priority to ensure ease of interpreting three-dimensional information on a two-dimensional medium.
- Information content should be identified by at least two distinctive coding parameters (size, shape, colour, location); colour alone should not be used as a discriminator.
- An accurate, easy, quick-glance interpretation of aircraft attitude shall be possible for all expected but unusual attitude situations and command guidance display configurations.
- Attitude, airspeed, altitude and heading must reside in the 'Basic T' arrangement. Heading and attitude must be presented on the same display. Airspeed and altitude should be arranged so that the present value is located as close as possible to a horizontal line extending from the centre of the attitude indicator.
- Flicker can cause mild fatigue and should not be perceptible day or night either foveally or in peripheral vision. Refresh rate is a major determinant of flicker; to avoid flicker, image refresh rates should be in excess of 70 Hz.
- Information update rates for analogue signals used in direct airplane control tasks should be equal to or better than 15 Hz.
- Any lag between input signal and display should be consistent with the airplane control task associated with that parameter. In particular, display system lag for attitude should not exceed 100 ms.
- For those elements of the display that are normally in motion, any jitter, jerkiness or ratcheting should be neither distracting nor objectionable. Conversely, smearing of moving symbols should be neither discernible nor objectionable.

12.10.3   Flight Deck Geometry

Figure 12.33 shows the viewing geometry of a typical flight deck. The instrument panel is at a comfortable distance for the pilot to reach all the controls, which gives a normal display

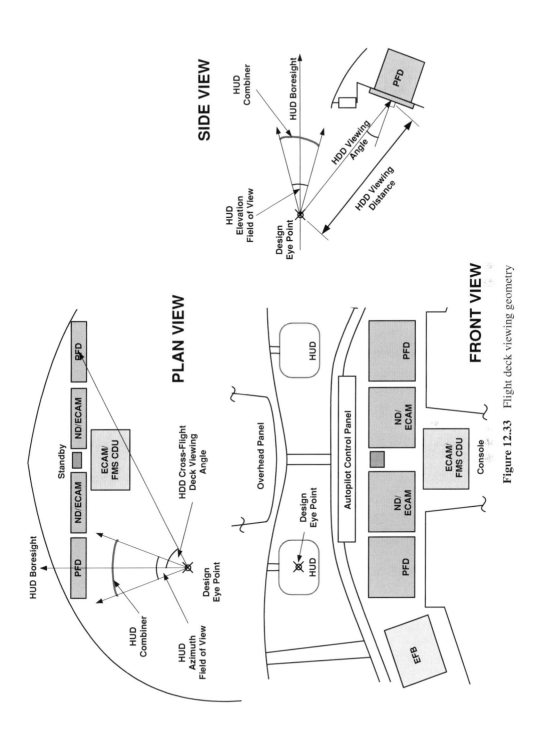

Figure 12.33 Flight deck viewing geometry

Table 12.3 Typical avionics display sizes and resolution

		Resolution (pixels)	
Typical display size	Format	Horizontal	Vertical
3 ATI standby instrument (3 inch square)	VGA (square)	480	480
5 ATI ADI/HSI (5 inch square)	SVGA (square)	600	600
6.25 × 6.25 inch square PFD/ND	XGA (square)	768	768
10 × 8 inch (portrait) PFD/ND	SXGA	1280	1024
15.1 inch diagonal PFD/ND/MFD	HDTV	1980	1080

viewing distance of around 600 mm (24 inches). The pilot over-the-nose vision line is usually about 15 degrees. The displays are normally tilted towards the pilot, but are generally viewed from above the display normal. The primary flight display is normally positioned so that the attitude ball is directly in front of the pilot. The HUD is harmonised to aircraft boresight to ensure the HUD image is conformable to the real world. The pilot will need to adjust his seat position vertically and fore–aft to ensure he is comfortably positioned within the HUD eyebox.

12.10.4   Legibility: Resolution, Symbol Line Width and Sizing

Under ideal conditions a normal human eye with 20/20 vision can discriminate two lines or two pixels separated by about one minute of arc (1/60th of a degree). Thus for a digital display media such as an AMLCD viewed at a distance of 600 mm, the ideal (or retinal) display resolution would be around 5.7 pixels per mm (140 pixels per inch) with a degree of anti-aliasing to blur the edges of lines and reduce any potential for ratcheting or 'staircasing' of rotating structures.

Table 12.3 indicates some common COTS display resolutions, and the avionics display sizes each will support at a viewing distance of around 600 mm.

It is generally recognised that symbol line widths should be of the order of 1 mr (3.5 arc minutes). A display object should subtend at least 12 minutes of arc and preferably 20 minutes of arc for any symbol that needs to be quickly assimilated. Symbol line width should be commensurate with font size.

12.10.5   Colour

The human eye contains three classes of photoreceptors (cones) that differ in the photo-pigments they contain. The perception of colour depends upon the excitation ratios of the cones, the tri-stimulus values. Thus from three colours suitably mixed it is possible to perceive all colours. A colour display such as an AMLCD has three primary colours, red, green and blue, with a spectral response shown in Figure 12.34. From a suitable mix of the ratios of these primary colours, the eye will perceive all colours within the inscribed triangle of the chromaticity chart.

FAA AC 25-11 advises adoption of the colour-coding scheme shown in Table 12.4 to facilitate information discrimination. The two option sets arise from the evolutionary nature

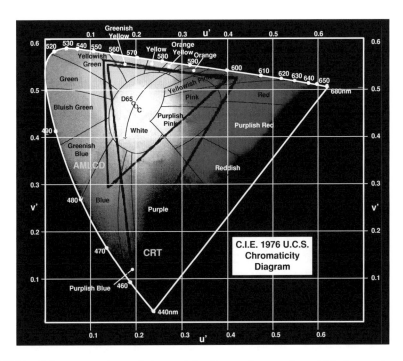

Figure 12.34 Colour perception and chromaticity. For a colour version of this figure, please refer to the colour plates

Table 12.4 Standardised colour-coding

Information	Colour-coding	
	Option set 1	Option set 2
Warnings	Red	
Flight envelope and system limits	Red	
Cautions, abnormal sources	Amber	
Earth	Tan/brown	
Sky	Cyan/blue	
Scales and associated figures	White	
Flight director	Magenta/green	
ILS deviation pointer	Magenta	
Autopilot engaged modes	Green	
Fixed reference symbols	White	Yellow*
Current data, values	White	Green
Autopilot armed modes	Cyan	White
Selected data, values	Green	Cyan
Selected heading	Magenta**	Cyan
Active route plan	Magenta	White

* Use of yellow other than for cautions is discouraged.
** Magenta used for 'fly-to' or 'keep-controlled' parameters.

of the development of display formats, and simply recognise different origins. Option set 1 is preferred for new designs.

12.10.6   Ambient Lighting Conditions

The ability of the crew to detect information presented on a flight deck display will depend upon the visual difference or contrast between the foreground image and the luminous background. FAA Advisory Circular AC 25-11 identifies four lighting conditions that should be considered in the display design:

- Direct sunlight falling on the display through a side window.
- Sunlight through a front window illuminating a shirtfront which is reflected in the display.
- Sun above the forward horizon and above a cloud deck reflecting into the pilots' eyes (known as veiling glare).
- Night or dark environment in which the display brightness should be dimmable such that outside vision is not impaired.

Head-Down Display: High Ambient – Sun Rear

This scenario is shown in Figure 12.35. The display is bathed in direct sunlight through a side window. The image contrast is degraded (washed out) primarily by diffuse (unfocused) reflectance from the display surface, grease/fingerprints or dust on the display surface or any

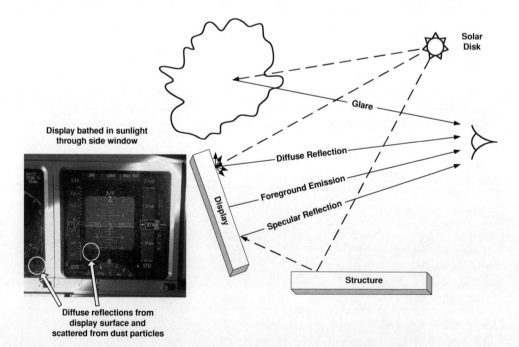

Figure 12.35   Head-down display: high ambient – sun rear

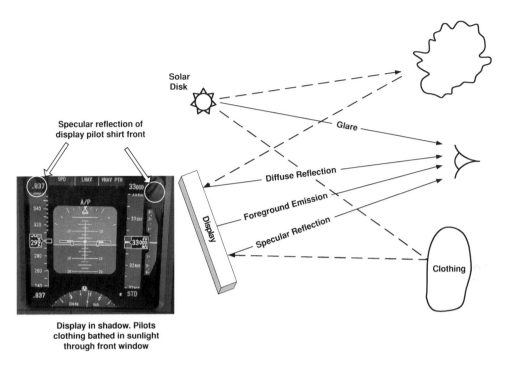

Figure 12.36 Head-down display: high ambient – sun forward

structures within the display device itself (e.g. the matt black matrix and colour filters of an AMLCD).

Head-Down Display: High Ambient – Sun Forward

This scenario is shown in Figure 12.36. The display is in shadow and illuminated by diffused skylight. The principal affect of the ambient illumination is to produce 'solar veiling glare', which degrades the pilots' perceptual capability to read the image against the bright outside world scene, plus an element of specular (focused) reflections from the display surface of the pilot's clothing, flight deck structure and lighting (at night).

Head-Up Display: High Ambient – Sun Forward

Practical flight experience indicates that the worst-case ambient illumination condition for the HUD is when the display symbology is presented against a cloud face, with a clear high-luminance solar disk producing additional glare just outside the display field of view. This scenario is shown in Figure 12.37.

FAA Advisory Circular AC 25-11 advises that:

- maximum display luminance should not be less than 75 fL (257 cd/m^2);
- minimum display contrast ratio should be greater than 3.0:1;
- display dark ambient contrast ratio should be better than 20:1;
- the display should be dimmable to 0.1 fL (0.343 cd/m^2).

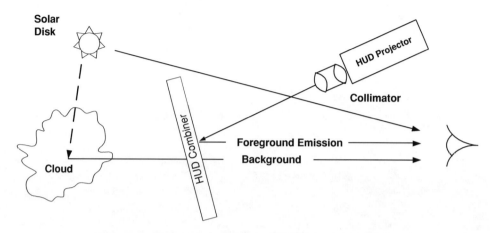

Figure 12.37   Head-up display: high ambient – sun forward

These requirements should be regarded as minimum performance standards, as their definition was somewhat constrained by the technology available at the time the document was prepared. SAE ARP 4256 provides some additional formulations specifically for AMLCD displays which include a measure of chrominance contrast.

Photometric Definitions

- **Visible spectrum**: The human eye is sensitive to all wavelengths in the electromagnetic spectrum from 380 (blue) to 760 nm (red).
- **Illumination** (or illuminance) is the density of radiant energy incident upon a receiving surface evaluated throughout the visible spectrum (the human eye response). The SI unit of illumination is the lux = 1 lumen/m^2.
- **Luminance** (or photometric brightness) is the luminous flux radiated from a surface. The SI unit of measurement is the candela per square meter (cd/m^2), also known as the nit.
- **Contrast ratio** is the ratio of the foreground luminance (the desired image content) against the background luminance (the undesired image content).

Further explanation of the optical terminology can be found in Chapter 12 of Reference [1].

12.11   Display Technologies

This section discusses current and emerging flight deck display technologies. It includes:

- direct view display technologies:
  - active matrix liquid crystal displays (AMLCD)
  - plasma panels
  - organic light-emitting diodes (O-LEDs)
  - electronic paper (e-ink)

- indirect view microprojection technologies:
  - transmissive LCD
  - reflective LCD
  - digital micro-mirror (DMD), also known as digital light projector (DLP)
- head-up display technologies:
  - pupil-forming holographic optics
  - optical waveguide optics.

12.11.1   Active Matrix Liquid Crystal Displays (AMLCD)

Liquid crystal displays (LCDs) do not emit light; instead they can control the transmission of light passing through them and should therefore be described as 'light modulators'. Because LCDs do not generate light, they must be back-lit.

Liquid crystals are a group of materials made up of organic molecules which are significantly long in comparison with their width. The presence of small electric dipole fields associated with individual molecules gives each a tendency to align itself with its immediate neighbour. In the liquid form the molecules are free to move, enabling them to be aligned by external influences. Because of their inherent self-alignment, the substances exhibit a phenomenon known as 'optical birefringence' in which the molecules interfere with light waves in a highly directional manner, thus causing the optical properties of the fluid to vary with direction. It is this feature of liquid crystal materials that is of use in display applications.

The most common form of LCD uses twisted nematic material. A cell is formed in which the long, thin, liquid crystal molecules are sandwiched between two glass plates. Microscopic grooves are formed on the inner surfaces of the plates, inducing the adjacent liquid crystal molecules to become aligned with the grooves. However, the grooves in one plate are aligned at 90° to those in the other, and this constrains the molecules to adopt a spiral orientation as shown in Figure 12.38. A polariser is attached to each glass plate such that its plane of polarisation is aligned with the grooves in the glass. Thus the planes of polarisation of the two polarisers are crossed; however, the spiral molecular orientation of the liquid crystal material causes the plane of polarisation of the light from one polariser to be rotated by 90° as it passes through the liquid, thus aligning it with the second polariser and allowing the light to be transmitted through the cell.

When an electric potential is applied between transparent electrodes on the inner glass surfaces, the resulting electric field across the liquid crystal material causes the molecules to become aligned along the electric field in a direction normal to the glass surfaces. This disrupts the spiral molecular structure of the fluid, inhibiting the rotation of polarised light as it passes through the material. The light is now blocked by the second polariser and is therefore prevented from passing through the liquid crystal cell. The cell is a light valve; light is transmitted through the cell in the de-energised condition but blocked when the cell is energised.

For the display of video scenes containing dynamic, complex pictorial information, nematic liquid crystal displays have been developed containing more than a million discrete pixels arranged as a matrix of rows and columns. The matrix is addressed via row and column selection circuits and the entire display is scanned in a sequential manner in which all the elements in one row are updated simultaneously, each row being addressed in turn. When a cell's row is energised, a transistor switch associated with each cell is turned on, thus applying

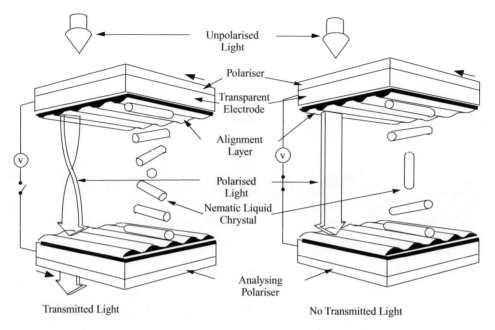

Figure 12.38   Operation of twisted nematic LCD

an electric field determined by the column voltage (a function of the scene luminance at that point) to the cell's electrodes and charging up the cell capacitance. During the period (most of the time) when other rows are being addressed, the switch is turned off, preventing the charge in the cell from leaking away. This arrangement is shown in Figure 12.39. Because each element of the array is operated by an active semiconductor switch, this type of display is referred to as an active matrix liquid crystal display, or AMLCD.

To achieve colour, a pattern of red, green and blue dyes is printed onto the surface of one of the glass sheets, such that each small area of dye coincides with an AMLCD cell. Thus each cell passes only red, green or blue light. The dye pattern can be arranged in various ways, of which the most common is the vertical stripe pattern shown in the figure.

The light transmissivity of a typical colour AMLCD is low, in the region of 3–6%. For the display to be viewable in direct sunlight a powerful backlight is required. In most current arrangements this is provided by one or more fluorescent tubes mounted together with a diffuser immediately behind the AMLCD panel. For maximum light transmission, the dyes used to colour the LCD cells are roughly spectrally matched to the red, green and blue emissions of the white fluorescent tube phosphor. The lamp must be dimmable to track the varying flight deck ambient light conditions, and this is achieved by pulse-width modulating the fluorescent tube drive waveform at a frequency above which the human eye perceives flicker (nominally 70 Hz).

Recent advances in LED technology make it possible to use very bright LEDs as an alternative backlight technology. The LED backlight comprises a dense array of red, green and blue LEDs. Matched to the colour filter transmission spectra, this tri-colour backlight can

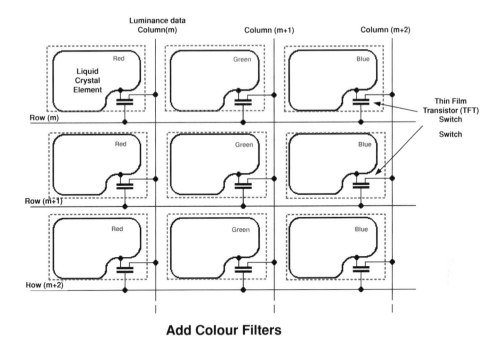

Add Colour Filters

Figure 12.39   Active matrix LCD arrangement

achieve better saturated primary colours and hence a wider range of colours (the colour gamut) can be produced.

12.11.2   Plasma Panels

Plasma displays use an electrical discharge in gas to produce light. As in other forms of flat panel, the displays are organised as rows and columns; the discharge is created by an electric field applied between the row and column electrodes. Pixels are addressed on a line-by-line basis, with the information in each individual video line being applied to all display columns whilst the appropriate row electrode is energised with a 'priming' voltage sufficient to strike a discharge in the addressed pixels.

However, although initially popular for large-size domestic TV, plasma panels have been overtaken by ever-increasing size, higher resolution and better quality AMLCD panels. Limited brightness and contrast have inhibited the use of plasma panels in airborne flight deck applications.

12.11.3   Organic Light-Emitting Diodes (O-LED)

The operation of an LED is based upon the fact that semiconductors can be of two types, p-type or n-type, depending upon whether dopants pull electrons out of the crystal, forming 'holes', or add electrons. An LED is formed when p-type and n-type materials are joined. When a voltage

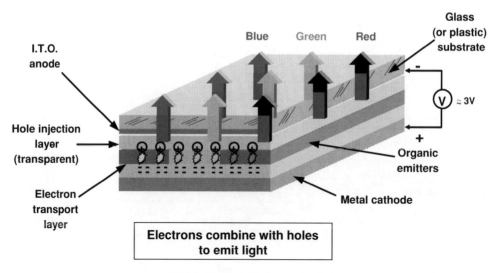

Figure 12.40   O-LED operation

is applied, electrons flow into the p-type material, and holes flow into the n-type material. An electron–hole combination is unstable and they recombine to release energy in the form of light. This can be a very efficient way to convert electricity to light.

Light-emitting diodes, based upon semiconductors such as gallium arsenide, have been around since the late 1950s. These crystalline LEDs are expensive, and it is difficult to integrate them into small high-resolution displays. However, there is a wide class of organic compounds that have many of the characteristics of semiconductors. They have energy gaps of about the same magnitude, and they can be doped to conduct by electrons (n-type) or holes (p-type). Electrons and holes recombine at the interface of the n-type and p-type materials and light is emitted – see Figure 12.40.

Initially, O-LED materials were used to replace inorganic LEDs in copiers. While the early diodes did not have sufficient efficiency or life to be commercially attractive, progress in improving these factors has been astounding. High-resolution full-colour displays have been fabricated in prototype quantities for domestic TV applications, although they are not yet available for general sale. Display brightness, life and robustness in the airborne environment have yet to be evaluated.

12.11.4   Electronic Paper (e-paper)

E-paper is a name that applies to a family of technologies. The intention is that ink-on-paper applications such as books and newspapers can be replaced by a 'similar', flexible material that permits the information contained to be re-programmed. E-paper displays make use of ambient illumination, so no backlight is required.

E Ink™ is the technology used in today's e-readers (the Kindle, etc.). Electrophoretic, or e-ink is made up of millions of tiny micro-capsules, about the diameter of a human hair. Each micro-capsule contains positively charged white particles and negatively charged black particles suspended in a clear fluid. When a positive or negative electric field is applied, corresponding particles move to the top of the micro-capsule where they become visible to the user. This makes the surface appear white or black at that spot. The disadvantages for aerospace use are the requirement to erase the entire page before writing a new image, poor dynamic performance, and limited grey-scale rendition.

An alternative electro-wetting technology has possibly greater potential for use in aerospace applications, for example for the electronic flight bag. An electro-wetting surface normally repels water. However, the surface attracts water when a voltage is applied. Each pixel comprises an oil droplet suspended in water. The cell is positioned between two electrodes and backed by a white substrate. In the unenergised state the electro-wetting surface repels water and the oil droplet will spread over the surface to act as a barrier. White light incident on the surface will be absorbed by the oil droplet and the cell will appear dark. In the energised state the electro-wetting surface attracts water and the oil droplet is forced out. The cell will now reflect the white light incident upon it. Subtractive colour is possible using different colour oil droplets.

12.11.5   Micro-Projection Display Technologies

Micro-miniature LCDs and digital micro-mirror devices (DMDs) have recently revolutionised digital projector technology. Figure 12.41 shows a conceptual implementation of a wide integrated instrument panel implemented using three rear projectors. Whilst superficially this looks very attractive, considerable issues would have to be overcome concerning viewability under high ambient sun-rear illumination conditions.

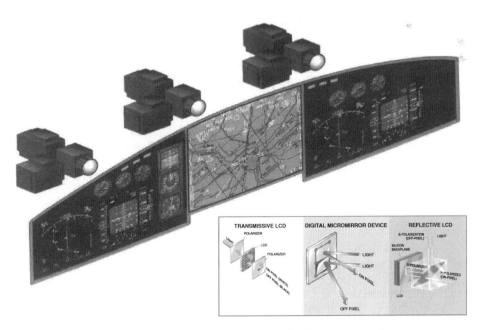

Figure 12.41   Wide-screen rear-projection instrument panel

There are currently three micro-display technologies employed in micro-projectors:

- **Transmissive LCDs**: These have been the dominant display device for conference room projection applications. They are a mature product, but suffer from resolution limitations in the sizes relevant for airborne display applications.
- **Reflective LCDs**: Also known as liquid crystal on silicon (LCoS) displays, they are rapidly emerging as the technology of choice for medium-end applications. These devices utilise a backplane of conventional crystalline silicon technology that is aluminised to increase the reflectivity. An LCD material is applied and sealed with a cover glass.
- **The digital micro-mirror device (DMD)**: This is the primary competitor to the LCD for the image modulating component of a projector, appearing more widely in medium- to high-end applications (e.g. digital cinemas). The digital micro-mirror device (or digital light projector, DLP) is a matrix array of tiny mirrors built into a CMOS static RAM chip. The mirrors can be matrix-addressed and rotated through a small angle using small electric fields internal to the device in such a way as to spatially modulate the mirror array so that it generates a picture when illuminated by an external light source. The DMD is essentially monochrome, though the mirror switching speeds are sufficiently high that a red–green–blue field-sequential colour display can be constructed by introducing a rotating colour wheel into the light path.

12.11.6 Head-Up Display Technologies

The power of the head-up display (HUD) comes from its collimated image, that is, an image focused at infinity. Symbology written onto the imaging device (CRT or LCD) is collimated by an optical system and then projected into the pilot's sightline by a partially reflective combiner so that the symbology appears to the pilot to be superimposed on the outside world. Being collimated, the projected image remains in perfect registration with the outside world irrespective of the pilot's head motion (provided his eyes remain within the HUD projection porthole). In civil transport aircraft the HUD is installed overhead and the combiner attached to the roof. The combiner can be folded (stowed) when not required.

Pupil-forming HUD

Traditionally HUDs applied to civil transport aircraft have used a pupil-forming optical arrangement (Figure 12.42). The optics of this type of HUD are classed as a 'compound' system. The optics may be thought of as being similar to a telescope in which light from the imaging source is brought to a focus at a plane in the space between the objective lens and the eyepiece. The real image formed at this plane is subsequently magnified by the eyepiece before being transmitted to the eye. In the case of the HUD, the refractive lens elements close to the imaging source are similar to the telescope objective lens; because these elements act as a relay between the CRT and a real image, they are collectively known as the 'relay lens assembly'. The combiner element takes the place of the telescope eyepiece, magnifying and collimating the real image.

Usually it is necessary to obtain the optical power of the combiner by holographic means. The hologram is tuned to the narrow-band spectral emission of the image source which it reflects and magnifies with high efficiency, while transmitting the wide spectrum of the real scene with low loss.

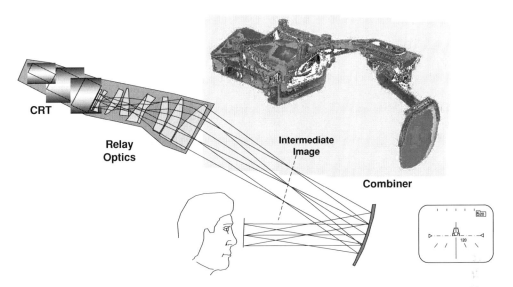

Figure 12.42 Pupil-forming HUD optical arrangement. Reproduced with permission of BAE Systems Electronic Systems in Rochester, Kent

Optical Waveguide HUD

The traditional pupil-forming HUD optics arrangement described above is complex, comprising many components, and is large, expensive and heavy. BAE Systems and Rockwell Collins are independently developing a revolutionary new optical technology that dramatically reduces size and mass by moving light without the need for complex conventional lenses.

The BAE Systems Q-HUD uses a specially designed hologram with a carefully tailored set of optical properties embedded within the combiner substrate (Figure 12.43). The image is formed conventionally by a reflective LCD illuminated with a high-brightness LED and collimated with a miniature lens. The pupil of image-bearing light is constrained to follow a path through the substrate using the optical principle of total internal reflection (TIR). A hologram introduced into the glass plate diffractively interacts with the pupil of light each time it is incident upon the hologram, allowing some energy to escape in a carefully controlled manner, and reforming the image that was injected into the substrate. The remainder of the incident energy continues wave-guiding within the plate. A similar waveguide structure is arranged to extend the pupil in the orthogonal axis to make a very large exit pupil display optic [35].

12.11.7 Inceptors

Cursor Control

Cursor control has been used for some time. A roller ball for each crew member mounted on the centre console is the more usual implementation, although a joystick may also be used.

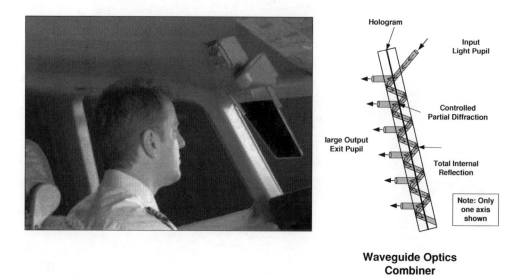

Figure 12.43 BAE Systems waveguide optics HUD. Reproduced by permission of Alex Cameron, Product Manager for BAE Systems Electronic Systems, Rochester, Kent

Touch Screen

Touch screen technology is now the norm for tablets and mobile phones. It is easy to use and intuitive. The next generation of pilots will have grown up expecting this technology. The most obvious applications for touch screen are associated with navigation and route planning: reviewing the flight plan and familiarisation of the departure and arrival procedures and aircraft performance calculations. Other applications include interrogating system synoptics and check lists, and navigating the electronic flight bag. Jeppesen charts are available as an iPad app.

It is not just the touch screen technology that makes it a success, but the whole human–machine software/graphics interface that goes with it. Technologies are capacitive, resistive and surface acoustic wave. However, there are some issues that will need to be carefully addressed to make it a success on the flight deck. Touch activation leaves fingerprints on the screen which will degrade viewability in high ambient illuminations. Careless frequent cleaning has the potential to 'buff' the surface, degrading diffuse reflectance.

12.12 Flight Control Inceptors

The modern flight deck incorporates a high degree of automation designed to reduce pilot workload. Whereas pilots originally flew aircraft manually throughout the flight, today they may perform only the take-off manually, commanding the rest of the flight, including the landing, through the autopilot and flight management systems. However, there are times when pilots will assume manual control and then it is essential that the pilot is able to precisely control the aircraft.

12.12.1   Handling Qualities

To ensure good handling qualities it is necessary that the aircraft's responses to the pilot's inputs are appropriate and predictable, and that all the information that he receives from the aircraft is clear and non-contradictory. When controlling the aircraft manually the pilot makes inputs through the primary flight control inceptor (wheel column, centre-stick or side-stick) and observes the aircraft response, primarily through visual cues. An aircraft that possesses good handling qualities exhibits desired, stable and predictable responses to pilot inputs.

12.12.2   Response Types

For most tasks, the pilot actually wants to control the flight path. For a conventional aircraft with a fixed mechanical linkage between stick and elevator, the pilot does not have direct command of flight path, nor can he see the flight path directly. When he pulls back on the stick he commands the elevator to deflect up which reduces lift on the tail and causes the aircraft to rotate around its centre of gravity until it settles with the wing meeting the oncoming air at a greater angle of attack. With this new angle of attack the wing will produce more lift, which results in an increase in normal acceleration and a concomitant increase in flight path. Hence, to control flight path, the pilot commands changes in angle of attack, and observes the response through pitch attitude.

Historically the aircraft responses were determined by the aircraft's geometric and aerodynamic design. However, with the introduction of electronic flight control systems it is possible to 'tailor' the response of an aircraft to pilot inputs. A number of alternative control strategies have been developed that allow the pilot directly to command a specific aircraft response. Examples of response types include angle of attack (a conventional aircraft), pitch rate, pitch attitude and normal acceleration.

In most applications, different response types will be implemented in different parts of the flight envelope, with blending between the response types depending upon flight condition. For example, response types that exhibit speed stability are better suited to the low-speed regime and the landing flare, since they provide natural tactile cues of an impending stall. Similarly, a pitch attitude response type may be suited to tasks that require accurate control to a specific pitch attitude. A pitch rate response type may be more appropriate when manoeuvring to greatly different pitch attitudes. Figure 12.44 contains time history responses of these response types to a boxcar input (an instantaneous aft movement of the stick, held constant for a few seconds and then instantaneously returned to its original position).

- **Conventional**: The response is a rapid increase to the new steady-state angle of attack, followed by a return to the original trimmed angle of attack when the input is removed.
- **Pitch rate**: A constant input produces a constant pitch rate response. The aircraft will maintain the pitch attitude at release.
- **Pitch attitude**: A constant input results in a new pitch attitude which will be held indefinitely, provided sufficient thrust is available. The aircraft will return to its original pitch attitude at release.
- **Normal acceleration**: A constant control inceptor input produces a constant normal acceleration, also referred to as load factor. The aircraft will maintain the current flight path when the inceptor is released.

Response Type

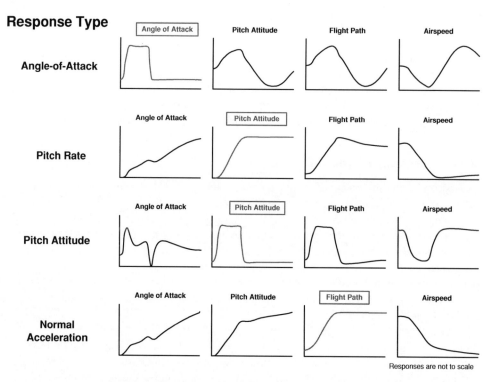

Figure 12.44   Flight control inceptor response types

12.12.3   Envelope Protection

While the characteristics for the angle of attack response type are conventional, clearly they are not so for the other response types. The natural cues associated with a conventional aircraft may be absent in an aircraft that utilises alternative response types when it is near the edge of the envelope. Thus, envelope protection or limiting may be essential for alternative response types, and onset of protection must be clearly communicated to the pilot.

12.12.4   Inceptors

It is essential that the flight deck inceptors and displays are designed to be compatible with all response types that are implemented throughout the aircraft envelope. Other important considerations in a multiple pilot flight deck include the means to transfer control between crew members, cross-feed between multiple inceptors and feed back to the inceptors of the autoflight system commands.

- **Wheel column:** The wheel column is suited to all response types and permits precise control around the datum trim position, while also providing larger displacements for larger control demands. It allows adjustment of the trim position datum for different flight conditions and configurations. It also provides the greatest peripheral visual cues of all the inceptors, but it takes up appreciable space and can obscure the displays. Boeing adopt a wheel column inceptor with cross-feed and tactile feedback.

- **Side-stick:** The characteristics of side-sticks are sensitive to their mechanisation. Many side-sticks employ only a limited displacement, and are operated through wrist action only. These are best suited to rate response types that only require occasional inputs from the pilot, and can lead to wrist fatigue if operated for long periods. The side-stick, however, is smaller than the wheel column and due to its location the problems of obscuration of the main flight deck displays are absent. Airbus adopt a side-stick controller with no cross-feed and no tactile feedback [36].

References

[1] Jukes, M. (2004) *Aircraft Display Systems*. John Wiley & Sons, Ltd., Chichester.

[2] Jukes, M. (1999) *Aircraft Cockpit Evolution*, Royal Aeronautical Society, Information Displays Conference, London 1999.

[3] Chorley, R.A. (1976) *Seventy Years of Instruments and Displays*, 3rd Folland Lecture, Royal Aeronautical Society, February 1976.

[4] Coombs, L.F.E. (1990) *The Aircraft Cockpit*. Patrick Stephens Ltd., Sparkford, UK.

[5] Gilchrist, P. (2001) *Boeing Classic 747*. Airlife Publishing Ltd., Shrewsbury.

[6] *Jane's All the World's Aircraft*, Jane's Information Group Ltd.

[7] TSO-C3d, *Technical Standard Order: Turn and Slip Instruments*, Federal Aviation Authority.

[8] TSO-C4c, *Technical Standard Order: Bank and Pitch Instruments*, Federal Aviation Authority.

[9] TSO-C52b, *Technical Standard Order: Flight Director Equipment*, Federal Aviation Authority.

[10] TSO-C6d, *Technical Standard Order: Direction Instrument, Magnetic (Gyro Stabilised Type)*, Federal Aviation Authority.

[11] Carr, G.P. and Montemerlo, M.D. (1984) *Aerospace Crew Station Design*. Elsevier Science Publishers, Amsterdam.

[12] Birtles, P. (2001) *Boeing 757*. Airlife Publishing Ltd.

[13] Bennett, S. (2012) Out of the fog – the Air France 447 accident. *The Aerospace Professional*, RAeS, October 2012.

[14] Gardner, R. (2010) Towards Tomorrow's Cockpit. *Aerospace International,* September 2010.

[15] Wisely, P.L. and Bartlett, C.T. (2000) *Gate to Gate Operation of a Civil Transport Head Up Display*. BAE Systems Avionics Ltd, Rochester, Kent.

[16] Alpin, J. (2001) Precise approach – every time. *Flight Deck International*, September 2001.

[17] AC 20-167 Advisory Circular: *Airworthiness Approval of Enhanced Vision Systems, Synthetic Vision Systems, Combined Vision Systems and Enhanced Flight Vision Systems*, Federal Aviation Authority.

[18] AC 25.1329A Advisory Circular: Transport Category, *Airplane Flight Guidance Systems*, Federal Aviation Authority.

[19] RTCA-DO-315 and subsequent releases currently -315A and -315B: *Minimum Aviation System Performance Standards for Enhanced Vision Systems, Synthetic Vision Systems, Combined Vision Systems and Enhanced Flight Vision Systems*, Radio Technical Commission for Aeronautics.

[20] AC 90-106 Advisory Circular: *Operational Requirements for EFVS, conducted below DA/DH or MDA down to 100ft above TDZE*, Federal Aviation Authority.

[21] AC 23-26 Advisory Circular: *Synthetic Vision and Pathway Depictions on the Primary Flight Display for Part 23 aircraft (small airplanes)*, Federal Aviation Authority.

[22] Carbaugh, D., *et al.* (2003) Vertical situation display for improved flight safety and reduced operational cost, Boeing Commercial Airplanes. *The Aerospace Professional*, February 2003.

[23] Brooks, P.E. (1998) *The Promise of Four-Dimensional Flight Displays*. Flight Deck International '98.

[24] Kramer, L., *et al.* (2011) *Enhanced and Synthetic Vision for Terminal Maneuvering Area Next Gen Operations*. NASA Langley Research Centre.

[25] Warwick, G. (2012) Advancing vision. *Aviation Week and Space Technology*, May 2012.
[26] *Flight International* (2011) Double vision on display. *Flight International*, October.
[27] JAR 25.1309, Joint Airworthiness Regulation: *Equipment Systems and Installation*, CAA (UK).
[28] AMJ 25.1309, Advisory Joint Material: *System Design and Analysis*, Federal Aviation Authority.
[29] TSO-C113, Technical Standard Order: *Airborne Multipurpose Electronic Displays*, Federal Aviation Authority.
[30] AC 25-11, Advisory Circular: Transport Category, *Airplane Electronic Display Systems*, Federal Aviation Authority.
[31] SAE AS 8034, *Minimum Performance Standards for Airborne Multipurpose Electronic Displays*, Society of Automotive Engineers Inc.
[32] SAE ARP 1068, *Flight Deck Instrumentation and Display Design Objectives for Transport Aircraft*, Society of Automotive Engineers Inc.
[33] SAE ARP 1874, *Design Objectives for CRT Displays for Part 25 (Transport) Aircraft*, Society of Automotive Engineers Inc.
[34] SAE ARP 4256, *Design Objectives for Liquid Crystal Displays for Part 25 (Transport) Aircraft*, Society of Automotive Engineers Inc.
[35] Cameron, A. (2009) *The Application of Holographic Optical Waveguide Technology to Q-Sight Family of Helmet Mounted Displays*, presented at SPIE Head and Helmet-Mounted Displays XIV Conference, BAE Systems.
[36] Field, E. (2004) Handling qualities and their implications for flight deck design, in *Human Factors for Civil Flight Deck Design* (ed. D. Harris), Ashgate Publishing Ltd., Farnham, Surrey.

13

Military Aircraft Adaptations

13.1 Introduction

There are two scenarios in which civil avionics are being used in military applications. The first is in the conversion of commercial aircraft to military roles, and the second is the updating of military types to make use of modern avionics where is economically viable to use commercial off-the-shelf (COTS) systems.

There are applications in the military field for which the civil aircraft platform together with its avionic systems is well suited. It may often be economically viable to convert an existing civil type rather than to develop a new military project. Much of the development costs of the structure and basic avionics will have been recouped from airline sales for a new platform. Alternatively, a used aircraft bought from an airline may also be an economic solution. In either case the basic avionics fitted will have been well tried and tested, and will be ideally suited for use in controlled airspace. The basic platform is often known as a 'green' aircraft before conversion.

There have been numerous examples of successful conversions serving the armed forces around the world. The successful Nimrod maritime patrol aircraft was based on the De Havilland Comet airframe and has seen service with the Royal Air Force as the MR1 and MR2 for over 40 years. The type was extensively refurbished by BAE Systems with the designation of MRA4 and trials aircraft flew until the project was cancelled as a result of the 2011 Strategic Defence Review in the UK. All the development and production aircraft were destroyed but the front fuselage of one production aircraft (PA05) is at Cranfield University. The refurbishment included a new wing, new engines and new avionics and mission systems being fitted to the existing MR2 fuselage. The Comet airframe was also the platform for the Nimrod R Mk 1 signals intelligence aircraft in service with the RAF, as well as the ill-fated Airborne Early Warning AEW Mk3. Similarly the Lockheed P-3 Orion maritime patrol aircraft was based on the Lockheed Electra and has seen many years of service with the US Navy and many other operators worldwide. The HS 748 passenger aircraft was successfully converted to the Andover troop and cargo transporter for use by the RAF, whereas the VC10 and Tristar have been used for personnel transport only. Tanker aircraft have been developed from the Boeing 707 (KC-135), the VC10 and the Lockheed L1011 Tristar, and proposals are in place for use

Civil Avionics Systems, Second Edition. Ian Moir, Allan Seabridge and Malcolm Jukes.
© 2013 John Wiley & Sons, Ltd. Published 2013 by John Wiley & Sons, Ltd.

of the Boeing 767 and Airbus A330 in this role. Commercial types are also used for carrying passengers, troops and VIPs, such the 125 Dominie, VC10, Tristar and 707.

There are also instances where civilian agencies have made use of aircraft for surveillance. These include police, customs and excise, fisheries protection and drug enforcement agencies. These applications are often civilian helicopters or small commercial aircraft. Typical applications for re-use of commercial aircraft include:

- personnel, materiel and vehicle transport;
- air-to-air refuelling tanker;
- maritime patrol;
- airborne early warning;
- ground surveillance;
- electronic warfare;
- flying classroom;
- range target/security aircraft.

For these roles, a large, slow-moving platform capable of transit over long distances and loitering for long periods of time, together with internal space for a mission crew and their workstations, is ideal. Many commercial aircraft are capable of conversion to these roles, retaining the basic structure, avionics and flight deck with the installation of additional avionics or mission systems to tailor the aircraft to a specific role. Retaining the basic avionic systems architecture makes good sense, since military aircraft make extensive use of civilian-controlled airspace during peacetime, and will often transit to theatres of operation, for training and defence purposes, using commercial routes.

The conversion to a military role, especially if the carriage of weapons is included, requires a different approach to safety and qualification, challenging the aircraft design teams to make the best use of civil and military certification rules. This often poses interesting problems in the mixing of design standards and processes.

The basic avionic systems are complemented by a set of sensors and systems to perform specific surveillance tasks. This is a situation in which the basic navigation and communication systems become part of the role-specific systems, and in which there are particular issues of accuracy, integration and security. These are especially important in instances where a commercial aircraft platform forms the basis of the military vehicle. In such instances there may be conflicts between the characteristics of the embedded systems on the commercial vehicle, and the requirements of the military vehicle. These issues may affect the approach to design and certification of the resultant aircraft. Some of the issues can be seen in Table 13.1.

13.2   Avionic and Mission System Interface

The basic avionics provide information to the flight deck crew for safe flying, handling and monitoring the behaviour of the aircraft and its systems, as described in the previous chapters in this book. In addition to this, the mission crew require assistance with their task of performing the military task. For this they will have their own sensors and systems that will be supplemented by information from the basic avionics suite. There are a number of sources

Table 13.1   Comparison of commercial and military certification issues

	Commercial	Military
Operating conditions	Worldwide	Worldwide
Design standards	RTCA	MIL-SPEC, Def Stan, Air Reglement
Certification	JAR/FAR 25, EASA ES 25	Military
Data bus	ARINC429/629, AFDX	MIL-STD-1553, IEEE1394
Data protocols	ARINC	Project defined
Availability	High – business-critical	High – mission-critical
Survivability	Systems failures	Battle damage
Battle damage repair	Not applicable	Essential
Software standards	RTCA	MIL-STD, Def Stan
Software language	C, C++	Ada, C++

of information required by the crew to build up a picture of the tactical scenario as illustrated in Figure 13.1.

The data that the crew have to mentally assimilate include:

- **Real-time data from avionics** – typical information such as position, speed, height, attitude and heading are provided from the navigation and air data sensors to aid tactical navigation. Information on the quantity of fuel remaining is essential to allow the mission crew to determine how long they can safely remain on task. The flight deck crew will monitor all these parameters as an independent check that the aircraft is being handled safely.
- **Sensors information** – information from the mission sensors will be prepared by sensor operators and provided to form a single tactical picture from which information can be gathered to conduct the mission.

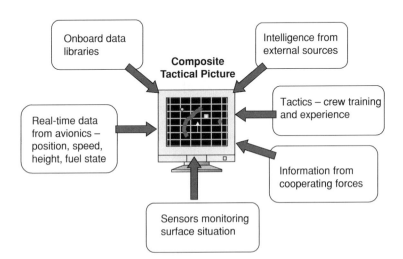

Figure 13.1   Contributors to the tactical picture

- **Onboard data libraries** – data are stored to enable the sensor operators to estimate what type of target they have detected. This information may be historical and may be supplemented by data obtained on each mission. Experienced operators are able to identify a target type, and sometimes even an individual ship or submarine by its combined sensor signature.
- **Tactics** – all operating crews are taught the appropriate tactics to detect and track their targets. These tactics are built up from many years of operational experience and are often highly classified, since their disclosure would enable an enemy to evolve tactics to avoid detection.
- **Information from cooperating forces** – most surveillance aircraft operate in collaboration with other forces and platforms such as fighter aircraft, helicopters, and ground forces operated by allies. All intelligence gained is made available by secure communications to add to the tactical picture.
- **Intelligence from external sources** – the headquarters teams may have access to other information in large databases or gleaned from the emerging scenario, especially if information is being provided by many sources. This is either sent to the aircraft by secure communications for inclusion in their own database, or decisions made at headquarters after analysis of the tactics may be sent, again by secure communications.

The requirement to provide a comprehensive system for capture, analysis and presentation of this information must be clearly understood so that the system architecture is fit for purpose. The architecture will then be designed to ensure that the basic avionics is able to provide the appropriate data with the appropriate accuracy and data rate, and with the required level of integrity. The mission system will be designed to meet its own internal requirements for sensors, processing and data display. The interfaces between the systems must then be designed to respect the need for information by the mission system, and to preserve the integrity of the basic avionics so that aircraft safety is not impaired. Figure 13.2 shows some examples of interfaces that may be required between the avionics and the mission system

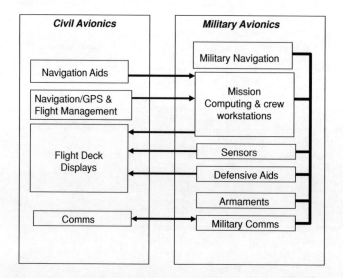

Figure 13.2 Generic interfaces between avionics and mission systems

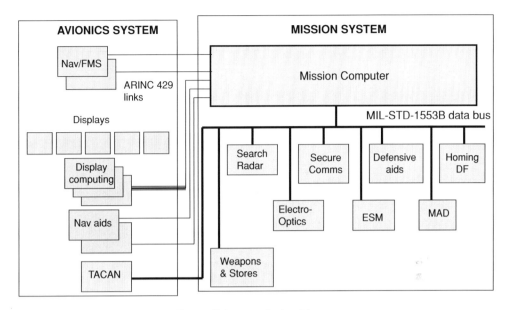

Figure 13.3   A typical architecture

Figure 13.3 shows an example of an architecture that separates the commercial and military data bus structures. One mechanism for doing this is to retain much of the basic avionics architecture, including the ARINC bus structure and data protocols. The mission computer can be used to accept the ARINC interfaces, convert the data formats for its own use, and provide the interface and data formatting and the bus control function for the MIL-STD-1553 data bus. Any military avionics that need to be incorporated into the flight deck can be included in the MIL-STD-1553 architecture by extending the bus or by providing a long stub connection. More information on ARINC and MIL-STD data buses can be found in Chapter 2.

13.2.1   Navigation and Flight Management

Unlike a commercial aircraft, which routinely flies on well-planned and pre-ordained commercial routes programmed into its flight management system, the military aircraft may only use such routes during training or transit to an operational area. Once on task the aircraft may be following a target, such as tracking a marine vessel, surface vehicles or submarine. In these circumstances its route becomes unplanned or determined by the target it is following, and an accurate knowledge of position is vital for a number of reasons:

- The long-range maritime patrol aircraft may spend many hours on patrol over open ocean, up to 1000 nautical miles from base. It is important that its position is known for coordination during rescue operations.
- To understand the position of the aircraft relative to its last known planned position, that is, when it entered the operational area, and its location relative to other forces, both friendly and hostile.

- To ensure that the aircraft does not enter hostile aircraft or missile engagement zones.
- To report the position of the target to other forces in order to assemble an accurate surface picture.
- To hand over the target to another aircraft if required.
- To attack the target if required and to provide accurate position information to the weapon.
- To ensure that the return route home is known and to determine when to break off an engagement and return safely to base with adequate fuel reserves to cover a diversion at home base. This also requires an accurate knowledge of fuel quantity.
- Commercial global positioning system (GPS) based navigation systems (see Chapter 12) are sufficiently accurate for most purposes. A military code enables access to even more accurate navigation data, and an active GPS antenna ensures that access to the right number of the most suitable satellites is guaranteed in most areas of the world.

The flight management system (FMS) will be used to store planned routes to and from operational areas, often using commercial air lanes and routes for fuel economy reasons in peacetime. In such cases the use of navigation aids is mandatory. It may also be used in conjunction with the mission computing system to allow the aircraft to fly automatically on stored search patterns. This enables the aircraft to fly for long periods of time doing a systematic search of the land or sea surface. The flight management system also eases crew workload during normal operations, making it possible to reduce the flight deck crew numbers with consequent savings in operating costs. For long duration missions, in which there may be long periods of flying at low level in adverse weather conditions, this reduction in workload is valuable in reducing crew fatigue and contributing to the safety of the mission. This is especially the case when flying repetitive search or surveillance patterns at low level for long periods of time. These patterns can be input into the FMS as routes and called up by the crew to start and finish at defined geographical points. The FMS and autopilot are not normally expected to perform this sort of task, and additional safety analysis will be required to ensure that the systems can be certificated to operate at low level without endangering the aircraft.

13.2.2   Navigation Aids

Modern navigation aids complement the role of many military aircraft. Apart from their use in normal airways flying and in using commercial airports, the systems may be used as follows:

- **Radar altimeter** (or rad alt, also known as radio altimeter) – to provide warning of deviation from a set height or to assist in the maintenance of height at very low levels (see Chapter 8). The radar altimeter measures absolute height above land or sea. It operates by sending either continuous or pulsed radio signals from a transmitter, and receives signals bounced back from the surface. The time taken for the signal to travel to the surface and back is converted to an absolute altitude that can be used by other systems, as well as provided as an indication on the flight deck. Aircraft flying low for long periods of time will use two radar altimeters, enabling the mission to be completed even if one altimeter fails. By positioning the transmitting and receiving antennas for each system on either side of the aircraft ventral surface, the radar altimeters continue to measure true height above the ground even when the aircraft is pulling a tight banking turn, or when the aircraft is flying in a steep-sided valley. This is especially used by maritime patrol aircraft where a height above the sea surface as little as 200 ft must be maintained even when operating in very tight turns (up to 2G). For such applications a dual radar altimeter is used, of a type that has a low probability of intercept type signal using frequency hopping or pulse modulation to avoid detection by enemy assets.

- **Landing aids** – using a multi-mode receiver (MMR) to allow the use of a microwave landing system or differential GPS in addition to the usual instrument landing system (ILS) (see Chapter 9). This allows the aircraft to operate in remote areas where conventional aids are not available or may have been destroyed. Mobile MLS or differential GPS can be set up by occupying ground troops to provide a service to friendly aircraft.
- **Ground proximity warning systems** (GPWS) and enhanced GPWS or Terrain Avoidance Warning System (TAWS) (see Chapter 11) will be used over foreign terrain, and may be connected to a military digital map to improve accuracy and confidence in the system. Such a system is known as terrain profile mapping (Terprom). Terprom correlates stored terrain data against inputs from the aircraft's navigation and radio altimeters to achieve a highly accurate, drift-free navigation solution. This is used to provide predictive terrain awareness with no detectable forward emissions, such as would be produced by terrain-following radar. This is especially useful in transport aircraft making covert approaches to dropping zones whilst using the terrain as defensive cover and wishing to avoid detection by hostile electronic support measures systems. The system can be updated with military intelligence to designate hostile areas and newly erected obstacles.
- **Traffic Collision Avoidance System** (TCAS) – is used not only in crowded airspace in the vicinity of airfields, but also in training where there may be a concentration of aircraft in a small space, for example, a designated low-flying area (see Chapter 9).
- **TACAN** is a chain of military navigation beacons used to obtain range and bearing. It can also be used to provide homing to a tanker aircraft (see Chapter 9).

13.2.3   Flight Deck Displays

A problem with early aircraft was that there was no space on the flight deck to provide tactical information to the pilots, mainly because of the proliferation of single function indicators and control panels. This meant that the flight deck crew were only made aware of tactical situations by the intercomm. This had an impact on their effectiveness and the safety of the aircraft for the following reasons:

- Their reaction time to commands from the tactical crew was delayed.
- Lack of exterior view in poor light conditions reduced the effectiveness of rescue efforts.
- If aircraft manoeuvres, especially violent manoeuvres or in poor weather conditions, were commanded by the rear crew with the flight deck crew 'hands off', the pilots tended to suffer disorientation or air sickness (rather similar to car passengers compared with the driver of a rally car).

The availability of modern multifunction display suites, as described in Chapter 12, allows information from the mission system to be more readily displayed within the existing display suite. This means that the flight deck crew can see the tactical map and can also see IR images that will enable them to conduct rescue missions more effectively. This also means that they can make a major contribution to the mission by flying the aircraft as if they were part of the mission crew. In this case the pilots will elect to alternate the role of flying pilot and tactical monitoring at regular intervals. This technique keeps both pilots alert and reduces the probability of both getting immersed in the tactical scenario and neglecting the flying task. The ability to switch the tactical picture from one side of the cockpit to the other is an advantage.

The flight deck displays can also display weather radar pictures obtained from the mission radar and are also able to display threat information produced by the defensive aids system. This enables the pilots to take evasive action rapidly if they believe that the threat is serious and endangers the aircraft.

13.2.4   Communications

The aircraft must have a set of communications to fly in controlled airspace – VHF and HF are part of the commercial aircraft fit, and both are essential for military use. However, most military communications use UHF so that additional UHF sets will be installed or combined V/UHF radios fitted. It may also be necessary to include encryption devices into the aircraft to provide secure communications, in which case the majority of the radios will need to be replaced. Maritime patrol aircraft that operate in close cooperation with naval and marine assets may need short-wave or marine band radios.

For most secure communication the military use satellite communications and data links. These systems allow encrypted standardised format messages to be sent and received that contain data as well as speech. This allows the aircraft to download its tactical information to other aircraft or to headquarters with a high probability of sound reception and minimum risk of interception. The most common data link in use in NATO is Link 16 or JTIDS (Joint Information Tactical Information Distribution System), although naval operations employ Link 11. Both links enable the mission crews to compile messages using preferred, defined formats and to transmit and receive using their normal communications transceivers.

Military applications may include the intermittent use of high-power transmitters which may interfere with, or damage, the aircraft communications and radio navigation aids. To reduce the likelihood of this, the sites of the antennae are carefully selected to prevent mutual interference. An active form of interference prevention known as blanking and suppression often reinforces this method of careful antenna location. This is accomplished by ensuring that equipment transmitting and receiving in the same frequency band are connected so that receiving equipment temporarily stops receiving, whilst another equipment is transmitting at high power.

13.2.5   Aircraft Systems

The aircraft electrical systems will be compatible with military avionics, although the civil and military specifications for power supply quality may differ in their method of defining quality. There may be a need to provide additional load conditioning such as filtering to ensure compatibility. Another issue is that some military applications may impose transient and intermittent high-power loads that may cause power surges and induced noise. In the event that the military loads exceed the capacity of the aircraft generation system, then it will be necessary to install larger generators; this will have an impact on the engine off-take loads and may lead to a re-certification programme.

The aircraft cooling systems will be designed to deal with a large passenger heat load, so their use in a military system is usually adequate – although the equipment cooling requirement may be higher, the number of passengers (crew) is considerably less. Very high-power loads such as radars or jamming transmitters may need their own integral liquid cooling systems.

One complication, however, is the need to operate and survive in conditions where biological and chemical contaminants may have been encountered. Rather than modify the environmental conditioning system of the aircraft, the usual method is to provide the crew with their own personal survival equipment in the form of a survival suit and a portable filtration unit. This can be complemented by installing an on-board oxygen generation system (OBOGS) if the aircraft role demands that it needs to remain on station for long periods of time in contaminated conditions [1]. However the installation of an OBOGS is a costly option.

The impact on the aircraft hydraulics systems is usually minimal, although some sensors may need hydraulic power for rotation or for extension and retraction.

13.3   Applications

13.3.1   Green Aircraft Conversion

A 'green' aircraft is the term given to an airframe purchased new and used to form the basis of a military platform. The main reasons for doing this include:

- reduced development cost and risk;
- basic avionics are integrated and qualified;
- basic aircraft systems are mature and it may be possible to use them unmodified;
- able to make use of established design and reliability as a mature product;
- able to make use of existing test evidence for basis type.

The process for adapting the green aircraft can take many forms but a typical sequence of events could be:

- Obtain the airframe, aircraft systems, flight deck and minimally fitted cabin from the production line.
- Cabin design to include:
  - galley facilities and toilets to suit crew size and mission duration
  - passenger section for resting crew, supernumerary crew, passengers
  - mission cabin space
  - installation of mission workstations and associated avionics.
- Fuselage and wings design to include:
  - design installation of antennae, pylons and external equipment
  - location and release envelope of weapons if required
  - installation of associated wiring harnesses.

- Avionics design to include:
  - incorporation of role equipment sensors and systems;
  - updates to basic systems to include military items, e.g. comms;
  - updates to flight deck to integrate with mission system;
  - integration of military standards with baseline aircraft commercial standards;
  - load analysis of electrical system for new mission systems increment
  - load analysis of cooling systems for new mission systems increment.
- Test and qualification to include:
  - integration test
  - ground test
  - EMH test
  - flight test
  - flight test
  - role equipment flight trials.
- Qualification as a military type with the new equipment and excrescences.

An example conversion is illustrated in Figure 13.4, which shows how the fuselage can be partitioned into areas to suit a particular task and also shows appropriate locations for antennae. This is often carried out by the military prime contractor at their own premises after delivery of the green aircraft from the commercial aircraft manufacturer, although any modifications or omissions that are cost-effective can be arranged to be performed before delivery. Examples of this are omission of overhead bins, basic primer paint scheme, wiring harnesses specified and laid in during build, blanking of windows, and so on.

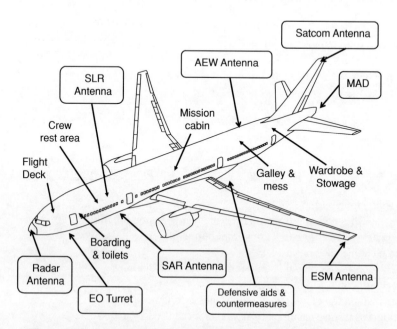

Figure 13.4   An example conversion of a green aircraft to a military role

13.3.2 Personnel, Material and Vehicle Transport

For large-scale military operations requiring troops or material to be transported into a war zone, specially designed aircraft such as C-130 or C-17 are used. However, there are instances where a less sophisticated platform will suffice. Typical candidates for such roles are passenger-carrying aircraft such as the Boeing 707, L1011 Tristar, VC10 or Airbus series. For personnel or troop transportation, very little change to the basic aircraft is required. Typical additional military avionics include:

- Communications systems may need to include military IFF with its own cryptos, as well as HF or satellite communication.
- TACAN is used to gain access to the military chain of homing and navigation beacons.
- If the aircraft is to be modified to enable paratroops to be carried and dispatched in large numbers, then the flight deck needs to be informed of the door or ramp opening to avoid the impact of pressurisation changes.
- Station keeping – to enable the aircraft to fly safely in large formations and to maintain a safe distance from others in the fleet, especially during inclement weather conditions.
- Landing aids such as a microwave landing system (MLS) or differential GPS – to allow the aircraft to fly safely into remote military landing zones using mobile ground stations set up in the field.
- Defensive aids system – to detect threats and to provide a defensive system using chaff and flare dispensing equipment, decoys or electronic countermeasures.

13.3.3 Air-to-Air Refuelling

Air-to-air refuelling has been used to extend the range and endurance of military aircraft and, in some instances, to extend test flying as successfully demonstrated in the Lockheed Joint Strike Fighter X-35 demonstration aircraft. The cost of designing and building special-to-type refuelling aircraft is extremely high since the numbers used are relatively small. The majority of aircraft in use today are developed from commercial airliners:

- The Royal Air Force uses tankers based on the VC10 and L1011 Tristar with drogues fitted to the tanker, and probes fitted to the receiving aircraft – the typical UK and European refuelling method.
- The USAF use tankers based on the Boeing 707, known as the KC-135, with a boom fitted to the tanker and a socket fitted to the receiver – the typical US refuelling method.
- Future candidates for conversion include airframes from Airbus (A330) and Boeing (767). The cost of maintaining a fleet of tankers at operational readiness at all times is causing air forces to consider an alternative in the form of the provision of a service by industry. Serious consideration has been given to modification kits to allow rapid conversion of commercial aircraft in times of tension.

The tanker is equipped with fuel tanks and modifications to its fuel system to allow pressurised refuelling of up to three receiver aircraft. The tanker must be able to navigate to remote locations, to loiter and to maintain station with its receiving aircraft. The difference in speed between the large tanker and fast jets can be problematic, with the tanker flying fast and the fighters flying slow at high angles of incidence. The differences in weight resulting from the transfer can alter this speed differential, resulting in a high workload task. Typical additional military avionics include the following:

- Communications systems may need to include military IFF with its own cryptos as well as HF or satellite communication.
- TACAN is used to gain access to the military chain of homing and navigation beacons. TACAN also includes a function that enables receiving aircraft to home onto the tanker.
- Data link in order to maintain secure communications with headquarters and the receiving aircraft and their controllers – the location of these aircraft and the refuelling zone is of vital tactical significance. Link 16 also includes a facility to allow aircraft to home on the tanker.
- Defensive aids system – to detect threats and to provide a defensive system using chaff and flare dispensing equipment.
- A means of monitoring the receiving aircraft on station and controlling the flow of fuel.
- A means of monitoring the correct deployment of the drogues or booms.
- A means of cutting and jettison if the drogue/boom fails to retract.

A number of different studies have been conducted to find a replacement for the current in-service tanker fleets, including multirole tanker transports and the provision of a contracted tanking service by commercial organisations. The Airbus A-400M also has a tanking capability, but the strong contender for a future replacement is the KC-46 based on the Boeing B767 aircraft.

13.3.4 Maritime Patrol

Maritime patrol aircraft (MPA) are used for a number of purposes by nations with extensive coastlines, those who may be dependent on commercial shipping for import and export, or those who have extensive naval fleets. Such aircraft are most often used in support of the homeland, although remote deployments are not uncommon in times of tension or war. Typical roles are a mixture of civilian, police and military duties:

- fisheries protection and monitoring of illegal fishing;
- offshore assets protection (e.g. oil installations);
- anti-drug/armaments/smuggling/terrorist operations;
- search and rescue;
- protection against hostile forces – anti-submarine/surface vessel threats;
- command and control for emergency operations (scene of action command).

To perform these roles the aircraft requires a number of sensors and a mission crew to gather and interpret data, identify vessels, report their location and prosecute an attack if necessary. Figure 13.5 provides a summary of aircraft roles, and the sensors necessary to perform these roles are shown in Figure 13.6 and described below.

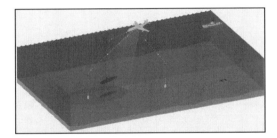

**Anti-Submarine
Warfare**

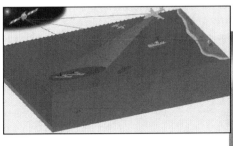

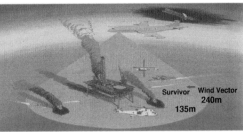

Anti-Surface Unit Warfare

Search and Rescue

Figure 13.5   Maritime patrol aircraft roles

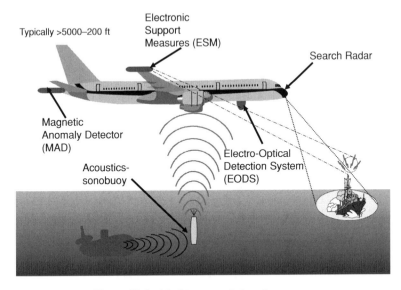

Figure 13.6   Maritime patrol aircraft – sensors

Maritime search radar

The maritime search radar's beam is optimised to conduct a wide-area surface search to discriminate small and fleeting targets against a background of sea clutter. The range of targets that an MPA must detect includes large capital ships, small surface craft, the periscope or antenna of a submarine, or human survivors and life rafts. The MPA must be able to detect and track such targets in sea states ranging from flat calm (sea state 0) to rough waves (often sea state 4). This is no easy task, and the typical search radar must be designed to detect such targets, and also to minimise side lobes so that the MPA is not detected by enemy sensors.

The radar is typically located in the nose of the aircraft so that the downward-looking search mode is complemented by weather mode and some degree of look-up mode to detect missiles, enemy aircraft or cooperating aircraft. This location restricts the search to about 270 degrees about the nose of the aircraft, and search patterns are designed to make maximum use of this limitation. 360-degree search coverage can be achieved by installing a radar antenna in the tail of the aircraft and fusing the two signal returns. This is expensive and complicated. An alternative method is to install the radar beneath the aircraft, which maximises the circular search at the expense of the look-up capability and is also a drag penalty.

Electro-optics

In conditions of poor visibility or night-time operations, electro-optical sensors are used to provide an image of targets that can be sensed, detected and tracked manually or automatically by a turret located on the ventral surface of the aircraft. Such sensors will include infra-red, low-light TV and digital cameras. They can be used singly or in combination to provide an image to the mission crew or the flight crew. This is especially helpful in locating survivors in the sea, whilst for military and policing operations the images can be annotated with position and time and may be used as evidence in legal proceedings.

Electronic support measures

The electronic support measures (ESM) system is able to intercept, locate, record and analyse radiated electromagnetic energy for the purpose of gaining tactical advantage. It is a completely passive system, its function being to 'listen' to signals radiated from other assets. The system is capable of covering the whole frequency spectrum of interest: radars, communications, missile guidance signals, laser emissions and infra-red emissions.

For monitoring radio and radar signals the intercept receiver is a versatile conventional radio receiver operating over a wide bandwidth. A signal analyser enables an operator to determine those parameters of interest to the mission. A direction-finding facility is provided so that range and bearing of the signal can be determined. A library of typical threat profiles allows an experienced operator to identify an emission to a particular emitter type, and also to identify the type of platform most likely to be equipped with that emitter.

Acoustics

A key role of the MPA is to detect and track submarines. Its main tool for doing this is the sonobuoy, which it dispenses into the water in patterns to enable the range, track and bearing of the submarine to be detected. The sonobuoys listen for acoustic signals from the submerged vessel and transmit the signal to the aircraft. Submarines are operated to avoid detection, maintaining silence by reducing the operation of machinery. Various types of sonobuoy are used:

- passive (listening only);
- active (using a sonar emission or 'ping' and monitoring the return echo).

Magnetic anomaly detector

The magnetic anomaly detector (MAD) is used to detect large metallic objects in the water, such as a submarine. The MAD is usually mounted at the end of a tail boom to keep the sensitive detector head away from magnetic objects in the airframe that could affect its performance. Once a maritime patrol aircraft has detected and pinpointed what it believes to be a large underwater mass that could be a submarine, it will confirm the target by overflying and looking for the MAD to signal that the mass is magnetic. This provides an extremely accurate fix and incidentally ensures that the target is not a shoal of fish or a whale.

The range of the MAD is very limited so the aircraft must fly very low over the suspected target, often as low as 200 ft. To do this the pilot will usually fly manually, disconnecting the autopilot to avoid pitch runaways causing a catastrophic descent. To do this automatically would require an autopilot designed and cleared to operate at very low levels; this is not a normal operation for a commercial aircraft.

Stores

The MPA is equipped as a hunter-killer and carries weapons to enable it to attack and destroy surface and subsurface vessels. For this purpose it will carry anti-ship missiles and torpedoes. The weapons are often carried internally in a bomb bay. It is also possible to carry missiles on wing-mounted pylons.

For rescue operations the aircraft will carry flares, life rafts and survival stores that can be dropped near to survivors.

Mission crew

Modern navigation systems based on inertial navigation with satellite-aided global navigation systems and flight management systems have led to the situation in which the flight deck crew members are able to perform all the navigation aspects of a mission. This has led to the replacement of the routine navigator in the Nimrod MRA4, leading to a reduction in the mission crew.

The mission crew act as a team, although each member may have a specific role such as radar operator, communications manager, acoustics operators and ESM operator. The data they each gather and refine are passed to the TACtical COmmander (TACCO) who compiles a composite picture of the surface, annotating friendly and enemy assets together with their tracks and identification if known. The avionic systems provide navigation data such as present position, heading, track, altitude, height above the surface and speed. These enable the TACCO to determine the position of the MPA with respect to other assets. Tactical information is also sent to the flight crew so that they are aware of any potential manoeuvres required and can take rapid action. The flight deck crew must monitor height above the surface at all times to ensure that the TACCO does not command an attack manoeuvre that would endanger the aircraft.

Examples

In the UK the Nimrod maritime patrol aircraft was originally developed from the de Havilland Comet and served the Royal Air Force as the MR1 and MR2 until 2011. A further development was started with the designation MRA4 and development flying was in progress when the project was cancelled in 2011. A block diagram of the MRA4 avionic and mission system is shown in Figure 13.7.

The Lockheed P3 was developed from the Lockheed Orion and is still in use in many countries. In Europe the French and German navies flew the Dassault Atlantic, a purpose-built maritime patrol aircraft originally designed by Avions Louis Breguet. Its successor is the Boeing P-8A Poseidon, a long-range anti-submarine warfare, anti-surface warfare, intelligence, surveillance and reconnaissance aircraft which first flew in 2009. It is equipped with an advanced mission system that ensures maximum interoperability in the future battle space. Capable of broad-area maritime and littoral operations, the P-8A will influence how the US Navy's maritime patrol and reconnaissance forces train, operate and deploy. The aircraft is designed to achieve a high degree of interoperability with other assets to exchange information and intelligence and to direct attacks from strike aircraft.

The P-8A is a derivative of the highly successful and reliable Next-Generation 737 with the fuselage of a 737-800 and the wings of a 737-900. Modifications to the baseline commercial aircraft are incorporated into the aircraft in-line. In the past, commercial aircraft were sent to modification centres where they were taken apart and rebuilt to meet military specifications. The P-8A departs from the description of the green aircraft above and is the first military derivative aircraft to incorporate structural modifications to the aircraft as it moves through the commercial production line.

The aircraft is equipped with the following systems and sensors:

- AN/APY-10 radar with weather, mapping, ISAR, SAR and periscope detection modes;
- electronic support measures (ESM–4);
- DIRCM;
- Northrop Grumman Nightfighter II E/O turret;
- common data link (CDL) × 2;
- SATCOM × 4;
- comms: V/UHF × 2; HF × 1; V/UHF/IFF/ATC × 2; UHF × 2;
- controlled reception GPS × 2;
- Store Management System (SMS) from GE Aviation;
- Mk54 torpedoes × 3;
- AGM84K SLAM-ER missiles × 3;
- AN/AWW-13 data link pod;
- LH & RH Sonobuoy Storage (× 48 each) and rotary launcher (10).

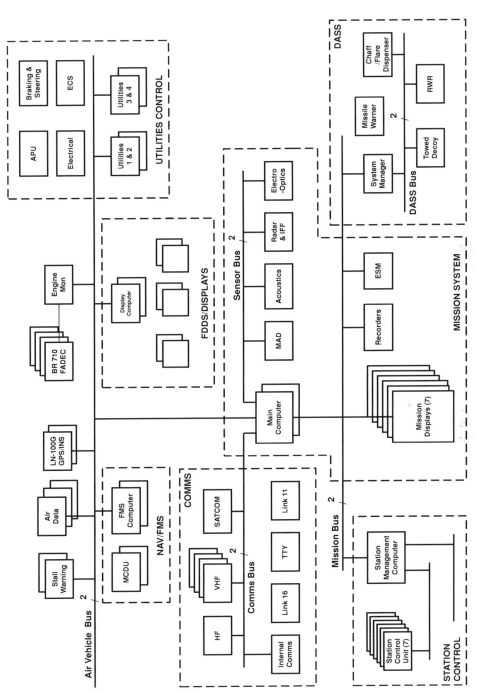

Figure 13.7 The Nimrod MRA4 avionics and mission system architecture

13.3.5 Airborne Early Warning

In many conflicts it is essential to gain air superiority to protect air, ground and maritime forces. A vital element of this strategy is constantly to monitor the hostile air threat. The airborne early warning aircraft can provide this capability by patrolling at height and observing aircraft movements beyond the sight of ground radars. The monitoring systems can provide range, bearing and speed of approaching air threats, and associated databases or libraries of threat profiles are able to provide intelligence on the type of threat. The aircraft crew are able to direct fighter interceptor aircraft to intercept targets of interest.

A typical aircraft is the E-3 Sentry Airborne Warning and Control System (AWACS), which is a modified Boeing 707 aircraft. A radar scanner is installed in a 9-metre diameter rotating dome mounted on struts above the aircraft rear fuselage. The radar is able to search for targets between the land or sea surface and up to the stratosphere. The radar has a range of over 250 miles for low-flying targets, and in excess of this for high-flying targets. The radar has an IFF interrogator which enables the crew to positively discriminate between friendly and hostile targets, and discriminate low-flying targets from ground clutter returns. The aircraft is also equipped with an electronic support measures system to allow detection and identification of radio frequency transmitters. Data link allows secure communication with external agencies and other forces. A mission computing system and a number of operator workstations provide facilities for a large crew of operators and analysts to work comfortably for long periods of time. The aircraft also includes facilities for rest and sleep.

Examples

The Boeing Airborne Early Warning and Control programme (AEW&C) is known as Wedgetail and is based on the B737-700 aircraft. The aircraft is designed to operate at an altitude of 30,000 to 40,000 ft with a maximum altitude of 41,000 ft.

A number of nations have expressed an interest in this project and the exact configuration of the aircraft will vary according to the customer's requirement. The aircraft will include in its capability a combination of the following:

- multi-role electronically scanned array (MESA) and IFF interrogator;
- multiple V/UHF radios, e.g. V/UHF × 4; V/UHF/TACAN × 2; UHF/TACAN × 1; V/UHF/IFF/ATC × 2;
- multiple HF;
- combinations of data links such as Link 4A, Link 11 and JTIDS Link 16;
- GPS controlled response antenna;
- electronic support measures (ESM) × 4;
- 2 × 180 kVA electrical generators (normally 2 × 90 kVA).

13.3.6 Ground Surveillance

An important aspect of air surveillance in times of tension or during conflict is to enable staff officers to gain an understanding of the disposition of assets, ground forces, installations and communications sites of opposing forces. This information completes the picture presented by maritime patrol, airborne early warning (AEW) and ground intelligence.

Surveillance of the battlefield can be conducted by aircraft flying high and deep inside friendly airspace. The aircraft carries a sideways-looking radar (SLR) or a synthetic aperture radar (SAR) that has a fixed or 'staring' antenna. The aircraft flies a fixed track that allows the best view of the area under surveillance. This enables the radar picture to be built in slices, which can be represented by a computer-generated picture. This can be supported by ESM to capture well-camouflaged or hidden transmitters that the radar may not detect.

An example of such a system is the Airborne Stand-Off Radar (ASTOR) system developed by Raytheon. The system is based on the Bombardier Global Express long-range executive jet fitted with a derivative of the Raytheon ASARS-2 sideways-looking radar (SLR). The radar is an upgrade of the SAR used on the U-2 aircraft and is capable of operation in all weathers at high altitudes to provide high-resolution images. The ASARS-2 includes dual synthetic aperture radar (SAR) which provides photographic quality images of the area being surveyed and a moving target indicator (MTI) radar that tracks moving vehicles over wide ranges, penetrating cloud and rainy conditions. The antenna location is illustrated in Figure 13.8.

The SAR can operate in spot mode to identify and track specific targets, or can be switched to swath mode which provides a large number of strips of pictures that can be joined together to form a detailed image of the battlefield. The SAR/MTI combination identifies the location of hostile forces and their quantity, direction and speed. Additional imagery can be provided by the use of electro-optical devices.

The image data are transmitted in real time by secure data links to ground stations, other forces or a headquarters command. The links are interoperable with existing U-2s and other surveillance platforms. The data can also be analysed and used by the onboard mission crew to direct ground operations and supporting land and air vehicles. At command headquarters the information is used by commanders and tacticians to understand the total battlefield scenario.

RAF Sentinel, Alan Radecki Akradecki

Figure 13.8 Location of SAR antenna on the Raytheon Sentinel aircraft. Copyright Alan Radecki. Image courtesy of Shutterstock.com

The aircraft is also equipped with a defensive aids system for self-protection, which includes missile warning, radar warning receiver, towed radar decoy and chaff/flare dispensers.

13.3.7   Electronic Warfare

Electronic warfare is a term that encompasses a number of aspects of intelligence-gathering and active jamming of enemy transmitting assets. Commonly used techniques of electronic warfare are:

- electronic counter measures (ECM);
- electronic support measures (ESM);
- SIGnals INTelligence (SIGINT);
- electronic counter-counter measures (ECCM).

The gathering of information is usually conducted by all available means including airborne, space satellite, naval and ground-based systems. Information gathered is transmitted to a headquarters intelligence cell where the data are analysed and used to conduct military operations. Part of the airborne element of this activity can be conducted by fast jets equipped with purpose-built photographic, receiving or jamming pods, making rapid transitions over enemy territory. However, for long- term intelligence gathering together with the ability to act as an airborne command post, a large long-endurance platform is preferred. Such a platform has the ability to fly at high altitude in friendly territory with its sensors observing targets many miles inside enemy territory. This operation is relatively safe, since the aircraft can 'stand-off' out of the range of surface-to-air missiles.

13.3.8   Flying Classroom

Mission crews or back-seat navigators will usually do their training in ground-based simulators. An additional degree of realism can be introduced by overflying ranges and operating in a live range environment. Although this can be performed in operational aircraft, the cost is high, especially in fast jets. An alternative that has been used is to install a number of crew workstations in the rear of a large aircraft. This allows a single instructor to train a number of crew members on a single mission. An example of this is the BAE Systems Jetstream equipped to train Tornado navigators.

13.3.9   Range Target/Safety

Military aircraft can conduct a lot of peace-time exercises on a military range. Targets can be simulated by specially equipped commercial aircraft equipped with radio frequency emitters that emulate enemy aircraft, so that crews can perform search and interdiction missions.

An example of this is the Falcon aircraft fleet operated on a commercial basis by Flight Refuelling. These aircraft can also patrol the range boundaries to provide safety and observation. This is especially useful for very large desert or maritime range exercises where the range area may have to be patrolled for long periods of time to restrict access to vessels or aircraft that may stray onto the range

Reference

[1] Moir, I. and Seabridge, A.G. (2008) *Aircraft Systems*, 3rd edition. John Wiley & Sons, Ltd., Chichester.

Further Reading

[1] Rihazcek, A.W. and Hershowitz, S.J. (2000) *Theory and Practice of Radar Target Identification*. Artech House.
[2] Hopkins, R.S. (1997) *Boeing KC-135 Stratotanker*. AeroFax.
[3] Morchin, W.C. (1990) *Airborne Early Warning Radar*. Artech House.
[4] Synafi, K. and Hovanessian, S.A. (1993) *Introduction to Electro-Optical Imaging and Tracking Systems*. Artech House.
[5] Gardner, W.J.R. (1996) *Anti-Submarine Warfare*. Brassey's (UK) Ltd.
[6] Oxlee, G.J. (1997) *Aerospace Reconnaissance*. Brassey's (UK) Ltd.
[7] Downing, T. (2011) *Spies in the Sky: The Secret Battle for Aerial Intelligence during WWII*. Little, Brown & Company.

Appendices

Introduction to Appendices

assist the reader in understanding how some of the analytical tools such as dependency grams, fault tree analysis (FTA) and Markov analysis may be applied to typical systems, ir Appendices are included. These appendices address the following systems:

Appendix A. Safety Analysis – Flight Control System
Appendix B. Safety Analysis – Electronic Flight Instrument System
Appendix C. Safety Analysis – Electrical System
Appendix D. Safety Analysis – Engine Control System

e analyses in the Appendices are presented in a simple mathematical fashion to provide reader with purely advisory and illustrative material. The failure rate probabilities offered for the purposes of illustrating the analysis methods and to draw simple conclusions; y are not representative of any real equipment or technologies. The analyses should not be nsidered as definitive of the standard that would be demanded during formal aircraft system ign. Nevertheless, it is hoped that they will aid the reader in appreciating some of the design ues that need to be considered early on in the design process. Refer to Chapter 4 for further evant sections.

During formal design, engineers utilise professional, qualified design tools to undertake the propriate analysis in a rigorous fashion. At the same time, these tools provide the required cumentation to the standard necessary to convince the certification authorities that the design afe.

l Avionics Systems, Second Edition. Ian Moir, Allan Seabridge and Malcolm Jukes.
)13 John Wiley & Sons, Ltd. Published 2013 by John Wiley & Sons, Ltd.

Appendix A

Safety Analysis – Flight Control System

This example evaluates the catastrophic failure case of total loss of the pitch axis of a fly-by-wire flight control system. It is a much simplified analysis, but serves to illustrate the principles and differences between the dependency diagram approach and the fault tree approach.

A.1   Flight Control System Architecture

The flight control system architecture is shown in Figure A.1. The primary means of controlling the aircraft in the pitch axis is by means of two elevators attached to either side of the rear of the tailplane horizontal stabiliser (THS). Each elevator section is operated by two hydraulic actuators powered from one of the three aircraft centralised hydraulic systems: yellow (Y), green (G) and blue (B), as indicated. Normally one actuator on each side is the controlling actuator, while the other is in damping mode, but for the purposes of this analysis we shall assume they are all the same, and that the aircraft is still controllable (albeit with some performance limitations) provided just one actuator is operational. We shall assume the actuators fail into a neutral position.

Elevator actuator demand is digitally signalled from four flight control computers. In normal mode the two elevator/aileron computers (ELACs) implement the control functions for the primary flight control surfaces. Should these computers fail, then reduced functionality is available from the spoiler/elevator computers (SECs) which normally control the secondary flight control surfaces. Each of these computers utilises a dual command:monitor architecture, but for the purposes of this analysis we shall assume that fly-by-wire demands in the pitch axis are available provided one of the four computers is operational.

In manual flight, the captain or first officer input their pitch commands through a side-stick controller, one per crew member. In autoflight, pitch commands are sourced from the autopilot. Envelope protection requires the provision of aircraft attitude and air data. For the purposes of this simplified analysis we shall assume the aircraft is being flown manually within its permitted flight envelope.

Civil Avionics Systems, Second Edition. Ian Moir, Allan Seabridge and Malcolm Jukes.
© 2013 John Wiley & Sons, Ltd. Published 2013 by John Wiley & Sons, Ltd.

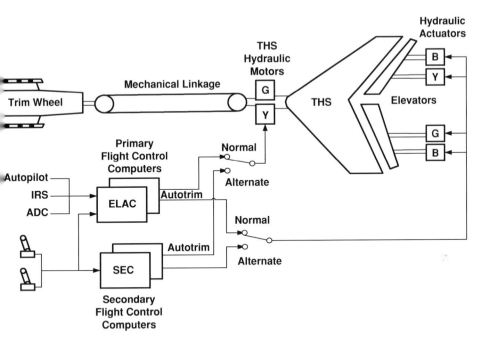

Figure A.1   Fly-by-wire flight control system – pitch axis

The aircraft is trimmed in the pitch axis by slow long-term movements of the tailplane izontal stabiliser (THS). In normal operation this function is commanded automatically by flight control computers via two hydraulic motors and a ball screw drive. Total loss of these nputers will leave the aircraft in a trim condition. Emergency backup in the pitch axis is vided by a mechanical linkage from the trim wheel located in the centre console between two crew members.

2   Dependency Diagram

ure A.2 provides the simplified dependency diagram for the architecture described. Con- ering first the primary flight control system:

itch axis control is lost if both left-hand AND right-hand elevators fail – indicated by the arallel nature of the branches. ach elevator is lost if both of its actuator channels are unavailable. An actuator channel is ost if either the actuator fails OR the hydraulic system that powers it fails OR the demand ignal is lost – indicated by the series nature of the each branch.

he actuator arrangement as shown is quadruplex, but there is a problem in that the aircraft only three hydraulic systems. The blue hydraulic system is used on both elevators. Loss he blue hydraulics system is a common mode failure. It is not easy to model this in a

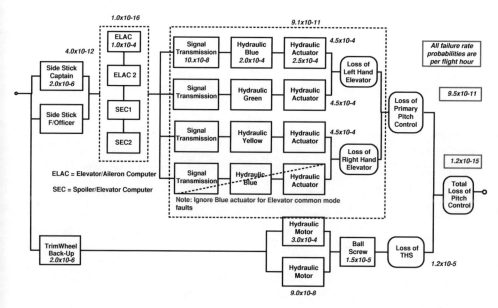

Figure A.2   Fly-by-wire flight control system – dependency diagram (simplified)

dependency diagram, so the simple approximation is to ignore the second blue channel and the treat the actuator channels as triplex. This will yield a pessimistic, but safe result.

- Demands to the actuators are computed by the four flight control computers.
- Crew input commands are sourced from two side-stick controllers (captain and first officer).

For the primary flight control function, if we assume the probability of failures per hour are as follows (note: these figures are for illustrative purposes only and do not represent the failure probabilities of actual equipment), then:

- **Side-stick controller**: 2.0×10^{-6} per flight hour (MTBF 500,000 hrs): duplex $= \mathbf{4.0 \times 10^{-12}}$ per flight hour.
- **Flight control computer**: 1.0×10^{-4} per flight hour (MTBF 10,000 hrs): quadruplex $= \mathbf{1.0 \times 10^{-16}}$ per flight hour.
- **Elevator actuator**:
  - signal transmission: $\mathbf{1.0 \times 10^{-8}}$ per flight hour
  - plus hydraulic system: $\mathbf{2.0 \times 10^{-4}}$ per flight hour (MTBF 5000 hrs)
  - plus hydraulic actuator: $\mathbf{2.5 \times 10^{-4}}$ per flight hour (MTBF 4000 hrs)
  So single actuator lane $= \mathbf{4.5 \times 10^{-4}}$ per flight hour: triplex $= \mathbf{9.1 \times 10^{-11}}$ per flight hour.

Thus the probability of loss of the primary flight control system is $\mathbf{9.5 \times 10^{-11}}$ per flight hour.

For the trim function, if we assume the probability of failures per hour are as follows:

Trim wheel: 2.0×10^{-6} per flight hour (MTBF 500,000 hrs).
Hydraulic motor: 3.0×10^{-4} per flight hour [3×10^{-6}/hr] (MTBF 3333 hrs): duplex $=$ 9.0×10^{-8} per flight hour.
Ball screw: 1.0×10^{-5} per flight hour (MTBF 100,000 hrs).

us the probability of loss of the mechanical backup trim function is 1.2×10^{-5} per flight ur.

Although we could compute the catastrophic loss of flight control in the pitch axis as $1.2 \times$ $^{-15}$ per flight hour, it is somewhat unrealistic as the aircraft would never be flown by the ergency backup system for anything other than a dire emergency, and then only to make an ergency landing at the nearest available airfield. The duty cycle of the emergency system somewhat uncertain and a judgement would need to be made as to what figure to use.

3   Fault Tree Analysis

ure A.3 provides the simplified fault tree for the architecture described. The fault tree roach facilitates a more complete model of the system architecture in respect of the

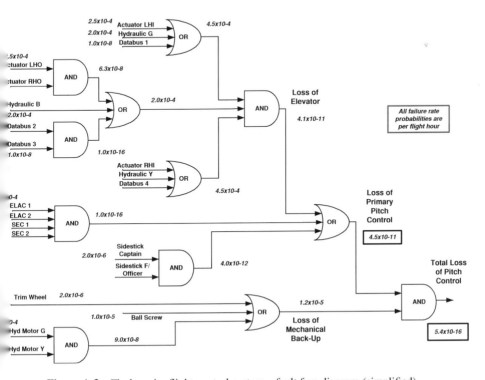

Figure A.3   Fly-by-wire flight control system – fault free diagram (simplified)

actuator power sources and the dual use of the blue hydraulic system by actuators driving both the left-hand and the right-hand elevators. In Figure A.3 the elevators are modelled as:

- Two inner sections (left hand and right hand), either of which is lost if either the associated actuator fails OR the associated hydraulic system fails OR the associated data bus signal fails: $= 4.5 \times 10^{-4}$ per flight hour each; and
- Two outer sections, both of which are lost if both actuators fail OR both data bus signals fail OR the associated hydraulic system fails: $= 2.0 \times 10^{-4}$ per flight hour.

Using the same probabilities as before:

- The probability of loss of all four elevator sections (left-hand inner AND right-hand inner AND both outer sections) becomes 4.1×10^{-11} per flight hour (c.f. 9.1×10^{-11} for the dependency diagram method);
- The probability of loss of the primary flight controls system (both controllers OR all four flight control computers OR the elevators) becomes 4.5×10^{-11} per flight hour (c.f. 9.5×10^{-11} for the dependency diagram method).

The difference between the two methods reflects the improved fidelity of the fault tree analysis, but it is not that significant in the overall scheme of things. It is the order of magnitude that is important when assessing the architecture against the safety objective, in this case to be less than 1×10^{-9} per flight hour for a catastrophic event.

The probability of loss of the primary flight control system for a three-hour flight time is obtained by factoring the probability of failure of each component by the time at risk. Thus, for example, the probability of failure of all four flight control computers during a three-hour flight time is $(3 \times 1.0 \times 10^{-4}$ per flight hour$)^4 = 8.1 \times 10^{-15}$ per flight. The probability of loss of the primary flight control system for a three-hour time of risk becomes 1.1×10^{-9} (fault tree method) which indicates that a more robust architecture may be needed for long haul and ETOPS flight operations, as is indeed the case in the flight control system architectures for the Airbus A340, the A380 and the Boeing 787.

Appendix B

Safety Analysis – Electronic Flight Instrument System

This example explores the flight safety aspects of the electronic flight instrument system (EFIS). The EFIS provides primary flight display (PFD) and navigation display (ND) functions to both crew members.

B.1 Electronic Flight Instrument System Architecture

The architecture of an early EFIS system is shown in Figure B.1. Typically, three symbol generators (SGs) source image formats on to four display units (DU). Each SG is able simultaneously to produce both PFD and ND formats. Each DU sources its display format from a normal or an alternate SG source. In normal operation SG#1 sources the images to the captain's displays, PFD#1 and ND#1; SG#2 sources the images to the first officer's displays, PFD#2 and ND#2. SG#3 is a 'hot spare' and can take over the function of either SG#1 or SG#2 in the event of their failure. Each SG sources its inputs from both left-hand and right-hand sensors. An independent integrated standby instrument system (ISIS) is installed on the flight deck to aid the crew in resolving discrepancies and to provide critical flight information (attitude, altimeter and airspeed) in the event of complete failure of the main display system.

Loss of primary flight data (attitude, airspeed and altitude) on the main displays is considered hazardous; and should have a probability of less than 1×10^{-7} per flight hour. Total loss of primary flight data from the flight deck is considered catastrophic, and should have a probability of less than 1×10^{-9} per flight hour.

The architecture is fault-tolerant and manages failures as follows:

- If one DU fails, then the SG driving it will reconfigure its outputs so that the image on the adjacent SG will be a composite PFD/ND display with all requisite primary flight data, including a compass.

Civil Avionics Systems, Second Edition. Ian Moir, Allan Seabridge and Malcolm Jukes.
© 2013 John Wiley & Sons, Ltd. Published 2013 by John Wiley & Sons, Ltd.

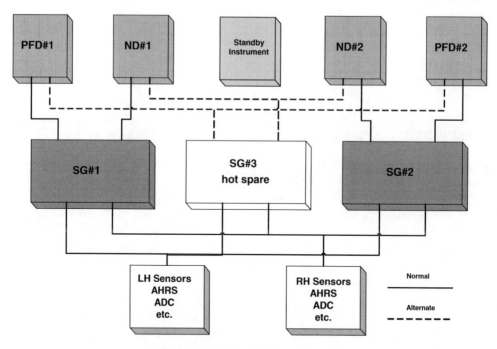

Figure B.1   Typical EFIS architecture

- In the event that a SG fails then the impacted DUs select their inputs from SG#3, the hot spare. The image formats sourced from SG#3 remain independent of those on the other side of the flight deck.
- If the entire primary display suit fails, then the crew revert to using the independent standby instrument (ISIS).

B.2   Fault Tree Analysis

Figure B.2 provides the simplified fault tree for the architecture described. Considering first the main displays, primary flight information is lost from the main displays if:

- all four DUs fail; OR
- all three SGs fail; OR
- both DUs fail on one side AND the SGs able to source images to the other side fail; OR
- both AHRSs fail, denying attitude data; OR
- both ADCs fail, denying speed and altitude data.

Assuming the probabilities of failure per hour are as follows (note: these figures are for illustrative purposes only and do not represent the failure probabilities of actual equipment):

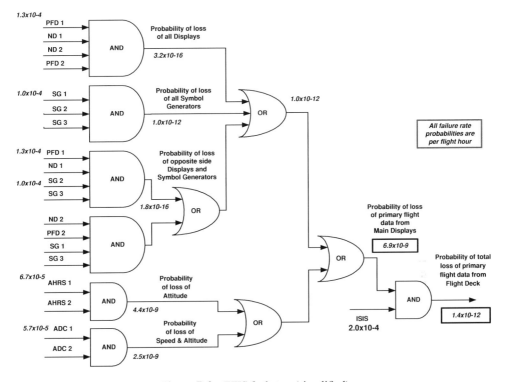

Figure B.2   EFIS fault tree (simplified)

- **Display unit**: 1.3×10^{-4} per flight hour (MTBF 7500 hrs): quadruplex $= 3.2 \times 10^{-16}$ per flight hour.
- **Symbol generator**: 1.0×10^{-4} per flight hour (MTBF 10,000 hrs): triplex $= 1.0 \times 10^{-12}$ per flight hour.
- **DUs and opposite side SGs**: quadruplex $= 1.8 \times 10^{-16}$ per flight hour.
- **AHRS**: 6.7×10^{-5} per flight hour (MTBF 15,000 hrs): duplex $= 4.4 \times 10^{-9}$ per flight hour.
- **ADC**: 5.0×10^{-5} per flight hour (MTBF 20,000 hrs): duplex $= 2.5 \times 10^{-9}$ per flight hour.

Thus the probability of loss of primary flight data from the main display system is 6.9×10^{-9} per flight hour, which satisfies the safety objective.

Assuming the probability of failure of the ISIS is 2.0×10^{-4} per flight hour, then the probability of total loss of primary flight data from the flight deck is $\sim 1.4 \times 10^{-12}$ per flight hour, which also satisfies the safety objective.

For a flight time of 5 hours, the probability of loss of primary flight data from the main displays becomes 1.6×10^{-7} per flight hour; a marginal outcome. (In this case each individual equipment failure rates are multiplied by five due to the increased risk of failure for a five-hour exposure time.) The weak link is not the display system itself, but the source data. The system architect would now review the design and consider augmenting the data sources for long haul flights or improving the technology of those equipments. This may include the provision of a triple air data module (ADM) architecture to replace the dual ADCs shown in the example.

The diligent reader will soon learn to master this analysis and will understand how to establish dispatch criteria to meet mission requirements while carrying various equipment failures at the outset. This can be achieved by negating the 'failed' unit(s) and executing the analysis in the usual manner.

Appendix C

Safety Analysis – Electrical System

This example analyses the total loss of aircraft electrical AC power on board an aircraft. The safety objective quantitative requirement established by FAR/JAR 25.1309 and as amplified in ARP 4754 will be such that this event could be catastrophic and the probability of occurrence shall be less than 1×10^{-9} per flight hour (or shall not occur more frequently than once per 1000 million flight hours). The ability of a system design to meet these requirements is established by a fault tree analysis (FTA) that uses the following probability techniques.

C.1   Electrical System Architecture

In this example it is assumed:

- that the aircraft has two independent but identical electrical power generation channels, the main components of which are the generator and the Generator Control Unit (GCU) which governs voltage regulation and system protection;
- the aircraft has an independent emergency system such as a ram air turbine (RAT);
- that the failure rates of these components may be established and agreed due to the availability of in-service component reliability data or a sound engineering rationale, which will provide a figure acceptable to the certification authorities.

The concept of this three-lane architecture is portrayed in Figure C.1 in a simplified form.

C.2   Fault Tree Analysis

The fault tree analysis – very much simplified – for this example is shown in Figure C.2.

The mean time between failures (MTBF) of a generator is 2000 hours – this means that the failure rate of Generator 1 is 1/2000 or 5.0×10^{-4} per flight hour. Similarly if the MTBF of the generator controller GCU 1 is 5000 hours then the failure rate of GCU 1 is 1/5000 or 2.0×10^{-4} per flight hour. The combined failure rate gives the probability of loss of electrical power

---

Civil Avionics Systems, Second Edition. Ian Moir, Allan Seabridge and Malcolm Jukes.
© 2013 John Wiley & Sons, Ltd. Published 2013 by John Wiley & Sons, Ltd.

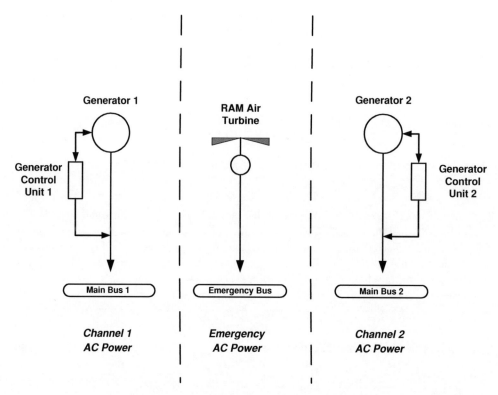

Figure C.1   Simplified electrical power system

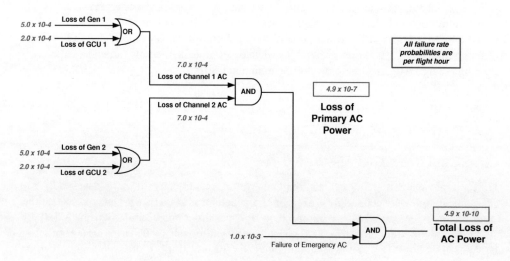

Figure C.2   Simplified FTA – AC power loss

to Main Bus 1. This is calculated by summing the failure rates of generator and controller, as either failing will cause the loss of Main Bus 1:

$$= 5.0 \times 10^{-4} + 2.0 \times 10^{-4} = \mathbf{7 \times 10^{-4}} \text{ per flight hour}$$
$$\text{(Generator 1)} \quad \text{(GCU 1)} \quad \text{(Main Bus 1)}$$

Similarly, assuming generator channels 1 and 2 are identical, the failure rate of Main Bus 2 is given by:

$$= 5.0 \times 10^{-4} + 2.0 \times 10^{-4} = \mathbf{7 \times 10^{-4}} \text{ per flight hour}$$
$$\text{(Generator 2)} \quad \text{(GCU 2)} \quad \text{(Main Bus 2)}$$

(Note that at this stage the experienced aircraft systems designer would be considering the effect of a common cause or common mode failure.)

The probability of two independent channels failing (assuming no common cause failure) is derived by multiplying the respective failure rates. Therefore the probability of both Main Buses failing is:

$$= 7 \times 10^{-4} \times 7 \times 10^{-4} = \mathbf{49 \times 10^{-8}} \text{ or } \mathbf{4.9 \times 10^{-7}} \text{ per flight hour}$$
$$\text{(Main Bus 1)} \quad \text{(Main Bus 2)}$$

Therefore the two independent electrical power channels alone will not meet the requirement of better than $\mathbf{1 \times 10^{-9}}$ per flight hour.

Assuming the addition of the ram air turbine (RAT) emergency channel as shown in the figure, with an MTBF of 1000 hours or a failure rate of $\mathbf{1 \times 10^{-3}}$ per flight hour, the probability of total loss of electrical power becomes:

$$4.9 \times 10^{-7} \times 1 \times 10^{-3} = \mathbf{4.9 \times 10^{-10}} \text{ per flight hour, which meets the requirements.}$$
$$\text{(Main Buses)} \quad \text{(RAT failure)}$$

This very simple example is illustrative of fault tree analysis, which is one of the techniques used during the PSSA and SSA processes. However, even this simple example outlines some of the issues and interactions that need to be considered. Real systems are very much more complex, with many more system variables and interlinks between a number of aircraft systems.

Appendix D

Safety Analysis – Engine Control System

The example chosen uses a Markov analysis to evaluate the likelihood of an engine in-flight shut down (IFSD), which would be a typical analysis necessary to determine the reliability of an engine prior to seeking Extended Twin Operations (ETOPS) clearance for a twin-engine aircraft.

D.1   Factors Resulting in an In-Flight Shut Down

Figure D.1 illustrates three main failure types, any of which could contribute to IFSD:

- a mechanical failure;
- an instrumentation failure that causes the crew to shut an engine down as a precautionary measure as they are unsure about engine health;
- a failure within the control portion of the engine that resides in the full authority digital engine control (FADEC) unit.

The first two items above may be readily calculated using past historical data gained from the experience of operating similar engine types. The FADEC is more difficult to assess as multiple failure states are possible and a detailed state analysis is required.

D.2   Engine Control System Architecture

A simple example will be used to illustrate the Markov analysis technique. In this case the dual-channel FADEC example outlined in Figure D.2 is used.

This simplified architecture is typical of many dual-channel FADECs using a command:monitor (COM:MON) implementation. There are two independent lanes: Lane A and Lane B. The FADEC undertakes the task of metering the flow of fuel to the engine, thereby controlling engine thrust. It also controls other valves on the engine related to bleed air and

Civil Avionics Systems, Second Edition. Ian Moir, Allan Seabridge and Malcolm Jukes.
© 2013 John Wiley & Sons, Ltd. Published 2013 by John Wiley & Sons, Ltd.

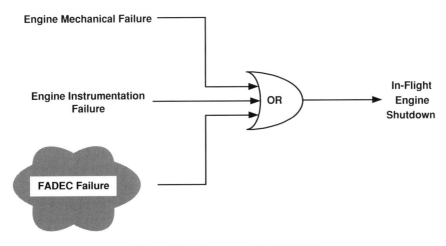

Figure D.1   Factors resulting in IFSD

cooling, and monitors key engine variables such as shaft speed, pressure ratio and engine temperatures.

Each lane comprises Command and Monitor elements that are interconnected for cross-monitoring purposes and which undertake the control and monitoring functions outlined above. The analysis required to decide upon the impact of certain failures in conjunction with others utilises a Markov model in order to understand the dependencies

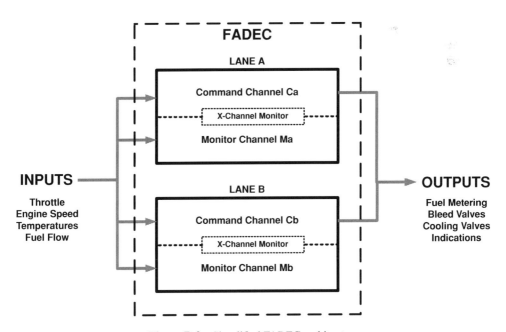

Figure D.2   Simplified FADEC architecture

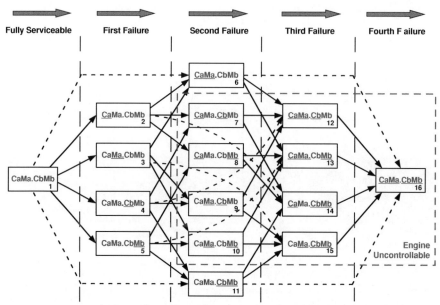

Figure D.3   Simple FADEC Markov model

D.3   Markov Analysis

Figure D.3 depicts a simple Markov analysis that models this architecture. By using this model, the effects of interrelated failures can be examined.

The model has a total of 16 states, as shown by the number in the bottom right-hand corner of the appropriate box. Each box relates to the serviceability state of the Lane A Command (Ca) and Monitor (Ma) channels, and Lane B Command (Cb) and Monitor (Mb) channels. These range from the fully serviceable state in box 1 through a series of failure conditions to the totally failed state in box 16. Clearly most normal operating conditions are going to be in the left-hand region of the model. Also represented on the diagram are the first, second, third and fourth failure areas.

Concentrating on the left-hand side of the model it can be seen that the fully serviceable state in box 1 can migrate to any one of six states:

- Failure of Command channel A (Ca) results in state 2 being reached.
- Failure of Monitor channel A (Ma) results in state 3 being reached.
- Failure of Command channel B (Cb) results in state 4 being reached.
- Failure of Monitor channel B (Mb) results in state 5 being reached.

These failures are represented by solid arrows and represent a single failure transition.

- Failure of the cross-monitor between Command A and Monitor A results in both functions being lost simultaneously and reaching state 6.

- Failure of the cross-monitor between Command B and Monitor B results in both functions being lost simultaneously and reaching state 11.

These failures are represented by dashed arrows. The failure of the cross-monitor represents a skip of two failures across the diagram and therefore has more effect than direct single command or monitor channel failure.

All of these failure states described above result in an engine that may still be controlled by the FADEC. However, further failures beyond this point will result in an engine that may not be controllable, either because both control channels are inoperative or because the 'good' control and monitor lanes are in opposing channels. The model shown above is constructed according to the following rules: an engine may be dispatched as a 'get-you-home' measure provided that only one monitor channel has failed. This means that states 3 and 5 are dispatchable, but not states 2, 4, 6 or 11, since subsequent failures could result in engine shut-down.

By knowing the failure rates of the command channels, monitor channels and cross-monitors, quantitative values may be inserted into the model and probabilities assigned to the various states. By summing the probabilities so calculated, numerical values may be derived.

Simplified Example (all failure rates per flight hour)

- Probability of engine mechanical failure $= 1 \times 10^{-6}$ per flight hour.
- Probability of instrumentation failure leading to engine shut-down $= 2 \times 10^{-6}$ per flight hour.
- The probability of determining shut-down due to failures within the FADEC is more complex and needs a Markov analysis.

If it is assumed that the failure rate of a command channel (Ca or Cb) is $\mathbf{8.3 \times 10^{-5}}$ per flight hour (MTBF of 12,000 hours); failure rate of a monitor channel (Ma or Mb) is $\mathbf{5.6 \times 10^{-5}}$ per flight hour (MTBF of 18,000 hours); and the failure rate of the cross-monitor (Ca.Ma or Cb.Mb) is $\mathbf{1.4 \times 10^{-5}}$ per flight hour; then the probability of all 16 logic states shown in the diagram may be calculated.

If it is further assumed that for the engine to remain controllable a functional command:monitor pair must be available, then the probability of the engine having to be shut down due to FADEC failures may be calculated. In fact, this probability is estimated by summing all of the relevant logic states shown in the dashed box shown on the diagram. This includes states 7 to 10, states 12 to 15, and state 16.

For the figures given above for the FADEC command and monitor channels this probability is $\mathbf{4.7 \times 10^{-8}}$ per flight hour.

The summation of these three factors yields the total probability of an engine shut-down as $\mathbf{3.1 \times 10^{-5}}$ per flight hour, and the contribution of FADEC failures to engine shut-down is in fact very small ($\sim 1.3\%$). This is despite the fact that the FADEC is not hugely reliable – the assumed figures used above equate to an overall FADEC MTBF of 3600 hours; it is the redundancy inherent in the architecture that leads to its availability. The fact that the second and subsequent failures that cause shut-down are relatively unlikely events is also important to consider.

Index

Civil Avionics Systems, Second Edition. Ian Moir, Allan Seabridge and Malcolm Jukes.
© 2013 John Wiley & Sons, Ltd. Published 2013 by John Wiley & Sons, Ltd.